Practical Transistors and Linear Integrated Circuits

Practical Transistors and Linear Integrated Circuits

Joseph D. Greenfield
Rochester Institute of Technology
Rochester, New York

JOHN WILEY & SONS
New York   Chichester   Brisbane   Toronto   Singapore

Interior and cover design: Dawn L. Stanley
Production Editor: Nancy Prinz
Cover photo: Arthur Singer/Phototake

Copyright © 1988, by John Wiley & Sons, Inc.

All rights reserved. Published simultaneously in Canada.

Reproduction or translation of any part of
this work beyond that permitted by Sections
107 and 108 of the 1976 United States Copyright
Act without the permission of the copyright
owner is unlawful. Requests for permission
or further information should be addressed to
the Permissions Department, John Wiley & Sons.

Library of Congress Cataloging in Publication Data:

Greenfield, Joseph D., 1930-
    Practical transistors and linear integrated circuits/Joseph D. Greenfield.
      p. cm.
    Bibliography: p.
    Includes index.
    ISBN 0-471-89097-9
    1. Transistor circuits. 2. Linear integrated circuits.
 I. Title.
TK7871.9.G67 1988
621.3815'30422—dc19

87-21708
CIP

Preface

This is an introductory book on transistors and linear electronics for electrical engineering or engineering technology students. It was written to acquaint them with the most commonly used transistor circuits and should allow them to understand, analyze, and design these circuits. A prime objective is to give the student a feeling of confidence when he or she encounters such a circuit. The text is designed for laboratory backup; most of the circuits discussed in it and the illustrative examples presented can be built and tested in the laboratory.

One of my primary objectives is to present the topics as clearly as possible. I have used the h parameters throughout and have refrained from distracting and confusing the student by introducing other parameters. I have tried to use as few equations as possible, but also to use them consistently throughout the book. The level of mathematics is appropriate for second or third year engineering or technology students, but long mathematical derivations have been relegated to the appendices because they slow down the pace of the presentation. Computer programs, written in BASIC, are used throughout the book wherever they are appropriate.

I also wrote the book because I felt that I had a unique and innovative approach to many of the topics that must be covered. There are identified in the chapter by chapter summary given in the following paragraphs.

In Chapter 1 diode circuits are introduced. The discussions of clippers, clampers, and photodiode circuits are supported by many specific examples to help the students comprehend these topics.

Chapter 2 introduces the BJT transistor. The examples are simple and easy to understand. Curve tracers are covered. A thorough discussion of the saturation and cutoff regions of a transistor, including the concept of forced Beta, is presented.

In Chapter 3, transistor biasing and stability are covered with many examples, including some that use the computer. Stability is also considered in detail with an example that illustrates the stability calculations.

Chapter 4 presents transistor equivalent circuits. This can be done in one of two ways: with and without using the Thevenin's equivalent. Both are presented here and I show that the results are the same. By solving the same problem two ways, and getting the same answer, the student can be confident that his results are correct.

I also use the Thevenin's equivalent circuit frequently. This will be helpful in later chapters. One of the transistor gain equations, 4-11, is very important and is not presented in any other textbook. Very little calculus is used in Chapters 3 and 4.

Chapter 5 covers JFETS. The analysis in this book relies heavily on computer programs written in BASIC. Many of these computer programs consider JFETS connected to a capacitively coupled load, and are unique to this book. The input and output files for the BASIC programs are presented so the student can examine the data being entered and the output of the program. The programs themselves will run on any computer or PC that has BASIC.

Chapter 6 covers most of the important multiple-transistor circuits. Particular attention is paid to Darlingtons and difference amplifiers.

Chapter 7 covers the frequency response of transistors. Equation 7-7 for the low-frequency response considers both the revelant zero and the pole and is unique to this book. The analysis of high-frequency response is made clear. The frequency response of oscilloscopes and scope probes, a topic omitted in many texts, is also covered.

Because I felt it would be less confusing to the student, Chapter 8, on feedback, is limited to two types of feedback rather than the four discussed by Millman. Also, when all four types of feedback are considered, it is often difficult to determine which types are being used in any particular circuit. Feedback is covered from several perspectives so the student can verify that the traditional analysis is correct. There is also a thorough discussion of feedback and frequency response.

Chapter 9 presents a very detailed discussion of power amplifiers, a subject treated lightly in many texts. I concentrate on showing how the power is dissipated in these amplifiers.

Chapter 10 covers regulated power supplies. The analysis is unique. It leads to an equation that relates the input voltage to the output voltage that is not in any other textbook. I cover switching power supplies thoroughly.

Chapter 11 covers tuned circuits and oscillators.

Chapter 12 is a comprehensive introduction to operational amplifiers.

I thank W. David Baker of RIT for his help and suggestions received when we taught the course together from the manuscript. I thank Jack Shanker of Scientific Radio, Inc., for his aid and advice. Hank Stewart, of John Wiley, was always helpful and encouraging. Finally, I must thank my wife, Gladys, for her love, help, typing, and support.

JOSEPH GREENFIELD

Contents

Introduction 1

CHAPTER ONE
Semiconductors and Diodes 3
1.1 Instructional Objectives 3
1.2 Self-evaluation Questions 3
1.3 Conductors and Insulators 4
1.4 Semiconductors 6
1.5 The p-n Junction 13
1.6 Diode Characteristics 16
1.7 Clippers and Rectifiers 24
1.8 Clamping Circuits 31
1.9 Zener Diodes 39
1.10 Schottky Diodes 45
1.11 Photodiodes 46
1.12 Applications of Photoemitters and Photodetectors 49
1.13 Light-emitting Diodes 52
1.14 Other Specialized Diodes 56
1.15 Summary 58
1.16 Glossary 58

1.17 References 59
1.18 Problems 59

CHAPTER TWO
Introduction To The Bipolar Junction Transistor 65
2.1 Instructional Objectives 65
2.2 Self-evaluation Questions 65
2.3 Introduction to the Junction Transistor 66
2.4 Transistor Operation 69
2.5 The Common-base Configuration 71
2.6 The Common-emitter Circuit 77
2.7 Curve Tracers 83
2.8 The Load Line 84
2.9 The Cutoff Region 87
2.10 The Transistor in Saturation 91
2.11 Manufacturers' Specifications 94
2.12 Summary 95
2.13 Glossary 95
2.14 References 96
2.15 Problems 96

CHAPTER THREE
Biasing The Bipolar Junction Transistor 101
3.1 Instructional Objectives 101
3.2 Self-evaluation Questions 101
3.3 Introduction to Biasing 102
3.4 The Fixed-bias Circuit 105
3.5 The Emitter-resistor Bias Circuit 110
3.6 The Load Line for Self-biased Circuits 120
3.7 Biasing the Common-base Amplifier 121
3.8 Biasing the Emitter follower 127
3.9 Bias Stability 128
3.10 Summary 134
3.11 Glossary 134
3.12 References 134
3.13 Problems 135

CHAPTER FOUR
Small Signal Analysis and AC Gain 141
4.1 Instructional Objectives 141
4.2 Self-evaluation Questions 141

4.3 Introduction to ac Analysis 142
4.4 The Hybrid Parameters 145
4.5 Calculating the Common-emitter Gain 151
4.6 The Common-base Amplifier 166
4.7 The Emitter-follower Circuit 169
4.8 Summary 178
4.9 Glossary 178
4.10 References 179
4.11 Problems 179

CHAPTER FIVE
Field Effect Transistors 187
5.1 Instructional Objectives 187
5.2 Self-evaluation Questions 187
5.3 Introduction 188
5.4 JFET Construction 189
5.5 Biasing the JFET 197
5.6 Small Signal Analysis of a JFET 206
5.7 The Source-follower 218
5.8 MOSFETs 222
5.9 VMOS 231
5.10 Summary 232
5.11 Glossary 232
5.12 References 233
5.13 Problems 233

CHAPTER SIX
Multiple Transistor Circuits 237
6.1 Instructional Objectives 237
6.2 Self-evaluation Questions 237
6.3 Introduction 238
6.4 RC Coupled Circuits 240
6.5 Direct-coupled Amplifiers 248
6.6 Darlingtons 251
6.7 Constant-current Sources 257
6.8 Comparators 259
6.9 Difference Amplifiers 263
6.10 The Cascode Amplifier 277
6.11 Summary 279
6.12 Glossary 280
6.13 References 280
6.14 Problems 280

CHAPTER SEVEN
The Frequency Response of Amplifiers 289
- 7.1 Instructional Objectives 289
- 7.2 Self-evaluation Questions 289
- 7.3 Capacitive Effects 290
- 7.4 The Low Frequency Response of a BJT Amplifier 293
- 7.5 The High Frequency Response of a BJT Amplifier 304
- 7.6 The BJT at High Frequencies 307
- 7.7 The Frequency Response of JFETs 316
- 7.8 The Frequency Response of Multistage Amplifiers 322
- 7.9 Transient Response of Amplifiers 327
- 7.10 Oscilloscope Probe Compensation 333
- 7.11 The Frequency Response Limitations of Oscilloscopes 341
- 7.12 Summary 343
- 7.13 Glossary 343
- 7.14 References 344
- 7.15 Problems 344

CHAPTER EIGHT
Feedback 353
- 8.1 Instructional Objectives 353
- 8.2 Self-evaluation Questions 353
- 8.3 Basic Feedback Circuits 354
- 8.4 Actual Voltage Feedback Circuits 361
- 8.5 Current Feedback 367
- 8.6 A New Look at Feedback 379
- 8.7 Feedback and Frequency Response 382
- 8.8 A More Rigorous Approach to Feedback Circuit Analysis 388
- 8.9 Summary 391
- 8.10 Glossary 391
- 8.11 References 391
- 8.12 Problems 392

CHAPTER NINE
Power Amplifiers 397
- 9.1 Instructional Objectives 397
- 9.2 Self-evaluation Questions 397
- 9.3 Introduction 398
- 9.4 Power Dissipation and Power Transistors 399
- 9.5 Thermal Resistance and Heat Sinks 411
- 9.6 Class A Power Amplifiers 419

9.7 Class B Amplifiers 431
9.8 Distortion and Class B Biasing 437
9.9 Complementary Symmetry Amplifiers 440
9.10 Integrated-Circuit Power Amplifiers 449
9.11 Summary 454
9.12 Glossary 454
9.13 References 454
9.14 Problems 455

CHAPTER TEN
Power Supplies 463
10.1 Instructional Objectives 463
10.2 Self-evaluation Questions 463
10.3 Introduction to Power Supplies 464
10.4 Rectifiers 465
10.5 The Capacitive Filter 472
10.6 Other Filters and the Voltage Doubler 479
10.7 Voltage Regulators 483
10.8 Transistor Voltage Regulators 487
10.9 Integrated Circuit Voltage Regulators 498
10.10 Switching Regulators 505
10.11 Summary 515
10.12 Glossary 516
10.13 References 516
10.14 Problems 517

CHAPTER ELEVEN
Tuned Amplifiers and Oscillators 523
11.1 Instructional Objectives 523
11.2 Self-evaluation Questions 523
11.3 The Tuned-circuit Amplifier 524
11.4 More Selective Amplifiers 531
11.5 Oscillator Theory 540
11.6 High-frequency Sinusoidal Oscillators 543
11.7 Phase-shift Oscillators 552
11.8 Digital Oscillators 558
11.9 Summary 562
11.10 Glossary 562
11.11 References 562
11.12 Problems 563

CHAPTER TWELVE
Operational Amplifiers 567
12.1 Instructional Objectives 567
12.2 Self-evaluation Questions 567
12.3 Introduction 568
12.4 Characteristics of the Op-amp 576
12.5 Integrators and Differentiators 590
12.6 Applications of Op-amps 600
12.7 Summary 613
12.8 Glossary 613
12.9 References 613
12.10 Problems 614

APPENDIX A
Characteristics of the 2N3903 and 2N3904 Transistors 619

APPENDIX B
Derivation of the Exact Hybrid Parameter Equations for a Transistor 627

APPENDIX C
A Program for Finding the Gain of a JFET with a Capacitively Coupled Load 631

APPENDIX D
Derivation of Equations for a Difference Amplifier 637

APPENDIX E
The Frequency Response Due to the Emitter Bypass Capacitor 641

APPENDIX F
Derivation of Feedback Circuit Equations 643

APPENDIX G
The Equivalence of a Series and a Parallel Coil 649

APPENDIX H
The Derivation of Tank Circuit Equations 651

Answers to Selected Problems 653

Index 667

Introduction

This book discusses transistors and the circuits that can be built from them. Transistors are the basic *amplifying element* in modern electronics. An amplifying element is a component of an electronic circuit that can take a small, varying electrical signal and increase its voltage, current, or power until it can drive a loudspeaker or perform some other useful function.

Historically, the first amplifying element, the triode vacuum tube, was invented in 1907. Before the triode, electricity was available; it could turn on lights and run motors. The triode and other vacuum tubes brought in *electronics*, the ability to transmit and control information in an electrical form. The first radios, television sets, and computers were built using vacuum tubes. Radios and TVs are examples of electronics at work. They use their antennas to pick minute signals out of the air. Then they process these signals and increase their strength to the point where they can drive a speaker for sound, or a cathode ray tube for picture. Computers are also built of electronic circuits, and are used to process information, perform calculations, and control external devices such as printers and disk drives.

Vacuum tubes were the dominant electronic amplifying devices until the mid-1950s. They had several major drawbacks, however. Tubes are much larger and bulkier than transistors. They also absorb much more energy and dissipate most of it as heat that must be removed from the circuit. Tubes require a heating element, called a filament, to heat their cathodes to the point where they can emit electrons. In addition, most tubes required high voltages for their operation. As electronics became more complex, and required more amplifying elements, the problems associated with tubes limited the growth of electronics. Although the first computers, which amazed many cynics by actually working, were built using tubes (18,000 of them in the UNIVAC I, circa 1950), cooling this vast array of tubes was a major problem. Keeping them operating was a second problem. Early computers were down, or out of service, several hours each day for routine maintenance. These computers dissipated several kilowatts of power, occupied an entire room (there were racks and racks of tubes), and cost several million dollars. Yet they were not as powerful as a modern Apple, TRS-80, or other small computers that can sit on a desk and cost less than $1000. Obviously something had to be done about all those tubes.

Fortunately, the transistor was invented at Bell Telephone Laboratories in 1947. It was based on the movement of electric charges in a solid semiconductor medium giving rise to the term "solid-state electronics." Because they did not use filaments, the power consumption dropped dramatically. The power supply voltages also dropped from about 300 volts for tubes to a safe 20 or 30 volts and virtually eliminated the danger of electrical shock.

Transistors dominated the electronics field from the mid-1950s until the early 1970s, when integrated circuits (ICs) became widely available. But an integrated circuit is really an array of transistors etched on a semiconductor wafer to perform a specific function. Of course discrete transistors are still used in many circuits, especially where the power requirements preclude the use of integrated circuits.

The purpose of this book is to explain the operation of transistors to technologists and engineers and concentrate on practical circuits that can be built using them. It is *not* a book for the *designer of transistors*, but for the designer and builder of *circuits with transistors*. Consequently, the book does not go into solid-state physics very deeply, leaving that for those books that emphasize the design and construction of transistors themselves. Instead, we concentrate on the characteristics and limitations of existing transistors and integrated circuits, so the reader can understand transistor circuits and be able to build, repair, and design circuits using them.

CHAPTER ONE

Semiconductors and Diodes

1-1 Instructional Objectives

This chapter introduces doped semiconductors, semiconductor junctions, and diodes made from these junctions. The reader must understand their operation before attempting to master transistors. Diodes are also used in a variety of important circuits that are discussed in this chapter. After reading it the student should be able to:

1. Calculate the resistance of a piece of material given its shape.
2. Calculate the conductivity or resistivity of a semiconductor.
3. Design clipping and clamping circuits to produce specified waveforms.
4. Analyze and design circuits using Zener diodes.
5. Design electric-eye or intrusion-detection circuits.
6. Use light-emitting diodes to indicate the status of a circuit.

1-2 Self-Evaluation Questions

Watch for the answers to the following questions as you read the chapter. They should help you to understand the material.

1. What chemical property of metals makes them good conductors?
2. What are donor and acceptor atoms? What effect do they have when they are added into a crystal lattice?
3. What are the major differences between a switching diode and a rectifier diode?

4. What is the important characteristic of a Zener diode?
5. What is the dynamic resistance of a diode and a Zener diode?
6. What is the principle of operation of a photodetector?
7. How do seven-segment displays work? What is the difference between common anode and common cathode?
8. How do varactor diodes function? What are they used for?

1-3 CONDUCTORS AND INSULATORS

The elements and compounds that comprise matter can be categorized in many different ways. One way is to subdivide matter into *conductors* and *insulators*. The ideal conductor provides no resistance to electric currents, while the ideal insulator allows no current to flow, or behaves as an infinite resistor. Although perfect conductors and perfect insulators do not exist, materials that approximate both of them are readily available.

The ability of a material to electrically conduct is measured by its resistivity (ρ, Greek letter rho). Resistivity is the resistance of a cube of the material one centimeter (cm) on a side, as shown in Figure 1-1. This is a characteristic of the material (resistivities for several materials are given in Table 1-1). The units of resistivity are ohms-cm (Ω-cm).

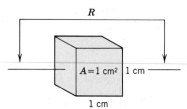

FIGURE 1-1
A cube of material. The resistance of a cube of this size equals the resistivity (ρ) of the material in Ω – cm.

Table 1-1 shows that a good conductor has a resistivity of about 2×10^{-6} ohm-cm, whereas insulators have a resistivity of about 10^{14} ohm-cm. Thus the difference between the resistivity of insulators and conductors is vast ($\approx 10^{20}$ ohm-cm).

The resistance of any object can be calculated, if its resistivity and shape are known, by Equation 1-1:

$$R = \frac{\rho L}{A} \qquad (1\text{-}1)$$

Table 1-1
Resistivities of Various Materials

	Material	ρ(Ω-cm)
Conductors	Aluminum	2.63×10^{-6}
	Copper	1.72×10^{-6}
	Gold	2.44×10^{-6}
	Iron	5.3×10^{-6}
	Silver	1.62×10^{-6}
Semiconductors	Carbon	.35
	Germanium	45
	Silicon	230,000
Insulators	Amber	5×10^{16}
	Mica	1.3×10^{14}
	Glass	10^{14}
	Wood	3×10^{10}

where ρ is the resistivity of the material, L is the length of the object in cm and A is its area in cm^2.

From Equation 1-1 we can see that the resistance between opposite faces of a 1-cm cube of copper is $\approx 10^{-6}$ ohms or about 0, whereas the resistance of a similar block of mica is $\approx 1.3 \times 10^{13}$ ohms, a reasonable approximation of infinity. These are extremes; copper is one of the best and most widely used conductors, and mica is often used as an insulator.

EXAMPLE 1-1

Wire size #0 has an area of 53.48 sq mm. Find the resistance of 100 m of this wire.

Solution

$$A = 53.48 \text{ mm}^2 = 0.5348 \text{ cm}^2$$

$$R = \frac{\rho L}{A} = \frac{1.72 \times 10^{-6} \times 10^4}{0.5348} = 0.03224 \text{ }\Omega\text{[1]}$$

---

[1] Wire tables give the resistance of #0 gauge wire as 0.3224 ohms per kilometer at 20°C, which agrees with our result.

1-3.1 Metals and Nonmetals

Elementary physics tells us that the atoms of any element consist of a number of protons and neutrons confined to the nucleus of the atom surrounded by number of electrons equal to the number of protons. The electrons exist in rings or energy levels, and the outermost level is called the *valence band*. A complete valence band consists of eight electrons and is very stable.

Metals consist of complete energy bands with one or two electrons in the outer, or valence, band. A diagram of the copper atom is shown in Figure 1-2.

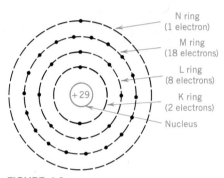

FIGURE 1-2
A simplified diagram of the copper atom.

It has complete rings of 2, 8, and 18 electrons called the K, L, and M rings, and one additional electron in its outer or N ring, which is the valence band. This electron in the N ring is very loosely bound to its parent atom and is really free to wander within the copper. Thus each copper atom contributes a free electron that moves randomly within the copper. Under the influence of an electric field, however, the free electrons move readily. This accounts for the high conductivity of copper and other metals.

Elements that are not metals usually have an incomplete valence band that lacks one or two electrons, and compounds are made up of elements that share electrons. In both nonmetals and compounds, the electrons are tightly held. Therefore, the best conductors are metals. Gold is a good conductor and does not tarnish, but it is prohibitively expensive. It is sometimes used in small quantities on printed circuit boards, especially on the fingers, which make contact with an external connector. Copper is much less expensive and is the most widely used conductor.

1-4 SEMICONDUCTORS

The use of *semiconductors* has allowed the electronics industry to progress by leading to the fabrication of solid-state devices such as diodes, transistors, and integrated circuits. Carbon, germanium, and silicon atoms all have four electrons in their valence bands. They straddle the line between metals and nonmetals and are called semiconductors. Table 1-1 shows that their resistivity is much higher than a metal and much lower than an insulator: another reason to label them semiconductors.

Germanium was the first semiconductor material to be used, but silicon has several advantages (see section 1-6.2) and is the major element used in most transistors and diodes.

1-4.1 The Crystal Lattice

As previously explained, both silicon and germanium have four valence electrons in their outer band. To complete the band, the atoms arrange themselves in a *crystal lattice* so that each atom *shares its valence electrons* with four other atoms in what is called a *covalent bond*. The actual arrangement of the atoms is a three-dimensional tetrahedral pattern, shown in Figure 1-3a, but the simplified two-dimensional diagram of Figure 1-3b is a clearer representation of the covalent bonds. Note that each nucleus is surrounded by eight electrons in the lattice structure due to the covalent bonding.

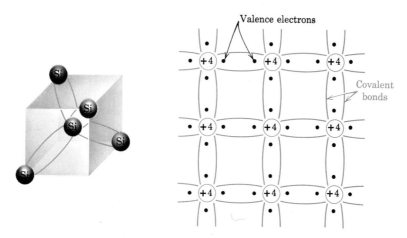

(a) Tetrahedral pattern (b) Two-dimensional representation

FIGURE 1-3
Arrangement of atoms in a silicon crystal. (From *Electronics Circuits & Devices* Third Edition by Ralph J. Smith. Copyright © 1987 by John Wiley & Sons, Inc. Reprinted by permission of John Wiley & Sons, Inc.)

1-4.2 Free Electrons and Holes

In the idealized representation of Figure 1-3, no free electrons are available to conduct current and the material is an insulator. At room temperatures, however, a few electrons acquire enough energy to break their covalent bond and escape from their parent atoms. When an electron does escape from a covalent bond, it is available to conduct electricity. It also leaves behind a

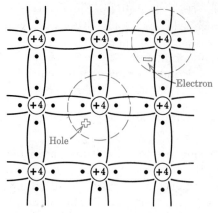

FIGURE 1-4
Silicon crystal with one covalent bond broken. (Quoted by Esther M. Conwell in "Properties of Silicon and Germanium II," *Proc. IRE*, June 1958, p. 1281.)

vacancy called a *hole*. A silicon crystal with a broken covalent bond is shown in Figure 1-4.

Holes are also conductors of electricity. This is because it requires very little energy for an electron in one covalent bond to jump to fill a hole in a neighboring covalent bond. Hole conduction is illustrated in Figure 1-5. Assume initially that there was a hole at point B and an electron at point A. Figure 1-5 shows the electron jumping from point A to point B. This effectively moves the hole from B to A. Note that the movement of holes is in the *opposite* direction to the movement of electrons.

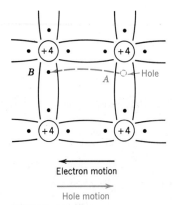

FIGURE 1-5
Hole conduction.

1-4.3 Conduction in Semiconductors

Under the influence of an electric field, a semiconductor will conduct current carried by both holes and free electrons. The electrons, of course, move opposite to the direction of conventional current, while the holes move in the direction of conventional current.

The total current in a semiconductor is therefore due to the sum of hole and electron conduction. *Conductivity is a measure of how well a material conducts electricity,* and is the *reciprocal* of the material's *resistivity*. The units of conductivity are Siemens per centimeter (S/cm). The conductivity (σ, Greek letter sigma) of any material depends on both the number of free electrons and their *mobility* (their ability to move in response to an electric field). Conductivity is given by equation 1-2, where n is the number of free electrons, μ_n is the mobility of these electrons, p is the number of holes, μ_p is the mobility of the holes, and e is the charge of an electron (1.6×10^{-19} coulombs). In a pure or intrinsic semiconductor

$$\sigma = (n\mu_n + p\mu_p)e \qquad (1\text{-}2)$$

the number of free electrons must equal the number of holes, because each electron that escapes from a covalent bond leaves a hole behind. The number of electrons or holes is called the intrinsic carrier density, n_i. Table 1-2 gives the pertinent characteristics for silicon and germanium. •

Table 1-2
Properties of Silicon and Germanium

Property	Si	Ge
Atomic number	14	32
Atomic weight	28.1	72.6
Density, g/cm³	2.33	5.32
Dielectric constant (relative)	12	16
Atoms/cm³	5.0×10^{22}	4.4×10^{22}
E_{GO}, eV, at 0 K	1.21	0.785
E_G, eV, at 300 K	1.1	0.72
n_i at 300 K, cm^{-3}	1.5×10^{10}	2.5×10^{13}
Intrinsic resistivity at 300 K, Ω-cm	230,000	45
μ_n, cm²/V · s at 300 K	1,300	3,800
μ_p, cm²/V · s at 300 K	500	1,800
D_n, cm²/s = $\mu_n V_T$	34	99
D_p, cm²/s = $\mu_p V_T$	13	47

Source: G. L. Pearson and W. H. Brattain, History of Semiconductor Research, *Proc. IRE,* vol. 43, pp. 1794–1806, December, 1955. ©1955 IRE (now IEEE).

EXAMPLE 1-2

Using Table 1-2, show why the resistivity of silicon is about 5000 times the resistivity of germanium.

Solution
The table indicates that the intrinsic carrier density of germanium exceeds silicon by a factor of 1700. Furthermore, the mobility of both electrons and hole is about 3 times greater in germanium, so the conductivity of germanium is $\approx 3 \times 1700 \approx 5100$ times the conductivity of silicon.

EXAMPLE 1-3

Using Table 1-2, find the resistivity of silicon.

Solution
From equation 1-2 the conductivity is:

$$\sigma_i = (n\mu_n + p\mu_p)e$$

since $n = p$ for an intrinsic semiconductor

$$\sigma_i = n(\mu_n + \mu_p)e = 1.5 \times 10^{10} \times (1300 + 500) \times 1.6 \times 10^{-19}$$
$$= 4320 \times 10^{-9} = 4.32 \times 10^{-6} \text{ S/cm}$$

$$\rho_i = \frac{1}{\sigma_i} = \frac{10^6}{4.32} = 232{,}000 \text{ }\Omega\text{-cm}$$

Note how close this answer is to the value given in Table 1-2.

1-4.4 Temperature Dependence

The resistivity of silicon and germanium in Table 1-2 is given at a specific temperature, 300°K, or 27°C, or about normal laboratory temperature. Electrons in the covalent bonds that gain enough energy to break these bonds do so primarily by extracting this energy from the heat of the environment. When an electron does break a bond, it becomes a conductor and leaves behind a hole, which is also a conductor. This is called a *thermally generated electron–hole pair*. Actually, the electron wanders around the crystal for a short time and then falls into a hole. Meanwhile, other electron–hole pairs are being created so that at any given temperature the concentration of carriers is in equilibrium.

As the ambient temperature increases, the number of electrons with sufficient energy to break out of the covalent bond and create a thermally generated electron–hole pair increases rapidly. A rule based on experience is that the conductivity of germanium or silicon at about 300°K doubles for each 10°C increase in temperature.

EXAMPLE 1-4

What is the resistivity of silicon at 330°K?

Solution
From Table 1-2 the resistivity of silicon at 300°K is 230,000 Ω-cm. 330°K is three intervals of 10°C, and since the conductivity doubles for each interval, it should increase by a factor of 8. Therefore the resistivity should decrease by a factor of 8 or:

$$\rho = \frac{230,000}{8} = 28,750 \text{ Ω-cm}$$

1-4.5 Doped Semiconductors

Intrinsic semiconductors are made much more useful by adding a small amount of an impurity called a *dopant* to the silicon lattice. Two types of dopants exist: donor atoms that produce *n*-type silicon or germanium, and acceptor atoms that produce *p*-type.

A *donor* impurity is an element with *five* electrons in the valence band, such as arsenic or antimony. The crystal lattice with a pentavalent donor atom is shown in Figure 1-6. It shows that four of the five valence electrons

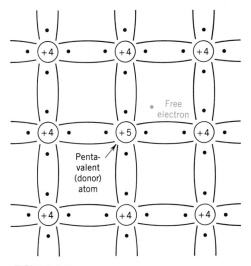

FIGURE 1-6
Effect of *n*-type doping. (From *Electronics Circuits & Devices* Third Edition by Ralph J. Smith. Copyright © 1987 by John Wiley & Sons, Inc. Reprinted by permission of John Wiley & Sons, Inc.)

combine in covalent bonds with neighboring silicon atoms. The fifth electron is very loosely held and becomes a free electron. In donor doped silicon almost all the carriers are electrons or negative charges, and this is called *n-type silicon*. We would like to emphasize that a very small amount of doping has a dramatic effect on the characteristics of the silicon. Millman (see References) shows that the addition of one impurity atom for every hundred million (10^8) silicon atoms reduces the resistivity of the silicon by a factor of 24,000, or from 230,000 to 9.6 Ω-cm. Even this sparse doping changes silicon from an insulator to a conductor.

Silicon or germanium can also be doped with *acceptor atoms*, such as boron, gallium, or indium. These elements have *three* electrons in their valence band. Figure 1-7 shows that the three valence electrons of the acceptor

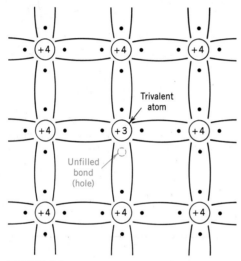

FIGURE 1-7
Effect of *p*-type doping. (From *Electronics Circuits & Devices* Third Edition by Ralph J. Smith. Copyright © 1987 by John Wiley & Sons, Inc. Reprinted by permission of John Wiley & Sons, Inc.)

impurity form covalent bonds with their neighboring silicon atoms, but *the missing electron creates a hole* in the covalent bonding. Thus in silicon doped with acceptor atoms, holes predominate, and this is called *p-type silicon*.

EXAMPLE 1-5

What effect on the resistivity of silicon should be expected if one acceptor atom is added for every 10^8 silicon atoms?

Solution
The number of holes increases in the same proportion as the number of electrons for the same concentration. The mobility of holes, however, is only 1/2.6 times the mobility of electrons, so the resistivity should drop to only 2.6 × 9.6 Ω-cm, or about 25 Ω-cm. Thus adding holes is not quite as effective as adding electrons because the mobility of holes is less.

We would like to emphasize that both *p*- and *n*-type silicon are *electrically neutral*. The number of protons and electrons are equal in both. In *n*-type silicon the free electrons are balanced by the occasional nucleus that contains five protons, and a similar situation exists in *p*-type silicon.

1-5 THE *p-n* JUNCTION

A diode is an electronic element that conducts current in only one direction. The symbol for a diode is shown in Figure 1-8. It is a two-terminal device

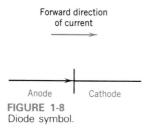

FIGURE 1-8
Diode symbol.

consisting of an *anode* and *cathode*. Conventional *current flows from anode to cathode*. The *ideal diode* exhibits *no resistance* to any current in the *forward direction* and acts as an infinite resistor or *open circuit* if a voltage is applied to it in the *reverse* direction (making the cathode positive with respect to the anode).

1-5.1 The Unbiased *p-n* Junction

Most diodes are made by connecting a region of *p*-type silicon or germanium material to an *n*-type, as shown in Figure 1-9. The boundary between the two regions becomes the *p-n junction*. First, imagine that a barrier is placed at the junction so that all the holes must remain on the *p* side and the free electrons are confined to the *n* side. Under these conditions, both sides are electrically neutral. Now assume, as in an actual junction, that the barrier is removed. Free electrons will move or diffuse into the *p* region and fill or occupy the holes near the junction.

The migration of electrons from the *n*- to the *p*-type material causes the *p* material to become *negatively charged* because it has gained electrons and the *n*

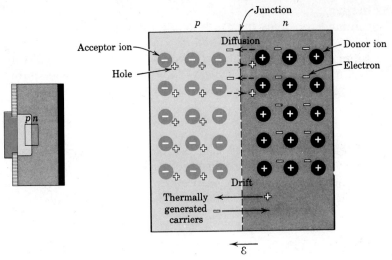

FIGURE 1-9
An unbiased *p-n* junction. (From *Electronics Circuits & Devices* Third Edition by Ralph J. Smith. Copyright © 1987 by John Wiley & Sons, Inc. Reprinted by permission of John Wiley & Sons, Inc.)

region to become *positively charged*. This, of course, creates a potential difference (note the $\mathscr{E}$ vector in Figure 1-9) that prevents any further migration of electrons, and a state of equilibrium is reached.[2] A second effect is that the electrons are occupying the holes at the junction. This creates a *depletion region* at the junction where there are very few free electrons or holes. The absence of free charges at the junction is also shown in Figure 1-9.

Figure 1-10 shows the charge density in the semiconductor. The charged region and the depletion region exist only near the junction and the potential increases in the *n* region because some electrons have migrated to the *p* region.

1-5.2 The Forward-Biased *p-n* Junction

A *p-n* junction can be *forward-biased* by applying a positive voltage to the *p* side as shown in Figure 1-11a. This voltage propels both electrons and holes, in their respective *n* or *p* regions, *toward* the junction. This action supports current flow. A current-limiting resistor has been added to this circuit to prevent excessive current.

1-5.3 The Reverse-Biased *p-n* Junction

If a voltage is applied to make the *p* side of the junction negative with respect to the *n* side, as shown in Figure 1-11b, both electrons and holes are *pulled*

---

[2]Under these conditions the drift current of thermally generated carriers equals the diffusion current. See *Microelectronics* by Jacob Millman.

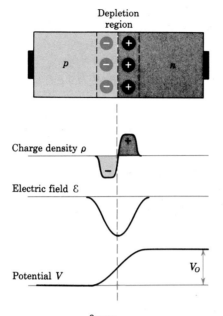

FIGURE 1-10
An open-circuited *pn* diode. (From *Electronics Circuits & Devices* Third Edition by Ralph J. Smith. Copyright © 1987 by John Wiley & Sons, Inc. Reprinted by permission of John Wiley & Sons, Inc.)

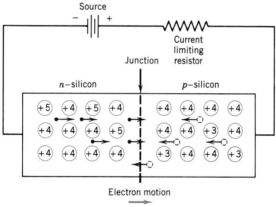

(a) Forward-biased *p-n* junction

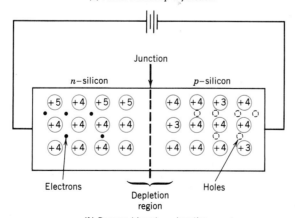

(b) Reverse-biased *p-n* junction

FIGURE 1-11
Biased *p-n* junctions.

1-5 THE p-n JUNCTION   **15**

away from the junction. This increases the depletion region. Because there are no carriers in the depletion region, the junction acts as an open circuit and prevents any current from flowing. Notice that there is no electron or hole motion in Figure 1-11b. Therefore the *p-n* junction is a good conductor in the forward direction and an insulator in the reverse direction.

Actually a small amount of current will flow in the reverse direction due to *thermally generated minority carriers*. When an electron–hole pair is created thermally in an *n*- or *p*-type semiconductor, one of the carriers (the electron in the *p*-type or the hole in the *n*-type) is a minority carrier. Minority carriers are propelled toward the junction in a reverse-biased diode and form a *leakage* or *reverse current*. Table 1-2 shows that silicon has a much lower carrier density, n_i, than germanium. Consequently, there are fewer thermally generated minority carriers and the reverse current is much less in a silicon diode. This is one of the major reasons for preferring silicon to germanium.

If the reverse voltage across the diode is too high, it will accelerate the free electrons to the point where they have enough energy to knock additional electrons out of the lattice structure when they collide with it. This results in a rapid increase in current and is known as *avalanche* breakdown. The inverse voltage applied to the diode must not be large enough to cause avalanche breakdown or the diode may be destroyed. Zener diodes (see section 1-9), however, make use of avalanche breakdown.

1-6 DIODE CHARACTERISTICS

Commonly available diodes are generally made of silicon *p-n* junctions and behave as described in section 1-5. Diodes can be divided into two main categories: switching diodes and rectifiers. Rectifier diodes are larger and are designed to handle high currents, usually from 1 to 10 amperes. Switching diodes are much smaller and are limited to a smaller forward current. They react to change in voltage more quickly, however, and are used in digital and other high-speed circuits where fast response is absolutely necessary.

The ideal diode has zero resistance in the forward direction. Real diodes are best represented as shown in Figure 1-12. In the forward direction they

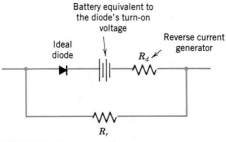

FIGURE 1-12
The equivalent circuit of a diode.

exhibit a *turn-on voltage,* or a certain forward voltage that must be attained in order for the diode to conduct. For silicon this voltage is about 0.7 V. This is represented by the battery in Figure 1-12. Below the turn-on voltage the diode still acts as an open circuit.

Above the turn-on voltage the current in a diode increases very rapidly. The voltage–current curves for a **1N4148,** a popular switching diode, are shown in Figure 1-13. The turn-on voltage at 25°C is about 0.4 V at very low

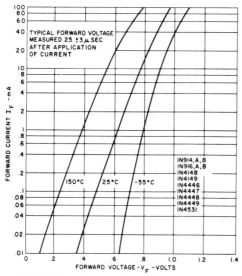

FIGURE 1-13
Forward characteristics of a family of diodes.
(Reprinted with permission of General
Electric Company.)

currents and the forward voltage increases to about 0.9 V at 100 mA of forward current. The figure also shows that the current increases as the temperature of the diode increases.

In the forward region the diode exhibits a small forward resistance called the *dynamic resistance,* r_d. It is defined as the slope of the curve at any point, dv/dI. The dynamic resistance is nonlinear; it decreases as the current through the diode increases, and this makes the equivalent circuit of Figure 1-12 difficult to use. Fortunately the dynamic resistance of most diodes is small, and in practice it is usually ignored. The rule followed by most engineers is that the forward voltage across a conducting silicon diode is 0.7 V and constant. Diodes operating in their normal range rarely have a forward voltage that deviates by more than 0.1 V from this value. R_r is the reverse resistance of the diode. It is nonlinear but normally very high and can be ignored. For a **1N4148,** the reverse resistance is calculated in Example 1-8.

EXAMPLE 1-6

A "clamp" circuit, sometimes used in digital circuits, is shown in Figure 1-14. Find the voltage at the clamp point and the current through the diodes.

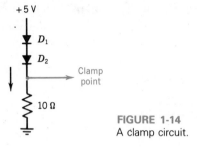

FIGURE 1-14
A clamp circuit.

Solution

Both diodes are forward biased so current will flow through them and the 10 kΩ resistor. Since the forward voltage of each diode is 0.7 V, the voltage at the clamp point will be 5 − 2 × 0.7 = 3.6 V. This is a logic 1 for TTL digital circuits.[3]

The current in the diodes cannot be calculated just by considering the diodes, because the diode voltage is always 0.7 V. In this problem and problems to follow, diode currents can be determined only by finding the currents in the connecting circuit elements. Here we observe that the current in the diodes is the same as the current in the resistor. Therefore:

$$I_{diode} = \frac{3.6 \text{ V}}{10 \text{ k}\Omega} = 0.36 \text{ mA}$$

EXAMPLE 1-7

Can 2 V be connected across a diode in the forward direction?

Solution

No. Of course it is possible to draw a circuit as shown in Figure 1-15, but it will not work. If the switch is ever closed, either the diode will burn out (perhaps explode) due to the excessive current through it, or the voltage of the supply will decrease. This problem was included to emphasize the fact

---

[3] Most digital circuits use TTL (Transistor–Transistor Logic), where a logic 0 is any voltage between 0 and 0.8 V, and any voltage between 2 and 5 V is a logic 1.

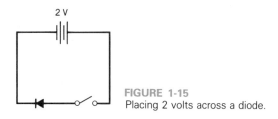

FIGURE 1-15
Placing 2 volts across a diode.

that forward voltage of a diode (or a transistor's base-to-emitter junction) can never be significantly more than 0.7 V. Students, however, often make the mistake of assuming that higher forward voltages may exist across a diode.

1-6.1 Diode Specifications

Table 1-3 is a list of the most important diode characteristics. The characteristics of a **1N4148** switching diode and a **1N4004** rectifier diode are shown. The table can also be used to compare switching and rectifier diodes. A photograph of the diodes is shown in Figure 1-16. Both diodes have bands at the cathode end, so the user can determine the current direction.

FIGURE 1-16
A **1N4148** switching diode and a **1N4004** rectifier diode.

In Table 1-3 the average rectifier forward current is the average current the diode can conduct if it is used as a half-wave rectifier (see sections 1-7.1 and 10-4.1). Here the rectifier diode is obviously superior to the switching diode. For large currents, the current capacity depends on the cooling or heat sinking of the diode (see section 9-5.3).

Table 1-3
Diode Characteristics

	1N4148 Switching diode	1N4004 Rectifier
Average rectified forward current	.15	1.0 amperes
Peak inverse voltage	75	400 volts
Typical reverse current	10	50 nA
Reverse recovery time	4 ns	6 μs

The *peak inverse voltage* (PIV) is the maximum voltage the diode can withstand in the reverse direction before breaking down. The larger rectifier diode can also withstand a higher PIV.

The reverse current is the diode current conducted in the reverse direction. Switching diodes have smaller reverse currents. Curves of the reverse current for a **1N4148** as a function of reverse voltage and temperature are shown in Figure 1-17. Note that the reverse current is not highly dependent on the reverse voltage. Figure 1-17a shows that it is about 10 nA for most reasonable reverse voltages. Figure 1-17b, however, shows that the reverse current does approximately double for each 10°C rise in temperature.

EXAMPLE 1-8

From Figure 1-17a determine the reverse resistance of a **1N4148** diode when the reverse voltage is 30 V.

Solution
At 30 V Figure 1-17a shows that the diode current is 10 nA. Therefore

$$R_r = \frac{30}{10 \times 10^{-9}} = 3 \times 10^9 \text{ ohms}$$

The *reverse recovery time* (t_{rr}) is a measure of how quickly the diode becomes an open circuit when the applied voltage abruptly switches from the forward to the reverse direction. The situation is illustrated in Figure 1-18. The circuit is shown in Figure 1-18a. Initially the voltage is V_F in the forward direction as shown in Figure 1-18b, and the current is V_F/R. When the voltage reverses to $-V_F$, the current also reverses to $-V_F/R$ for a short period known as the *storage time*. When the diode is conducting in the forward direction, both electrons and holes are propelled toward the junction, and an excess of

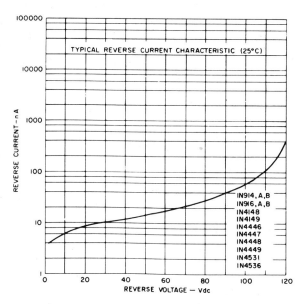

(a) Current versus reverse voltage

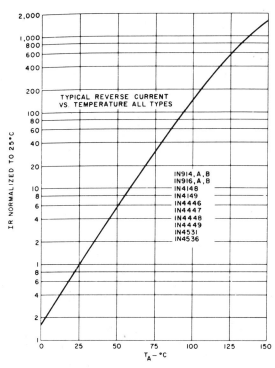

(b) Current versus temperature

FIGURE 1-17
Curves of diode reverse current. (Reprinted with permission of General Electric Company.)

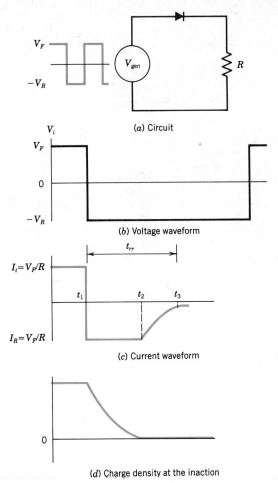

FIGURE 1-18
Reverse recovery time (t_{rr}) of a diode.

charged particles exist there. Figure 1-18d shows the charge density at the junction, which is positive when the circuit is forward biased.

When the voltage is suddenly reversed, it takes a finite time for the charges to be swept out of the junction, and the depletion region established. This is the storage time. During this time, the charges support a reverse current until they are all swept out of the junction. The current waveform is shown in Figure 1-18c.

When the depletion region is established, there are essentially two oppositely charged regions (the p material with a negative charge and the n material with a positive charge), separated by an insulator (the depletion layer). Two conducting regions separated by an insulator form a capacitor that must be charged up before conduction can stop. The time required to charge this capacitor is called the *transition time*.

In Figure 1-18, the storage time is the time between the voltage switching (t_1) and the excess charges decaying to approximately 0 (t_2). The storage time is added to the transition time (between t_2 and t_3) to give the reverse recovery time (t_{rr}) of the diode.

The **1N4148** is a very fast switching diode. It is specified to have a $t_{rr} \leq 4$ ns and a capacitance of ≤ 2 pf. We tested several samples in the laboratory and found they were too fast to show any reverse recovery time. We also tried the **1N4004** rectifier diode; the results are shown in Figure 1-19. V_{in} was a ± 5 V square wave. Figure 1-19 shows that when the input goes negative, the

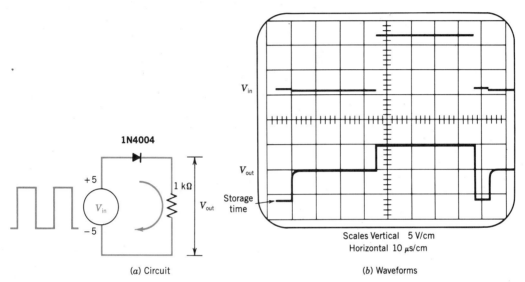

FIGURE 1-19
Oscilloscope traces showing reverse recovery time.

output voltage drops to -5 V for about 6 μs. During this time the excess charges support the reverse current. After the 6 μs storage time, the diode begins to operate and the current and output voltage become 0. In this circuit, the transition time is too small to be significant and the diode appears to recover instantly. This diode is far too slow to be used in any high-speed applications.

1-6.2 Silicon and Germanium Diodes

The first transistors and diodes were constructed using germanium as the base semiconductor. Germanium diodes have the advantage of a smaller forward voltage drop (0.3 V vs. 0.7 V for silicon), and some are still used in circuits where this is important.

Silicon, however, has become the predominant semiconductor in diodes and transistors. There are two main reasons for this:

1. Germanium transistors fail at about 100°C. Silicon can withstand temperatures up to about 200°C. This gives silicon a significant advantage when circuits that generate internal heat because of high current or power consumption are required. The **1N4148** silicon diode is very small. It can conduct a maximum current of 200 mA due mainly to heat considerations. The current capacity of a similar germanium diode would be much less.
2. The reverse leakage current is much less in silicon diodes. Millman and Taub (see References) have curves that show that the reverse current of a germanium diode at -40 V is 5 µA, whereas the silicon diode has a reverse current of 2 nA. This advantage, greater than three orders of magnitude, is even more important in transistors than in diodes.

Silicon diodes now predominate to such an extent that most manufacturers no longer make germanium diodes, and they are becoming expensive because they are not manufactured in large quantities.

1-7 CLIPPERS AND RECTIFIERS

A *clipping circuit* is used to clip off part of a waveform and transmit only the remaining portion of the waveform. Clipping circuits are made of diodes and resistors.

1-7.1 The Half-Wave Rectifier

A rectifier is a clipping circuit that converts an ac input into a dc output. It clips or removes the *negative* portion of the input ac waveform. The half-wave rectifier circuit, shown in Figure 1-20, is the simplest clipper circuit. It can be analyzed

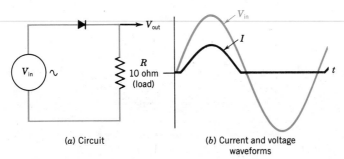

(a) Circuit

(b) Current and voltage waveforms

FIGURE 1-20
A half-wave rectifier.

using two different regions of the input. During the positive half cycle of the input sine wave when the input voltage is greater than 0.7 V, the voltage across the resistor is $V_{in} - 0.7$. During the negative half cycle of the input sine wave, or when the input voltage is less than 0.7, the diode can be approximated by an open circuit and the output voltage is 0.

EXAMPLE 1-9

If a 10 V peak sine wave is applied to the circuit of Figure 1-20:
a. Draw the current waveform.
b. What is the peak inverse voltage?

Solution
a. When the input voltage is positive, $I = (V_{in} - 0.7)/R$. At its peak $V_{in} = 10$ and $I = 93$ mA. When the input voltage is negative the current is 0, so the current is approximately a half pulse whose magnitude is 93 mA, as shown in Figure 1-20b.
b. The peak inverse voltage (PIV) occurs when the input is most negative. Since the voltage across the resistor is 0, the entire inverse voltage is across the diode at that time and the PIV for the diode is 10 V.

The circuit of Figure 1-20 produces dc because *the current flows in only one direction*. Note that it is *not* dc in the sense that the voltage is constant. In order to provide a more constant voltage, a capacitor is often placed across the load, as shown in Figure 1-21a. The capacitor acts as a filter and produces the smoother waveform of Figure 1-21b. When the input voltage is at its peak, between times t_1 and t_2 in Figure 1-21b, the capacitor charges. When the input

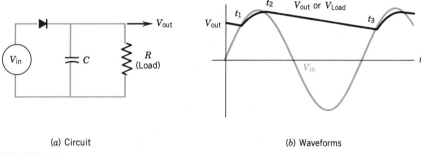

(a) Circuit   (b) Waveforms

FIGURE 1-21
A half-wave rectifier with a filter capacitor.

voltage falls, the capacitor discharges through the load resistance. This is between t_2 and t_3 in Figure 1-21b. At t_3 the input voltage is once more high enough to charge the capacitor again.

The output voltage waveform shows that there is always voltage across the load and hence always current in the load. This current is supplied mainly by the capacitor, which recharges at the peak of every cycle. This is an improvement over the circuit without the capacitor (Figure 1-20a), where no load current flows during the negative half cycle. The exact waveform depends on the value of R and C. This circuit and more advanced rectifier circuits are discussed in detail in Chapter 10.

Adding a capacitor affects the PIV. In the worst case the capacitor holds the voltage across the load at the peak input voltage, V, while the input is most negative, or at $-V$. At this instant the voltage across the diode is $2V$ and the diode must be able to withstand this voltage.

The circuit of Figure 1-21 can also be used as a *demodulator* for an *amplitude modulated* (AM) wave, such as is used in an AM radio. An AM wave is shown in Figure 1-22*a*. It consists of a radio frequency wave ($\approx$1 MHz) modulated by an *audio signal*. The function of the demodulator is to detect the audio information contained in the AM wave. First the circuit of Figure 1-21 rectifies the wave as shown in Figure 1-22*b*. Then the capacitor charges up to the peak of the wave and the output approximates the audio information. In a rectifier circuit the capacitor is very large to provide better filtering. In a demodulator circuit the capacitor cannot be too large; if it were, the output wave could not follow the audio input. More advanced texts on radios and communications have formulas that allow the user to calculate the optimum value of the capacitor.

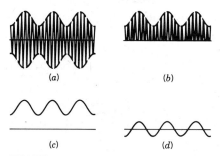

FIGURE 1-22
The mechanism of diode detection. (*a*) Modulated carrier wave. (*b*) Waveform resulting from half-wave rectification. (*c*) Envelope of the output of half-wave rectification. (*d*) Dc value removed by blocking capacitor. (From Fundamentals of Electronics by E. Norman Lurch. Copyright © 1960 by John Wiley & Sons, Inc. Reprinted by permission of John Wiley & Sons, Inc.)

1-7.2 Biased Clippers

A *biased* clipping circuit generally uses a battery or other voltage source to provide a *clipping voltage*. The output is clipped above or below the clipping voltage, depending on the direction of the diode. The half-wave rectifier of Figure 1-20 is a special case where the clipping voltage is 0 V. To simplify the discussion of clipping circuits, the forward diode voltage drop will be ignored.

The circuit of Figure 1-23*a* is a clipping circuit that clips all voltages below 5 V. The input voltage must be greater than 5 V for the diode to conduct. When the diode does conduct, it effectively shorts the input to the

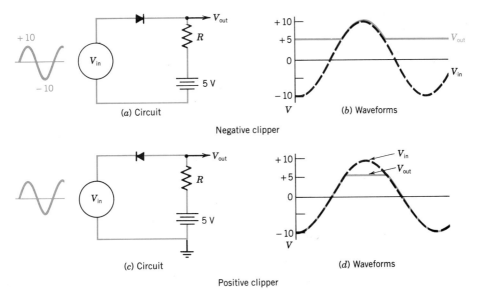

FIGURE 1-23
Clipping circuits.

output. When the input sine wave is below 5 V, the diode is open and the output is 5 V due to the battery. The output waveform is shown in Figure 1-23b.

If the diode is reversed, as shown in Figure 1-23c, it conducts whenever the input is less than 5 V. The output is shown in Figure 1-23d. Thus either the top or the bottom of the waveform can be clipped depending on the direction of the diode.

EXAMPLE 1-10

For the circuit of Figure 1-24 find the current waveform, the output voltage waveform, and the peak inverse voltage across the diode. Assume an ideal diode where the forward voltage drop and the reverse current are both 0. The input is a 30V sine wave as shown in Figure 1-24b.

Solution
The diode conducts only when the input voltage exceeds 10 V. The output voltage waveform is shown in Figure 1-24c. Note that when the input voltage is less than 10 V, the diode is open and the output is effectively equal to the battery voltage, 10 V.

Current flows only when the diode is conducting. The equation for current in this circuit is

$$I = \frac{V - 10 \text{ V}}{500}$$

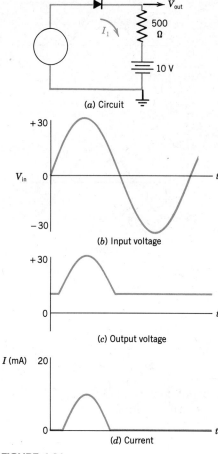

FIGURE 1-24
Circuit and waveforms for Example 1-10.

The current waveform is shown in Figure 1-24d. It is approximately a sinusoidal pulse that peaks when $V_{out} = 30$ V, or $I = 20/500 = 40$ mA.

The peak inverse voltage occurs when the input is -30 V. At this time the anode of the diode is at -30 V and the cathode is at $+10$ V, so the maximum inverse voltage across the diode is 40 V.

1-7.3 More Complex Clipping Circuits

More complex clipping circuits generally involve more components and often include several diodes. Generally the diodes are treated as ideal diodes, although if the actual 0.7 V forward drop were taken into account, it would not significantly increase the complexity of the analysis.

Circuits of this type are analyzed by dividing the input voltage into *ranges* and finding equations that relate the output to the input over each range. The boundary between the ranges occurs when any of the diodes in the circuit *change state* (from conducting to nonconducting or vice versa). We recommend that a three-column table be drawn up to analyze this type of circuit. The first column is the range of the input voltage. On the first line it starts at the lowest possible voltage and goes up until a diode changes state. The second column is the state of the diodes in the first range, and the third column is an equation relating the input voltage to the output voltage. When the range changes, another line is added to the table because both the condition of the diodes and the input–output equation will be different. The table continues, using one line for each different range until the input reaches its highest value. This procedure is best illustrated by Example 1-11.

EXAMPLE 1-11

a. For the circuit of Figure 1-25a, find the set of equations that relates the output voltage to the input voltage as the input voltage goes from -50 to $+100$ V.

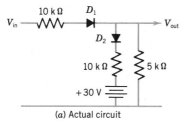

(a) Actual circuit

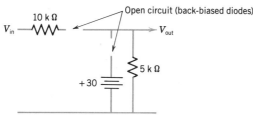

(b) Equivalent circuit when $-50 \text{ V} \leqslant V_{in} \leqslant 0 \text{ V}$

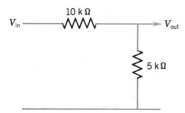

(c) Equivalent circuit when D_1 is on and D_2 is off

FIGURE 1-25
Circuit for Example 1-11.

b. Find the actual output voltage if the input voltage is

   1. 60 V
   2. 100 V

Solution

a. There are two diodes in this circuit, labeled D_1 and D_2 to distinguish them. Table 1-4 was constructed to help with this problem. The first step is to write $-50 \leq V_{in} \leq$ in the first column of Table 1.4. The most negative voltage, -50 V, is written on the left side of the equation. The number on the right side is left blank for now because the upper limit of the range has not yet been determined.

The second and third columns can now be written. With $V_{in} = -50$ V an inspection of the circuit reveals that both diodes are off or open circuits. The circuit looks as shown in Figure 1-25b and $V_{out} = 0$ V.

The crucial question is for what voltages does this situation apply? An examination of the circuit shows that it applies until $V_{in} = 0$ V. Just above 0 V, D_1 will turn on, but D_2 will remain off until V_{out} becomes 30 V or greater. The first row can now be completed, $-50 \leq V_{in} \leq 0$, D_1, D_2 off, and $V_{out} = 0$ V.

The second row of Table 1-4 starts with $0 \leq V_{in} \leq 90$. In this range D_1 is on and D_2 is off. The resulting circuit is shown in Figure 1-25c and the equation for V_{out} is $V_{out} = V_{in}/3$. When V_{out} becomes 30 V, D_2 turns on. The equation for this range shows that V_{in} must equal 90 V for V_{out} to be 30 V, so the limit of the range is 90 V.

The third row is for $90 \leq V_{in} \leq 100$ V. For this range, both diodes are on and the circuit is the original circuit (Figure 1-25a) with the diodes as short circuits. Many circuit theorems can be used to find V_{out} as a function of V_{in}, but for this type of circuit, node analysis is usually the best way. Using node analysis on this circuit gives

$$\frac{V_{out} - V_{in}}{10 \text{ k}\Omega} + \frac{V_{out}}{5 \text{ k}\Omega} + \frac{V_{out} - 30 \text{ V}}{10 \text{ k}\Omega} = 0$$

Table 1-4
Table for Example 1-11

Voltage Range	Diode Condition	Equation	Corner Point
$-50 \leq V_{in} \leq 0$	D_1 off, D_2 off	$V_{out} = 0$ V	$V_{in} = 0$, $V_{out} = 0$
$0 \leq V_{in} \leq 90$	D_1 on, D_2 off	$V_{out} = V_{in}/3$	$V_{in} = 90$, $V_{out} = 30$
$90 \leq V_{in} \leq 100$	D_1 on, D_2 on	$V_{out} = \dfrac{V_{in} + 30 \text{ V}}{4}$	

or

$$V_{out} - V_{in} + 2V_{out} + V_{out} = 30 \text{ V}$$
$$4V_{out} = 30 \text{ V} + V_{in}$$

or

$$V_{out} = \frac{V_{in} + 30 \text{ V}}{4}$$

Table 1-4 can now be completed. The places where the *range changes* are called the *corner points*. At the corner points *both* equations must be satisfied. There is a corner point here where $V_{in} = 90$ V, $V_{out} = 30$ V. Note that these values satisfy both the equation for $V_{in} < 90$ V and for $V_{in} > 90$ V.

b. If $V_{in} = 60$ V, the equation for this value is on the second line of the Table ($V_{out} = V_{in}/3$) and $V_{out} = 20$ V.

If $V_{in} = 100$ V, the equation is on the third line of the Table [$V_{out} = (V_{in} + 30$ V$)/4$] and $V_{out} = 130$ V$/4 = 32.5$ V.

1-8 CLAMPING CIRCUITS

The function of a *clamping circuit* is not to change the waveform of an incoming signal, as a clipping circuit does, but to *clamp the peak of the incoming waveform to a specific voltage*. Clamping circuits also change the *average* or *dc level* of a waveform, but *not its shape*. Thus the ideal clamping circuit can be visualized as taking the incoming waveform and moving it up or down in voltage to suit the application.

1-8.1 Unbiased Clamper

The simplest clamp circuit is shown in Figure 1-26. When the input is at +10 V, the diode is a short circuit and the capacitor charges up to +10 V. When the input goes to −10 V, the diode cuts off and the capacitor discharges very slowly through the 100 kΩ resistor. If we assume the capacitor discharge is negligible, then the voltage across the capacitor is 10 V constantly and $V_{out} = V_{in} - 10$. The input and output waveforms are shown in Figures 1-26b and 1-26c. The output is clamped to 0 V and the average or dc level of the input is at −10 V. As long as the capacitor can maintain the 10 V across it, the clamping circuit functions perfectly, producing an output voltage that is equal to the input voltage minus the clamp voltage.

The waveforms for a real circuit are shown in Figures 1-26d and 1-26e. Figure 1-26d shows that on the positive edge of the input there is a short positive voltage pulse while the capacitor recharges. Then the capacitor remains charged during the rest of the positive portion of the wave. When the input voltage goes negative, the capacitor discharges through the resistor and its voltage decays.

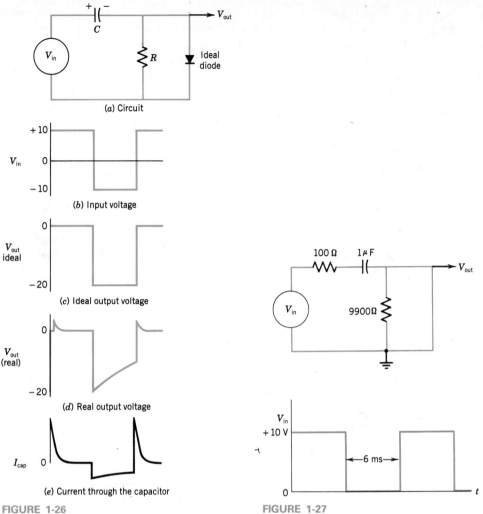

FIGURE 1-26
The simplest clamping circuit.

FIGURE 1-27
The clamp circuit for Examples 1-12 and 1-13.

The current waveform is shown in Figure 1-26e. It shows a large pulse of current when the capacitor charges and a negative current when the capacitor discharges.

EXAMPLE 1-12

Find the ideal output voltage waveform for the circuit of Figure 1-27a in response to the input shown in Figure 1-27b.

Solution

If this circuit were an ideal clamp, the capacitor would charge to 10 V and the output would be $V_{in} - 10$ V. The output would go from 0 V when the input is at $+10$ V, to -10 V when the input is 0 V.

32 SEMICONDUCTORS AND DIODES

It takes more detailed calculations to find out what the circuit actually does and how closely it approximates the ideal clamp. To calculate the precise response, we must realize that this is an R-C circuit subject to step changes in the input voltage. The formula for such a circuit is

$$V(t) = V_F + (V_I - V_F)e^{-t/RC} \qquad (1\text{-}3)$$

Formula 1-3 is one of the most basic and important formulas in the book. The following definitions apply to it:

- t — **Time in seconds.** Generally the time following a change in input voltage.
- **R-C** — The circuit resistance and capacitance. The product, RC is the *time constant* of the circuit.
- $V(t)$ — The voltage at any point in the circuit, as a function of time.
- V_I or $V_{(initial)}$ — The voltage at the point of interest the instant after a step change in voltage has occurred.
- V_F or $V_{(final)}$ — The voltage the point of interest would assume if it were given infinite time and no change in input voltage. Practically speaking, any time greater than 5 time constants is considered infinite because if $t \geq 5\ RC$, $e^{-5} \leq 0.0067\ (\approx 0)$ and $V(t) \approx V_F$.

The use of this formula is best illustrated by an example.

EXAMPLE 1-13

Using formula 1-3, analyze the circuit of Figure 1-27 more precisely.

Solution

Assume the input voltage is $+10$ V. We must first determine whether the capacitor has sufficient time to charge. When the voltage is $+10$ V, the diode is forward-biased and the time constant is $RC = 100\ \Omega \times 10^{-6}\ F = 10^{-4}$ s. The pulse width is 6 ms, so the circuit has $6 \times 10^{-3}/10^{-4} = 60$ time constants to charge the capacitor. The capacitor has far more than enough time to fully charge to 10 V, and the current will be 0 at the end of the interval. If the diode impedance were also considered, it would increase the time constant slightly, but not enough to make a significant difference.

The situation at the instant that the input goes to 0 V is shown in Figure 1-28. The fully charged capacitor drives current through the 100 Ω and the 9900 Ω resistors. They act as a voltage divider and the initial value of V_{out} is -9.9 V. To find the final value of V_{out}, consider what would happen if the input voltage to the circuit of Figure 1-28 were allowed to remain at 0 V indefinitely. Then the capacitor would completely discharge and $V_{out\ (final)} = 0$ V.

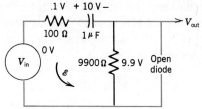

(a) Circuit immediately after the input voltage has gone to 0 V

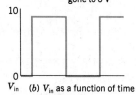

(b) V_{in} as a function of time

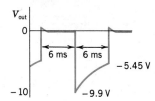

(c) V_{out} as a function of time

(d) Capacitor voltage as a function of time

FIGURE 1-28
Circuit and waveforms for Example 1-13.

Now that V_I and V_F have been determined to be -9.9 V and 0 V, respectively, equation 1-3 can be written as

$$V_{out}(t) = -9.9\, e^{-t/RC}$$

For this circuit the input voltage does not remain at 0 indefinitely, but only for 6 ms, and the time constant is 10 kΩ times 1 µf = 10^{-2} s. Using this value we find

$$V_{out}\,(6\text{ ms}) = -9.9\, e^{-0.6} = -5.45 \text{ V}$$

Similarly

$$V_{cap} = +10\, e^{-0.6} = +5.52 \text{ V}$$

The waveforms for the output and capacitor voltages are shown in Figures 1-28c and 1-28d. The spike of voltage where V_{out} goes above 0 lasts for at most 5 time constants when the diode is forward biased, or 500 μs. Observe that there are no jumps in the capacitor voltage waveform because the voltage across a capacitor cannot change instantaneously.

The waveform of V_{out} is a mediocre approximation to the ideal clamped waveform. There are three ways it can be improved.

1. Shorten the period of the driving square wave.
2. Increase C.
3. Increase R.

Shortening the period will keep the output closer to 10 V, but the capacitor must have enough time to recharge when the voltage goes positive. In this circuit the period should not be less than 500 μs (5 time constants in the forward direction) or the capacitor may not recharge fully.

Increasing the capacitance will improve the waveform but will also lengthen the recharge time. In this circuit the capacitance cannot be increased to the point where $5\,RC = 6$ ms.

Increasing the resistance will improve the waveform without affecting the recharge time. The resistance, however, often represents any *load* attached to the output. Raising the resistance decreases the current available for any load. This circuit is an example of a typical engineering problem where improving the circuit in one area may present problems in another area. Selecting the optimum circuit for the application requires good engineering judgment.

1-8.2 Clamping Sine Waves

If the clamping circuits of Figure 1-26 or 1-27 were driven by a sine wave input, they would clamp the input to ground. On every positive peak of the sine wave, the capacitor would charge to the peak value of the sine wave. It would then discharge throughout the rest of the cycle until the sine wave was again near its peak. The waveforms are shown in Figure 1-29. The top oscilloscope trace is a ± 10 V input sine wave. The bottom trace is the output clamped to ground. The flattening of the output voltage near ground is clearly seen. The trace shows that the output goes from 0 to about -18 V, so the capacitor has discharged somewhat in this circuit. Because of the discharge, the capacitor starts to recharge before the sine wave reaches its peak. This results in a flattening of the top of the clamped waveform, as shown in Figure 1-29. If a good clamping circuit is used, the capacitor charging time and waveform flattening will be small.

Consider the circuit of Figure 1-30a with a 10 V sine wave applied. The capacitor will charge to 10 V and the output will be $V_{in} - 10$ V or between 0 and -20 V. Now assume that the input voltage drops to 6 V peak, as shown

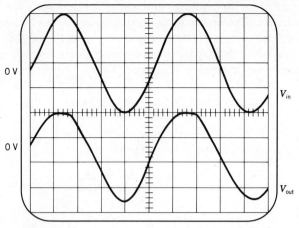

FIGURE 1-29
Oscilloscope traces of a clamped sine wave.

FIGURE 1-30
Clamping sine waves.

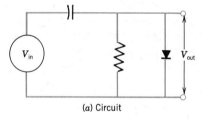

(a) Circuit

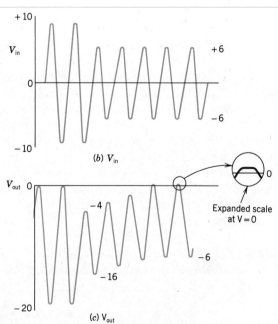

(b) V_{in}

(c) V_{out}

in Figure 1-30b. With the capacitor charged to 10 V, the output during the first few cycles will never be positive. At first it will go between -4 and -16 V. But in these circumstances the diode will always be reverse biased and the capacitor can get no recharging pulses. As a result, it will discharge until its voltage drops to 6 V, as shown in Figure 1-30c. Then it starts getting recharge pulses and $V_{out} = V_{in} - 6$. Notice that the clamping circuit automatically adjusts itself to keep the peak of the output waveform at 0 V, regardless of the amplitude of the input voltage, and will maintain the peak at 0 V despite any changes in the input. This is also true for square wave inputs.

Figure 1-30c also shows the slight flattening we can expect when the voltage reaches 0 V. This is similar to the waveforms shown in Figure 1-29.

1-8.3 Biased Clamping Circuits

The circuits considered in the previous sections clamped the output voltage to ground. By adding a voltage source, such as a battery, the output waveform can be clamped to any voltage. The important points are that $V_{out} = V_{in} + V_{cap}$ and V_{cap} will generally be the *maximum voltage the capacitor can charge to* when the diode is forward biased.

EXAMPLE 1-14

Given: $V_{in} = 15 \sin \omega t$. Design circuits to produce the outputs shown in Figures 1-31a and 1-31b.

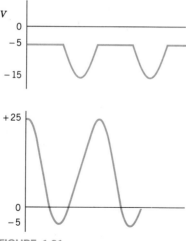

FIGURE 1-31
Waveforms for Example 1-14.

Solution

a. This circuit is the result of a biased clipper. Looking at the output waveform, we realize the diode must short the output to -5 V whenever the input is more positive than -5 V. Therefore, a 5 V battery or voltage source is required. The clipping circuit is shown in Figure 1-32.

b. This waveform is the output of a clamp circuit that adds 10 V to the ± 15 V range of the input. Thus V_{out} must equal $V_{in} + 10$ or there must be 10 V across the capacitor. The start of the circuit is shown in Figure 1-33a. The

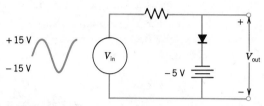

FIGURE 1-32
Biased clipping circuit of Example 1-14a.

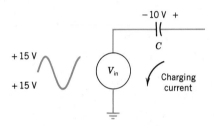

(a) Path of charging current

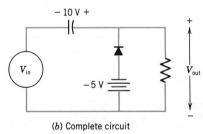

(b) Complete circuit

FIGURE 1-33
Biased clamp circuit for Example 1-14b.

direction of the charging current determines the direction of the diode. The maximum charging current occurs when V_{in} is minimum or -15 V. At this time we require 10 V across the capacitor so it must be charging from a -5 V source to -15 V. The complete clamping circuit is shown in Figure 1-33b. Note that the diode is reverse biased for most of the input waveform. It conducts only when the input is close to -15 V to recharge the capacitor.

1-9 ZENER DIODES

When diodes are subjected to an inverse voltage, there is a point where they will "breakover" or provide a *large reverse current*. This point or voltage depends upon the construction and doping of the diode and can be controlled.

A very strong electric field can break electrons out of the covalent bonds and thus cause a significant current. This is called a *Zener breakdown* and occurs only at low inverse voltages. When higher inverse voltages are applied, any free electrons in the semiconductor are accelerated by the resulting electric field. At some point they acquire enough energy to break a covalent bond when they collide with it, resulting in the generation of additional free carriers. This is called an *avalanche breakdown*.

In many circuits these breakdown phenomena are very useful and have resulted in the design and production of a class of diodes called Zener diodes. This is the name given to diodes that are used because of their *breakover* characteristics, regardless of whether the breakover is caused by the Zener or the avalanche effect.

1-9.1 Zener Characteristics

The typical characteristic for a Zener diode is shown in Figure 1-34a. The curve can be divided into three parts:

1. **The forward characteristic.** The forward characteristic of the Zener diode is much like that of a normal diode. Because Zener diodes are usually reverse-biased, the reverse characteristic is more important.
2. **The reverse characteristic for voltages less than the Zener voltage, V_Z.** The reverse-biased Zener has a small leakage current for voltages less than V_Z. Although the Zener leakage current is somewhat larger than the leakage current for a normal diode, it is still small enough to be insignificant in most circuits. In the vicinity of V_Z, the reverse current begins to rise slowly until the current reaches what is known as the *knee* of the Zener curve. Below the knee the Zener impedance is very high. At the knee it is moderate, and it falls rapidly as the current increases beyond the knee. The dynamic impedance is the slope of the Zener diode curve (Figure 1-34), which is very flat at low voltages (high impedance) and very steep at V_Z (low impedance).
3. **The reverse characteristics for a reverse voltage of V_Z.** When the reverse Zener voltage becomes V_Z, the current increases very rapidly and the dynamic impedance, the slope of the Zener curve, $\Delta V_Z/\Delta I_Z$, becomes small. Because of its tendency to conduct large currents at V_Z, the reverse voltage across a Zener diode cannot practically be more negative than V_Z.

An equivalent circuit for a Zener diode is given in Figure 1-34b. In the reverse direction it consists essentially of a "battery" (V_Z) opposing the reverse current flow. The figure is not perfectly accurate because the re-

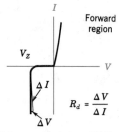

(a) Voltage–current characteristic

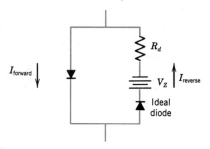

(b) Equivalent circuit

FIGURE 1-34
A typical Zener diode.

sistance varies with temperature. In most practical cases the resistance is ignored and only the Zener voltage is used in calculations.

EXAMPLE 1-15

For the circuit of Figure 1-35, find the current in the resistor and the voltage across the Zener diode.
a. Ignore the Zener resistance.
b. Assume the Zener resistance is 10 Ω.

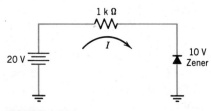

FIGURE 1-35
Circuit for Example 1-15.

Solution
a. If the Zener impedance is ignored, the diode has a constant 10 V drop across it and:

$$I_R = \frac{V_{source} - V_Z}{R} = \frac{20 \text{ V} - 10 \text{ V}}{1000 \text{ }\Omega} = 10 \text{ mA}$$

b. Including the impedance means the Zener diode acts like a 10 V battery in series with a 10 Ω resistor.

$$I = \frac{10 \text{ V}}{1010 \text{ }\Omega} = 9.9 \text{ mA}$$

The voltage across the diode equals the Zener voltage, plus the drop across the Zener resistance.

$$V_Z = 10 \text{ V} + 9.9 \times 10^{-3} \text{ mA} \times 10 \text{ }\Omega = 10.099 \text{ V} \approx 10.1 \text{ V}$$

The drop across the 1 kΩ resistor is $9.9 \times 10^{-3} \times 10^3 = 9.9$ V and the sum of the voltages adds up to 20 V. Taking the Zener impedance into account in this circuit changed the result by only 1% or 0.1 V.

1-9.2 Manufacturers' Specifications

The manufacturers' specifications for a set of Zener diodes are listed in Figure 1-36. The specifications cover JEDEC type diodes **1N4728** through **1N4764** with the equivalent Motorola numbers **1M3.3ZS10** through **1M200ZS10**. The Zener voltages of these diodes range from 3.3 V to 200 V. The Motorola numbers have the advantage of including the Zener voltage as part of the number.

The power dissipation of Zener diodes is a more important characteristic than the power dissipation of an ordinary diode. In either case the diode must absorb power equal to the product of its current and voltage. An ordinary diode conducts in the forward direction only, and its power is $0.7 \times I_{forward}$, because the forward voltage is approximately 0.7 V. The Zener, however, conducts in the reverse direction, and its voltage can be significant. All the diodes in Figure 1-36 are 1 watt diodes.

EXAMPLE 1-16

What is the maximum current that can be passed through a 10 V Zener diode (**1N4740** or **1M10ZS10**)?

Solution
Because these are 1 W diodes

$$P = V_Z I$$

or

ELECTRICAL CHARACTERISTICS ($T_A = 25°C$ unless otherwise noted) *$V_F = 1.5$ V max, $I_F = 200$ mA for all types

JEDEC Type No. (Note 1)	Motorola Type No. (Note 2)	*Nominal Zener Voltage V_Z @ I_{ZT} Volts (Note 2 & 3)	*Test Current I_{ZT} mA	*Max Zener Impedance (Note 4)			*Leakage Current		*Surge Current @ $T_A = 25°C$ i_r — mA (Note 5)
				Z_{ZT} @ I_{ZT} Ohms	Z_{ZK} @ I_{ZK} Ohms	I_{ZK} mA	I_R μA Max	V_R @ Volts	
1N4728	1M3.3ZS10	3.3	76	10	400	1.0	100	1.0	1980
1N4729	1M3.6ZS10	3.6	69	10	400	1.0	100	1.0	1260
1N4730	1M3.9ZS10	3.9	64	9.0	400	1.0	50	1.0	1190
1N4731	1M4.3ZS10	4.3	58	9.0	400	1.0	10	1.0	1070
1N4732	1M4.7ZS10	4.7	53	8.0	500	1.0	10	1.0	970
1N4733	1M5.1ZS10	5.1	49	7.0	550	1.0	10	1.0	890
1N4734	1M5.6ZS10	5.6	45	5.0	600	1.0	10	2.0	810
1N4735	1M6.2ZS10	6.2	41	2.0	700	1.0	10	3.0	730
1N4736	1M6.8ZS10	6.8	37	3.5	700	1.0	10	4.0	660
1N4737	1M7.5ZS10	7.5	34	4.0	700	0.5	10	5.0	605
1N4738	1M8.2ZS10	8.2	31	4.5	700	0.5	10	6.0	550
1N4739	1M9.1ZS10	9.1	28	5.0	700	0.5	10	7.0	500
1N4740	1M10ZS10	10	25	7.0	700	0.25	10	7.6	454
1N4741	1M11ZS10	11	23	8.0	700	0.25	5.0	8.4	414
1N4742	1M12ZS10	12	21	9.0	700	0.25	5.0	9.1	380
1N4743	1M13ZS10	13	19	10	700	0.25	5.0	9.9	344
1N4744	1M15ZS10	15	17	14	700	0.25	5.0	11.4	304
1N4745	1M16ZS10	16	15.5	16	700	0.25	5.0	12.2	285
1N4746	1M18ZS10	18	14	20	750	0.25	5.0	13.7	250
1N4747	1M20ZS10	20	12.5	22	750	0.25	5.0	15.2	225
1N4748	1M22ZS10	22	11.5	23	750	0.25	5.0	16.7	205
1N4749	1M24ZS10	24	10.5	25	750	0.25	5.0	18.2	190
1N4750	1M27ZS10	27	9.5	35	750	0.25	5.0	20.6	170
1N4751	1M30ZS10	30	8.5	40	1000	0.25	5.0	22.8	150
1N4752	1M33ZS10	33	7.5	45	1000	0.25	5.0	25.1	135
1N4753	1M36ZS10	36	7.0	50	1000	0.25	5.0	27.4	125
1N4754	1M39ZS10	39	6.5	60	1000	0.25	5.0	29.7	115
1N4755	1M43ZS10	43	6.0	70	1500	0.25	5.0	32.7	110
1N4756	1M47ZS10	47	5.5	80	1500	0.25	5.0	35.8	95
1N4757	1M51ZS10	51	5.0	95	1500	0.25	5.0	38.8	90
1N4758	1M56ZS10	56	4.5	110	2000	0.25	5.0	42.6	80
1N4759	1M62ZS10	62	4.0	125	2000	0.25	5.0	47.1	70
1N4760	1M68ZS10	68	3.7	150	2000	0.25	5.0	51.7	65
1N4761	1M75ZS10	75	3.3	175	2000	0.25	5.0	56.0	60
1N4762	1M82ZS10	82	3.0	200	3000	0.25	5.0	62.2	55
1N4763	1M91ZS10	91	2.8	250	3000	0.25	5.0	69.2	50
1N4764	1M100ZS10	100	2.5	350	3000	0.25	5.0	76.0	45
—	1M110ZS10	110	2.3	450	4000	0.25	5.0	83.6	—
—	1M120ZS10	120	2.0	550	4500	0.25	5.0	91.2	—
—	1M130ZS10	130	1.9	700	5000	0.25	5.0	98.8	—
—	1M150ZS10	150	1.7	1000	6000	0.25	5.0	114.0	—
—	1M160ZS10	160	1.6	1100	6500	0.25	5.0	121.6	—
—	1M180ZS10	180	1.4	1200	7000	0.25	5.0	136.8	—
—	1M200ZS10	200	1.2	1500	8000	0.25	5.0	152.0	—

*Indicates JEDEC Registered Data

FIGURE 1-36
Electrical characteristics of a set of Zener diodes. (Copyright by Motorola, Inc. Used by permission.)

$$I = \frac{P}{V_Z} = \frac{1\ \text{W}}{10\ \text{V}} = 100\ \text{mA}$$

In Figure 1-36 columns 4 and 5 indicate the typical dynamic impedance. Column 4 lists the test current in mA and is approximately one-quarter of the maximum diode current. For the 10 V Zener, for example, it is 25 mA. Column 5 gives the maximum dynamic resistance at this current. For the 10 V Zener it is 7.0 Ω, which is quite small. For higher currents the resistance generally gets smaller, whereas lower currents move the operation closer to the knee of the curve and the dynamic resistance generally increases. Columns 6 and 7 give the maximum dynamic resistance around the knee of the curve. For the 10 V Zener this is 700 Ω at the small current of 0.25 mA.

Columns 8 and 9 give the maximum or worst-case leakage current at about 75% of the Zener voltage. If the 10 V Zener diode has 7.6 V across it, it will allow a maximum leakage current of 10 µA, which is still very low.

1-9.3 Uses of Zener Diodes

Zener diodes can be used whenever it is necessary to set a voltage at a certain point in a circuit. They are used in most power supply and voltage regulator circuits. These circuits are described in Chapter 10.

The following examples illustrate some other uses of Zener diodes.

EXAMPLE 1-17

Given a circuit where the input can vary from 45 to 55 V, design a circuit so that the output is 10 V when the input is 50 V.
a. Use a resistive voltage divider.
b. Use a 30 V Zener diode.

Solution

The simple resistive divider circuit is shown in Figure 1-37a. The output voltage is $V_{out} = V_{in} \times 1\ \text{k}\Omega / 5\ \text{k}\Omega = V_{in}/5$. If $V_{in} = 50$ V, $V_{out} = 10$ V.

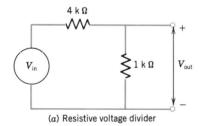

(a) Resistive voltage divider

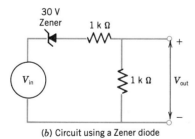

(b) Circuit using a Zener diode

FIGURE 1-37
Voltage divider circuits for Example 1-17.

The circuit using the Zener diode is shown in Figure 1-37b. The output voltage is $V_{out} = (V_{in} - 30)/2$. Again if $V_{in} = 50$ V, $V_{out} = 10$ V.

In Example 1-17 the reader may wonder what the advantages of the Zener circuit are. Consider what happens if V_{in} goes to 51 V. For the resistive divider the output voltage becomes 10.2 V, but for the Zener circuit it becomes 10.5 V. Thus in the Zener circuit the output is more *sensitive* to a change of input voltage. By comparison, we see that a change of input voltage produces a much greater (2.5 times) change of output voltage. If the function of this circuit is to monitor the input voltage and cause something to respond to its changes, the Zener circuit is superior.

EXAMPLE 1-18

In the circuit of Figure 1-38, find the current I
a. When the switch is open.
b. When the switch is closed.

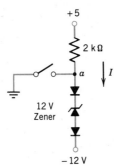

FIGURE 1-38
Circuit for Example 1-18.

Solution
a. With the switch open, the voltage drops across the two diodes and the 12 V Zener provide a total of 13.4 V to oppose the 17 V source (from +5 V to −12 V) driving the current through the resistor. Therefore

$$I = \frac{17 - 13.4}{2 \text{ k}\Omega} = 1.8 \text{ mA}$$

b. With the switch closed, the voltage at point a is 0 V. It requires 13.4 V to drive current through the diodes and the Zener, but the negative voltage is only −12 V. Because the voltage is insufficient, no current will flow.

This circuit was used in the laboratory as part of a TTL-to-20-mA current loop converter used to drive a teletype. Actually, the switch was an open-collector TTL gate and the lower diode was the base-to-emitter junction of an NPN transistor. When the input was at +5 V, the transistor saturated and drove 20 mA through the teletype.

1-10 SCHOTTKY DIODES

A diode can be formed by bringing a metal into contact with an *n*-doped semiconductor, as shown in Figure 1-39. When the contact is made, electrons from the *n*-doped semiconductor migrate into the metal, resulting in a net

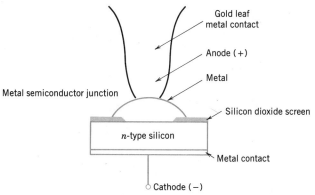

FIGURE 1-39
Construction of a Schottky or hot carrier diode. (Robert Boylestad/Louis Nashelsky, *Electronic Devices & Circuit Theory*, 3rd ed., © 1982. Reprinted by permission of Prentice-Hall, Inc., Englewood Cliffs, New Jersey.)

negative charge in the metal and a positive charge in the semiconductor. The resulting electric field prevents any further migration of electrons and sets up a depletion region at the junction, causing diode action. If a *positive voltage* is applied to the *metal contact*, it attracts the excess electrons in the metal and results in *forward conduction*. A negative potential of the metal increases the *p-n* barrier and no current flows. These diodes are called *Schottky* diodes, or *surface barrier diodes* (because of the electric field barrier at the junction), or *hot carrier diodes* (because of the high energy levels of the electrons injected into the metal from the semiconductor).

The Schottky phenomenon also creates a problem when a metal wire for a lead must make contact with the *n* material in an ordinary *p-n* junction diode. To prevent a barrier at the metal junction, the *n* material is heavily doped at the contact point to provide a sufficient number of electrons to prevent a *p-n* junction from forming. The n^+ symbol in a diode indicates a region of heavy doping.

The situation is illustrated in Figure 1-40. In Figure 1-40*a* there is n^+ doping and no barrier is formed. This is for a *p-n* junction diode. In Figure 1-40*b* the entire body of the diode is made up of *n* material. There is no extra doping at the anode lead so it forms a metal-to-semiconductor Schottky diode where it contacts the *n*-type silicon. Note, however, the n^+ doping at the cathode lead so only one barrier junction is formed.

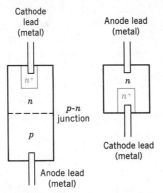

FIGURE 1-40
Schottky diodes and ordinary diodes.

1-10.1 Uses of Schottky Diodes

Schottky diodes have two main advantages over silicon *p-n* diodes:

1. Their forward voltage is lower for the same curent.
2. Conduction is by majority carriers (electrons) only. Consequently there is *no minority carrier storage* and the *reverse recovery time is very low* because storage time is practically zero.

Their low forward voltage and fast recovery time has led to their use in rectifier and switching power supply circuits (see Chapter 10). They are widely used to speed up transistor conduction in digital circuits. The Schottky TTL series is one of the fastest series of digital gates currently available.

Schottky diodes also have two drawbacks:

1. Their leakage current is higher than that of *p-n* diodes, although still small (approximately 40–100 nA at 10 V reverse voltage).
2. Their peak inverse voltage is less than that of silicon diodes.

Because of these disadvantages, *p-n* junction diodes are still used in most common applications, but Schottky diodes are being increasingly used in specialized circuits.

1-11 PHOTODIODES

Photodiodes are diodes that are affected by *light*. They can be classified in two broad categories: *emitters* of light and *detectors* of light. They have many practical uses, such as light-emitting diodes and seven-segment displays, "electric eyes," and bar code readers. These will be discussed in the following paragraphs that explain the operation of these devices.

1-11.1 Photodetectors

A photodetecting diode, commonly called a photodiode, is essentially a *p-n* diode with a flat glass window cap, as shown in Figure 1-41. The glass window allows light to impinge upon the *p-n* junction.

(*a*) Photograph

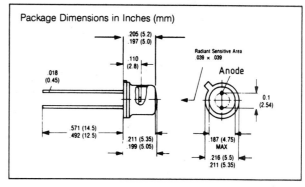
(*b*) Dimensions

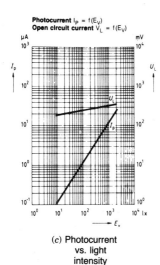
(*c*) Photocurrent vs. light intensity

FIGURE 1-41
A Litronix tape BPX 66 PIN photo diode. (Courtesy of Seimens Components, Inc., Optoelectronics Division.)

Photodiodes are set up as reverse-biased diodes. In the absence of light, only a very small reverse current, called the *dark current*, flows. When light hits the *p-n* junction, the photons in the light create electron–hole pairs, just as additional heat would. These photo-generated electron–hole pairs become a source of minority carriers and the reverse current increases rapidly. Figure 1-41*c* shows the diode current versus light intensity. The scale is logarithmic, so the current does increase greatly as the junction is illuminated.

EXAMPLE 1-19

For the circuit of Figure 1-42, find V_{out} if
a. The diode is dark.
b. The diode is illuminated so that 1 µA flows.

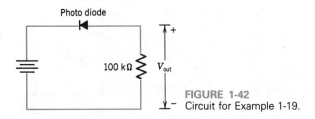

FIGURE 1-42
Circuit for Example 1-19.

Solution
a. When the diode is dark, Figure 1-41c shows that the dark current is approximately 10^{-2} µA. This is the current at the lowest illumination. Therefore

$$V_{out} = IR = 10^{-8} \text{ A} \times 10^5 \text{ } \Omega = 0.001 \text{ V}$$

b. When the diode is illuminated and the current is 1 µA, $V_{out} = IR = 10^{-6} \text{ A} \times 10^5 \text{ } \Omega = 0.1 \text{ V}$

Example 1-19 shows that this photodiode will produce only a small change of voltage when illuminated. Photodiode circuits of this type are often connected to high-gain, high-input impedance amplifiers, such as op-amps (see Chapter 12), to increase the output voltage.

The construction of a typical photodiode is shown in Figure 1-43. Most commercially available photodiodes are PIN diodes, which means they have a

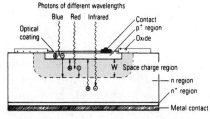

FIGURE 1-43
Construction of a photodiode. (Courtesy of Seimens Components, Inc., Optoelectronics Division.)

48 SEMICONDUCTORS AND DIODES

p and an *n* region, separated by a region of *intrinsic* (undoped) semiconductor, the I region or *space charge region* shown in Figure 1-43. The additional separation results in lower diode capacity and increases the speed of response of photodiodes.

The principle of photodiodes has also been incorporated in *phototransistors*. These are transistors that conduct only in response to a light input, but use the transistor to amplify the output.

1-11.2 Photoemitters

When a diode is forward biased, the power it dissipates is the product of the forward voltage times the current. Most of this energy is dissipated as heat, but some is dissipated as light. Silicon and germanium generate a negligible amount of light, but diodes using gallium arsenide (GaAs) or gallium arsenide phosphorus (GaAsP) can generate significant light in response to a forward current and are called emitters or *light-emitting diodes*. They are used in optical couplers and electric eyes (see section 1-12).

1-12 APPLICATIONS OF PHOTOEMITTERS AND PHOTODETECTORS

In this section several applications of photoemitters and photodetectors will be discussed.

1-12.1 Electric Eye or Intrusion Detection

Intrusion detection is used to determine whether an object is present or absent. It can determine when a person enters or leaves a room or whether a box is on a conveyor belt.

The general method of detection is to allow light to shine on a photodetector. If an object comes between the light and the detector, the detector current drops and this indicates the presence of the object.[4]

Photodetection must often be accomplished in the presence of ambient light that is always shining on the detector. This problem can sometimes be solved by using *infrared emitters* and *detectors* that are available from Litronix and other companies. In these systems the detector responds only to infrared light and is not sensitive to ambient light. In an infrared system the following precautions must be observed:

1. The *spectral response* (the light frequencies emitted by the emitter and the frequencies the detector responds to) must match. Curves of spectral response are available from the manufacturer.

---

[4]One example is measuring the speed of slot cars. At RIT we purchased a set of slot cars that we connected to an Apple computer. We drilled a small hole in the track and placed a photodetector under it. Ambient light activated the photodetector. When the slot car passed over the hole, however, it blocked the light, thus signaling its presence to the computer. This system is capable of determining the time for the car to complete one lap to the thousandth of a second.

2. The emitter must be properly *focused* so the detector responds to it and not to ambient light. In most detection circuits the focusing is done by trial and error until a satisfactory response is obtained.

1-12.2 Optical Couplers

An *optical coupler* or *opto-isolator* consists of a light-emitting diode and phototransistor in the same package, as shown in Figure 1-44. When current is

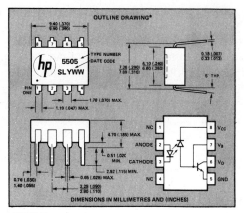 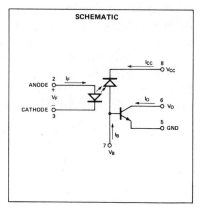

FIGURE 1-44
The Hewlett Packard SL5505 high-speed optical coupler. (Courtesy of Hewlett Packard Co.)

driven through the diode (pins 2 and 3 in Figure 1-44), it produces light that activates the photodiode and transistor in the receiving circuit. The result is a transistor current (from pins 6 to 5). The coupling between the transmitting circuit (the circuit driving the diode) and the receiving circuit is via light.

Opto-isolators are used when it is necessary to *isolate* one system from another, including the system grounds, while transferring information between them. This occurs often in industrial environments where the system that originates the signal is in an electrically noisy environment that may change the information before it is received. The originating system drives current through the emitter diode, which turns the phototransistor on or off. Thus the coupling between the two systems is by light in the encapsulated package (not visible to the user), and the two systems can be totally isolated from each other so that there are no electrical connections between them.

1-12.3 Bar Code Reading

A *bar code* is the array of bars placed on packages of food and other items. It contains information describing the contents of the package. At present some supermarkets are using optical scanners to read the bar code, calculate the buyer's bill, and also help the market control its inventory.

Bar codes are read by a reflective system. Light is sent out by an *optical emitter*, and *reflected* off the object to a *sensor*, a photodiode or phototransistor.

With a bar code as the object, the light will be reflected off the white areas but not the black bars, and the sensor can thus detect the bar pattern. Bar codes can be detected by moving a special wand across the code, as shown in Figure 1-45, or by sliding the package over a glass plate with an optical sensor embedded in it.

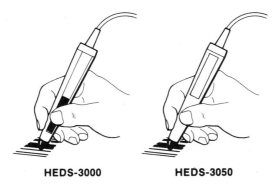

HEDS-3000   **HEDS-3050**

FIGURE 1-45
Reading a bar code with a photodetecting wand.
(Courtesy of Hewlett Packard Co.)

The **HEDS-1000** is a device made by the Hewlett-Packard Corporation for reflective sensing. It combines a photoemitter and a photodiode. The photodiode can be connected to a transistor for greater amplification. The efficiency of the **HEDS-1000** is highly dependent on the distance between the glass cap and the reflecting surface. Figure 1-46 shows this relationship. Clearly the reflecting surface should be about 4.5 mm from the cap for maximum effect.

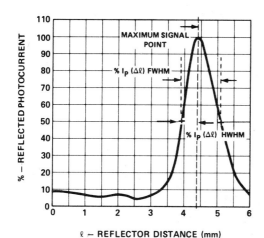

FIGURE 1-46
Depth of field vs. maximum signal point for a HEDs 1000. (Courtesy of Hewlett Packard Co.)

1-13 LIGHT-EMITTING DIODES

The primary function of the photoemitters discussed previously was to supply light for a photodetector. Another and more important use of a photodiode is as a *light-emitting diode* (LED). LEDs are made of gallium arsenide phosphide (GaAsP) and most emit a red light, although yellow and green LEDs are also available.

LEDs are used primarily as indicators, especially in digital circuits. If there is a current through them, causing them to light, it usually indicates the presence of a logic 1 (high voltage) at a point in the circuit. Conversely, if there is a logic 0 at the point in the circuit, the LED will receive no current and remain off.

The dimensions and typical characteristics for the **HLMP-3000** series of LEDs, manufactured by Hewlett-Packard, are shown in Figure 1-47. Notice

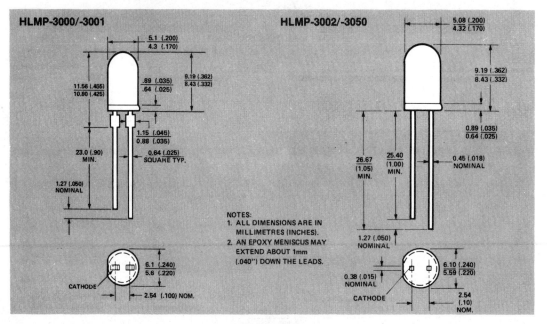

(a) Dimensions

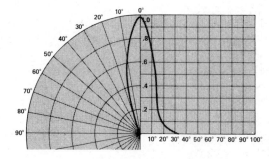

(b) Voltage–current curve

FIGURE 1-47
Characteristics of the HLMP-3000 series of LEDs manufactured by Hewlett Packard, Inc. (Courtesy of Hewlett Packard Co.)

that these diodes are constructed so that the anode lead is longer than the cathode lead. For these diodes the forward voltage is almost constant at 1.6 V (other LEDs may have forward voltages of about 2 V). As the current through the diode increases, so does the brilliance of the emitted light, but the light is not very sensitive to the diode current. A typical value of current for reasonable brilliance is 20 mA.

EXAMPLE 1-20

An LED of the **5082-4480** series is being driven by an open-collector TTL gate whose output is 0.4 V when the gate is on and can absorb current, and an open circuit when the output is off. Design a circuit to drive the LED from the TTL gate and a $+5$ V source.

Solution

The LED can be connected to the output of the TTL gate. When the gate is on, it will drop 0.4 V and the LED will drop 1.6 V. Consequently, a resistor must be placed in the circuit to absorb the other 3 V. The circuit is shown in Figure 1-48. Because 20 mA should flow through the diode, the resistor value must be

$$R = \frac{3 \text{ V}}{20 \text{ mA}} = 150 \text{ }\Omega$$

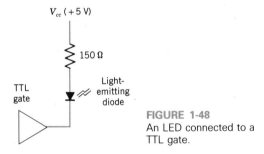

FIGURE 1-48
An LED connected to a TTL gate.

When the gate turns on, it provides a path for the current and the LED lights. If the gate is off, there is no current path and the LED remains dark.

The circuit of Figure 1-48 is called a *common anode* circuit because the LED is between the voltage source and the TTL output. Another method of connecting LEDs is the *common cathode* circuit shown in Figure 1-49. Using this configuration gates can be driven directly from TTL integrated circuit outputs.

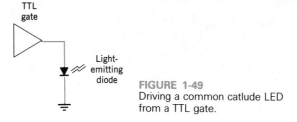

FIGURE 1-49
Driving a common catlude LED from a TTL gate.

External resistors are not required because there is an internal current limiting resistor within the TTL gate.[5]

1-13.1 Seven-Segment Displays

LEDs are often arranged in groups of seven to form a *seven-segment display*, as shown in Figure 1-50. The seven LEDs are used to display a single decimal

FIGURE 1-50
Seven-segment displays. (Courtesy of Hewlett Packard Co.)

digit. An eighth LED is added to most seven-segment displays to function as a decimal point. Hand calculators, microwave ovens, and many other devices use seven-segment displays. The inputs for the seven segments and the decimal point are generally derived from digital circuits or microprocessors.[6]

Both common anode and common cathode seven-segment displays are readily available. In a common anode display, all the anodes are connected together and are meant to be tied to the supply voltage, V_{CC}. Current-limiting resistors are usually required, as shown in Figure 1-51a. Common cathode seven-segment displays have the cathodes of each segment tied together. The common cathodes are generally tied to ground, as shown in Figure 1-51b.

---

[5]TTL gates driving LEDs should not also be used to drive other TTL gates because the output voltage will be out of specification.

[6]A thorough discussion of the methods of interfacing seven-segment displays to digital circuits is given in Greenfield, *Practical Digital Design Using ICs*, 2nd Edition, 1983, John Wiley & Sons, Inc., New York.

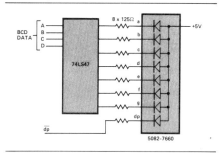
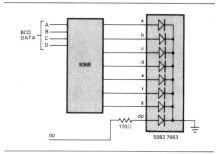

(a) DC drive circuit for the **5082-7660** common anode display

(b) DC drive circuit for the **5082-7663** common cathode display

FIGURE 1-51
Driving seven-segment displays. (Courtesy of Hewlett Packard Co.)

Both the common anode and common cathode circuit of Figure 1-51 are shown being driven by digital interface circuits.

Seven segments are enough to display numbers but not alphabetic characters. Hewlett-Packard and other manufacturers make 16- and 18-segment displays, as shown in Figure 1-52, that can display alphanumeric information. Dot matrix displays, where the characters are formed from a series of dots, are also available.

FIGURE 1-52
The HDSP 6504 and 6508 16-segment displays. (Courtesy of Hewlett Packard Co.)

Magnified Character Font Description

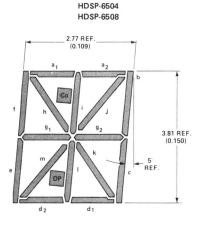

Device Pin Description

Pin No.	Function			
	HDSP-6504		HDSP-6508	
1	Anode	Segment g_1	Anode	Segment g_1
2	Anode	Segment DP	Anode	Segment DP
3	Cathode	Digit 1	Cathode	Digit 1
4	Anode	Segment d_2	Anode	Segment d_2
5	Anode	Segment l	Anode	Segment l
6	Cathode	Digit 3	Cathode	Digit 3
7	Anode	Segment e	Anode	Segment e
8	Anode	Segment m	Anode	Segment m
9	Anode	Segment k	Anode	Segment k
10	Cathode	Digit 4	Cathode	Digit 4
11	Anode	Segment d_1	Anode	Segment d_1
12	Anode	Segment j	Cathode	Digit 6
13	Anode	Segment C_0	Cathode	Digit 8
14	Anode	Segment g_2	Cathode	Digit 7
15	Anode	Segment a_2	Cathode	Digit 5
16	Anode	Segment i	Anode	Segment j
17	Cathode	Digit 2	Anode	Segment C_0
18	Anode	Segment b	Anode	Segment g_2
19	Anode	Segment a_1	Anode	Segment a_2
20	Anode	Segment c	Anode	Segment i
21	Anode	Segment h	Cathode	Digit 2
22	Anode	Segment f	Anode	Segment b
23			Anode	Segment a_1
24			Anode	Segment c
25			Anode	Segment h
26			Anode	Segment f

(a) Magnified character font description

(b) Device pin description

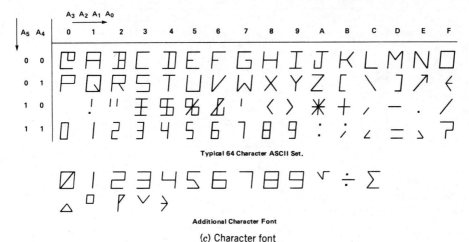

Typical 64 Character ASCII Set.

Additional Character Font

(c) Character font

FIGURE 1-52c

1-14 OTHER SPECIALIZED DIODES

In this section specialized diodes manufactured for specific purposes will be introduced. Space and the scope of this book preclude a detailed discussion of their circuits, but they are mentioned so the reader with a particular need can pursue them further in more advanced literature.

1-14.1 Varactor Diodes

Varactor or *varicap* diodes are reverse biased so there is very little current through them. The *capacitance* of these diodes varies with the amount of reverse voltage placed across them. The curve of capacitance versus reverse voltage for a **BB139** varactor diode manufactured by Fairchild, Inc., is shown in Figure 1-53. The capacity of the diode varies from 50 pf to 5 pf as the reverse voltage changes from 0.5 V to 20 V.

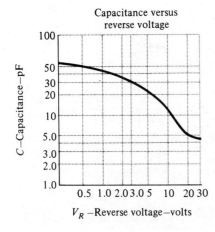

FIGURE 1-53
Capacitance vs. reverse voltage for a BB139 varactor diode. (Courtesy of Fairchild Inc.)

One use of the varactor diode is as an FM modulator. If the varactor is placed in a circuit with an inductor, the resonant frequency that is generated depends on the capacitance of the varactor. Therefore, if an audio signal is used to modulate the diode's reverse voltage, its capacitance will vary accordingly and the output frequency will be proportional to audio input signal.

1-14.2 Tunnel Diodes

Tunnel diodes are doped much more heavily than ordinary diodes. As a result these diodes have a peculiar forward volt-ampere characteristic, as shown in Figure 1-54, where the current rises to a peak, then drops to a valley, before exhibiting normal diode characteristics. Specifications for tunnel diodes include the voltages and currents at both the peak and valley areas.

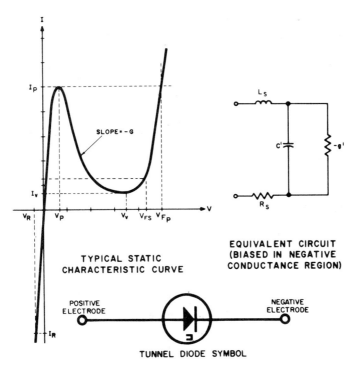

FIGURE 1-54
Tunnel diodes. (Reprinted with permission of General Electric Company.)

In the region between the peak and valley, the slope of the curve is negative. This is called the negative resistance region (or negative conduction, $-g$, as shown in Figure 1-54). Devices that have a negative resistance region can be used in oscillators (see Chapter 11). These diodes are also very fast and find some use in computer circuits because of their speed.

1-14.3 Liquid-Crystal Displays

Liquid-crystal displays (LCDs) are actually a form of capacitor rather than a diode. They can be used as indicators instead of light-emitting diodes (LEDs) when low power consumption is important and, like LEDs, are often arranged in seven-segment displays.

LCDs operate by polarizing light so that it does or does not reflect ambient light. They have one outstanding advantage over LEDs: they essentially act as a capacitor and consume almost no power. For this reason they are now being used in watches and as readouts for handheld computers and electronic games. This is a significant advantage because in many circuits LEDs consume more power than all the ICs in the rest of the system. Unfortunately, LCDs reflect, rather than generate, light and must be in a well-lit environment to be seen clearly. They must also be driven by ac voltage, but this can be generated by EXCLUSIVE-OR circuits.

1-15 SUMMARY

In this chapter the concepts of doping and the operation of the p-n junction were introduced. The ordinary diode was discussed, as were a number of specialized diodes, such as rectifiers and Zener diodes.

Circuits depending on diodes, such as clamping circuits and clippers were also discussed. Finally, the use of light-emitting diodes as displays and other particular diode circuits were considered. One must understand p-n junctions and diodes before proceeding to a study of transistor circuits.

1-16 GLOSSARY

Acceptor. A trivalent material used to dope silicon to create a p-type semiconductor.

Clamping circuit. A circuit that changes the dc level of a waveform without changing its shape.

Clipping circuit. A circuit that clips off part of an input waveform above or below a certain voltage.

Conductivity. See Resistivity.

Covalent bond. A pair of electrons shared between two atoms to help complete a valence ring.

Depletion region. A semiconductor area devoid of free electrons or holes. Usually found at a p-n boundary.

Diode. A device that conducts current in one direction only.

Donor. A pentavalent material used to dope silicon to create an n-type semiconductor.

Doping. Adding p- or n-type material to a semiconductor to give it the characteristics desired.

Dynamic resistance. The resistance of a diode in the vicinity of the point where it is conducting.

Hot carrier diode. See Schottky diode.

Light-emitting diodes (LEDs). Diodes that emit light when current is passed through them.

n-Type silicon (germanium). Silicon doped with donor atoms so that it has many free electrons.

p-Type silicon (germanium). Silicon doped with acceptor atoms so that it has many holes.

Peak inverse voltage (PIV). The maximum reverse voltage that can be applied across a diode.

Rectifier. A circuit that transforms ac into dc.

Resistivity (ρ). A measure of the resistance of a material. Its inverse is conductivity (σ).

Schottky diode. A diode formed by a metal n-type junction. It is a very fast diode.

Semiconductor. A material (generally silicon) whose conductivity is between those of an insulator and a conductor. It forms the base material for diodes and transistors.

Seven-segment display. An arrangement of LEDs to display a decimal digit.

Storage time. The time required for the charges to be swept out when a diode goes from forward to reverse bias.

Surface barrier diode. See Schottky diode.

Varactor diode. A diode used primarily as a voltage-controlled variable capacitor.

Zener diode. A diode used to maintain a fixed voltage (V_Z) across it when subject to a reverse voltage.

1-17 REFERENCES

ROBERT BOYLESTAD and LOUIS NASHELSKY, *Electronic Devices and Circuit Theory*, 4th Edition, Prentice-Hall, Englewood Cliffs, N.J., 1987.

JOSEPH D. GREENFIELD, *Practical Digital Design Using ICs*, 2nd Edition, Wiley, New York, 1983.

E. NORMAN LURCH, *Fundamentals of Electronics*, 3rd Edition, Wiley, New York, 1981.

JACOB MILLMAN, *Microelectronics*, McGraw-Hill, New York, 1979.

JACOB MILLMAN and HERBERT TAUB, *Pulse, Digital, and Switching Waveforms*, McGraw-Hill, New York, 1965.

J. F. PIERCE and T. J. PAULUS, *Applied Electronics*, Charles E. Merrill, Columbus, Ohio, 1972.

RALPH J. SMITH, *Electronics: Circuits and Devices*, 2nd Edition, Wiley, New York, 1980.

1-18 PROBLEMS

1-1 Find the resistance of a bar of aluminum 10 cm long with a cross section of 2 cm by 3 cm.

1-2 Using wire tables[7] find the resistance of 10 feet of #22 gauge copper wire. Repeat for 10 feet of #30 gauge copper wire.

1-3 Using wire tables find the resistance of 20 cm of #22 gauge copper wire.

1-4 Find the resistivity of germanium at 300°K using formula 1-2.

1-5 Find the resistivity of germanium at 350°K.

1-6 Pierce and Paulus (see References) give the following formula for the number of free electrons in silicon:

$$n_i = p_i = AT^{3/2} \exp[-qV_{g/2KT}]$$

---

[7]Wire tables are available in most handbooks for electrical engineers.

where A is a constant $\approx 6 \times 10^{15}$, V_g is the energy gap voltage ≈ 1.1 V and $KT/q = 25.8$ mV at 300°K. Find
a. The free electron concentration at 300°C. How does this compare with Table 1-2?
b. The free electron concentration at 320°C.
c. The free electron concentration at 330°C.
d. How do b and c compare with the rule that conductivity doubles for each 10°C rise in temperature?

1-7 From Figure 1-13 find the forward resistance of the diode at 25°C if the current varies between 1 and 2 mA.

1-8 Find the current in the diode of Figure P1-8.

FIGURE P1-8

1-9 Using Figure 1-17a, find the reverse resistance of a **1N4148** diode when the reverse voltage is
a. 20 V.
b. 40 V.

1-10 Using Figure 1-17a, find the dynamic resistance of a **1N4148** diode at
a. 40 V.
b. 80 V.

1-11 Using Figure 1-17b, find the normalized reverse current at 45° and 75°C. How does this compare to the rule that the current doubles for every 10°C increase in temperature?

1-12 Find the output voltage and current waveforms for the circuits of Figure P1-12 if
a. $V_B = 0$.
b. $V_B = 5$ V.
c. $V_B = -5$ V.

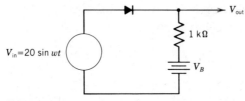

FIGURE P1-12

1-13 For the circuit of Figure P1-13 write equations relating V_{in} and V_{out} as V_{in} goes from -50 to $+100$ V. Be sure to indicate the range over which each equation is valid.

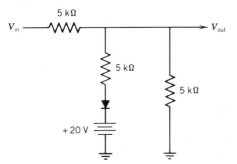

FIGURE P1-13

1-14 For the circuit of Figure P1-14, find the output voltage as the input varies from -50 to $+50$ V.

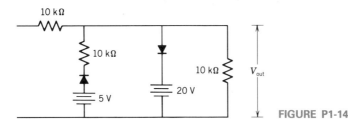

FIGURE P1-14

1-15 For the circuit of Figure P1-15, assume the diodes are ideal. Find V_{out} as V_{in} goes from -50 to $+100$ V.
  a. Find V_{out} when $V_{in} = 100$ V.
  b. Find V_{out} when $V_{in} = 200$ V.

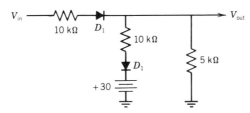

FIGURE P1-15

1-16 Find the response of the clamping circuit of Figure P1-16a to the input waveform of Figure P1-16b.

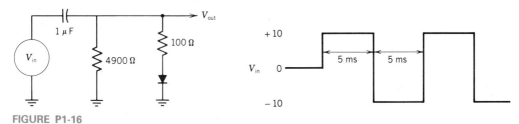

FIGURE P1-16

1-18 PROBLEMS **61**

1-17 Find the response of Figure P1-17 to the waveform of Figure 1-16b.

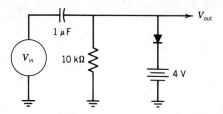

FIGURE P1-17

1-18 Calculate the response of the circuit of Figure P1-18 to a 10 V symmetric square wave with a period of 12 M seconds.

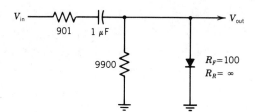

FIGURE P1-18

1-19 Given $V_{in} = 20 \sin \omega t$. Design a circuit to produce the outputs of Figure P1-19.

(a)   (b)   FIGURE P1-19

1-20 For the circuit of Figure P1-20, find the current in the resistor and the voltage across the Zener diode.
    a. Ignore the Zener resistance.
    b. Assume the Zener resistance is 20 ohms.

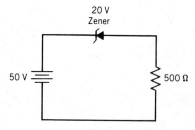

FIGURE P1-20

1-21 For a **1M24ZS10**, find
    a. The maximum current through the Zener diode.
    b. How does the test current compare with the maximum current?
    c. What is its maximum Zener impedance
        1. At the test current?
        2. At the knee?

1-22 Design a voltage divider circuit to take a V_{in} of 100 V and reduce it a V_{out} of 10 V:
   a. Use resistors.
   b. Use resistors and a 50 V Zener diode.
   c. What is $\Delta V_{out}/\Delta V_{in}$ in each case?

1-23 The circuit of Figure P1-23 is a voltage regulator circuit. Find the current in the Zener diode, R_1 and R_2 if
   a. $V_{in} = 16$ V.
   b. $V_{in} = 20$ V.
   c. $V_{in} = 30$ V.
   How much power does the Zener absorb in each case?

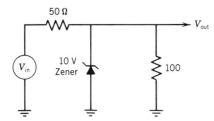

FIGURE P1-23

1-24 The circuit of Figure P1-24 is used to test an LED. If 20 mA is to flow through the LED to light it properly, what must the resistance be?

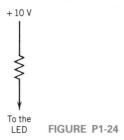

FIGURE P1-24

1-25 In the circuit of Figure P1-25, the manufacturer's specifications on the Zener diode are that the Zener voltage is 20 V, and the maximum power dissipation is 400 W.
   a. Find the value of R_1.
   b. Over what range of values of R_2 will the voltage be controlled by the diode?

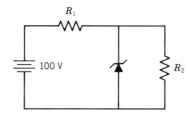

FIGURE P1-25

1-26 A BPX-66 is connected to a 50 kΩ resistor. What is the output voltage if the luminance is 10^3 electron-volts?

1-18 PROBLEMS   63

1-27 For the circuit of Figure P1-27 find the capacitance of the **BB139** varactor diode and the resonant frequency of the circuit if
  a. $V_{in} = -5$ V.
  b. $V_{in} = -10$ V.

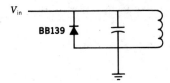

FIGURE P1-27

CHAPTER TWO

Bipolar Junction Transistor

2-1 Instructional Objectives

This chapter presents the operating principles of the BJT (bipolar junction transistor), the most popular type of transistor in current use, and discusses its basic parameters. After reading it the student should be able to:

1. List the differences between a transistor's base, emitter, and collector.
2. Calculate the base and collector current in a transistor, given the emitter current and α.
3. Calculate the gain of a common-base circuit.
4. Use common-base and common-emitter characteristic curves to find the relationship between the transistor's voltages and currents.
5. Calculate the β of a transistor given α and vice versa.
6. Use curve tracers to find transistor characteristics.
7. Plot load lines on transistor's characteristic curves.
8. Find transistor limitations and parameters on a manufacturer's specification sheet.

2-2 Self-Evaluation Questions

Watch for the answers to the following questions as you read the chapter. They should help you to understand the material presented.

1. Why must the collector-base junction withstand a higher reverse voltage than the emitter-base junction? What effect does this have on the physical construction of the transistor?
2. Why is the emitter doped more heavily than the base?
3. Why is the current gain of a common-base circuit less than 1? Is this also true for a common-emitter circuit?
4. Why is BV_{CEO} less than BV_{CBO}?
5. What are the characteristics of a transistor in saturation?
6. Why does a transistor dissipate very little power when in saturation and cutoff?

2-3 INTRODUCTION TO THE JUNCTION TRANSISTOR

In Chapter 1 many uses for *pn* type diodes were considered, but they have a basic limitation. They cannot amplify a signal; amplification requires a transistor. The main function of the transistors considered in this book is to increase the voltage, current, or power level of an ac signal. Transistors must do this in radios, for example, where the antenna picks an extremely small, high-frequency ac signal out of the air, recovers (or detects) the audio information in it, and amplifies it until it has enough power to drive the speaker. The circuits must absorb dc power to increase the ac power, but dc power is usually readily available (from batteries or power supplies), and the information, the music, or sound of interest, is contained in the ac signal. It is, therefore, important to amplify the ac signal, and the dc power required to do this is a secondary consideration.

A transistor is formed by connecting three semiconductor regions called the *emitter*, *base*, and *collector*. The emitter and collector are always doped with the same type of impurity atoms. They are separated by the base region, which is oppositely doped. Thus a transistor consists of a semiconductor sandwich of two possible types, *npn*, where the collector and emitter are *n*-doped, and the base is *p*-doped, and *pnp*, where the reverse is true.

The construction and transistor symbols for both transistors are shown in Figure 2-1. The direction of the arrow on the emitter lead is in the direction of *conventional current flow* and indicates whether the transistor is *npn* or *pnp*. Transistors are generally operated with forward-biased base-to-emitter junctions and reverse-biased base-to-collector junctions. For an *npn* transistor, the base is *p*- and the emitter *n*-type semiconductor. The current flows from the base to the emitter, as the arrow indicates. For a *pnp* transistor, the reverse is true and the arrow goes from the emitter to the base.

2-3.1 The Base-to-Emitter Junction

The base-to-emitter junction of a transistor is a diode and controls the action of the transistor. There are two possibilities:

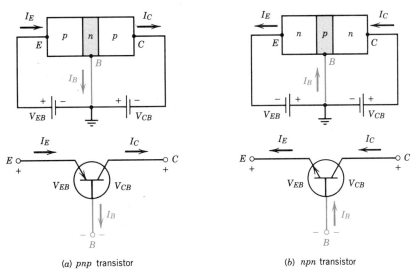

FIGURE 2-1
Basic transistors and their symbols.

1. The transistor is *on*. Here the junction is conducting current and the base-to-emitter voltage (V_{BE}) is approximately 0.7 V.
2. The transistor is *off*. In this case the base-to-emitter voltage is less than 0.7 V (it may be negative) and there is no significant current flow. Note that V_{BE} can never be greater than 0.7 V (see example 1-7).

Transistors are designed so that the emitter is much more heavily doped than the base. When the emitter–base junction is forward biased, almost all the current consists of the *emitter carriers* (electrons for an *npn* transistor, and holes for a *pnp* transistor). Pierce and Paulus (see References) give an example for a *pnp* transistor where the hole current is 5600 µA and the electron current is 3.5 µA.

2-3.2 Transistor Construction

Transistors are designed to provide high gain or amplification. These criteria determine the physical construction of the transistor. The base, which separates the collector and emitter, is very thin and lightly doped. This keeps the base current small compared to the emitter and collector currents. The collector and emitter are doped in the same way, but in most transistors there are differences between them. The collector–base junction is designed to be reverse-biased, whereas the emitter–base junction is forward-biased, so the collector base junction must withstand a greater inverse voltage. Collectors generally are larger than emitters for this reason. Also, collectors are not usually doped as heavily as emitters, although they are doped more heavily than bases.

The fabrication or construction of a transistor consists essentially of growing a silicon ingot, slicing it into wafers, and then adding impurities (*p*-

or *n*-dopants) in the proper concentrations and at the proper places. Figure 2-2 shows approximate cross sections of three types of transistors: grown junction, alloy, and diffused planar. In all cases the thin base is clearly shown and the larger collector is shown for the alloy and diffused planar transistors. In Figure 2-2c the aluminum metalization areas are the electrical connections to the base, emitter, and collector, and the silicon dioxide is used as an insulator. The methods of fabricating transistors and the advantages and disadvantages of each are beyond the scope of this book.

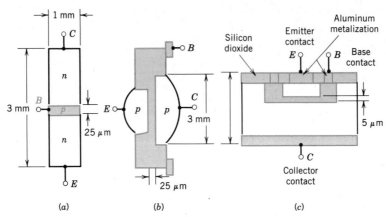

FIGURE 2-2
Construction of transistors. (a) Grown (*n-p-n*), (b) alloy (*p-n-p*), and (c) diffused planar (*n-p-n*) types. (The dimensions are approximate, and the figures are not drawn to scale.) (From *Microelectronics* by Jacob Millman, published by McGraw-Hill, Inc., 1979.)

2-3.3 The Actual Transistor

Figure 2-3 is a photograph and drawing of a **2N3904** transistor. This is a common transistor used for both amplification and switching and will be used in many examples in this book. The body of the transistor is about 0.2 inches ($\approx$5 mm) high and the leads extend for another 0.5 inches (12.7 mm).

The drawing shows the emitter, base, and collector leads from a bottom view. Holding the transistor upright and facing the flat part of the body, the emitter is the left lead, the base lead is in the center, and the collector lead is on the right.

Other transistors may have different configurations and come in different packages. The **2N3904** is packaged in the small TO-92 type case. Power transistors are packaged in larger cases, which are often metal (usually aluminum), and are designed to be mounted on heat sinks (see Chapter 9). The heat sink absorbs the heat by conduction and allows the transistor to dissipate more power.

Some transistors come in an integrated-circuits package. One example is the **Q2T2905,** manufactured by Texas Instruments, and shown in Figure 2-4. It comes in a dual-in-line package (DIP) and contains four *pnp* transistors.

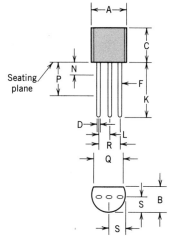

FIGURE 2-3
The physical construction of a **2N3904** transistor. (Copyright by Motorola, Inc. Used by permission.)

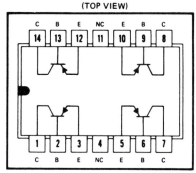

FIGURE 2-4
Type **Q2T2905** Texas Instruments quad *pnp* silicon transistors: (*a*) appearance; (*b*) pin connections. (Courtesy of Texas Instruments, Inc.)

2-4 TRANSISTOR OPERATION

The basic operation of an *npn* transistor is shown in Figure 2-5. The blue lines indicate the direction of conventional current flow and hole flow. The small black line on Figure 2-5 is the electron component of I_{CBO} and the blue line is its hole component. The heavy black line is the emitter electron flow.

2-4.1 The Collector–Base Junction

If switch S_1 is open, V_{CC} reverse-biases the base–collector junction. The only current that flows is a leakage current called I_{CBO} (collector to base with the emitter open circuited).[1] Most manufacturers will specify maximum values of

---

[1]Millman (see References) defines I_{CO} as the collector–base leakage current and I_{CBO} as I_{CO}, plus leakage current across the insulators separating the junction, etc. The differences between them are small enough to be ignored here.

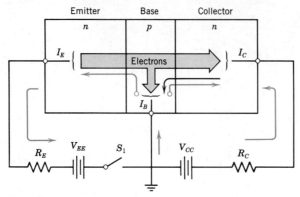

FIGURE 2-5
The basic operation of an *npn* transistor.

I_{CBO} in their data sheets. In the early days, when there were many germanium transistors, I_{CBO} was occasionally significant. Now that silicon transistors dominate, I_{CBO} is so small that it can generally be ignored. For most common silicon transistors, I_{CBO} is less than 10 nA. Of course, I_{CBO} is a leakage current and doubles for every 10°C rise in junction temperature, but it is still very small compared to the normal values of collector current.

2-4.2 The Emitter–Base Junction

When switch S_1 in Figure 2-5 is closed, the emitter–base junction becomes forward-biased, as it normally is. This forward bias produces a heavy base-to-emitter current, which, in an *npn* transistor, consists almost entirely of electrons moving from the emitter into the base because the emitter is more heavily doped than the base. Once the electrons enter the base region, they become minority carriers in the *p*-type base region. Minority carriers are swept across a reverse-biased junction (see section 1-5.2), so most of these electrons are swept into the collector. The base is also deliberately made thin so that the electrons entering the base can easily continue into the collector.

With the switch closed the transistor is fully functional. There are three currents, I_E, I_B, and I_C; the emitter, base, and collector currents, respectively. By Kirchhoff's current law, the current entering the transistor must equal the current leaving it. Therefore, for any transistor:

$$I_E = I_B + I_C \qquad (2\text{-}1)$$

As we saw before (and neglecting I_{CBO} and the base hole current as insignificant), the electrons from the emitter move into the base, where they are attracted both by the positive base potential and the bias across the base–collector junction. Because of the thin base, most of the electrons continue into the collector. We define a constant α for any transistor as the ratio of collector current to emitter current.

$$\alpha = \frac{I_C - I_{CO}}{I_E} \approx \frac{I_C}{I_E} \tag{2-2}$$

By equation 2-1, I_E is always greater than I_C; therefore α must always be less than 1. For modern high-gain transistors, this value ranges between 0.98 and 0.998. With these values of α, the following approximation is often used:

$$I_E \approx I_C \tag{2-3}$$

2-5 THE COMMON-BASE CONFIGURATION

Figure 2-5 is an example of the transistor connected in the common-base configuration,[2] where the base is grounded. Figure 2-6 shows the common-base circuits for both *npn* and *pnp* transistors. Note that in the *npn* configuration the emitter is negative with respect to ground and in the *pnp* configuration the collector is negative.

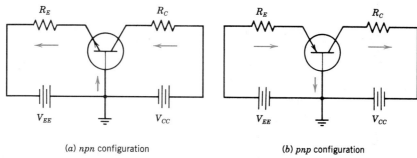

(a) *npn* configuration   (b) *pnp* configuration

FIGURE 2-6
Common base circuits.

2-5.1 Common-Base Amplification

Amplification or gain means that the circuit can provide a greater output signal than its input signal. In the common-base configuration, the input signal is applied to the emitter, as shown in Figure 2-7, and the output is taken at the collector. The common-base circuit can never provide current amplification because I_E is always greater than I_C, but it can provide voltage amplification, as Example 2-2 shows.

---

[2]The word *common* used in the context of common base, common emitter, or common collector (see section 3-8) means the associated terminal is connected to a dc or unchanging voltage. For many common-base and common-emitter circuits, including those discussed in this chapter, that voltage is ground.

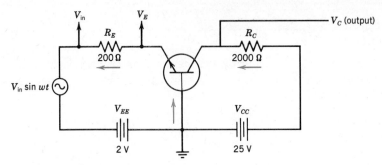

FIGURE 2-7
Circuit for Example 2-1.

EXAMPLE 2-1

a. For Figure 2-7, find the dc operating conditions. Assume $\alpha = 0.99$ and $V_m = 0$ so the sine wave generator is a short circuit. Find I_E, I_B, I_C and V_C.
b. How large can R_C become?

Solution

a. The value of I_E can be found by considering the emitter loop. Here $V_E = -0.7$ V because of the base-to-emitter drop. Therefore:

$$I_E = \frac{2\text{ V} - 0.7\text{ V}}{200\ \Omega} = \frac{1.3\text{ V}}{200\ \Omega} = 6.5\text{ mA}$$

$$I_C = \alpha I_E = 0.99 \times 6.5\text{ mA} = 6.435\text{ mA}$$

$$I_B = I_E - I_C = (1 - \alpha)I_E = 0.01 \times 6.5\text{ mA} = 0.065\text{ mA}$$

The voltage at the collector, V_C, is

$$V_C = V_{CC} - I_C R_C$$
$$= 25\text{ V} - 6.435\text{ mA} \times 200\ \Omega = 12.13\text{ V} \qquad (2\text{-}4)$$

b. As the value of R_C increases, the gain tends to increase (see section 4-5.1). This would indicate that R_C should be as large as possible. There is a limitation, however. For linear amplification, the collector-to-base voltage must remain positive to reverse-bias the collector–base junction. Since $V_C = V_{CC} - I_C R_C$, the voltage across the collector resistor, $I_C R_C$, must be less than V_{CC} for V_C to remain positive. In this example:

$$I_C R_C \leq 25\text{ V}$$

or

$$R_C \leq \frac{25\text{ V}}{6.435\text{ mA}} \leq 3885\ \Omega$$

Actually when bias and clipping considerations are taken into account (see section 3-6.2), the selected R_C is typically half the value calculated by this method.

2-5.2 ac and dc Nomenclature

An ac signal can be injected into the emitter of Figure 2-7 by turning on the sine wave generator. The voltage V_{in} will then be an ac voltage superimposed upon the -2 V dc level. If V_m is set to 1 V, V_{in} will vary between -1 and -3 V as Figure 2-8a shows.

In voltage or current expressions it is often necessary to differentiate between ac and dc components or value. Table 2-1 gives the most commonly used nomenclature that will be followed in this book. For the circuit of Figure 2-7 and the voltage waveform of Figure 2-8a, for example:

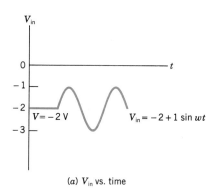

(a) V_{in} vs. time

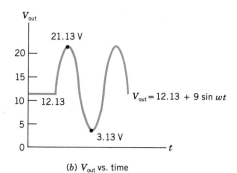

(b) V_{out} vs. time

FIGURE 2-8
V_{in} and V_{out} for the circuit of Figures 2-7 and 2-9.

Table 2-1
Transistor Nomenclature

Symbol	Example	Definition
Capital letter Capital subscript	I_B, I_E, V_C	dc or quiescent value
Capital letter Lowercase subscript	$V_m \sin \omega t$	peak ac value
Lowercase letter Lowercase subscript	v_c, i_c, i_e	instantaneous ac value
Lowercase letter Capital subscript	v_C, i_E	total instantaneous value of a signal
Capital letter Double capital subscript	V_{CC}, V_{EE}, V_{BB}	a fixed dc voltage generally from a battery or power supply

$$V_E = V_{EE} = -2 \text{ V}$$
$$V_{in} = 1 \text{ V}$$
$$V_{in} = -2 \text{ V} + 1 \sin \omega t$$

When $\sin \omega t = 1$, $V_{in} = -1$ V. Note the difference in nomenclature between the peak ac voltage, $V_{in} = 1$ V, and the peak of the total voltage, $V_{in} = -1$ V.

2-5.3 Common-Base ac Voltage Gain

A primary characteristic of transistor circuits is the ac gain or ratio of output voltage to input voltage. Gain is symbolized by the letter A. Four different gains will be used in this book.

$A_{v(ck)}$ The voltage gain of the transistor circuit; the ratio of the output voltage of the circuit to the input voltage of the circuit.

$A_{v(tr)}$ The voltage gain of the transistor itself; the ratio of the output voltage to the input voltage at the transistor terminals.

$A_{i(ck)}$ The ac current gain of the transistor circuit.

$A_{i(tr)}$ The ac current gain of the transistor itself.

When analyzing a transistor circuit for its ac values, the dc values or constant voltage sources are ignored. This is done in accordance with the superposition theorem of circuit analysis. A dc analysis can be made ignoring the varying components of the input and an ac analysis can be made that ignores the power supplies and other dc voltage sources. The total analysis is the addition or superposition of the dc and ac analyses.

In the circuit of Figure 2-7, ac analysis will ignore both batteries and the 0.7 V across the base-to-emitter p-n junction. There is, however, an ac impedance between base and emitter. For the common-base configuration this impedance, called h_{ib} (see section 4.6), is usually quite small.

EXAMPLE 2-2

For the circuit of Figure 2-7, find $A_{i(ck)}$, $A_{i(tr)}$, $A_{v(ck)}$, and $A_{v(tr)}$. Assume $h_{ib} = 20$ ohms.

Solution
The ac equivalent circuit for Figure 2-7 is shown in Figure 2-9. Note the use of the lowercase i for the ac component of current. The emitter current is a sine wave given by

$$I_e = \frac{V_{in} \sin \omega t}{R_E + h_{ib}}$$

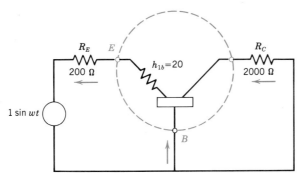

FIGURE 2-9
The ac equivalent circuit for Figure 2-5.

The peak value of i_e occurs when $\sin \omega t = 1$ and is

$$I_e = \frac{V_m}{R_E + h_{ib}} = \frac{1 \text{ V}}{220 \, \Omega} = 4.54 \text{ mA}$$

Similarly, the peak value of collector current is

$$I_c = \alpha I_e = 0.99 \times 4.54 \text{ mA} = 4.5 \text{ mA}$$

In Figure 2-9 the current gain of the transistor I_c/I_e, and the current gain of the circuit are the same, 0.99. A gain of less than one indicates that the output is smaller than the input.

To find the voltage gains the ac component of the output voltage must first be calculated.

$$V_c = I_c R_c = 4.5 \text{ mA} \times 2000 \, \Omega = 9 \text{ V}$$

The gain of the circuit is simply

$$A_{v(ck)} = \frac{V_c}{V_m} = 9$$

The gain of the transistor is V_c/V_e where the emitter voltage $V_e = h_{ib} \, I_e = 20 \, \Omega \times 4.54 \text{ mA} = 0.091$ V. The transistor gain is therefore

$$A_{v(tr)} = \frac{V_c}{V_e} = \frac{9 \text{ V}}{0.091 \text{ V}} = 98.9$$

This example is typical; a common-base circuit can provide significant voltage gain, but no current gain.

The ac output voltage has a 9 V peak around a 12.13 V (see example 2-1), and is shown in Figure 2-8b. In accordance with the nomenclature convention of Table 2-1, $V_{CC} = 25$ V, $V_C = 12.13$ V, $V_c = 9$ V and v_C varies between 21.13 and 3.13 V.

2-5.4 Common-Base Characteristic Curves

The common-base characteristics are a set of curves that show the relationship between V_{CB}, the collector-to-base voltage, I_C, the collector current, and I_E, the emitter current. The characteristic curves for a **2N3904** transistor are shown in Figure 2-10. They were obtained on an **HP 4145A** curve tracer (see section 2-7).

FIGURE 2-10
Common base characteristics for a **2N3904** transistor. (Courtesy of Hewlett Packard Co.)

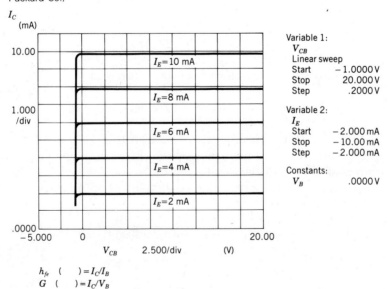

Notice that the curves are practically flat for $V_{CB} > 0$. We deliberately allowed V_{CB} to go negative to show some curvature, but transistors are never operated in that region. For the active region, where $V_{CB} > 0$ and the transistor can function as an amplifier, it is easy to see that $I_C \approx I_E$. Indeed, until $I_C = 10$ mA, there is almost no discernible difference between I_C and I_E. This is consistent with the fact that α for this transistor is typically between 0.99 and 0.995.

2-6 THE COMMON-EMITTER CIRCUIT

In the common-base circuit an injected emitter current produces a collector current of approximately the same value and a very small base current. In the *npn* common-emitter circuit, shown in Figure 2-11a, the emitter is grounded and the base is forward-biased. The positive voltage across the base-emitter *p-n* junction causes a large number of electrons to enter the base from the emitter. A small percentage of them leave through the base lead and constitute the base current. Most of the electrons flow into the collector, so the collector current is larger than the base current. Therefore the common-emitter circuit has a current gain greater than 1. This is one of the reasons that the common-emitter circuit is used much more frequently than the common-base circuit.

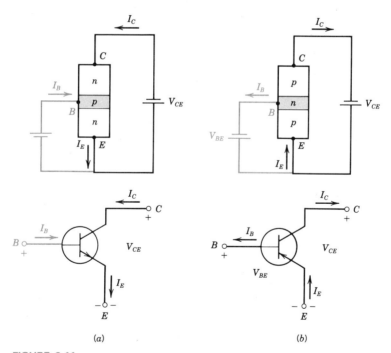

FIGURE 2-11
Notation and symbols used with the common-emitter configurations:
(a) *npn* transistor; (b) *pnp* transistor.

The relationship between the base and collector currents can be calculated:

$$I_C = \alpha I_E + I_{CO} = \alpha(I_B + I_C) + I_{CO}$$
$$I_C(1 - \alpha) = \alpha I_B + I_{CO}$$
$$I_C = \frac{\alpha I_B}{1 - \alpha} + \frac{I_{CO}}{1 - \alpha} \qquad (2\text{-}5)$$

The term $\alpha/(1 - \alpha)$ is called β (beta), and is the ratio of collector current to base current, or the current amplification of a common-emitter transistor. β and α are both current ratios and have no units.

EXAMPLE 2-3

If $\alpha = 0.99$ for a transistor, find β.

Solution

$$\beta = \frac{\alpha}{1 - \alpha} = \frac{0.99}{0.01} = 99$$

Most transistors have high βs and consequently high current amplification. For a **2N3904**, β is specified to be between 100 and 300. This shows that the range of βs for a particular transistor is broad, and designers of transistor circuits have to carefully consider this fact.

With the approximation that

$$\frac{\alpha}{1 - \alpha} \approx \frac{1}{1 - \alpha} \approx \beta$$

equation 2-5 becomes

$$I_C = \beta I_B + \beta I_{CBO} \qquad (2\text{-}6)$$

Thus the small common-base leakage current is amplified by β in the common-emitter configuration. This leakage current is called I_{CEO}. Although magnified it is still quite small for silicon and generally can be ignored. For a **2N3904** this current is specified to be 50 nA maximum.

Most transistor specification sheets do not use β, but give the hybrid parameter term (see section 4-4), h_{FE} for dc conditions, and h_{fe} for ac conditions. The terms β and h_{FE} are used interchangeably. There is a dc β or h_{FE} defined simply as

$$h_{FE} = \frac{I_C}{I_B} \tag{2-7}$$

and an ac β, which is the amplification of an alternating current. The ac β is defined as the *change* in collector current divided by the change in emitter current.

$$\beta_{ac} = h_{fe} = \left. \frac{\Delta I_c}{\Delta I_b} \right|_{V_{CE}\,\text{constant}} \tag{2-8}$$

The β of a transistor does vary somewhat depending on the point of operation of that transistor. Fortunately, for most transistors, the dc and ac βs are approximately equal.

EXAMPLE 2-4

For the circuit of Figure 2-12, find V_C if the *npn* transistor has a β of 155.

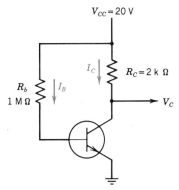

FIGURE 2-12
Circuit for Example 2-4.

Solution

In this circuit $V_C = V_{CC} - I_C R_C$. To find the collector current, I_C, the base current must first be found. It flows from V_{CC} through R_B and into the base–emitter junction. Because $V_{BE} = 0.7$ V, I_B can be determined:

$$I_B = \frac{V_{CC} - V_{BE}}{R_B} = \frac{20 - 0.7 \text{ V}}{1 \text{ M}\Omega} = 19.3 \text{ μA}$$

$$I_C = \beta I_B = 155 \times 19.3 \text{ μA} = 3 \text{ mA}$$

$$V_C = V_{CC} - I_C R_C = 20 - 3 \text{ mA} \times 2 \text{ k}\Omega = 14 \text{ V}$$

2-6.1 Common-Emitter Transistor Characteristics

Common-emitter volt-ampere characteristic curves can be plotted similarly to common-base characteristics, and these plots are often very valuable. A common-emitter plot for a 2N3904 (again using an **HP 4145A** curve tracer) is shown in Figure 2-13. Notice the small increments of base current (10 µA) compared to the collector current scale of 1 mA.

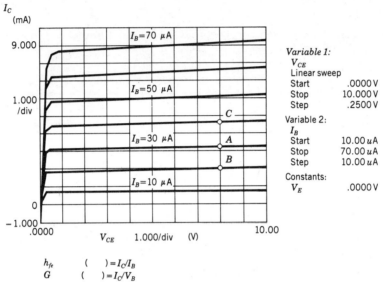

FIGURE 2-13
The common emitter characteristics of a **2N3904** transistor.
(Courtesy of Hewlett Packard Co.)

EXAMPLE 2-5

For the transistor of Figure 2-11, find h_{FE} and h_{fe} if $V_C = 8$ V and $I_B = 30$ µA.

Solution
a. At 30 µA and 8 V (point A on Figure 2-13), I_C is found to be 3.2 mA. Therefore

$$h_{FE} = \frac{I_C}{I_B} = \frac{3.2 \text{ mA}}{30 \text{ µA}} = 106.7$$

b. h_{fe}, the ac value is defined as $\Delta I_C / \Delta I_B$ with V_C constant. To find h_{fe} at the 30 µA point, it is best to take the values above and below it for the same V_C, or at points B and C. The values are:

Point	A	B	C	Units
V_C	8	8	8	V
I_B	30	20	40	μA
I_C	3.2	1.9	4.5	mA

$$h_{fe} = \left.\frac{\Delta I_C}{\Delta I_B}\right|_{V_C = 8\text{ V}} = \frac{4.5 - 1.9 \text{ mA}}{40 - 20 \text{ μA}} = \frac{2.6 \text{ mA}}{20 \text{ μA}} = 130$$

Note that h_{fe} and h_{FE} differ by about 25% at this point. For larger values of I_B, the two are closer (see problem 2-9).

Figure 2-14 is a plot of h_{FE} versus V_{CE} for 10 μA increments of base current for the same **2N3904** transistor using the **HP 4145A**. The 30 μA, 8 V point is at point A on the plot. It shows an h_{FE} of 105, which agrees very well with the results of Example 2-5.

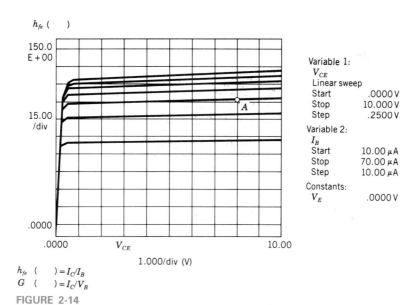

h_{fe} ( ) = I_C/I_B
G ( ) = I_C/V_B

FIGURE 2-14
A plot of h_{FE} for the **2N3904** transistor. (Courtesy of Hewlett Packard Co.)

EXAMPLE 2-6

Describe how the common-emitter curves of a transistor can be obtained in the laboratory.

Solution

To obtain curves similar to Figure 2-13, a three-step procedure can be followed.

a. Set I_B to a fixed value (for example, the first step could be 10 μA).
b. Vary V_C and measure I_C.
c. Now set I_B to the second step (20 μA, for example) and repeat step 2. Continue until a sufficient number of steps have been taken.

Figure 2-15 is a circuit that works in the laboratory. The first step is to fix I_B.

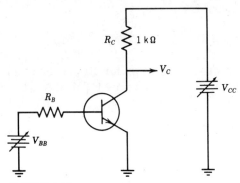

FIGURE 2-15
A circuit for obtaining the common emitter characteristics of an *npn* transistor.

$$I_B = \frac{V_{BB} - V_{BE}}{R_B}$$

As a first approximation $V_{BE} = 0.7$ V. If V_{BB} is a digital power supply (+5 V) and I_B is selected at 10 μA, then

$$R_B = \frac{4.3 \text{ V}}{10 \text{ μA}} = 430 \text{ k}\Omega$$

Now V_{CC} is varied, starting from 0, and V_C monitored. When $V_C = 1, 2, 3$ volts, the collector current can be determined from the equation

$$I_C = \frac{V_{CC} - V_C}{R_C}$$

Choosing a 1 kΩ collector resistor greatly simplifies this calculation.

After the 10 μA curve has been obtained, R_B can be halved and the 20 μA curve can be taken. To be extremely precise, V_{BB} and V_{BE} should be measured at each step of V_C, but normally this is not necessary, and V_{BE} can be assumed to be 0.7 V.

2-7 CURVE TRACERS

A *curve tracer* is an instrument that allows the user to plug in a diode or transistor, and displays that device's characteristics on a cathode ray tube (CRT) screen. Figures 2-10, 2-13, and 2-14 were taken from an **HP 4145A** curve tracer. Because one of the major components of a curve tracer is the CRT screen, it resembles an oscilloscope.

To use a curve tracer, the user must:

1. Plug in the transistor. There is always a switch or other method of telling the curve tracer whether an *npn* or *pnp* transistor is being tested.
2. Select a range of base currents.
3. Select a range of collector voltages.
4. Select the number of steps to be used.

The curve tracer will then cycle, injecting the first increment of the base current and varying the collector voltage, then injecting the second increment of base current, and so on. When it finishes all the base current steps, it starts over. This cyclical process is done so rapidly that the output appears as a set of curves on the CRT screen and the relationship of V_{CE}, I_B, and I_C can be determined directly from the screen.

EXAMPLE 2-7

How was the curve tracer set up to produce the plot of Figure 2-13?

Solution
An examination of the figure reveals

1. The base current steps were 10 µA.
2. The collector voltage was varied from 0 to 10 V.
3. There were seven steps of base current, from 10 µA to 70 µA.

All of these parameters can be changed by the operator of the different displays.

2-7.1 Tektronix Curve Tracers

The Tektronix Corporation is presently marketing two curve tracers, the **576** and **577**. The **576** is intended to test high-power transistors, such as the **2N3055** (see Chapter 9), whereas the **577** is intended for lower power transistors, such as the **2N3904**.

The **577** curve tracer and the **177,** a companion fixture, are shown in Figure 2-16.

The transistor or diode under test is plugged into the fixture, and the dials on the panel of the **577** are used to select the range of base currents, collector voltages, and number of steps. The current inputs range from 5 nA to 200 mA per step, and the test voltage can go as high as 1600 V.

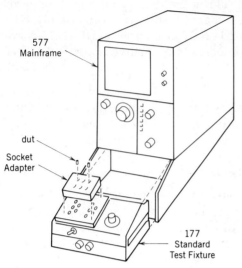

FIGURE 2-16
The Tektronix **577** curve tracer with the **177** standard test fixture. (Courtesy of Tektronix, Inc.)

Figure 2-16 shows the **577** displaying an emitter characteristic. It can also test diodes, FETs, SCRs and other devices.

2-7.2 Hewlett-Packard Curve Tracers

Figure 2-17 is a photograph of the **HP 4145A** Semiconductor Parameter Analyzer, which is an elaborate, programmable curve tracer. It contains a **68000** microprocessor to perform its calculations.

The **HP 4145A** is *programmable* because the current steps, voltage steps, and so on are entered via the keys on the front panel instead of by a dial. The parameters may also be entered via a computer or the cassette reader on the Parameter Analyzer.

The **HP 4145A** has many advanced features. It can measure transistor parameters such as h_{FE}, as well as displaying the transistor characteristics. It has a marker control on the front panel that allows the user to move a highlighted dot along a curve while the **HP 4145A** calculates the values of the parameters and slope of the curve at the marker position. It can also be tied to a plotter that will automatically plot the curves. Figures 2-10, 2-13, and 2-14 were obtained in that way.

2-8 THE LOAD LINE

Consider the circuit of Figure 2-18 where a current source is supplying the base current. Variations in the base current will cause the collector current, I_C, to vary, but I_C is subject to two constraints:

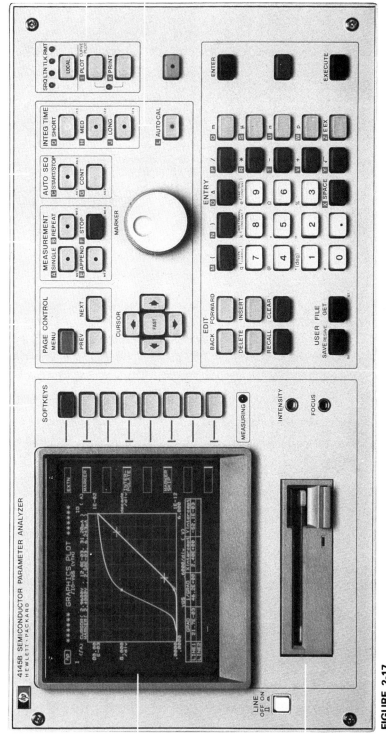

FIGURE 2-17
The HP **4145A** curve tracer and parameter analyzer. (Courtesy of Hewlett Packard Co.)

2-8 THE LOAD LINE

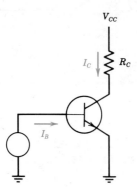

FIGURE 2-18
A basic common emitter *npn* transistor circuit.

$$I_C = \beta I_B \quad (2\text{-}9)$$
$$V_C = V_{CC} - R_C I_C \quad (2\text{-}10)$$

The characteristic curves for the transistor are plots with I_C as the vertical or y axis and V_C as the horizontal or x axis (see Figure 2-13). Equation 2-10 is the equation of a straight line on these axes called the *load line*. Figure 2-19 shows the load line plotted on a set of characteristic curves for a typical *npn* transistor with assumed values of $V_{CC} = 20$ V, $R_C = 2$ kΩ.

EXAMPLE 2-8

For the circuit whose load line is shown as the solid line in Figure 2-19, find I_C and V_C if $I_B = 40$ μA.

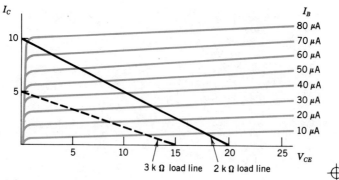

FIGURE 2-19
Load lines on the characteristic curves for a **2N3904** transistor.

86 BIPOLAR JUNCTION TRANSISTOR

Solution

With the load line and curves available, the simplest solution is to follow the $I_B = 40\ \mu A$ curve until it intersects the load line. From the curves, it can be seen that $I_C = 4.5$ mA and $V_C = 11$ V. This problem can also be solved using equations 2-9 and 2-10 (see example 2-4), but with the curves and the load line available, this method is simpler.

Equation 2-10 can be transposed to

$$I_C = \frac{-V_C}{R_C} + \frac{V_{CC}}{R_C}$$

This is in the familiar slope-intercept ($y = mx + b$) form of the straight line equation. It shows that the slope of the load line is $-1/R_C$ and the y intercept is V_{CC}/R_C.

Having considered equation 2-10 mathematically, it will now be considered physically. At one extreme, $I_C = 0$; then no current flows and $V_C = V_{CC}$. The other extreme is when the transistor is a short circuit. Then $V_C = 0$, and $I_C = V_{CC}/R_C$. The simplest way to plot a load line for this type of circuit is to draw a line between V_{CC} on the horizontal axis and V_{CC}/R_C on the vertical axis.

EXAMPLE 2-9

If $V_{CC} = 15$ V and $R_C = 3$ kΩ, draw the load line on Figure 2-19.

Solution

Two points on the load line are $V_C = 15$ V (V_{CC}) on the x or V_{CE} axis and 5 mA (V_C/R_C) on the y or I_C axis. The load line is drawn between them, as shown by the dashed line. Note that the slopes of the two load lines on Figure 2-19 are different due to the different resistors.

The load lines for *pnp* transistors can be plotted on their characteristic curves (see problem 2-18) in a similar manner. The only differences are that V_{CE} is negative and all the currents are reversed.

2-9 THE CUTOFF REGION

Figure 2-20 shows that there are three regions where a transistor can operate, called *active*, *cutoff*, and *saturation*. The *active* region is where *amplification* takes place and will be discussed in Chapters 3 and 4. The saturation region is discussed in section 2-10.

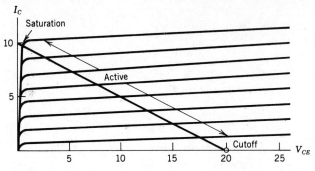

FIGURE 2-20
The saturation, active, and cutoff regions for a transistor.

As a first approximation, we can define the *cutoff* region as the region where *no significant current flows*. The circuit of Figure 2-21 was tested in the laboratory, and at $V_{BE} = 0.5$ V a collector current of 5 µA flowed. Only a very minute current will flow if $V_{BE} \leq 0.5$ V and this can be considered as cutoff. If the transistor's base is open, it is also cutoff since only I_{CEO} flows, and this current is so small that it is difficult to measure (see problem 2-13).

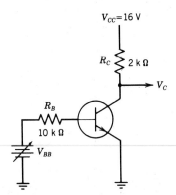

FIGURE 2-21
A test circuit to find the cutoff voltage of a transistor.

2-9.1 Voltage Limitations in Cutoff

In cutoff there are three voltage values that limit the operation of a transistor. These values are generally specified by the manufacturer and should not be exceeded. The *B* preceding the term specifies a *breakdown voltage*, which can destroy the transistor.

BV_{CBO} This is the breakdown voltage between the collector and base with the emitter open. If this voltage is exceeded, avalanche breakdown (see section 1-5.3) may occur and destroy the transistor.

BV_{CEO} This is the breakdown voltage between the collector and emitter if the base is left open. Because of the amplification of the leakage current it is somewhat smaller than BV_{CBO}. Since most circuits are common-emitter, this is the more important parameter. Practically, it means that the transistor power supply (V_{CC}) *should not exceed* BV_{CEO}, because if the base circuit is momentarily disconnected for any reason, V_{CC} appears between the collector and emitter and can destroy the transistor if it is greater than V_{CEO}.

BV_{EBO} To keep the transistor cutoff some circuits reverse bias the base-to-emitter junction, as in Figure 2-22. BV_{EBO} is the maximum reverse voltage that should be applied between the base and emitter. To be very precise the base voltage in Figure 2-22 is $-V_{BB} + I_{CBO}R_B$, but I_{CBO} is so small that in most cases the $I_{CBO}R_B$ term can be ignored.

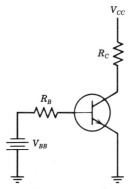

FIGURE 2-22
An *npn* transistor with the base reverse biased.

If BV_{EBO} is exceeded the transistor may be destroyed. In many transistors, however, the base-to-emitter junction tends to act as a Zener diode when BV_{EBO} is exceeded. This results in significant reverse base-to-emitter current without damaging the transistor.

In addition to the foregoing voltages, the manufacturers' specifications on I_{CBO} and I_{CEO} should be checked, but these currents generally will be small enough to be ignored.

2-9.2 Curve Tracer Tests of Transistor Characteristics

For any particular transistor, a curve tracer can be used to determine the limiting characteristics such as BV_{CEO}, BV_{CBO}, and I_{CEO}. To measure a breakdown voltage, the voltage applied to a transistor is slowly increased until significant current starts to flow. A typical curve tracer display is shown in Figure 2-23. It was taken from Tektronix Application Note 4 W-4021-1 entitled "Testing the **2N3904** Bipolar Transistor," which shows how to measure the cutoff voltages and currents for a **2N3904**. Other transistors would be measured similarly.

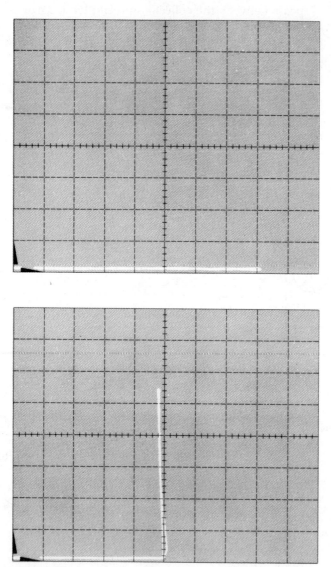

FIGURE 2-23
Testing BV_{CEO} on a curve tracer. (Courtesy of Tektronix, Inc.)

2-10 THE TRANSISTOR IN SATURATION

The *saturation* region is at the top of the load line in Figure 2-20. It is characterized by relatively high base and collector currents and a very low V_{CE}.

To examine the saturation region further, consider the circuit of Figure 2-24 and let us try to calculate the collector voltage.

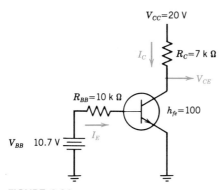

FIGURE 2-24
A saturated transistor.

As before

$$I_B = \frac{V_{BB} - 0.7 \text{ V}}{R_B} = \frac{10.7 \text{ V} - 0.7 \text{ V}}{10 \text{ k}\Omega} = 1 \text{ mA}$$

$$I_C = h_{FE}I_B = 100 \text{ mA}$$

$$V_C = V_{CC} - I_C R_C = 20 \text{ V} - 100 \text{ mA} \times 1 \text{ k}\Omega = -80 \text{ V}$$

The mathematics is impeccable, but the answer is wrong. What is really happening must be determined. In this circuit an increasing base current produces an increasing collector current that lowers the collector voltage. But the collector voltage can never go below 0 V; actually it gets as low as 0.1 or 0.2 volt.

The saturation region is entered by forcing so much base current through the transistor that the collector current reaches its maximum possible value:

$$I_{C(max)} \approx \frac{V_{CC}}{R_C}$$

Once in saturation any increase or decrease in base current will not change I_C, unless the base current decreases sufficiently to bring the transistor into the active region.

Transistors in saturation are said to be operating with a *forced* β that is less than the inherent h_{FE} of the transistor. For the circuit of Figure 2-24, $I_C = 20$ mA, $I_B = 1$ mA and

$$\beta_{(\text{forced})} = \frac{I_C}{I_B} = \frac{20 \text{ mA}}{1 \text{ mA}} = 20$$

Furthermore, in the saturation region, V_{CE} can drop to 0.1 or 0.2 V, while V_{BE} remains at 0.7 V. Thus the base voltage is positive with respect to the collector and the *collector–base junction is forward-biased*.

EXAMPLE 2-10

If $V_{CE} = 0.1$ V, how much is the forward bias of the collector–base junction?

Solution
Because $V_{BE} = 0.7$ V the forward bias is 0.7 − 0.1 or 0.6 V.

To sum up, there are three basic characteristics of a transistor in saturation:

1. I_C equals its maximum possible value, generally V_{cc}/R_c as in Figure 2-24.
2. The transistor is operating with a forced

$$h_{FE(\text{forced})} = \frac{I_C}{I_B} < h_{FE(\text{transistor})}$$

3. The base-to-collector junction is forward-biased.

A transistor in saturation exhibits all of these characteristics.

2-10.1 Saturation Curves

Saturation curves for the **2N3904** transistor are shown in Figure 2-25. These are curves of collector voltage versus base current for various values of

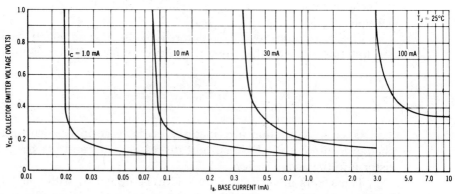

FIGURE 2-25
Saturation curves for a **2N3904** transistor. (Copyright by Motorola, Inc. Used by permission.)

collector current. At low values of I_B the curves are almost vertical because the transistor is not in saturation and I_C/I_B is the inherent h_{FE} of the transistor. At higher values of I_B the curves flatten out as the transistor goes into saturation.

EXAMPLE 2-11

Using the $I_C = 10$ mA curve of Figure 2-25, determine
a. The inherent h_{FE} of the transistor.
b. The h_{FE} and V_{CE} if $I_B = 0.5$ mA.

Solution
a. To determine the inherent h_{FE} of the transistor, it is best to get as far out of the saturation region as possible. $V_{CE} = 1$ V is as far as one can go on these curves, and at this point the collector–base junction is reverse-biased. From the curve at $V_{CE} = 1$ V, we find $I_B = 0.08$ mA. When $I_C = 10$ mA, therefore

$$h_{FE} = \frac{10 \text{ mA}}{0.08 \text{ mA}} = 125$$

b. At $I_B = 0.5$ mA we have

$$h_{FE(forced)} = \frac{10 \text{ mA}}{0.5 \text{ mA}} = 20$$

At this point V_{CE} is 0.16 V so the collector–base junction is forward-biased. This point is well below the knee of the curve and the transistor is well into saturation.

Another set of saturation curves supplied by the manufacturers of the **2N3904** are the *on* voltages shown in Figure 2-26. These are taken for values of $I_C/I_B = 10 = h_{FE(forced)}$, and clearly show the collector and base voltages for the transistor operating with a forced β of 10 for a range of collector currents.

2-10.2 Digital Operation

For amplification the transistor must not be operated in cutoff or saturation. In the *switching mode*, however, the transistor is either *fully on* (saturated) or *fully off* (cutoff). In many circuits this causes the output voltage to be either approximately V_{CC} or approximately 0. Switching mode is used in digital circuits, and the speed of the transition from cutoff to saturation and back determines the speed of the digital circuit (see section 7-9).

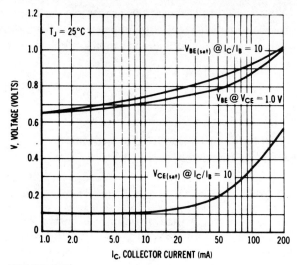

FIGURE 2-26
"On" voltages for a **2N3904** transistor. (Copyright by Motorola, Inc. Used by permission.)

In addition to having the output voltages at V_{CC} or 0, transistors in switching mode have a power advantage. The power absorbed by any device is the product of the voltage across it and the current in it. For a transistor

$$P_D = V_{CE} I_C \tag{2-11}$$

where P_D is the power dissipation.

P_D is significant if the transistor is in the active region, but when the transistor is cutoff $I_C \approx 0$, and when the transistor is saturated $V_{CE} \approx 0$. Thus in both saturation and cutoff the transistor dissipates very little power, and this fact makes switching mode operation very attractive. Power dissipation in transistors is discussed in Chapter 9.

2-11 MANUFACTURERS' SPECIFICATIONS

To find the characteristics and parameters for any transistor, the manufacturers' specifications should be consulted. Appendix A contains the manufacturers' specifications for the 2N3903 and 2N3904 transistors that were frequently used throughout this chapter.

The first page of the specifications gives the maximum ratings (V_{CBO}, V_{CEO}, V_{EBO}, $I_{C(max)}$, Power Dissipation and operating temperature) for the transistors. Exceeding these ratings may damage or destroy the transistor.

The electrical characteristics on the second page of the Appendix repeat some of the maximum ratings. They also give ranges of h_{FE} for various operating conditions and many other transistor characteristics that will be discussed in future chapters.

Under small-signal characteristics, the small signal ac current gain, h_{fe}, is specified as a minimum of 100 and a maximum of 400 at $V_{CE} = 10$ V and $I_C = 1$ mA, a typical operating point for this transistor.

The remaining pages are curves of a typical transistor for many characteristics. The variation of current gain versus collector current are shown, as are the saturation curves and *on* voltages discussed in section 2-10. The other curves and parameters specified for the **2N3904** will be discussed in the later chapters of this book.

The **2N3903/04** specifications also mention the *complementary* types, the **2N3905** and **2N3906**. These are *pnp* transistors with the same characteristics and limitations as the **2N3903** and **2N3904**. Some circuits are both *npn* and *pnp* transistors operating together to achieve the desired results. These circuits are often called complementary circuits (see section 9-9).

2-12 SUMMARY

This chapter started by explaining the basic principles of transistor operation. The common-base and common-emitter configurations were discussed and their characteristics and curves were developed. Curve tracers that display these characteristics were presented. Transistor operation in the cutoff and saturation regions was discussed and the manufacturers' specifications for transistors were introduced.

This chapter provides a base of facts and insights the student will need as he or she proceeds with the more detailed and advanced study presented in the chapters to follow.

2-13 GLOSSARY

Active region. The region where the transistor amplifies an ac signal.

A_v. The voltage gain of a transistor.

BV_{CEO}, BV_{CBO}, BV_{EBO}. Breakdown voltages for a transistor. These voltages should not be exceeded or the transistor may be damaged or operate improperly.

Common-base circuit. A transistor circuit where the base is grounded.

Common-emitter circuit. A transistor circuit where the emitter is grounded.

Curve tracer. An instrument for displaying the characteristic curves and other parameters of a transistor.

Cutoff region. The region of operation where the transistor conducts no or very little current.

Forced β. The transistor's β in the saturation region. Here the β is really determined by the voltages and resistors connected to the transistor, rather than by the transistor itself.

I_{CBO}. The current flow from collector to base with the emitter open and the collector–base junction reverse-biased as it normally is. This is a leakage current and is usually very small.

I_{CEO}. The current flow from collector to emitter with the base junction open. This is also a very small leakage current. There is very little current flow in a transistor under these conditions.

The region where $V_{CE} \approx 0$ and $I_C \approx V_{CC}/R_C$.

2-14 REFERENCES

ROBERT BOTOS, "Automated Curve Tracer Handles Complex Tests," Hewlett-Packard Co., Loveland, Colorado, *Electronics*, June 16, 1982.

ROBERT BOYLESTAD and LOUIS NASHELSKY, *Electronic Devices and Circuit Theory*, 4th Edition, Prentice-Hall, Englewood Cliffs, N.J., 1987.

4145A Semiconductor Parameter Analyzer, Hewlett-Packard, Palo Alto, Ca., May, 1982.

JACOB MILLMAN and HERBERT TAUB, *Pulse, Digital, and Switching Waveforms*, McGraw-Hill, New York, 1965.

J. F. PIERCE and T. J. PAULUS, *Applied Electronics*, Charles E. Merrill, Columbus, Ohio, 1972.

DONALD L. SCHILLING and CHARLES BELOVE, *Electronic Circuits, Discrete and Integrated*, 2nd Edition, McGraw-Hill, New York, 1979.

Testing the 2N3904 Bipolar Transistor, Tektronix, Inc., Beaverton, Ore., 1983.

2-15 PROBLEMS

2-1 A transistor in the common-base configuration has an emitter current of 3 mA.
  a. If $\alpha = 0.995$, find the base and collector currents.
  b. If $V_{CC} = 20$ V and $R_C = 5$ kΩ, find the collector voltage.

2-2 For the circuit of Figure 2-7, assume $\alpha = 0.99$. Find I_E, I_B, I_C, and V_C if $V_{EE} = 3$ V. Under these conditions what is the largest value of R_C, if the transistor is to remain in the active region?

2-3 For the circuit of Figure 2-9, find $A_{i(ck)}$, $A_{i(tr)}$, $A_{v(ck)}$, and $A_{v(tr)}$ if $R_C = 1500$ Ω, $\alpha = 0.995$, and $h_{ib} = 50$.

2-4 From Figure 2-10, estimate α if $I_E = 10$ mA.

2-5 Find the β of a transistor whose $\alpha =$
  a. 0.98.
  b. 0.995.
  c. 0.996.

2-6 If I_{CBO} for a transistor at 25°C is 10 nA, find I_{CEO} if the junction temperature is 105°C and the α of the transistor is 0.99.

2-7 Find the α of a transistor whose β is
  a. 100.
  b. 150.
  c. 200.

2-8 For the circuit of Figure P2-8, find I_C and V_C if $\beta = 150$ for the *pnp* transistor.

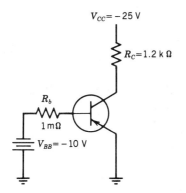

FIGURE P2-8

2-9 Find h_{FE} and h_{fe} from Figure 2-13 if
  a. $V_C = 5$ V and $I_B = 50$ μA.
  b. $V_C = 9$ V and $I_B = 60$ μA.
  How do these values compare with those found in Figure 2-14?

2-10 The characteristics of a transistor with an h_{fe} of 100 are to be displayed on a curve tracer. If I_C is to vary from 0 to 12 mA, what is a reasonable I_B per step and what is the corresponding number of steps?

2-11 The transistor whose characteristic curves are given in Figure 2-19 operates with $V_{CC} = 18$ V, $R_C = 2$ kΩ.
  a. Draw the load line.
  b. Find V_{CE} and I_C if $I_B = 40$ μA.

2-12 What is the h_{FE} and h_{fe} of the transistor of Figure 2-19 when $V_{CE} = 11$ V and $I_B = 40$ μA?

2-13 If the transistor of Figure 2-21 has an I_{CBO} of 50 nA at 25°C with the base open, find V_C. Repeat if the ambient temperature rises to 75°C.

2-14 a. For Figure 2-25, find the inherent h_{FE} of the **2N3904** for $I_C = 1.0$ mA, 30 mA, and 100 mA.
  b. For each curve find V_{CE} if I_B is three times its value at $V_{CE} = 1.0$ V.

2-15 For Figure 2-25, find the power dissipation of the transistor if
  a. $I_B = 0.08$ mA, $I_C = 10$ mA.
  b. $I_B = 0.5$ mA, $I_C = 10$ mA.

2-16 a. Plot the load line on the curve of Figure P2-16 for $V_{CC} = 32$ V, $R_C = 80$ Ω.
  b. For $I_B = 5$ mA find V_{CE}, I_C, h_{FE}, and h_{fe}.

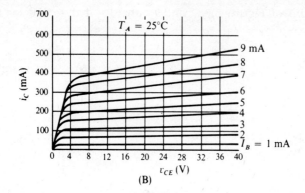

FIGURE P2-16

2-17 Using the curves of the *pnp* transistor shown in Figure P2-17, find
  a. I_c at the point $I_E = -3$ mA, $V_C = -20$ V.
  b. I_C at the point $V_C = -20$ V, $I_B = -150$ mA.

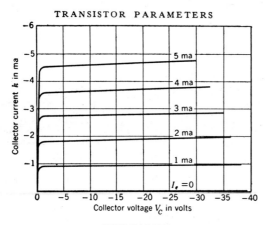

FIGURE P2-17

COMMON BASE

Output Characteristics

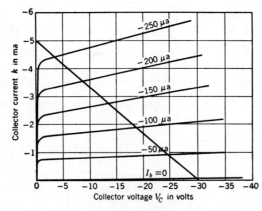

COMMON EMITTER

Output Characteristics

2-18 The load line on the common-emitter characteristics of Figure P2-17 was drawn for a circuit like Figure 2-18. Sketch the circuit and find the values of V_{CC} and R_C.

2-19 Using a circuit similar to Figure 2-15 with $V_{BB} = 10$ V, describe how to take the curves for the transistor of Figure 2-15. State specifically the values of R_B you would use.

2-20 For the circuit of Figure P2-8, R_B is changed to 10 kΩ. If the intrinsic β of the transistor is 80, what is its forced β under these conditions?

2-21 In Problem 2-20, what is the smallest value R_B can have if the transistor is not to saturate?

2-22 For the circuit of Figure P2-22, the β of the transistor is 120. Find V_C and I_C if the value of R_B is
  a. 1 MΩ.
  b. 100 kΩ.
  c. 10 kΩ.
  If the transistor saturates, find the forced β.

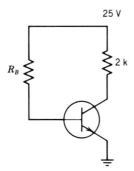

FIGURE P2-22

CHAPTER THREE

Biasing the Bipolar Junction Transistor

3-1 Instructional Objectives

This chapter shows how to analyze and design bias circuits for the bipolar junction transistor (BJT). Bias circuits for the common-emitter, common-base, and common-collector circuits are considered. After reading the chapter the student should be able to:

1. Perform a dc analysis of a fixed biased or H-biased common emitter amplifier to find the quiescent values of I_B, I_C, and V_{CE}.
2. Design a fixed bias or H-bias circuit to provide the optimum values of I_B, I_C, and V_{CE}.
3. Find the Thevenin's equivalent of the H-bias circuit.
4. Use a computer program to analyze an H-bias circuit.
5. Analyze and design bias circuits for a common-base amplifier.
6. Design and analyze bias circuits for an emitter follower.
7. Calculate stability factors and determine the stability of a transistor circuit.

3-2 Self-Evaluation Questions

Watch for the answers to the following questions as you read the chapter. They should help you to understand the material presented.

1. What is a blocking capacitor? Why is it necessary?
2. What is clipping? Why is it to be avoided?

3. What is the function of an emitter bypass capacitor? Why must it be large?
4. Why does the presence of an emitter-resistor stabilize the gain?
5. Where should V_{CEQ} be set in a fixed-bias circuit? In an H-bias circuit? Why?
6. Why don't common-base or common-collector circuits need bias stabilization as badly as common-emitter circuits?

3-3 INTRODUCTION TO BIASING

To design an amplifier, the proper dc conditions must be set up first. This is called *biasing* or setting the *quiescent point* of the amplifier. The quiescent point of an amplifier means the dc values of base current, collector current, and collector voltages at that point. These are determined by the power supply voltage and the resistors connected to the transistor. If a load line is drawn on a set of characteristic curves, as it is in Figure 2-19, the quiescent point is often placed near the middle of the load line.

After the amplifier is properly biased, the ac signal is applied to the input and the amplified signal is taken at the output. This chapter is concerned with biasing or setting up of the dc conditions. The ac analysis of amplifiers is the subject of Chapter 4.

A common-emitter transistor amplifier operates by injecting an ac signal into the base, as shown in Figure 3-1. This ac input voltage varies the base

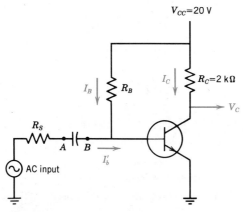

FIGURE 3-1
A fixed bias circuit.

current, which causes a larger variation of the ac output voltage. The variation in ac output voltage is sometimes called the *output swing* (it is shown later in

Figure 3-3). The ac input voltage usually comes in through a large capacitor. Its function is to isolate the ac source from the dc voltage at the base, and vice versa.

EXAMPLE 3-1

In Figure 3-1, what is the dc voltage at point A if the ac source is a pure sine wave generator? What is the dc voltage at point B?

Solution

If the ac source is a pure sine wave, it has no dc component and its average value is 0. This is the dc voltage at point A. Point B is at the base of the transistor, so its dc value must be 0.7 V. The capacitor allows us to couple the two signals together despite the differences in their dc voltages. If it were not present (R_S connected directly to the base), the sine wave generator would drag the dc base voltage down toward cutoff and the output voltage would be distorted.

Example 3-1 shows that the voltage across the input capacitor is 0.7 V. For best operation the capacitor voltage should remain constant. Actually the voltage across the capacitor depends on the frequency of the ac input and the size of the capacitor, and is discussed in section 7-4.3 on low-frequency response. Until then, it is assumed that the capacitor is large enough to hold its voltage constant. This means that the capacitor acts *ideally*; it is a short circuit for ac voltages and an open circuit for dc voltages.

3-3.1 Setting the Operating or Q Point

The circuit of Figure 3-1 is called a *fixed-bias* circuit because the base current is determined by the voltage source driving the base current and the resistance between the voltage source and the base, R_B. The voltage source for driving the dc base could be another supply different from the collector voltage source, V_{CC}, but only one power supply is needed if V_{CC} is used to drive both the base and collector currents. Once the base voltage source and R_B are specified, the base current can be calculated. If R_C is also specified, the load line can be drawn on the characteristic curves and the *operating point*, which is the dc values of collector voltage and current, can be determined.

EXAMPLE 3-2

The load line for the circuit of Figure 3-1 has already been plotted in Figure 2-19 (repeated here for convenience). What value should R_B have so that $I_B = 40$ μA?

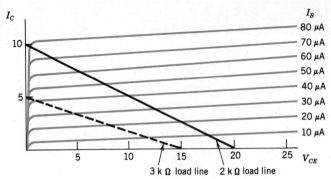

FIGURE 2-19

Solution

In the circuit of Figure 3-1, the dc base current also flows through the base resistor. Therefore

$$I_B = \frac{V_{CC} - V_{BE}}{R_B}$$

$$R_B = \frac{20 \text{ V} - 0.7 \text{ V}}{40 \text{ μA}} = 482 \text{ k}\Omega$$

EXAMPLE 3-3

If the ac generator signal is $V_m = 5$ V, how large should the source resistance, R_S, be if the base current is to vary by 20 μA?

Solution

For ac conditions both the capacitor voltage and V_{BE} are constant and can be assumed to be 0. There is an ac input impedance, h_{ie} (see Chapter 4), but for this circuit we will consider it to be negligible. Neglecting R_B as very high compared to the ac input impedance, we have

$$V_m \sin \omega t = \frac{I_m \sin \omega t}{R_S} \qquad R_S = \frac{V_m}{I_m} = \frac{5 \text{ V}}{20 \text{ μA}} = 250 \text{ k}\Omega$$

Examples 3-2 and 3-3 show that by selecting the proper values of R_B and R_S the desired values of dc quiescent current and ac swing can be obtained (see problem 3-1).

3-4 THE FIXED-BIAS CIRCUIT

In section 3-3 the question of selecting the best bias point was not discussed. Actually any bias point will do if it keeps the quiescent or operating point in the active region and does not allow it to go into cutoff or saturation (see sections 2-9 and 2-10). The best point, however, is that dc voltage that allows the *largest ac swing* without leaving the active region. Generally this leads to a dc quiescent voltage that is approximately one-half the power supply voltage, V_{CC}.

3-4.1 Clipping in Saturation and Cutoff

Usually the ac voltages to be amplified by a transistor are sine waves. If the ac signal is too large, or the circuit is biased badly, the operating point will enter the saturation or cutoff regions. In either case, a flattening of the peaks occurs, as shown in Figure 3-2, and the ac output is distorted. This is called *clipping*. For most applications it is unacceptable.

Clipping at cutoff is shown in Figure 3-2a. It occurs when the varying ac input drives the transistor below the point where it can conduct. When clipping occurs $I_C = 0$, $V_{out} = V_{CC}$, and the transistor cannot amplify because no current is flowing in it. Figure 3-2a shows the flattening at the tops of the waveform when the transistor is driven into cutoff and $V_{CE} = 12 \text{ V}$ (V_{CC}).

Clipping in saturation is shown in Figure 3-2b. It occurs when the ac input signal drives too much current into the base. In saturation $V_{CE} \approx 0$ and $I_C \approx V_{CC}/R_C$. Any further increase in base current has no effect on these values.

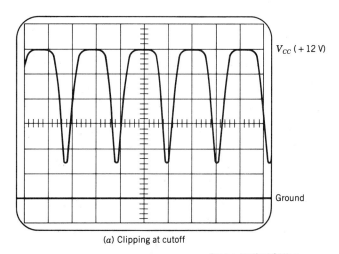

(a) Clipping at cutoff

Scales: Vertical 2 V/cµ
Horizontal 500 µs/cµ

FIGURE 3-2
Clipping in cutoff and saturation.

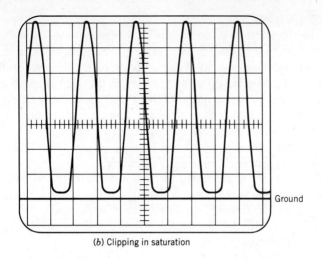

(b) Clipping in saturation

Scales: Vertical 1 V/cµ
Horizontal 500 µs/cµ

FIGURE 3-2
continued

The vertical scale in Figure 3-2b has been changed to show the effects of saturation more clearly. The collector voltage, V_{CE}, never quite gets to 0. When the transistor saturates, the collector voltage becomes $V_{CE(sat)}$, the collector-to-emitter saturation voltage, which is slightly above ground, as shown in Figure 3-2b.

3-4.2 The Optimal Bias Point

The transistor of Figure 2-19, with $V_{CC} = 20$ V, $R_C = 2$ kΩ, can be biased at any point in the active region by properly selecting the bias point. Table 3-1 shows the quiescent conditions for bias points of 20, 40, and 60 µA.

Table 3-1
Three Operating Points for the Transistor of Figure 2-19

Parameter	A	B	C	Units
I_B	20	40	60	µA
I_C	1.9	4.5	7.2	mA
V_{CE}	16.1	11	5.5	V

EXAMPLE 3-4

For each of the operating points in Table 3-1, what is the maximum ac output swing?

Solution

The quiescent value of V_{CE} at point A is 16.1 V. The ac voltage can swing up by only 3.9 V until $V_{CE} = 20$ V, $I_C = 0$, and the transistor cuts off. Thus the maximum ac swing at this point is 3.9 V.

For point B the quiescent V_{CE} is 11 V. It can swing up about 9 V to reach cutoff and down about 10 V before it reaches saturation. The 9 V swing to cutoff will cause clipping first, so 9 V is the maximum swing.

Point C, where the quiescent V_{CE} is 5.5 V, is closer to saturation. If saturation is assumed to start at a V_{CE} of 0.5 V, 5 V is the maximum swing before the transistor starts to clip in saturation.

Figure 3-3 shows the circuit with a quiescent base current of 20 μA and an ac swing of 40 μA. The figure shows that the output voltage swings up by

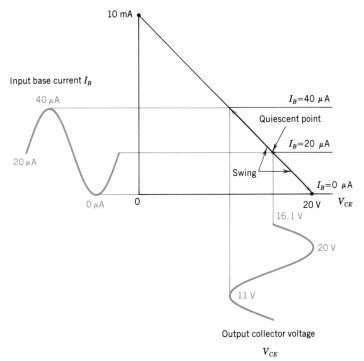

FIGURE 3-3
Curves and swing for point A of Example 3-4.

3.9 V from its quiescent value of 16.1 V (to 20 V), but down by 5.1 V (to 11 V) in response to the 20 μA ac input swing. Since the upward and downward swings are *not equal*, the output is a *distorted* sine wave. This distortion is caused by the inherent nonlinearity of the curves, the large swing, and operating too close to cutoff. If point B were used as the quiescent point, the distortion would not be as bad. Many circuits, however, can tolerate this amount of distortion with no ill effects. The quality of the sound in a radio, for example, might still be satisfactory despite this distortion.

3-4.3 Setting the Bias Point

If the β of a transistor is known, it is not difficult to set the quiescent point properly. Ideally, we would like to design for a particular quiescent point, plug any transistor in, and have it operate at that point. Unfortunately, transistors, even of the same type, can have widely varying βs, and this will effect the quiescent point. The **2N3904** can have βs anywhere from 100 to 300. Characteristic curves do not help either, because they are for a *typical transistor*, and transistors with different βs have different curves.

To set up the bias conditions, the following steps could be followed:

1. Select the supply voltage, V_{CC}. Chapter 2 showed that V_{CC} must be less than V_{CEO}. Often V_{CC} is determined by the available power supplies.
2. Select the collector resistor, R_C.
3. Select the desired quiescent collector voltage, V_{CEQ}. This selection determines the quiescent collector current, I_{CQ}. In most circuits, V_{CEQ} will be half of V_{CC} to allow maximum swing.
4. Use a reasonable value of h_{FE}.
5. Calculate the base biasing resistor.

Using a reasonable value of h_{FE} depends on the specifications and some experience. The βs of the **2N3904** are specified to be between 100 and 300, so it might seem reasonable to assume a β of 200. Experience shows, however, that most **2N3904**s have βs closer to 100 than to 300, so assuming a β near 100, say 150, is closer to what we would expect in an actual circuit.

EXAMPLE 3-5

For the circuit of Figure 3-4, $V_{CC} = 25$ V, $R_C = 2$ kΩ and V_{CEQ} is to be 13 V (approximately $V_{CC}/2$), find the value of the biasing resistor R_B. Assume the transistor has a β(h_{FE}) of 150.

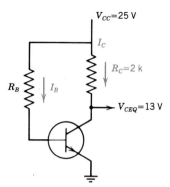

FIGURE 3-4
Circuit for Example 3-5.

Solution
With the given condition, I_{CQ} can be calculated, then I_B and then R_B.

$$I_{CQ} = \frac{V_{CC} - V_{CEQ}}{R_C} = \frac{25\text{ V} - 13\text{ V}}{2\text{ k}\Omega} = 6\text{ mA}$$

$$I_B = \frac{I_{CQ}}{h_{FE}} = \frac{6\text{ mA}}{150} = 40\text{ }\mu\text{A}$$

$$R_B = \frac{V_{CC} - V_{BE}}{I_B} = \frac{25 - 0.7}{40\text{ }\mu\text{A}} = 607\text{ k}\Omega$$

EXAMPLE 3-6

If the circuit of Figure 3-4 is set up with $R_B = 607\text{ k}\Omega$, as calculated in Example 3-5, find V_{CEQ} if the transistor used has an h_{FE} of

a. 100.
b. 300.

Solution
With V_{CC}, V_{BE}, and R_B fixed, the base current will be constant at 40 μA, regardless of the transistor's h_{FE}.

a. If $h_{FE} = 100$

$$I_C = 100\, I_B = 4\text{ mA}$$
$$V_{CEQ} = V_{CC} - I_C R_C = 25\text{ V} - 4\text{ mA} \times 2\text{ k}\Omega = 17\text{ V}$$

b. If $h_{FE} = 300$

$$I_C = 300\,I_B = 12\text{ mA}$$
$$V_{CEQ} = 25\text{ V} - 12\text{ mA} \times 2\text{ k}\Omega = 1\text{ V}$$

The results of this example are disturbing. They indicate that the transistor's quiescent point can vary from 17 V to 1 V, due to changes in the transistor's h_{FE}. This, of course, limits the swing. Indeed, the transistor whose h_{FE} is 300 operates with a V_{CEQ} of 1 V, and is practically in saturation. This has led to the design of other biasing circuits where the quiescent point is not as dependent on h_{FE}.

3-5 THE EMITTER-RESISTOR BIAS CIRCUIT

As the previous section shows, the large variation in the Q point (quiescent point) of the transistor as its β varies is a serious drawback of the fixed-bias circuit. For this reason, fixed-bias circuits are not often used for amplifiers, although they are used for switching circuits.

The emitter-resistor bias circuit shown in Figure 3-5 reduces the dependence of the Q point on h_{FE}. It contains an emitter resistor, R_E, whose value is generally between 10% and 20% of R_C. If R_E is larger, the emitter voltage will be relatively high and will reduce the possible output swing. The current flow through this resistor produces the bias.

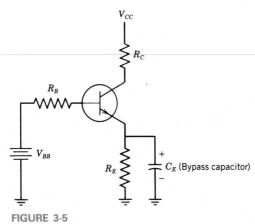

FIGURE 3-5
The self-bias circuit.

3-5.1 The Bypass Capacitor

The presence of an emitter resistor by itself, however, would seriously reduce the ac gain. Therefore, a large value, small working voltage capacitor (50 µF,

10 V typical) is almost always placed in parallel with R_E as shown. This capacitor should be an open circuit to dc, a short circuit to ac, and hold the voltage across the emitter resistor absolutely constant. At low frequencies it does not function perfectly; this is considered in Chapter 7, on low-frequency response. Until then, the capacitor is assumed to be ideal. In any case, the capacitor has *no* effect on the dc voltages and is *ignored* in the bias calculations.

3-5.2 Analysis of the Emitter-resistor Bias Circuit

The circuit of Figure 3-5 can be analyzed to reveal how it works and why it stabilizes the gain. Figure 3-6 clarifies the analysis. It shows that the base current, I_B, causes a collector current of $h_{FE}I_B$, and they both flow into the emitter. If the base current were to increase for any reason, the voltage across the emitter resistor would also increase. This voltage opposes the increase in base current and limits it.

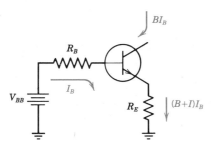

FIGURE 3-6
A simplified view of the self-bias circuit.

The bias circuit can be analyzed as follows:

$$I_E = I_C + I_B = (h_{FE} + 1)I_B$$

Using Kirchhoff's voltage law around the base loop we find

$$V_{BB} = I_B R_B + V_{BE} + (h_{FE} + 1)I_B R_E$$

Solving for I_b, we obtain

$$I_B = \frac{V_{BB} - V_{BE}}{R_B + (h_{FE} + 1)R_E} \tag{3-1}$$

$$\approx \frac{V_{BB} - V_{BE}}{R_B + h_{FE}R_E} \tag{3-2}$$

Equation 3-2 is simpler than 3-1 and was derived from it by assuming

$h_{FE} + 1 = h_{FE}$, which is true for large values of h_{FE}.[1] If $h_{FE}R_E \gg R_B$, then equation 3-2 becomes

$$I_B \approx \frac{V_{BB} - V_{BE}}{h_{FE}R_E}$$

Notice that I_B is now inversely proportional to h_{FE}, and

$$I_C = h_{FE}I_B = \frac{V_{BB} - V_{BE}}{R_E} \tag{3-3}$$

Equation 3-3 is independent of h_{FE}. Therefore, with the foregoing assumptions, the collector current and the quiescent output voltage will be constant regardless of h_{FE}.

Unfortunately, R_B cannot be made too small or it will reduce the gain and increase the power consumption of the circuit, and R_E cannot be made too large. A reasonable choice is to set $R_B = h_{FE}R_E/10$. This satisfies the condition of $h_{FE}R_E \gg R_B$ fairly well and still allows a significant value of R_B.

In practice, the circuit of Figure 3-5 is rarely used because it requires two power supplies. Figure 3-7 shows a circuit that uses V_{CC} to generate the bias

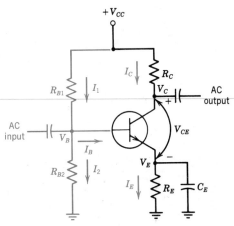

FIGURE 3-7
The H-bias circuit.

---

[1] If $h_{FE} = 100$, the assumption is off by only 1%. If h_{FE} is higher or lower, the assumption is more or less accurate.

and requires only the main power supply. This circuit is sometimes called the *H-bias* circuit because the circuit with the collector resistor, emitter resistor, and two bias resistors resembles the letter H. Fortunately the H-bias circuit can be reduced to the circuit of Figure 3-5 by using Thevenin's theorem applied at the base of the transistor.

EXAMPLE 3-7

Using Thevenin's theorem find V_{BB} and R_B for the circuit of Figure 3-7.

Solution

If the base is open-circuited, R_{B1} and R_{B2} form a simple voltage divider and

$$V_{\text{(open circuit)}} = V_{BB} = V_{CC} \frac{R_{B2}}{R_{B1} + R_{B2}} \tag{3-4}$$

The Thevenin's resistance is seen looking back from the base and is the parallel combination of R_{B1} and R_{B2}.

$$R_{Th} = R_B = \frac{R_{B1} R_{B2}}{R_{B1} + R_{B2}} \tag{3-5}$$

Using the Thevenin's equivalent circuit, the quiescent point for a given circuit can be found.

3-5.3 Analysis of the H-bias Circuit by the Approximate Method

There are three ways to analyze an H-bias circuit to determine its quiescent point. The approximate method is discussed in this section. It can be performed quickly, with a minimum of mathematics, and gives answers that are correct to within 10% for most circuits. Section 3-5.4 presents a much more accurate method of analysis, which requires more mathematics.

A computer program to analyze the H-bias circuit is given in section 3-5.5. It requires only that the user enter the circuit values and then produces the results. If a computer is available, this is the easiest method, but does little to help the reader understand the circuit.

The approximate analysis method assumes that the voltage at the base of the transistor is V_{BB} and is determined by resistors R_{B1} and R_{B2} acting as a voltage divider. Then the emitter voltage can be found and the current determined. Example 3-8 demonstrates this method.

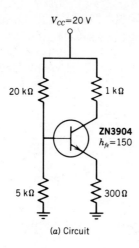

FIGURE 3-8
Circuit for Example 3-8.

(a) Circuit

(b) Thevenin's equivalent input circuit

EXAMPLE 3-8

For the circuit of Figure 3-8a, find the quiescent values of I_B, I_C, I_E, and V_{CEQ}.

Solution
First the approximate base voltage is found by assuming R_{B1} and R_{B2} form a voltage divider.

$$V_B = \frac{R_{B2}}{R_{B1} + R_{B2}} \cdot V_{CC} = 4 \text{ V} = V_{BB}$$

Then the emitter voltage can be found.

$$V_E = V_{BB} - V_{BE} = 3.3 \text{ V}$$

Now I_C and I_E can be found.

$$I_E = \frac{3.3 \text{ V}}{300 \text{ }\Omega} = 11 \text{ mA}$$

$$I_C \approx I_E = 11 \text{ mA}$$

114 BIASING THE BIPOLAR JUNCTION TRANSISTOR

Finally the voltage at the collector can be found.

$$V_C = V_{CC} - I_C R_C = 20\text{ V} - 11\text{ mA} \times 1\text{ k}\Omega = 9\text{ V}$$

V_{CEQ} is the quiescent collector to emitter voltage.

$$V_{CEQ} = V_C - V_E = 9\text{ V} - 3.3\text{ V} = 5.7\text{ V}$$

3-5.4 Exact Analysis of the H-bias Circuit[2]

The exact analysis depends on the Thevenin's equations (equations 3-4 and 3-5) and requires more effort.

EXAMPLE 3-9

Repeat Example 3-8 using exact analysis.

Solution

First the Thevenin's equation of the base circuit must be found.

$$V_{BB} = \frac{V_{CC} R_{B2}}{R_{B1} + R_{B2}} = 20\text{ V} \times \frac{5\text{ k}\Omega}{25\text{ k}\Omega} = 4\text{ V}$$

$$R_B = 5\text{ k}\Omega \parallel 20\text{ k}\Omega = 4\text{ k}\Omega$$

The Thevenin's equivalent circuit is shown in Figure 3-8b.
Now the circuit of Figure 3-8b can be analyzed. Using the exact formula 3-1, we find:

$$I_B = \frac{V_{BB} - V_{BE}}{R_B + (h_{FE} + 1)R_E} = \frac{4\text{ V} - 0.7\text{ V}}{4\text{ k}\Omega + 151 \times 300} = \frac{3.3\text{ V}}{49.300} = 66.9\text{ }\mu\text{A}$$

$$I_C = h_{FE} I_B = 150 \times 66.9\text{ }\mu\text{A} = 10.035\text{ mA}$$

$$V_C = V_{CC} - I_C R_C = 20\text{ V} - 10.035\text{ mA} \times 1\text{ k}\Omega = 9.965\text{ V}$$

Note V_C is the collector-to-ground voltage. In this circuit the emitter is not grounded, so V_C does not equal V_{CE}. Actually, $V_C = V_{CE} + V_E$.

$$I_E = I_B + I_C = 10.035\text{ mA} + 0.0669\text{ }\mu\text{A} = 10.1\text{ mA}$$

$$V_E = I_E R_E = 10.1\text{ mA} \times 300\text{ }\Omega = 3.03\text{ V}$$

---

[2] This section may be omitted at first reading.

Finally,

$$V_{CEQ} = V_C - V_E = 9.965 \text{ V} - 3.03 \text{ V} = 6.935 \text{ V}$$

Check: These calculations can be checked in two ways. First the voltages around the base loop can be calculated.

$$V_{BB} = I_B R_B + V_{BE} + V_E = 66.9 \times 10^{-6} \times 4000 + 0.7 + 3.03$$
$$= 0.267 + 3.03 + 0.7 = 3.997 \text{ V}$$

This checks with the calculated value of V_{BB}, 4 V.

The second check involves calculating the base current. The base voltage of the actual circuit is

$$V_B = V_{BE} + V_E = 0.7 \text{ V} + 3.03 \text{ V} = 3.73 \text{ V}$$

Now the currents in R_{B1} and R_{B2} can be calculated, and the base current is the difference between them.

$$I_{R_{B2}} = \frac{V_B}{R_{B2}} = \frac{3.73 \text{ V}}{5 \text{ k}\Omega} = 746 \text{ }\mu\text{A}$$

$$I_{R_{B1}} = \frac{V_{CC} - V_B}{R_{B1}} = \frac{16.27 \text{ V}}{20 \text{ k}\Omega} = 813.5 \text{ }\mu\text{A}$$

$$I_B = I_{R_{B1}} - I_{R_{B2}} = 813.5 \text{ }\mu\text{A} - 746 \text{ }\mu\text{A} = 67.5 \text{ }\mu\text{A}$$

This compares well with the previously calculated value of I_B, 66.9 μA.

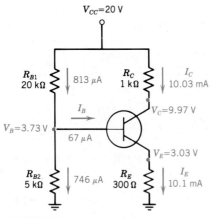

FIGURE 3-9
The results of Example 3-8.

Example 3-9 was calculated precisely so that the checks would be very close. This is especially necessary when checking I_B because I_B is a small difference between two larger currents ($I_{R_{B1}}$ and $I_{R_{B2}}$) and accuracy is very difficult in that kind of calculation. Figure 3-8 is redrawn in Figure 3-9 to show the results of the calculations.

3-5.5 Computer Analysis of the H-bias Circuit

A computer program, written in BASIC, to analyze the H-bias circuit of Figure 3-8 is given in Figure 3-10. It was run on Digital Equipment's VAX computer, but can also be used on Apples, TRS-80s, and other small computers. It uses the exact method. The values of the resistors, V_{CC} and h_{FE} are given in the input file (INFIL). The results appear in the output file (OUTB) and check with the results of Example 3-8 almost exactly. Those readers without access to a VAX computer can still use the BASIC programs presented in this book, but they cannot use INFIL or OUTB. Instead they must type in their input data, and the output data will appear on the screen.

```
10 REM  PROGRAM TO CALCULATE THE QUIESCENT POINT OF AN H BIAS CIRCUIT
15 INPUT "RB1,RB2,RE,RC";RB1,RB2,RE,RC
20 INPUT "VCC, BETA";VCC,HFE
30 VBB = VCC*RB2/(RB1+RB2)
40 RB = RB1*RB2/(RB1+RB2)
50 IB = (VBB - 0.7)/(RB + (HFE+1)*RE)
60 IC = HFE*IB
70 VCG = VCC - IC*RC
80 VEG = IB*(HFE+1)*RE
90 PRINT "IB = ";IB,"IC = ";IC
100 PRINT "COLLECTOR VOLTAGE = ";VCG
110 PRINT "EMITTER VOLTAGE = ";VEG
120 END
```

(a)

```
20000,5000,300,1000
20,150
```

(b)

```
IB =   .669371E-04           IC =   .100406E-01
COLLECTOR VOLTAGE =   9.95943
EMITTER VOLTAGE  =   3.03225
```

(c)

FIGURE 3-10
A BASIC computer program for analyzing the bias circuit.

3-5.6 Design of the Self-bias Circuit

In the previous section a given self-bias circuit was analyzed. The next step is to be able to design a circuit so the user may select the optimum quiescent point. There are two methods of designing for the proper bias point; the first is discussed next and we recommend it. The second method is discussed in section 3-6.1 on load lines.

An H-bias circuit can be designed by following the steps listed below:

1. Select V_{CC} and R_C. Generally V_{CC} is determined by the available power supply and R_C is determined by the required gain and swing (see Chapter 4).
2. Decide on the emitter and collector voltages. The collector voltage is typically $V_{CC}/2$ or slightly higher to place it halfway between V_E and V_{CC}. The emitter voltage is typically between 10% and 20% of V_{CC}. These decisions determine R_E.
3. Select a reasonable value of h_{FE} for the transistor.
4. Set $R_B = h_{FE}R_E/10$. This allows the designer to determine R_B and V_{BB}.
5. Formulas 3-4 and 3-5 gave the relationship of V_{BB} and R_{B1} and R_{B2}. They can be algebraically manipulated to give

$$R_{B1} = R_B \cdot \frac{V_{CC}}{V_{BB}} \tag{3-6}$$

$$R_{B2} = \frac{R_B}{1 - \frac{V_{BB}}{V_{CC}}} \tag{3-7}$$

This completes the design.

EXAMPLE 3-10

Design an H-bias circuit similar to the fixed-bias circuit of Example 3-5, where $V_{CC} = 25$ V, $R_C = 2$ kΩ, and h_{FE} is again assumed to be 150.

Solution

First the values of V_C and V_E must be selected. In accordance with step 2 above, a V_C of 13 V and a V_E of 3 V are chosen. Now the calculations can be started:

$$V_C = V_{CC} - I_C R_C \quad \text{or} \quad I_C = \frac{V_{CC} - V_C}{R_C} = \frac{25 \text{ V} - 13 \text{ V}}{2 \text{ k}\Omega} = 6 \text{ mA}$$

$$I_E \approx I_C = 6 \text{ mA}$$

$$R_E = \frac{V_E}{R_E} = \frac{3 \text{ V}}{6 \text{ mA}} = 500 \text{ }\Omega$$

$$I_B = \frac{I_C}{h_{FE}} = \frac{6 \text{ mA}}{150} = 40 \text{ μA}$$

$$^3R_B = \frac{h_{FE}R_E}{10} = \frac{100 \times 500 \text{ Ω}}{10} = 5000 \text{ Ω}$$

$$V_{BB} = V_E + V_{BE} + I_B R_B = 3 \text{ V} + 0.7 \text{ V} + 0.2 \text{ V} = 3.9 \text{ V}$$

$$R_{B1} = R_B \times \frac{V_{CC}}{V_{BB}} = 5000 \text{ Ω} \times 25 \text{ V}/3.9 \text{ V} = 32 \text{ kΩ}$$

$$R_{B2} = \frac{R_B}{1 - V_{BB}/V_{CC}} = \frac{5000 \text{ Ω}}{1 - \frac{3.9}{25}} = \frac{5000 \text{ Ω}}{0.844} = 5924 \text{ Ω}$$

The final circuit is shown in Figure 3-11.

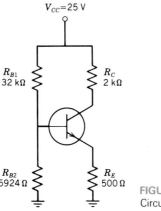

FIGURE 3-11
Circuit for Example 3-10.

After all these calculations it is wise to consider how the self-bias circuit has improved the stability of the circuit.

EXAMPLE 3-11

For the circuit of Figure 3-11, find I_C if h_{FE} is
a. 100.
b. 300.

---

[3] To be conservative the lowest h_{FE} for the transistor is used in this calculation.

Solution
Either by calculation (see section 3-5.2) or by computer, an analysis of Figure 3-11 gives $I_C = 5.81$ mA when $h_{FE} = 100$ and $I_C = 6.2$ mA when $h_{FE} = 300$. Thus, even for large changes in h_{FE}, the values of I_C and V_C remain almost constant. This circuit is a great improvement over the similar fixed-bias circuit.

3-6 THE LOAD LINE FOR THE SELF-BIASED CIRCUIT

Using Kirchhoff's voltage law around the collector circuit of the self-biased transistor, we find

$$V_{CC} = R_C I_C + V_{CE} + R_E I_E \qquad (3\text{-}8)$$

If we assume $I_C \approx I_E$, then equation 3-8 becomes $V_{CC} = V_{CE} + I_C(R_C + R_E)$ or

$$I_C = \frac{V_{CC} - V_{CE}}{R_C + R_E}$$

This is similar to equation 2-10 and allows us to plot the load line on the characteristic curves for the self-biased circuit. The difference is that the slope of the load line is $-1/(R_C + R_E)$ instead of $-1/R_C$, so the load line is flatter because of the emitter resistor.

The load line can be plotted by connecting the point where $I_C = 0$ ($V_{CC} = V_{CE}$) to the point where $V_{CE} = 0$ ($I_C = V_{CC}/[R_C + R_E]$).

EXAMPLE 3-12

For the characteristic curves of Figure 2-19, plot the load line for a self-biased circuit if $V_{CC} = 25$ V, $R_C = 2$ kΩ and $R_E = 500$ Ω.

Solution
The load line is plotted in Figure 3-12. It was plotted by connecting the points $I_C = 0$, $V_{CE} = V_{CC} = 25$ V to the point on the V axis ($V_{CE} = 0$, $I_C = V_{CC}/(R_E + R_C) = 25$ V/2500 Ω = 10 mA).

3-6.1 Designing the Bias Circuit from the Load Line

The bias circuit can be designed from the load line as follows:

1. Select a reasonable operating point (generally $V_{CEQ} = V_{CC}/2$ or slightly less). The reason for it being less is discussed in section 3-6.2.

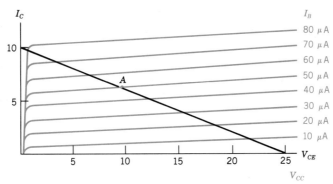

FIGURE 3-12
The dc load line for Example 3-12.

2. Find I_B and I_C from the curves.
3. Using these values, V_{BB} and R_B can be found. Then R_{B1} and R_{B2} can be calculated using equations 3-6 and 3-7.

EXAMPLE 3-13

Calculate the bias circuit for the curves of Figure 3-12. Assume the selected quiescent point is point A on Figure 3-12 where $V_{CE} = 10$ V, $I_C = 6$ mA, and $I_B = 50$ μA. This point satisfies the criteria of step 1.

Solution
First R_B can be found. Again:

$$R_B = \frac{h_{FE(min)}R_E}{10} = \frac{100 \times 500 \, \Omega}{10} = 5000 \, \Omega$$

Now V_{BB} can be found by

$$V_{BB} = I_B R_B + V_{BE} + I_E R_E$$

Here $I_B = 50$ μA and $I_E \approx I_C = 6$ mA
Therefore $V_{BB} = 50$ μA $\times$ 5000 Ω + 0.7 V + 6 mA $\times$ 500 Ω

$$V_{BB} = 3.95 \text{ V}$$

Now by use of equations 3-6 and 3-7 R_{B1} and R_{B2} can be calculated. The results are

$$R_{B1} = 31,645$$
$$R_{B2} = 5938$$

In Example 3-10 we found that $I_B = 40\ \mu A$ when $I_C = 6$ mA. Using the curves in Example 3-13, we found that $I_B = 50\ \mu A$ when $I_C = 6$ mA for a circuit with the same collector and emitter resistors. The differences are due to variations in h_{FE}. In Example 3-10 we assumed $h_{FE} = 150$, whereas the h_{FE} at point A on Figure 3-10 is 120 (6 mA/50 μA). The results of Example 3-13 can be checked by computer using this value of h_{FE}.

The characteristic curves are for a transistor with a particular h_{FE}, and will vary widely with large changes in h_{FE}. Fortunately, the self-bias circuit greatly reduces the dependence of the quiescent point on h_{FE} (as Example 3-11 shows), so that the quiescent point will remain stable.

3-6.2 The ac Load Line

The load line of Figure 3-12 is called the *dc load line* because it applies to dc bias conditions. When an ac signal is superimposed on the circuit, however, the load line changes because the emitter resistor is in parallel with an emitter capacitor whose function is to *short out R_E* for ac signals and also to hold the emitter voltage constant. Consequently, another load line called the *ac load line* can be plotted on Figure 3-12. When an ac signal is applied, the voltages and currents vary along the ac load line. The ac load line must go through the quiescent point (if $V_{ac} = 0$, the circuit will operate at its quiescent point), but its slope is $-1/R_C$, so it is somewhat steeper than the dc load line.

EXAMPLE 3-14

Plot the ac load line on the curves of Figure 3-12.

Solution

The ac load line must go through the quiescent point ($V_{CE} = 10$ V, $I_C = 6$ mA) and have a slope of $-1/2000$. The slope is $\Delta I_C/\Delta V_C$. If a ΔI_C of -5 mA is selected then

$$\frac{\Delta I_C}{\Delta V_{CE}} = \frac{-1}{2000} \qquad \Delta V_C = -2000\ \Delta I_C = 10\ V$$

Thus, if I_C decreases by 5 mA it will be 1 mA and V_{CE} will increase by 10 V and become 20 V. The point $V_{CE} = 20$ V, $I_C = 1$ mA is a second point on the load line, and the load line is plotted by connecting this point to the quiescent point. The plot with both the ac and dc load lines is shown in Figure 3-13, where the ac load line is in blue.

Notice that the ac load line in Figure 3-13 is slightly steeper than the dc load line and intersects the V axis at $V_{CE} = 22$ V. This indicates that the highest ac voltage this transistor can swing to is 22 V because the emitter

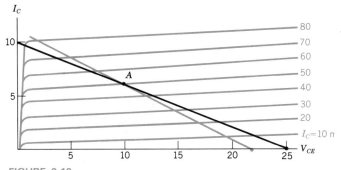

FIGURE 3-13
The dc and ac load lines plotted on the characteristic curves for a **2N3904**, using the values given in Example 3-12.

voltage is constant at 3 V. Thus the 22 V ac swing is from $V_C = 3$ V (where the transistor is saturated) to $V_C = 25$ V (where the transistor is cut off). Note that the quiescent V_{CEQ} of 10 V actually means that the quiescent collector-to-ground voltage is 13 V, and the maximum ac voltage swing is 10 V down and 12 V up. For maximum swing the quiescent voltage should be halfway between V_{CC} and V_E. This means that V_C should be slightly more than $V_{CC}/2$ and V_{CEQ} should be slightly less than $V_{CC}/2$.

The equation for this load line can be developed from the point-slope equation for a line from algebra. This equation is $y - y_1 = m(x - x_1)$ where x_1, y_1 is a point on the line and m is its slope. In this example, it becomes:

$$I_C - 6 \text{ mA} = \frac{-1}{2000}(V_{CE} - 10 \text{ V})$$

Note that if I_C is set to 0 mA, $V_{CE} = 22$ V as expected.

EXAMPLE 3-15

For the circuit of Example 3-12, assume $R_E = 500\ \Omega$ and $I_C = 6$ mA. Find V_C and V_{CEQ} for maximum swing. How can R_C be selected to give this swing?

Solution
If $R_E = 500\ \Omega$ and $I_C = 6$ mA, V_E will be 3 V. For maximum swing, the bias point should be halfway between 3 V and 25 V or at 14 V. This is V_C. At this point $V_{CEQ} = V_C - V_E$ or 14 V $-$ 3 V $= 11$ V.

To set the quiescent values at this point, the collector resistor can be selected as

$$R_C = \frac{V_{CC} - V_C}{I_C} = \frac{11 \text{ V}}{6 \text{ mA}} = 1833\ \Omega$$

3-7 BIASING THE COMMON-BASE AMPLIFIER

The common-base amplifier was introduced in section 2-5. It is an amplifier for which the base instead of the emitter is at a fixed voltage (usually ground), and the signal is injected into the emitter instead of into the base.

A classic common-base circuit is shown in Figure 3-14, where the base is grounded, the ac signal is connected to the emitter via R_G and C_1, and the dc bias current flows through R_E. C_1 is a blocking capacitor used to isolate the dc emitter voltage from the ac source. In this section the biasing of a common-base circuit will be analyzed.

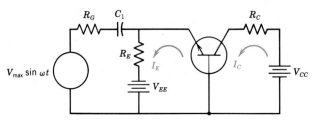

FIGURE 3-14
The common-base circuit.

The gain considerations for a common-base circuit are discussed in section 4-6. Fortunately, if the reasonable assumption $I_C = I_E$ is made, the analysis is rather simple, as Example 3-16 shows.

EXAMPLE 3-16

The circuit of Figure 3-14 has the following values:

$$V_{EE} = -5 \text{ V}$$
$$R_E = 1 \text{ k}\Omega$$
$$R_C = 2 \text{ k}\Omega$$
$$V_{CC} = 25 \text{ V}$$

Find the quiescent collector current and collector voltage.

Solution
The emitter current can be calculated easily because the base is at ground, V_{BE} is 0.7 V, and therefore the dc voltage at the emitter must be -0.7 V.

$$I_E = \frac{-0.7 - V_{EE}}{R_E} = \frac{4.3 \text{ V}}{1000 \text{ }\Omega} = 4.3 \text{ mA}$$

To find I_C

$$I_C \approx I_E = 4.3 \text{ mA}$$

The collector voltage is

$$V_C = V_{CC} - I_C R_C = 25 \text{ V} - 4.3 \text{ mA} \times 2000 \text{ }\Omega = 17.4 \text{ V}$$

Example 3-16 shows that the bias current I_E is $(V_{EE} - V_{BE})/R_E$. This is independent of h_{FE}, so circuits like the common-emitter self-bias circuit that stabilized the common-emitter circuit against variations in h_{FE} are not necessary in the common-base circuit.

3-7.1 Design of the Common-base Circuit

Like analysis, design of a common-base circuit is relatively simple and straightforward. The steps are as follows:

1. Select V_{CC} and V_{EE}. Generally these are the power supplies readily available to the user.
2. Select reasonable values of collector current and collector voltage. Generally V_C is set at about $V_{CC}/2$ for maximum swing.
3. Calculate R_C and R_E.

EXAMPLE 3-17

Design the bias circuit for the common-base circuit of Figure 3-14. V_C must be 10 V and I_C = 5 mA, V_{EE} = 5 V, and V_{CC} = 25 V.

Solution
With these specifications, only R_C and R_E need be calculated.

$$V_C = V_{CC} - I_C R_C$$

$$R_C = \frac{V_{CC} - V_C}{I_C} = \frac{25 \text{ V} - 10 \text{ V}}{5 \text{ mA}} = 3 \text{ k}\Omega$$

$$R_E = \frac{V_{EE} - 0.7}{I_E} \quad \text{and} \quad I_E \approx I_C = 5 \text{ mA}$$

$$R_E = \frac{4.3 \text{ V}}{5 \text{ mA}} = 860 \text{ }\Omega$$

3-7.2 The Single-source Common-base Circuit

Like common-emitter circuits, common-base circuits can be designed to work with a single power supply, as shown in Figure 3-15. Here R_1 and R_2 form a voltage divider. This circuit is similar to the common-emitter bias circuit (Figure 3-7), except for the capacitor. The large capacitor across R_2 is used to keep the base voltage constant, which is the essential characteristic of a common-base circuit. As before, the ac signal is injected at the emitter. The steps to design the bias circuit for Figure 3-15 are shown in Example 3-18.

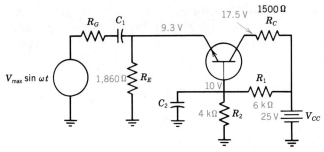

FIGURE 3-15
The single-source common-base circuit.

EXAMPLE 3-18

Design a self-biased common-base circuit for Figure 3-15 where

$$V_B = 10 \text{ V}$$
$$V_{CC} = 25 \text{ V}$$
$$I_{CQ} = 5 \text{ mA}$$

Solution
In order to bias the base at 10 V we can assume that

$$V_B = \frac{R_2 V_{CC}}{R_1 + R_2} \quad (3\text{-}9)$$

Equation 3-9 neglects the fact that the base current flows through R_1 but not through R_2, and is valid as long as the currents in R_1 and R_2 are large compared to I_B (see problem 3-7). This occurs if R_1 and R_2 are not too large.

$$V_B = 10 \text{ V} = \frac{4}{10} V_{CC}$$

If we choose $R_1 = 6\ \text{k}\Omega$, $R_2 = 4\ \text{k}\Omega$, this requirement will be satisfied.

The emitter current approximately equals the collector current of 5 mA. Now R_E can be calculated:

$$R_E = \frac{V_B - V_{BE}}{I_E} = \frac{10 - 0.7}{5\ \text{mA}} = \frac{9.3\ \text{V}}{5\ \text{mA}} = 1860\ \Omega$$

The collector resistor, R_C, should be selected for maximum swing. In this circuit the limits on the swing are 10 V and 25 V, because a collector voltage of less than 10 V means the transistor is saturated since the base is held at 10 V.

The midpoint of the range is 17.5 V and

$$V_{CQ} = V_{CC} - I_{CQ}R_C \qquad R_C = \frac{V_{CC} - V_{CQ}}{I_{CQ}} = \frac{25\ \text{V} - 17.5\ \text{V}}{5\ \text{mA}} = 1500\ \Omega$$

The answers are shown in blue on Figure 3-15.

The circuit of Figure 3-15 has the advantage of using only one power supply, but the price paid for this is the smaller swing of the output voltage.

3-8 BIASING THE EMITTER FOLLOWER

The *emitter-follower* circuit, sometimes called the *common-collector* circuit, is shown in Figure 3-16. It has a gain of less than 1, but it has a high input impedance (see section 4-7) and is often used as a *buffer* or *isolating amplifier* for

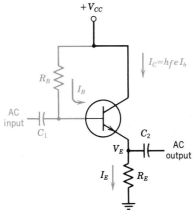

FIGURE 3-16
The emitter-follower or common-collector circuit.

this reason. The ac signal comes in at the base and the output is taken across the emitter resistor, R_E.

The bias for this circuit can easily be found and the analysis of this circuit is straightforward. Taking the voltages through the base and emitter gives

$$V_{CC} = I_B R_B + V_{BE} + (1 + h_{FE}) I_B R_E \qquad (3\text{-}10)$$

To design the bias for an emitter follower circuit, the designer must first select the emitter resistor. A value of 1 kΩ is typical. Then the designer should select the desired dc operating point (often this is $V_{CC}/2$). The value of I_B can then be found and this completes the design.

EXAMPLE 3-19

For the circuit of Figure 3-16, find the bias resistor if:

$$V_{CC} = 20 \text{ V}$$
$$R_E = 1 \text{ k}\Omega$$
$$V_E = 9 \text{ V}$$
$$h_{FE} = 150$$

Solution
This is really a design problem where V_E has been chosen to be 9 V.

$$I_E = \frac{9 \text{ V}}{1 \text{ k}\Omega} = 9 \text{ mA} \approx I_C$$

$$I_B = \frac{I_C}{h_{FE}} = \frac{9 \text{ mA}}{150} = 60 \text{ μA}$$

$$V_{CB} = V_{CC} - V_B = V_{CC} - (V_{BE} + V_E) = 20 \text{ V} - 9.7 \text{ V} = 10.3 \text{ V}$$

$$R_B = \frac{V_{CB}}{I_B} = \frac{10.3 \text{ V}}{60 \text{ μA}} = 172 \text{ k}\Omega$$

3-9 BIAS STABILITY

Stability is the ability of a transistor circuit to hold its quiescent current constant despite variations in the conditions that affect transistor operation. The quiescent collector current will vary due to changes in I_{CO}, h_{FE}, and temperature. The variations caused by unit-to-unit changes in h_{FE}, however, are the most significant.

The effects of instability are shown in Figure 3-17. Assume the circuit is

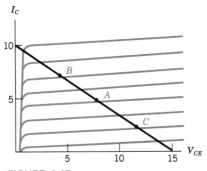

FIGURE 3-17
Effects of instability on a transistor.

properly designed to operate at point *A*. If variations in the parameters cause changes in the collector current so that the quiescent point moves to *B* or *C*, the transistor will be operating too close to saturation or cutoff and clipping will distort the output for all but the smallest ac output voltages. In the most unstable cases the quiescent point could actually wander into the saturation or cutoff regions and the transistor would not amplify. It is, therefore, necessary to keep I_C fairly constant to prevent this instability.

3-9.1 The Classic Calculation of Stability

The classic calculation of stability assumes that the change in I_C, ΔI_C, is due to changes caused by each parameter. From calculus we have

$$\Delta I_C = \frac{\partial I_C}{\partial I_{CO}} \Delta I_{CO} + \frac{\partial I_C}{\partial V_{BE}} \Delta V_{BE} + \frac{\partial I_C}{\partial \beta} \Delta \beta \tag{3-11}$$

Each of the partial derivatives in equation 3-11 is called a *stability factor*, and equation 3-11 can be written

$$I_C = S_I \Delta I_{CO} + S_V \Delta V_{BE} + S_\beta \Delta \beta \tag{3-12}$$

where S_I = the stability factor with respect to I_{CO}

$$S_I = \frac{\partial I_C}{\partial I_{CO}}$$

S_V = the stability factor with respect to V_{BE}

$$S_V = \frac{\partial I_C}{\partial V_{BE}}$$

and S_β = the stability factor with respect to

$$\beta(h_{FE})$$

$$S_\beta = \frac{\partial I_C}{\partial \beta}$$

A stability factor simply means the rate at which I_C changes due to a change in conditions in the transistor. For example, S_I is the rate at which I_C changes due only to changes in I_{CO} and ΔI_{CO} is the change in I_{CO}.

Small stability factors and the small changes in conditions lead to small changes in I_C and a stable circuit, so it is advantageous to minimize these factors if possible.

For the H-bias circuit of Figure 3-7, Pierce and Paulus (see References) have developed equation 3-13, which is very precise in that it takes I_{CO} into account.

$$I_C = I_{CO} + \frac{V_{CC}\beta (R_B/R_{B1}) - \beta V_{BE} + \beta I_{CO} R_B}{R_B + (\beta + 1) R_E} \tag{3-13}$$

where R_B is the parallel combination of the biasing resistors, R_{B1} and R_{B2}, as it was in section 3-5.2. By taking partial derivatives of I_C with respect to I_{CO}, V_{BE}, and β we find

$$S_I = \frac{\partial I_C}{\partial I_{CO}} = 1 + \frac{\beta(R_B/R_E)}{1 + \beta + R_B/R_E} \approx 1 + R_B/R_E$$

$$S_V = \frac{(-1)}{R_E} \frac{\beta}{1 + \beta + R_B/R_E} \approx \frac{-1}{R_E}$$

$$S_\beta = \left[\frac{I_C - I_{CO}}{\beta}\right] \left[\frac{1 + R_B/R_E}{1 + \beta + R_B/R_E}\right]$$

Fortunately, in most cases these changes are so small that they can be ignored. Pierce and Paulus give an example for the H-bias circuit, where they did get a 17% change in I_C, but they used germanium transistors to get a significant value of I_{CO}. With commonly used silicon transistors, ΔI_C would have been negligible.

3-9.2 Stability and Temperature Variations

I_{CO}, V_{BE}, and β all vary with temperature, and these cause variations in the quiescent point. I_{CO} approximately doubles for each 10°C rise in temperature and V_{BE} changes by 2 mV/°C. Changes of β with temperature can often be found by using the manufacturer's specifications. For the **2N3904,** for example, the specifications are given in Appendix A and Figure 15 in Appendix A shows how h_{FE} increases with rising temperatures.

EXAMPLE 3-20

For the circuit of Figure 3-8 (see Example 3-9), find the variation in I_C due to changes in temperature if I_{CO} at 25°C is 10 nA and the transistor operating temperature goes from 25°C to 125°C. Assume $h_{FE} = 150$ at 25°C.

Solution

The 100°C change in temperature means that I_{CO} increases by a factor of $2^{10} \approx 1000$, or from 10 nA to 10 μA. Therefore, $\Delta I_{CO} \approx 10$ μA.

$$S_I \approx 1 + \frac{R_B}{R_E} = 1 + \frac{4000}{300} = 14.3$$

V_{BE} changes by 2 mV/°C × 100°C or 200 mV.

$$S_V = \frac{-1}{R_E} = \frac{-1}{300}$$

For the β change we have, with a β of 150

$$S_\beta = \left[\frac{I_C - I_{CO}}{\beta}\right]\left[\frac{1 + R_B/R_E}{1 + \beta + R_B/R_E}\right]$$

From Example 3-9, $I_C = 6$ mA. I_{CO} is so small, compared to 6 mA, that it can be ignored.

$$S_\beta = \frac{6 \times 10^{-3}}{150} \times \frac{14.3}{164.3} = 3.5 \times 10^{-6}$$

To find $\Delta\beta$, Figure 15 of Appendix A can be used. It shows that the normalized h_{FE} at 6 mA increases from 1 to 1.5. Assuming h_{FE} at 25°C is 150 means that h_{FE} at 125°C will be 1.5 × 150 = 225 and $\Delta\beta$ = 225 − 150 = 75.

The total change of collector current is the product of the stability factors and the corresponding parameter changes.

$$\Delta I_C = S_I \cdot \Delta I_{CO} + S_V \cdot \Delta V_{BE} + S_\beta \Delta\beta$$

$$= 14.3 \times 10 \text{ μA} + \frac{200 \times 10^{-3}}{300} + 3.5 \times 10^{-6} \times 75$$

$$= 143 \text{ μA} + 667 \text{ μA} + 262 \text{ μA} = 1.07 \text{ mA}$$

Thus, even with the large temperature change, the quiescent value of I_C changes by slightly more than 1 mA for this circuit. Note that higher values of R_E would lower each stability factor and make the circuit more stable.

3-9.3 Stability and β Variations

Equation 3-11 does not really apply to β variations, because it considers small changes in values whereas β variations can be large, especially if one transistor is replaced by another. The best approach is to calculate I_C at the extreme values of h_{FE} to determine whether the stability is satisfactory.

In Examples 3-5 and 3-6 we started by finding I_C for a typical transistor with an h_{FE} of 150. Then we assumed it was replaced by transistors with extreme βs of 100 and 300. The results were that I_C varied from 4 mA to 12 mA, or all over the load line. When the fixed-bias circuit was replaced by the H-bias circuit (Example 3-11) for the same transistor, I_C varied from 5.8 mA to 6.2 mA. This small variation would be acceptable in almost all cases. If it is unacceptable, stability can be improved by raising R_E or lowering R_B, but this results in less available swing, less gain, and more power dissipation so the compromise

$$R_B = \frac{h_{FE} R_E}{10}$$

is usually used.

EXAMPLE 3-21

For the emitter-follower circuit of Figure 3-16 and the values given in Example 3-19, find S_β.

Solution
Equation 3-10 gives:

$$V_{CC} = I_B R_B + V_{BE} + (1 + h_{FE}) I_B R_E$$

First this equation must be solved for I_C.

$$I_B = \frac{V_{CC} - V_{BE}}{R_B + (1 + h_{FE}) R_E}$$

$$I_C = h_{FE} I_B = h_{FE} \frac{(V_{CC} - V_{BE})}{R_B + (1 + h_{FE}) R_E}$$

$$\frac{\partial I_C}{\partial h_{FE}} = S_\beta = \frac{[R_B + (1 + h_{FE}) R_E - h_{FE} R_E][V_{CC} - V_{BE}]}{[R_B + (1 + h_{FE}) R_E]^2}$$

$$= \frac{(R_B + R_E)(V_{CC} - V_{BE})}{[R_B + (1 + h_{FE}) R_E]^2}$$

$$= \frac{[R_B + R_E]I_B}{R_B + (1 + h_{FE})R_E}$$

$$= \frac{I_C}{h_{FE}\left(1 + \dfrac{R_E h_{FE}}{R_E + R_B}\right)}$$

For the circuit of Figure 3-16 using the values of Example 3-19, $I_C \approx I_E = 9$ mA

$$R_B = 172 \text{ k}\Omega \quad \text{and} \quad R_E = 1 \text{ k}\Omega$$

$$S_\beta = \frac{9 \times 10^{-3}}{150\left(1 + \dfrac{150{,}000}{173{,}000}\right)} = \frac{9 \times 10^{-3}}{150 \,(1.867)} = 32 \times 10^{-6}$$

Notice that the stability factor is greater for the emitter-follower than for the H-bias circuit. Consequently it is somewhat less stable.

EXAMPLE 3-22

Find I_C for the circuit of Example 3-19 if
a. $h_{FE} = 100$.
b. $h_{FE} = 300$.

Solution
From Example 3-21 we have

$$I_C = \frac{h_{FE}(V_{CC} - V_{BE})}{R_B + (1 + h_{FE})R_E}$$

For $h_{FE} = 100$ this becomes

$$I_C = \frac{100 \,(19.3 \text{ V})}{172{,}000 + 101{,}000} = \frac{1.930 \text{ V}}{273{,}000} = 7.07 \text{ mA}$$

For $h_{FE} = 300$ this becomes

$$I_C = \frac{300 \,(19.3 \text{ V})}{172{,}000 + 301{,}000} = \frac{5790}{473{,}000} = 12.24 \text{ mA}$$

The variation in collector and emitter current is larger for this circuit than for the H-bias circuit. This is consistent with the larger S_β found in Example 3-21.

3-10 SUMMARY

This chapter discussed the biasing of a BJT transistor. Biasing circuits for the common-emitter, common-base, and common-collector (emitter-follower) circuits were presented. Emphasis was on both analysis and design of these circuits. Stability of the biasing circuits was also discussed.

This chapter is necessary because biasing circuits affect both the ac gain and voltage swing. The ac analysis of BJT transistor circuits is presented in Chapter 4.

3-11 GLOSSARY

Common-base circuit. A circuit where the base is at a fixed voltage. The input is at the emitter and the output is at the collector.

Common-collector circuit. A circuit where the collector is at a fixed voltage, often V_{CC}. The input is at the base and the output is at the emitter.

Common-emitter circuit. A circuit where the emitter is either at ground or a fixed voltage. The input is at the base and the output is at the collector.

Emitter-follower. A circuit where the output is taken across the emitter resistor (see Common-collector circuit).

Fixed-bias circuit. A circuit where the bias is determined by a single base resistor (see Figure 3-1).

H-bias circuit. The same as the self-bias circuit.

R_B. Thevenin's resistance looking backward from the base. R_B is usually equal to the parallel combination of R_{B1} and R_{B2}.

Self-bias circuit. A circuit where the bias is determined by R_{B1}, R_{B2}, and R_E (see Figure 3-7).

Stability factor. A measure of the variation of the collector current with changes in other parameters of the transistor.

Swing. The variation of the operating point of a transistor about its quiescent point due to an ac input.

V_{BEQ}. The base-to-emitter voltage (quiescent).

V_{CEQ}. The dc collector-to-emitter voltage (quiescent).

V_{CGQ}. The dc collector-to-ground voltage (quiescent).

3-12 REFERENCES

ROBERT BOYLESTAD and LOUIS NASHELSKY, *Electronic Devices and Circuit Theory*, 4th Edition, Prentice-Hall, Englewood Cliffs, N. J., 1987.

JACOB MILLMAN, *Microelectronics*, McGraw-Hill, New York, 1979.

JACOB MILLMAN and HERBERT TAUB, *Pulse, Digital, and Switching Waveforms*, McGraw-Hill, New York, 1965.

J. F. PIERCE and T. J. PAULUS, *Applied Electronics*, Charles E. Merrill, Columbus, Ohio, 1972.

DONALD L. SCHILLING and CHARLES BELOVE. *Electronic Circuits, Discrete and Integrated*, 2nd Edition, McGraw-Hill, New York, 1979.

3-13 PROBLEMS

The characteristic curves for the transistor of Figure P3-1 are given in Figure 2-19. Find R_B and R_S if the quiescent current is to be 50 µA and the ac current swing is to be 30 µA.

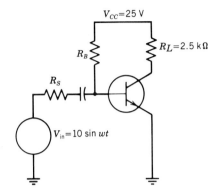

FIGURE P3-1

3-2 A fixed-bias circuit with $V_{CC} = 18$ V is biased at $V_{CEQ} = 13$ V. How far can the ac output swing? Repeat if $V_{CEQ} = 6$ V.

3-3 For the circuit of problem 3-1, what is the maximum output voltage if $I_{BQ} =$
  a. 10 µA.
  b. 30 µA.
  c. 50 µA.

3-4 Design the circuit for the transistor of Figure P3-1 using the curves of Figure 2-19 if $I_{BQ} = 30$ µA, the ac source is 2 sin ωt, and a 25 µA input current swing is to be used. What is the corresponding output voltage swing?

3-5 Sketch a drawing, similar to Figure 3-3, for point C of Example 3-4.

3-6 As the β of a transistor increases, do the lines on its characteristic curves become closer or tend to spread further apart?

3-7 The transistor whose characteristic curves are given in Figure P2-16, is used in a circuit similar to Figure P3-1 with $V_{CC} = 32$ V and $V_{in} = 10$ sin ωt. It is to be biased so that at its quiescent point $V_{CE} = 20$ V and $I_C = 250$ mA. Find
  a. R_B
  b. R_C
  c. The value of R_S if I_B is to swing 3 mA.

3-8 A transistor whose V_{CC} is 20 V is biased so that $V_{CE} = 15$ V.
  a. What is the maximum voltage swing of the transistor?
  b. If the transistor is overdriven, will it go into saturation or clipping?

3-9 For the circuit of Figure 3-4, $V_{CC} = 30$ V, $R_C = 2.5$ kΩ and V_{CEQ} is to be 15 V. Find the value of the biasing resistor if
   a. $h_{FE} = 100$.
   b. $h_{FE} = 150$.

3-10 For the circuit of Figure 3-4, $V_{CC} = 30$ V and the biasing resistor is 800 kΩ. Find V_{CEQ} if β equals
   a. 100.
   b. 150.
   c. 300.

3-11 For the circuit of Figure P3-11, find the collector voltage if h_{FE} is
   a. 100.
   b. 200.
   c. 300.

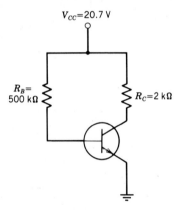

FIGURE P3-11

3-12 For the circuit of Figure P3-12, find I_B and I_C if the h_{FE} of the transistor equals
   a. 100.
   b. 150.

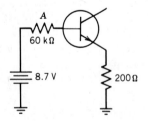

FIGURE P3-12

3-13 For the circuit of Figure P3-13, find I_1, I_2, I_B, I_C, I_E, V_C and V_E. Use the approximate method.

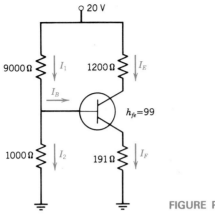

FIGURE P3-13

3-14 Repeat problem 3-13 using the exact method of Example 3-9.

3-15 Repeat problem 3-13 using the computer program.

3-16 For Figure P3-16, find R_1 and R_2 if V_C is to be 10 V.

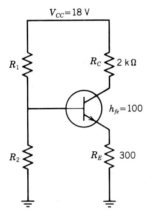

FIGURE P3-16

3-17 Design the H-bias circuit for a *pnp* transistor if
$V_{CC} = -20$ V
$R_C = 1500$
$h_{FE} = 100$
$R_E = 200$
and V_C should be -11 V. Sketch the final circuit.

3-18 Design an H-bias circuit for a transistor with the following specifications: $I_B = 50$ µA, $R_E = 400$ Ω, $V_{CC} = 20$ V, $h_{FE} = 200$. Find R_C so that V_{CQ} is halfway between V_E and V_{CC}. Also find R_{B1} and R_{B2}.

3-19 Using the curves and load line of Figure P3-19, design an H-bias circuit for an I_{CQ} of 3 mA. Design it so that $V_E = 3$ V. Find h_{FE} from the curves. Find V_{CC}, R_E, R_C, R_{B1}, and R_{B2}.

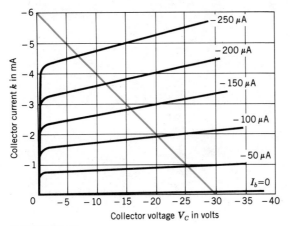

FIGURE P3-19
(Robert L. Riddle/Marlin P. Ristenbatt, *Transistor Physics & Circuits*, © 1958, p. 415. Reprinted by permission of Prentice-Hall, Inc., Englewood Cliffs, New Jersey.)

3-20 Using the curves of Figure 2-19, plot the dc and ac load lines if $V_{CC} = 20$ V, $R_C = 1600 \,\Omega$, $R_E = 400 \,\Omega$, and $I_B = 40 \,\mu A$. What is the maximum ac swing?

3-21 Using the curves of Figure P3-21, plot the dc and ac load lines if $V_{CC} = 28$ V, $R_C = 60 \,\Omega$, $R_E = 10 \,\Omega$, and $I_B = 5$ mA.

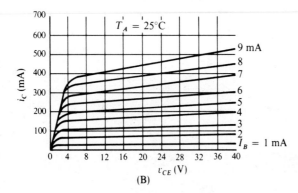

FIGURE P3-21
(Figure P2-16)

3-22 Prove mathematically that the ac load line for a transistor whose quiescent point is V_C, I_C intersects the V axis at $V_{CC} - I_C R_E$.

3-23 For the circuit of Figure P3-23, find I_C and V_C.

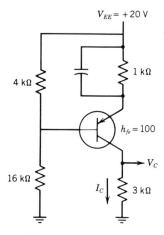

FIGURE P3-23

3-24 For the circuit of Figure 3-14, $V_{CC} = 15$ V, $R_C = R_E = 1$ kΩ $V_{EE} = -5$ V. Find I_C, I_E, and V_C.

3-25 For the circuit of Figure 3-14, $V_{CC} = 20$ V, $V_{EE} = -5$ V, and $R_C = 1$ kΩ. Find R_E if V_{CQ} is to be 12 V.

3-26 For Figure 3-15, calculate the actual base voltage taking into account the base current flowing in R_1. Assume $h_{FE} = 150$ and $I_C = 5$ mA.

3-27 Repeat problem 3-26 if $R_2 = 40$ kΩ and $R_1 = 60$ kΩ.

3-28 For the circuit of Figure P3-28, find V_{C1} and V_{C2}.

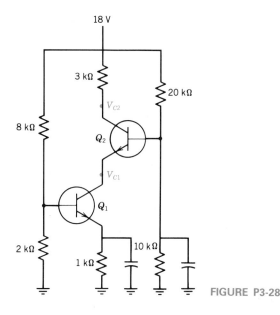

FIGURE P3-28

3-13 PROBLEMS

3-29 For the emitter follower of Figure 3-16, find I_E and V_E if $V_{CC} = 15$ V, $h_{FE} = 149$, $R_B = 100$ kΩ, and $R_E = 1$ kΩ.

3-30 Design the emitter follower of Figure 3-16 so that $V_E = 7$ V. If $V_{CC} = 18$ V and $R_E = 1$ kΩ, find R_B. Assume $h_{FE} = 99$.

3-31 Find the stability factors for the circuit of Figure P3-13.

3-32 Find I_C of Figure P3-13 if the transistor is replaced by a transistor whose $h_{FE} = 300$. How close is this to the answer of problem 3-9?

3-33 For the circuit of figure P3-33 find
  a. The current in R_{B1}
  b. The current in R_{B2}
  c. The current in the base of the transistor.
  d. If the transistor has an h_{FE} of 100, is it in saturation?
  e. How would you change the circuit to prevent saturation?

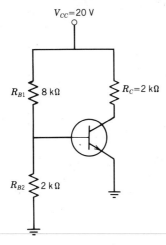

FIGURE P3-33

CHAPTER FOUR

Small Signal Analysis and ac Gain

4-1 Instructional Objectives

The gain of a circuit, the ratio of ac output to ac input, has been defined in Chapter 2. This chapter shows how to find the ac gain of the most common bipolar junction transistor (BJT) circuits. After reading it the student should be able to:

1. Find the gains of a circuit if the voltages at the input, base, and collector are given.
2. Calculate the gain of a common-emitter transistor using both the exact and approximate equations.
3. Calculate the gain and input impedance of a common-base circuit.
4. Calculate the gain and input and output impedances of an emitter-follower.
5. Calculate the gain of a phase-splitter circuit.

4-2 Self-Evaluation Questions

Watch for the answers to the following questions as you read the chapter. They should help you understand the material presented.

1. Why must the biasing circuit be specified before gains can be calculated?
2. What is an equivalent circuit? Why is it used?
3. How does a capacitively coupled load resistor affect the gain of transistor circuits? How does it affect the output swing?

4. What is the relationship between h_{fe}, h_{ie} and h_{fb}, h_{ib}?
5. The voltage gains of a common-base and common-emitter transistor are about the same. Why is the circuit gain of a common emitter transistor usually greater?
6. Why does an emitter-follower have a high input impedance?
7. What resistors can be used to represent the output impedance of an emitter-follower?

4-3 INTRODUCTION TO ac ANALYSIS

This chapter concentrates on finding the midfrequency gains of various bipolar junction transistor (BJT) circuits. The term *midfrequency* generally means that the effects of the capacitors in the circuit can be ignored, or that all capacitors act ideally as an open circuit to dc and a short circuit to ac. Practically, most circuits operating at an ac frequency between 1 and 10 kHz have these characteristics. Gain degradation due to capacitive effects is discussed in Chapter 7 on frequency response.

Figure 4-1a shows the classic H-biased single-stage transistor amplifier. This chapter concentrates on the ac signals throughout the circuit, but their values depend on the dc conditions set up by the bias circuit, so the bias resistors must be specified before an ac analysis can begin. The calculation of the biasing resistors has already been covered in Chapter 3, and the circuit is assumed to be properly biased.

In Figure 4-1a the ac signal, V_{max} sin ωt enters the base of the transistor through the source resistance, R_S, and the blocking capacitor, C_1. The transistor then amplifies the signal to produce an ac output voltage at its collector. Often the output voltage is taken from the circuit through a blocking capacitor (C_2 in Figure 4-1a), so as to isolate the dc level at the collector from the output or load.

One of the major problems in this circuit is to find the ac voltage and current gains. These have been defined in section 2-5.3 and are repeated here for convenience:

$A_{v(ck)}$ The voltage gain of the transistor circuit; the ratio of the output voltage to the input voltage.
$A_{v(tr)}$ The voltage gain of the transistor itself; the ratio of the output voltage to the input voltage at the transistor terminals.
$A_{i(ck)}$ The ac current gain of the transistor circuit.
$A_{i(tr)}$ The ac current gain of the transistor itself.

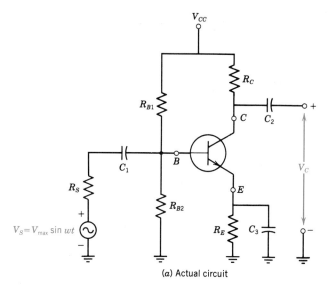

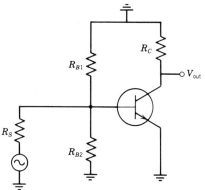

(b) The ac equivalent circuit

FIGURE 4-1
The H-biased, single-stage transistor amplifier.

The first step in calculating the gains is to construct an ac *equivalent circuit*. In such a circuit *all points that have a constant dc voltage are considered to be at ground potential*. This is because the constant dc voltage is usually maintained by a large capacitor, which is an ac short to ground. Thus in Figure 4-1 both the power supply voltage V_{CC} and the emitter voltage V_E are ground or 0 V from an ac viewpoint. Note that V_E is connected to ground by C_3, which should be very large, and the output of a dc power supply (V_{CC}) is usually paralled by a large capacitor. The ac equivalent circuit is shown in Figure 4-1b.

4-3 INTRODUCTION TO ac ANALYSIS   **143**

EXAMPLE 4-1

The circuit of Figure 4-1 has the following values:

$$R_S = 100 \text{ k}\Omega$$
$$R_{B1} = 20 \text{ k}\Omega$$
$$R_{B2} = 5 \text{ k}\Omega$$
$$R_C = 2.0 \text{ k}\Omega$$

The circuit points were measured with an ac voltmeter:

$$v_S = 2.5 \text{ V}$$
$$v_B = 25 \text{ mV}$$
$$v_C = 2 \text{ V}$$

where v_S, v_B, and v_C are the source, base, and collector voltages, with respect to ground. For this circuit find the voltage and current gains.

Solution

The voltage gains are very simple to find with the given data.

$$A_{v(ck)} = \frac{v_{out}}{v_{in}} = \frac{v_C}{v_S} = \frac{2 \text{ V}}{2.5 \text{ V}} = 0.8$$

$$A_{v(tr)} = \frac{v_C}{v_b} = \frac{2 \text{ V}}{25 \times 10^{-3} \text{ V}} = 80$$

To find the current gains, the currents must first be found.

$$i_{in} = \frac{v_S - v_b}{R_S} = \frac{2.5 \text{ V} - 25 \text{ mV}}{R_S} = \frac{2.475 \text{ V}}{100 \text{ k}\Omega} = 24.75 \text{ μA}$$

$$i_{out} = \frac{v_C}{R_C} = \frac{2 \text{ V}}{2 \text{ k}\Omega} = 1 \text{ mA}$$

$$A_{i(ck)} = 40.4$$

To find the current gain of the transistor, the current entering its base must be found. But this is the ac current in R_S minus the ac current in R_{B1} and R_{B2}.

$$i_b = i_{in} - (i_{RB1} + i_{RB2})$$

$$i_{RB1} = \frac{25 \text{ mV}}{20 \text{ k}\Omega} = 1.25 \text{ μA}$$

$$i_{RB2} = \frac{25 \text{ mV}}{5 \text{ k}\Omega} = 5 \text{ μA}$$

$$i_b = 24.75 \text{ μA} - 6.25 \text{ μA} = 18.5 \text{ μA}$$

$$A_{i(tr)} = \frac{i_c}{i_b} = \frac{1 \text{ mA}}{18.5 \text{ μA}} = 54$$

This is also the h_{fe} or ac β of the transistor.

In taking voltage measurements around a circuit similar to Figure 4-1, two points should be observed: the ac voltage on either side of the blocking capacitor should be the same, and the ac voltage at the emitter should be 0 (a measurement of a few millivolts (mV) is permissible). If these conditions do not exist, either the capacitors are too small or some other problem exists.

4-4 THE HYBRID PARAMETERS

The analysis of a circuit containing a transistor, such as that of Figure 4-1, requires the use of an *equivalent circuit*. An equivalent circuit uses a group of generators and resistors to represent a transistor. The most commonly used equivalent circuit for a BJT treats the transistor as a *two-port* network. The basic two-port network is shown in Figure 4-2a. It consists of an input port (terminals 1 and 2), and an output port (terminals 3 and 4). This is also called a *four-terminal* network, but many four-terminal networks reduce to *three-terminal* networks because terminals 2 and 4 are often both ground, as shown in Figure 4-2b. In this case, the three terminals are generally labeled input, output, and ground.

Any two-port circuit, no matter how complex, can be described by a set of four parameters. If the four parameters for the circuit are known, and any two of the input variables (v_1, i_1, v_2, i_2) are also known, the others can be determined.

There are three types of parameters in common use:

r or resistive parameters
y or admittance parameters
h or hybrid parameters

Unlike the *r* or *y* parameters, each *h* parameter is a different type of element, which is why they are called *hybrid* parameters. Because the *h* parameters are used to describe the characteristics of transistors and given in the manufacturers' specification sheets, they will be considered here. The other parameters are discussed in books on circuit theory.

The *h* parameter equivalent circuit is shown in Figure 4-2c. It consists of a Thevenin's type circuit, a resistor h_i in series with a voltage generator $h_r V_2$

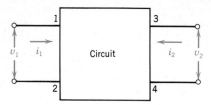

(a) Four-terminal two-port network

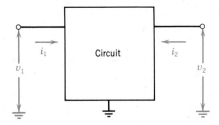

(b) Three-terminal two-port network

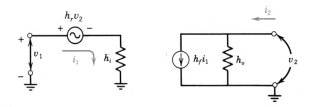

(c) The h parameter equivalent circuit

FIGURE 4-2
2-port networks.

---

on the input side, and a Norton's type circuit, a current generator $h_f i_1$ in parallel with an admittance h_o. The parameters for this circuit, h_i, h_r, h_f, and h_o, are used in the basic equations:

$$v_1 = h_i i_1 + h_r v_2 \tag{4-1}$$

$$i_2 = h_f i_1 + h_o v_2 \tag{4-2}$$

The four h parameters are:

h_i This is a resistance. Its value can be found by short-circuiting the output terminals, causing v_2 to be 0 V. Then $h_i = v_1/i_1$.

h_f the current generator in the output leg of the circuit produces a current whose value is $h_f i_1$. If the output is short-circuited, this current becomes i_2. Therefore h_f is the ratio of the output current to the input current when the output is short-circuited ($v_2 = 0$). It can be determined by supplying an input current and placing an

146   SMALL SIGNAL ANALYSIS AND ac GAIN

h_r The value $h_r v_2$ is a reverse voltage generator that appears in the input leg. If the input terminals are open-circuited, so that $i_1 = 0$, and a voltage is applied at v_2, h_r is the ratio of the voltage that appears at v_1 in response to v_2.

h_o This is an admittance on the output side. If the input is open ($i_1 = 0$), $h_o = i_2/v_2$.

4-4.1 Analysis of a Resistive Circuit Using h Parameters[1]

The simplest illustration of the h parameters is to analyze a resistive circuit, as Examples 4-2 and 4-3 show.

EXAMPLE 4-2

Find the parameters for the resistive circuit of Figure 4-3a.

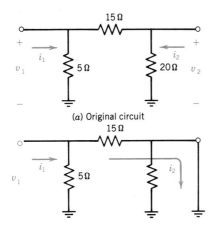

(a) Original circuit

(b) Circuit with the output shorted so that $V_2 = 0$

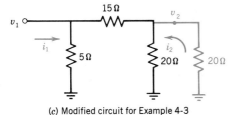

(c) Modified circuit for Example 4-3

FIGURE 4-3
Circuit for Examples 4-2 and 4-3.

---

[1]This section is mathematical and may be omitted at first reading.

Solution

Equations 4-1 and 4-2 show that the parameters h_i and h_f can be found simply if v_2 is set to 0. But v_2 can be forced to 0 by short-circuiting the output terminals, as shown in Figure 4-3b. This shorts out the 20 Ω resistor. Then

$$h_i = \left.\frac{v_1}{i_1}\right|_{v_2=0} = 5\,\Omega \parallel 15\,\Omega = 3.75\,\Omega$$

$$h_f = \left.\frac{i_2}{i_1}\right|_{v_2=0}$$

All the current in the 15 Ω resistor flows through the short circuit. This is $-i_2$, because it is opposite to the defined direction of i_2, and i_1 is the total current in the 5 Ω and 15 Ω resistors. By current division:

$$h_f = \frac{i_2}{i_1} = \frac{-5\,\Omega}{5\,\Omega + 15\,\Omega} = -0.25$$

The parameters h_r and h_o can be found if $i_1 = 0$. This simply means applying a voltage v_2 and leaving the input terminal open-circuited

$$h_r = \left.\frac{v_1}{v_2}\right|_{i_1=0} = 0.25$$

because the 5 Ω and 15 Ω resistors form a voltage divider if the input terminals are open-circuited.

$$h_o = \left.\frac{i_2}{v_2}\right|_{i_1=0}$$

Looking in at v_2 with the input side open, one sees two 20 Ω resistors in parallel.

$$h_o = \frac{1}{10} = 0.1\,\text{S}^2$$

EXAMPLE 4-3

Use the h parameters to determine i_1, v_2, and i_2 for the circuit of Figure 4-3c if v_1 is 10 V.[3]

---

[2]The unit of admittance is the Siemen (formerly the mho).

[3]This circuit can be analyzed easily by using elementary circuit theory. That method, however, would not help the reader understand the h parameters.

Solution

Figure 4-3c is the circuit of Figure 4-2, with an external 20 Ω resistor connected across the output terminals, as shown in blue on the figure. Because the *h* parameters have already been determined, the equations are

$$10 \text{ V} = 3.75\, i_1 + 0.25\, v_2$$

$$i_2 = -0.25\, i_1 + 0.1\, v_2$$

These are two equations with three unknowns, i_1, i_2, and v_2. The external 20 Ω resistor, however, creates a relationship between v_2 and i_2 so that $v_2 = -20\, i_2$. Now the equations become

$$10 \text{ V} = 3.75\, i_1 + 0.25\, v_2$$

$$-0.05\, v_2 = -0.25\, i_1 + 0.1\, v_2$$

or

$$10 \text{ V} = 3.75\, i_1 + 0.25\, v_2$$

$$0 = -0.25\, i_1 + 0.15\, v_2$$

Determinants can be used to find i_1.

$$i_1 = \frac{\begin{vmatrix} 10 & 0.25 \\ 50 & 0.15 \end{vmatrix}}{\begin{vmatrix} 3.75 & 0.25 \\ -0.25 & 0.15 \end{vmatrix}} = \frac{1.5}{0.5625 + 0.0625} = \frac{1.5}{0.625} = 2.4 \text{ A}$$

Similarly, v_2 can be found to be 4 V.

4-4.2 The Transistor Equivalent Circuit

As stated before, the *h* parameters are often used to represent a transistor. The circuit of Figure 4-2c can represent a transistor connected in either the common-base, common-emitter, or common-collector configurations. Even for the same transistor, however, the *h* parameters are different for each configuration. The parameters generally take a subscript that indicates the configuration they apply to. Thus the parameters h_{ie}, h_{fe}, h_{oe}, and h_{re} apply to common-emitter circuits; the parameters h_{ib}, h_{fb}, h_{ob}, and h_{rb} apply to the common-base circuit; and h_{ic}, h_{fc}, h_{oc}, and h_{rc} apply to the common-collector or emitter-follower circuits.

For the same transistor the three sets of parameters are interrelated, and if one set is known the others can be found. This is discussed in section 4-6.

4-4.3 The Common-emitter Parameters

Because the common-emitter circuit is the most popular, most manufacturers tend to specify the emitter parameters. In Appendix A, the specifications for the **2N3904**, Figures 11, 12, 13, and 14, show the variation of h_{fe}, h_{oe}, h_{ie}, and h_{re} as the dc collector current, I_C, varies.

EXAMPLE 4-4

Find h_{fe}, h_{oe}, h_{ie}, and h_{re} from the **2N3904** specifications if $I_C = 5$ mA.

Solution
From the figures at 5 mA we find

$$h_{fe} = 150$$
$$h_{oe} = 30 \times 10^{-6} \text{ S}$$
$$h_{ie} = 900 \text{ }\Omega$$
$$h_{re} = 1.5 \times 10^{-4}$$

4-4.4 The Relationship Between h_{ie} and h_{fe}

The figures in Appendix A show that the h parameters vary with collector current, with h_{fe} and h_{oe} rising slowly and h_{ie} decreasing as I_C increases. Equation 4-3 is a relationship between h_{ie}, h_{fe}, and the dc collector current that we will use in this book:

$$h_{ie} = \frac{30 \, h_{fe}}{I_E} \qquad (4\text{-}3)[4]$$

where I_E is in mA and $I_E \approx I_C$. Equation 4-3 shows that the ac h parameters do depend on the dc biasing.

EXAMPLE 4-5

a. Using only the curves of Figure 11 of Appendix A, find h_{ie} if $I_E \approx I_C = 5$ mA.
b. Repeat if $I_C = 1$ mA.
c. How do the answers compare with h_{ie} as found in Figure 13 of Appendix A?

---

[4] Other authors using theoretical considerations give $h_{ie} = 26 \, h_{fe}/I_E$. Our experience with silicon transistors after testing several of them, indicates that equation 4-3 is more accurate. This is also verified by the curves for the parameters in the specifications.

Solution
a. From Figure 11 with $I_C = 5$ mA, $h_{fe} = 150$.

$$h_{ie} = \frac{30\, h_{fe}}{I_E} = \frac{30 \times 150}{5} = 900\ \Omega$$

b. When $I_C = 1$ mA, $h_{fe} = 120$.

$$h_{ie} = \frac{30 \times 120}{1} = 3600\ \Omega$$

c. The curves of Figure 13 of Appendix A shows that h_{ie} is 900 Ω at 5 mA and 3600 Ω at 1 mA. Thus the answers to this example are verified by the curves.

4-5 CALCULATING THE COMMON-EMITTER GAIN

Figure 4-4 shows the entire equivalent circuit for a typical transistor circuit such as that of Figure 4-1. It includes the voltage source and impedance and the load resistance, R_C. In the figure, v_{Th} is the Thevenin's equivalent voltage source and R_{Th} is the Thevenin's equivalent resistance looking outward from the base of the transistor.

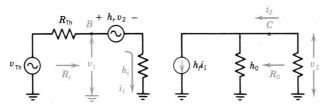

FIGURE 4-4
The transistor equivalent circuit with a source and load connected.

Figure 4-4 can be used as a generalized circuit for any transistor configuration (common-emitter, common-base, or common-collector). Since it is being used here in the common-emitter configuration, the base (B) and collector (C) junctions are identified on the figure. The emitter is connected to ac ground.

EXAMPLE 4-6

For the circuit of Figure 4-1, find R_{Th} and v_{Th}.

Solution

R_{Th} is the Thevenin's equivalent resistance looking back from the base with the voltage source short-circuited. This places the three resistors connected to the base in parallel, between the base and ground. Therefore:

$$R_{Th} = R_S \parallel R_{B1} \parallel R_{B2}$$

Note that this is the ac Thevenin's impedance. The dc impedance, as shown in Chapter 3, is $R_B = R_{B1} \parallel R_{B2}$.

The Thevenin's source voltage is

$$V_{Th} = V_S \frac{R_B}{R_B + R_S}$$

because R_S and R_B effectively form a voltage divider circuit, as shown in Figure 4-5.

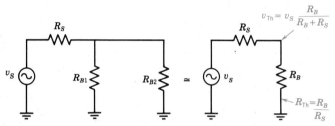

FIGURE 4-5
The Thevenin's equivalent of the input circuit to a transistor

Figure 4-4 also shows the input and output impedances for the circuit of Figure 4-1. The transistor's input impedance, R_i, is the impedance seen looking directly into the base, or V_1/i_1 in Figure 4-4. The output impedance, R_o, is the impedance seen looking into the collector or V_2/i_2.

A word of caution might be in order here. In laboratory situations a voltage generator is often connected directly to the base of the transistor. This effectively sets $R_S = 0$, which shorts out the input resistors, and makes $R_{Th} = 0$. This is because the input voltage can be measured only at the terminals of the generator, which is after the generator's source impedance.

For laboratory experiments, we object to connecting a voltage generator directly to the base, instead of through a resistor, because a direct connection makes it difficult to measure the input current, sets $R_{Th} = 0$, and shorts out the biasing resistors so they have no direct effect on the gain of the circuit.

4-5.1 Gain Calculations Using Exact Equations

The gains and input impedances for the common-emitter circuit of Figure 4-1 can be calculated from Figure 4-4. Equations 4-4 through 4-7 give exact formulas for the voltage gain, current gain, input impedance, and output impedance of a transistor based on its h parameters.[5]

$$A_v = \frac{v_2}{v_1} = \frac{-h_f R_C}{h_i + \Delta^h R_C} \tag{4-4}$$

$$A_i = \frac{i_2}{i_1} = \frac{h_f}{1 + h_o R_C} \tag{4-5}$$

$$R_i = \frac{v_1}{i_1} = \frac{\Delta^h R_C + h_i}{h_o R_C + 1} \tag{4-6}$$

$$R_o = \frac{v_2}{i_2}\bigg|_{v_1 = 0} = \frac{h_i + R_{Th}}{\Delta^h + h_o R_{Th}} \tag{4-7}$$

where Δ^h is the value of the determinant of the h parameters. Note also that R_i depends upon R_C, and R_o depends on R_{Th}.

EXAMPLE 4-7

Find the gains and impedances of Figure 4-4 using the exact formulas if the h parameters are those found from the curves at 5 mA (see Example 4-4), $R_C = 1800\ \Omega$, and $R_{Th} = 2292\ \Omega$.

Solution

First the Δ^h for the common-emitter circuit must be found.

$$\Delta^h = \begin{vmatrix} h_{ie} & h_{re} \\ h_{fe} & h_{oe} \end{vmatrix} = \begin{vmatrix} 900 & 1.5 \times 10^{-4} \\ 150 & 30 \times 10^{-6} \end{vmatrix}$$

$$\Delta^h = 900 \times 30 \times 10^{-6} - 150 \times 1.5 \times 10^{-4}$$

$$= 27 \times 10^{-3} - 22.5 \times 10^{-3} = 4.5 \times 10^{-3}.$$

Now the gains and impedances can be calculated.

---

[5] The derivation of these equations and an example using the exact equivalent circuit is given in Appendix B.

$$A_v = \frac{-h_{fe}R_C}{h_{ie} + \Delta^h R_C} = \frac{-150 \times 1800 \, \Omega}{900 \, \Omega + 1800 \, \Omega \times 4.5 \times 10^{-3}}$$

$$= \frac{-270{,}000 \, \Omega}{908.1 \, \Omega} = -297.3$$

The minus sign here indicates that there is a 180° phase shift between the input and output voltages.

$$A_i = \frac{h_{fe}}{1 + h_{oe}R_C} = \frac{150}{1 + 30 \times 10^{-6} \times 1800} = \frac{150}{1.054} = 142.3$$

$$R_i = \frac{\Delta^h R_C + h_{ie}}{1 + h_{oe}R_C} = \frac{908.1 \, \Omega}{1.054} = 862 \, \Omega$$

$$R_o = \frac{h_{ie} + R_{Th}}{\Delta^h + h_{oe}R_{Th}} = \frac{900 + 2292 \, \Omega}{4.5 \times 10^{-3} + 30 \times 10^{-6} \times 2292} = \frac{3192 \, \Omega}{73.3 \times 10^{-3}}$$

$$= 43547 \, \Omega$$

The output impedance in this circuit is quite high. In most circuits, however, it is in parallel with R_C, which is a much lower resistance. Therefore R_o is often ignored.

4-5.2 Gain Calculations Using the Approximate Circuit

The use of the exact equations and circuits are cumbersome, as shown by Example 4-7. Fortunately, they are rarely used. Practically, it has been found that the assumption $h_{re} = h_{oe} = 0$ can be made without affecting the calculations significantly. As a rule of thumb, we have found that the gains calculated after ignoring h_{re} will be about 6% too high. This greatly simplifies the calculations, especially because using $h_{re} = 0$ allows us to omit the feedback voltage generator that causes most of the complications. If h_{oe} must be taken into account, it can be combined in parallel with R_C.

The circuit of Figure 4-4 is redrawn in Figure 4-6 with the voltage

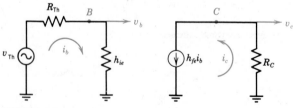

FIGURE 4-6
A transistor circuit using the simplified equivalent circuit.

generator and output admittance omitted, according to our simplifying assumptions. Formulas for the gain can be derived very simply from this circuit.

For the transistor:

$$A_{v(tr)} = \frac{v_c}{v_b}$$

But

$$v_b = h_{ie} i_b$$

and

$$v_c = -i_c R_C = -h_{fe} i_b R_C$$

so that

$$A_{v(tr)} = \frac{v_c}{v_b} = \frac{-h_{fe} R_C i_b}{h_{ie} i_b}$$

$$A_{v(tr)} = \frac{-h_{fe} R_C}{h_{ie}} \qquad (4\text{-}8)$$

The transistor current gain for the simplified circuit is simply

$$A_{i(tr)} = \frac{i_c}{i_b} = h_{fe} \qquad (4\text{-}9)$$

The input impedance is the impedance seen looking in at the base. This is now only h_{ie}

$$R_i = h_{ie} \qquad (4\text{-}10)$$

The voltage gain of the Thevenin's equivalent circuit is v_c/v_{Th}. It can be derived as follows

$$i_b = \frac{V_{Th}}{R_{Th} + h_{ie}}$$

$$v_c = -R_c i_c = -R_C h_{fe} i_b = \frac{-R_C h_{fe} v_{Th}}{R_{Th} + h_{ie}}$$

$$A_{v(ck)} = \frac{v_c}{v_{in}} = \frac{v_c}{v_{Th}} \times \frac{v_{Th}}{v_{in}}$$

where v_{in} is the actual input voltage and v_{Th} is the Thevenin's equivalent input voltage.

$$A_{v(ck)} = \frac{-h_{fe} R_C}{R_{Th} + h_{ie}} \times \frac{v_{Th}}{v_{in}} \quad (4\text{-}11)$$

To calculate the circuit gain by this method, the ratio of the Thevenin's input voltage to the actual input voltage must be found (see Example 4-8).

If a voltage generator is connected directly to the base of the transistor, $R_{Th} = 0$ (see section 4-5) and the Thevenized input voltage becomes the input voltage at the base or $V_{Th}/V_{in} = 1$. Connecting a generator directly to the base causes the circuit gain to be the same as the transistor gain and equation 4-11 reduces to equation 4-8.

EXAMPLE 4-8

For the values of Example 4-7 ($h_{fe} = 150$, $h_{ie} = 900\ \Omega$, $R_{Th} = 2292\ \Omega$, $R_C = 1800\ \Omega$), find the transistor voltage and current gains and the voltage gain of the circuit.

Solution
The gains can be calculated using equations 4-8 through 4-11.

$$A_{v(tr)} = \frac{-h_{fe} R_C}{h_{ie}} = \frac{-150 \times 1800\ \Omega}{900\ \Omega} = -300$$

$$A_{i(tr)} = h_{fe} = 150$$

$$A_{v(ck)} = \frac{-h_{fe} R_C}{R_{Th} + h_{ie}} \times \frac{v_{Th}}{v_{in}} = \frac{-150 \times 1800}{3192} \times \frac{v_{Th}}{v_{in}} = -84.6 \frac{v_{Th}}{v_{in}}$$

Table 4-1 shows a comparison of the gains as calculated by the exact and approximate methods and as measured in the lab (see section 4-5.3). It can be seen that the values calculated by the exact and approximate methods are quite close.

Table 4-1
Comparison of Exact and Approximate Calculations
and Measured Results

Parameter	Exact calculation (Example 4-7)	Approx. calculation (Example 4-7)	Laboratory measurement
$A_{v(tr)}$	−297.3	−300	|274|
$A_{i(tr)}$	142.3	150	143
R_i	862 Ω	900 Ω	940 Ω
$A_{v(ck)}$		19.4[a]	18.2

[a] Assuming $v_{Th}/v_{in} = 0.23$ (see Example 4-11).

The approximate formulas can be used if only h_{fe} and h_{ie} are known. In many transistors, however, only values for h_{fe} are given. But the quiescent current can be found and h_{ie} can then be approximated by the formula $h_{ie} = 30h_{fe}/I_E$. Example 4-9 is a complete analysis of this type of circuit.

EXAMPLE 4-9

For the circuit of Figure 4-7, find
a. $A_{v(tr)}$.
b. $A_{i(tr)}$.
c. $A_{v(ck)}$.
d. $A_{i(ck)}$.

Solution
Because only h_{fe} is given, the first step is to find the quiescent current and h_{ie}.

$$V_{BB} = \frac{V_{CC} \times R_{B2}}{R_{B1} + R_{B2}} = \frac{20 \text{ V} \times 2.5 \text{ k}\Omega}{12.5 \text{ k}\Omega} = 4 \text{ V}$$

$$R_B = 10{,}000 \text{ }\Omega \parallel 2500 \text{ }\Omega = 2 \text{ k}\Omega$$

$$V_{BB} = R_B I_B + h_{fe} R_E I_B + V_{BE}$$

$$4 \text{ V} = 2000 \, I_B + 64{,}000 \, I_B + 0.7 \text{ V}$$

$$3.3 \text{ V} = 66{,}000 \, I_B$$

$$I_B = 50 \text{ μA}$$

$$I_E \approx I_C = h_{fe} I_B = 5 \text{ mA}$$

$$h_{ie} = \frac{30 \, h_{fe}}{I_E} = \frac{30 \times 100}{5} = 600 \text{ }\Omega$$

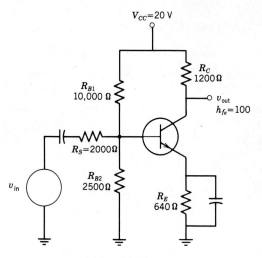

(a) Actual circuit

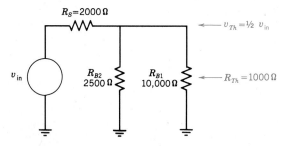

(b) Therenin's circuit at the base of the transistor

FIGURE 4-7
Circuit for Example 4-9.

Now the gains can be found:

$$A_{i(tr)} = h_{fe} = 100$$

$$A_{v(tr)} = \frac{-h_{fe}R_C}{h_{ie}} = \frac{100 \times 1200\ \Omega}{600\ \Omega} = -200$$

The voltage gain of the circuit, $A_{v(ck)} = v_c/v_{in}$, can be found by either one of two methods
Method 1:

$$A_{v(ck)} = \frac{v_b}{v_{in}} \times \frac{v_c}{v_b}$$

158 SMALL SIGNAL ANALYSIS AND ac GAIN

The second term in this equation is simply the voltage gain of the transistor, $A_{v(tr)} = -200$. The first term can be found by voltage division. R'_{in} is the resistance seen looking into the circuit. It is R_B in parallel with h_{ie}.

$$R'_{in} = 10{,}000\ \Omega \parallel 2500\ \Omega \parallel 600\ \Omega = 462\ \Omega$$

By voltage division

$$v_b = \frac{R'_{in}\, v_{in}}{R_S + R'_{in}} = \frac{462\ \Omega \times v_{in}}{2462\ \Omega} = 0.1876\, v_{in}$$

$$A_{v(ck)} = \frac{v_b}{v_{in}} \times \frac{v_c}{v_b} = 0.1876 \times 200 = 37.52$$

Method 2:

The second method is to use equation 4-11:

$$A_{v(ck)} = \frac{h_{fe} R_C}{h_{ie} + R_{Th}} \times \frac{v_{Th}}{v_{in}}$$

Here the Thevenin's resistance looking back from the base is two 2000 Ω resistors in parallel or 1000 Ω.

$$A_{v(ck)} = \frac{100 \times 1200}{1600} \times \frac{v_{Th}}{v_{in}} = 75 \times \frac{v_{Th}}{v_{in}}$$

Now the relationship of the Thevenin's input voltage to the actual input voltage must be calculated. With the base open the input voltage generator is looking at the two 2000 Ω resistors in series, as shown in Figure 4-7b. Therefore $v_{Th}/v_{in} = 0.5$ and:

$$A_{v(ck)} = 0.5 \times 75 = 37.5$$

The current gain of the circuit is the output current divided by the input current, at the generator.

$$A_{i(ck)} = \frac{i_{out}}{i_{in}} = \frac{i_{out}}{i_b} \times \frac{i_b}{i_{in}}$$

But $i_{out}/i_b = h_{fe}$, so only i_b/i_{in} must be found. However, i_{in} flows into the parallel combination of the biasing resistors and h_{ie}. Therefore by current division

4-5 CALCULATING THE COMMON-EMITTER GAIN

$$\frac{i_b}{i_{in}} = \frac{2000}{2600} = 0.77$$

$$A_{i(ck)} = \frac{i_{out}}{i_b} \times \frac{i_b}{i_{in}} = 100 \times 0.77 = 77$$

EXAMPLE 4-10[6]

Check the results of Example 4-9 by assuming that $v_{in} = 100$ mV. Find the voltage at the base and collector and the currents in each of the resistors.

Solution
This example can be started by simply stating that $v_c = 3.75$ V because this is the input voltage times $A_{v(ck)}$. Unfortunately, this doesn't give the reader a "feel" for what is happening within the circuit. Instead, we can start by calculating the currents.

$$i_{in} = \frac{v_{in}}{R_S + R'_{in}} \frac{100 \text{ mV}}{2462 \text{ }\Omega} = 40.6 \text{ }\mu\text{A}$$

By voltage division

$$v_b = \frac{462 \text{ }\Omega}{2462 \text{ }\Omega} \times v_{in} = 18.76 \text{ mV}$$

$$i_b = \frac{v_b}{h_{ie}} = \frac{18.76 \text{ mV}}{600 \text{ }\Omega} = 31.26 \text{ }\mu\text{A}$$

$$i_c = h_{fe}i_b = 3.126 \text{ mA}$$

$$V_{out} = -3.126 \text{ mA} \times 1200 \text{ }\Omega = -3.751 \text{ V}$$

Figure 4-8 shows the final circuit with all the ac voltages and ac currents. Note that the circuit current gain is $i_{out}/i_{in} = 3.12$ mA/40.6 μA = 77 as predicted.

4-5.3 Laboratory Analysis of the H-bias Circuit

In Examples 4-7 and 4-8 we found the gains for a transistor with the values given in Example 4-7 ($h_{fe} = 150$, $h_{ie} = 900$ Ω, $R_{Th} = 2292$ Ω, $R_C = 1800$ Ω,

---

[6]This example may be omitted on first reading.

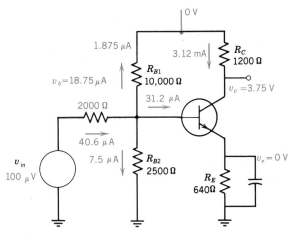

FIGURE 4-8
Final check circuit for Example 4-9.

$I_C = 5$ mA). A circuit with these values is shown in Figure 4-9, and was set up in the laboratory at RIT.[7] The dc voltages were measured and are shown in black; the ac voltages were also measured and are shown in blue on the circuit.

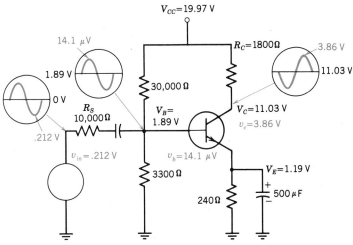

FIGURE 4-9
The laboratory circuit for Examples 4-7 and 4-8.

---

[7]Rochester Institute of Technology.

EXAMPLE 4-11

From the values given in Figure 4-9, find $A_{v(tr)}$, $A_{v(ck)}$, h_{ie}, and h_{fe}.

Solution

$$A_{v(tr)} = \frac{-v_c}{v_b} = \frac{-3.86 \text{ V}}{1.14 \text{ mV}} = -274$$

$$A_{v(ck)} = \frac{v_c}{v_{in}} = \frac{-3.86 \text{ V}}{0.212 \text{ V}} = -18.2$$

The ac currents in each of the resistors can be found from the given voltages.

$$i_{R_s} = \frac{212 \text{ mV} - 14 \text{ mV}}{10{,}000 \text{ }\Omega} = 19.8 \text{ } \mu\text{A}$$

$$i_{R_{B1}} = \frac{14.1 \text{ mV}}{30{,}000 \text{ }\Omega} = 0.47 \text{ } \mu\text{A}$$

$$i_{R_{B2}} = \frac{14.1 \text{ mV}}{3300 \text{ }\Omega} = 4.3 \text{ } \mu\text{A}$$

$$i_c = \frac{3.86 \text{ V}}{1800 \text{ }\Omega} = 2.145 \text{ mA}$$

The current flowing into the transistor is:

$$i_b = i_{R_s} - (i_{RB1} + i_{RB2}) = 15 \text{ } \mu\text{A}$$

Now h_{ie} and h_{fe} for the transistor in this circuit can be found.

$$h_{ie} = \frac{v_b}{i_b} = \frac{14.1 \text{ mV}}{15 \text{ } \mu\text{A}} = 940 \text{ } \Omega$$

$$h_{fe} = \frac{i_c}{i_b} = \frac{2.145 \text{ mA}}{15 \text{ } \mu\text{A}} = 143$$

A list of the calculated values and the experimental values is given in Table 4-1, where they can be compared easily. To obtain the calculated value for $A_{v(ck)}$, the value of v_{Th}/v_{in} was assumed to be 0.23, which applies to this circuit.

All calculated values are within 6% of the expected values. They would have been even closer if the measured h_{fe} of the transistor, 143, were used instead of 150.

4-5.4 The Capacitively Coupled Load

In the circuits of Figure 4-8 or 4-9, the ac output voltage is riding on a dc level (11 V for Figure 4-9). In many cases this output cannot be used directly because the dc level at the collector must be isolated from the load. The load might be another circuit, a loudspeaker, a light, or whatever the transistor circuit is intended to drive.

A capacitor is often used to block the dc voltage at the collector from the load, as shown in Figure 4-10. In this figure R_L represents the resistance of

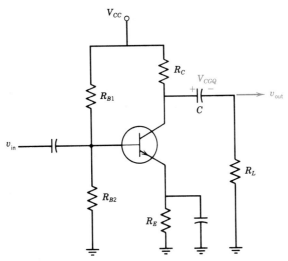

FIGURE 4-10
The H-bias circuit with a capacitively coupled load.

the load. The capacitor charges up to the quiescent dc voltage, and causes the average output voltage to be 0. The capacitor should be large enough to act as an ac short circuit, so the ac voltage at the output and at the collector are equal. Because the capacitor blocks dc, R_L does *not* affect the biasing, but the capacitor also acts as an *ac short*, so for gain calculations, the load resistor is in *parallel* with the collector resistor. This equivalent resistance is R'_L and effectively lowers the gain of the circuit. It also changes the slope of the load line. The effect of R_L on the ac load line is shown in Figure 4-11.

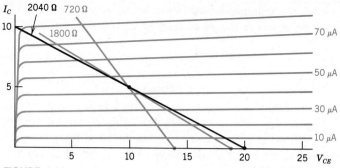

FIGURE 4-11
Ac and dc load lines for the circuits of Figures 4-9 and 4-10.

EXAMPLE 4-12

Find the $A_{v(tr)}$ for the circuit of Figure 4-10 if $R_C = 1800\ \Omega$, $R_L = 1200\ \Omega$, $h_{fe} = 150$, and $h_{ie} = 900\ \Omega$.

Solution

$$R'_L = R_L \parallel R_C = 1800\ \Omega \parallel 1200\ \Omega = 720\ \Omega$$

$$A_{v(tr)} = \frac{-h_{fe}R'_L}{h_{ie}} = \frac{-150 \times 720\ \Omega}{900\ \Omega} = -120$$

Example 4-12 shows that the additional 1200 Ω load resistor lowers the gain of the circuit from 300 (see Example 4-7) to 120.[8] The fact that it also reduces the maximum possible swing is often overlooked, because the addition of R_L does not affect the quiescent point.

EXAMPLE 4-13

Assuming the bias circuit of Figure 4-10 is the same as Figure 4-9, $I_{CQ} = 5$ mA and only the 1800 Ω resistor is connected, what is the maximum possible swing of the collector voltage?

Solution
There are two methods of handling this problem. The first is to realize that $V_C = 11$ V and $V_E = 1.2$ V. Thus the collector can swing down by 9.8 V and up by 9 V (to V_{CC}) and the maximum swing must be 9 V.

---

[8]The following material in this section may be omitted at first reading.

The second method is to use the point-slope formula (see section 3-6.2). The quiescent point is V_{CEQ} = 9.8 V (11 V at the collector, minus 1.2 V at the emitter) and I_{CQ} = 5 mA. The equation for this line becomes

$$(I_C - 5 \text{ mA}) = \frac{-1}{1800} (V_{CE} - 9.8 \text{ V})$$

When I_C = 0, the load line intersects the V axis and V_{CE} = 18.8 V. Therefore V_{CE} can swing down by 9.8 V and up by 9 V (from 9.8 to 18.8 V) before clipping occurs.

EXAMPLE 4-14

Repeat Example 4-13 if the capacitor and 1200 Ω load resistor are added.

Solution

For the ac load line, R_L is in parallel with R_C. The load lines are shown in Figure 4-11. Because the addition of R_L and the capacitor does not affect the bias, the new ac load line must go through the same quiescent point, but its slope is $-1/R'_L = -1/720$ Ω. The point slope formula becomes

$$I_C - 5 \text{ mA} = \frac{-1}{720} (V_{CE} - 9.8)$$

When I_C = 0, V_{CE} = 13.4 V, so the ac load line intersects the V axis at 13.4 V and the maximum swing is 3.6 V (from 9.8 V to 13.4 V).

The solution of Example 4-14 is a mathematical rather than a physical explanation of the situation. It is still sometimes difficult to understand why the addition of a resistor that does not affect the quiescent point reduces the maximum swing.

To explain the phenomenon physically, we must consider the coupling capacitor. In any circuit it charges to V_C (11 V for Figure 4-10) and the capacitor must be large enough so its voltage remains constant. Now consider what happens when the peak of the input signal is just large enough to cut off the transistor, as shown in Figure 4-12a. The circuit at this instant is shown in Figure 4-12b. By being cut off at this instant, the transistor is effectively not there (i_c = 0) and all current flows through the resistors and capacitor. But because of the constant 11 V across the capacitor, there is only 9 V available to drive the circuit. The 1200 Ω and 1800 Ω resistors act as a voltage divider and the output, across the 1200 Ω resistor, is (1200 Ω/3000 Ω) × 9 V = 3.6 V. This agrees with the mathematical solution. This analysis shows that ac coupling will reduce the output swing and may cause clipping if the designer is not

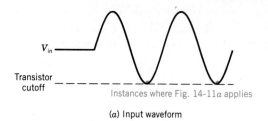

(a) Input waveform

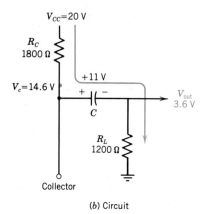

(b) Circuit

FIGURE 4-12
Current flow in the circuit of Figure 4-9 at the instant of cutoff.

careful. For this type of circuit, the designer can gain slightly more swing by biasing at a lower V_{CEQ} (see problem 4-16).

4-6 THE COMMON-BASE AMPLIFIER

The common-base amplifier was introduced in section 2-5 and its biasing was discussed in section 3-7. Actually, the common-base circuit can be represented by the same equivalent circuit (Figure 4-6) as the common-emitter circuit and the voltage and current gains can be calculated by the same equations, 4-8 and 4-9. In order to use them, however, it is necessary to transform the common-emitter parameters, h_{fe} and h_{ie}, into the corresponding common-base parameters, h_{fb} and h_{ib}. As before, we will assume that h_{re} and h_{oe} make no significant contributions. For most practical circuits, this assumption is valid.

The transposition of parameters from common-emitter to common-base is shown in Figure 4-13. Figure 4-13a shows the standard common-emitter configuration with v_{in} connected to the base. The same transistor can be operated as a common-base circuit by connecting the input to the emitter and grounding the base, as shown in Figure 4-13b. Note that the direction of the

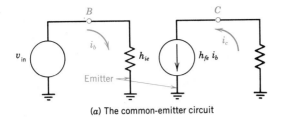

(a) The common-emitter circuit

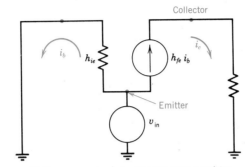

(b) The common-base circuit for the same transistor

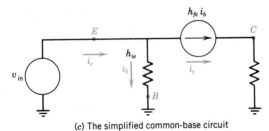

(c) The simplified common-base circuit

FIGURE 4-13
Transforming the common-emitter circuit to a common-base circuit.

emitter-to-collector current generator has been reversed because the direction of base current has been reversed. Figure 4-13c is simply Figure 4-13b redrawn for simplicity.

From Figure 4-13c, we can see that $i_e = h_{fe}i_b + i_b$. For this circuit h_{ib} is the voltage at the emitter divided by the emitter current:

$$h_{ib} = \frac{v_{in}}{i_e}$$

but $v_{in} = h_{ie}i_b$ so we have

$$h_{ib} = \frac{v_{in}}{i_e} = \frac{h_{ie}i_b}{i_b(h_{fe} + 1)}$$

or

$$h_{ib} = \frac{h_{ie}}{h_{fe} + 1} \approx \frac{h_{ie}}{h_{fe}} \tag{4-12}$$

The parameter h_{fb} is analogous to h_{fe} in that it is the relationship of the collector current to the input current at the emitter, instead of the base. Because the direction of the current generator in Figure 4-13 has been reversed, h_{fb} is a negative number.

$$h_{fb} = \frac{-i_c}{i_e} = \frac{-h_{fe}i_b}{(1 + h_{fe})i_b} = \frac{-h_{fe}}{1 + h_{fe}} \approx -1 \tag{4-13}$$

As we found in Chapters 2 and 3, h_{fb}, the current gain of a common-base circuit, is very close to, but slightly less than, 1. If h_{ib} and h_{fb} are known, the voltage gain of a common-base circuit can be found easily. The gain of a common-base circuit is positive; the output is in phase with the input.

EXAMPLE 4-15

Design a common-base amplifier with the base grounded, $R_C = 1800\ \Omega$, $V_{CC} = 20\ V$, and $V_{EE} = -5\ V$. Choose R_E so that $I_{CQ} = 5\ mA$. Find its voltage and current gain if $h_{fe} = 150$.

Solution
To bias the circuit at 5 mA the emitter resistor must be

$$\frac{-0.7\ V - (-5\ V)}{5\ mA} = 860\ \Omega$$

The circuit is shown in Figure 4-14. The ac voltage and current gains can be found as follows:

$$h_{ie} = 30\frac{h_{fe}}{I_E} = 900\ \Omega \quad \text{(as before)}$$

$$h_{fb} = \frac{-150}{151} \approx -1$$

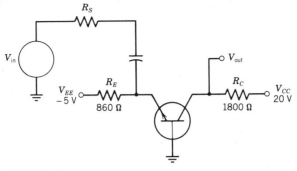

FIGURE 4-14
The common-base circuit for Example 4-15.

$$h_{ib} = \frac{h_{ie}}{h_{fe}} = \frac{900\ \Omega}{150} = 6\ \Omega$$

$$A_{v(\text{tr})} = -\frac{h_{fb}R_L}{h_{ib}} = -\frac{-1 \times 1800\ \Omega}{6\ \Omega} = 300$$

This circuit was built in the laboratory and the results confirm the calculations (see problem 4-18). Accurate measurements are difficult because of the low input impedance of the circuit, and the input capacitor has to be huge.

The equations for the common-base circuit and the results of Example 4-15 show that the voltage gain of a common-base circuit is about the same as a similar common-emitter circuit, although the current gain is less than 1. The low input impedance of the common base circuit, however, tends to drag down the source voltage so that the voltage gain of the circuit is usually less than that of a common-emitter circuit. For these reasons the common-base circuit is not used very often. It is used, however, in some radio frequency (RF) amplifiers that require a high-frequency response and a low input impedance to match a 50 ohm coaxial cable or microstrip line. It is also used in some multiple-transistor circuits, such as the cascode amplifier (see section 6-10).

4-7 THE EMITTER-FOLLOWER CIRCUIT

The *common-collector* or *emitter-follower* circuit is often used as a buffer amplifier because of its high input impedance. It was introduced in section 3-8, where its biasing was discussed. In this section its characteristics as an amplifier are considered.

The emitter-follower circuit is shown in Figure 4-15a and the equivalent circuit is shown in Figure 4-15b. The common-collector equivalent circuit is simply the common-emitter equivalent circuit with resistor moved from the collector to the emitter.

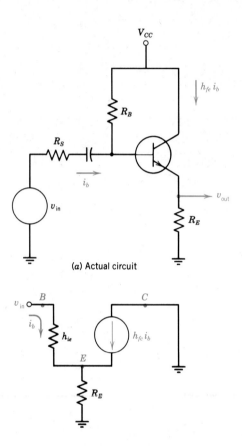

FIGURE 4-15
The emitter-follower circuit.

Figure 4-15a shows that the input signal is applied to the base, and this causes a much larger collector current. The currents flow through the emitter resistor and create an emitter voltage that *opposes* the input voltage. Consequently only a small current flows. Therefore the input impendance of the emitter-follower is very high.

The circuit of Figure 4-15 can be analyzed by converting the common-emitter parameters (h_{ie}, h_{fe}) into common-collector parameters (h_{ic}, h_{fc}), but it is simpler to analyze the circuit of Figure 4-15b directly. From it we find

170 SMALL SIGNAL ANALYSIS AND ac GAIN

$$i_e = i_b + h_{fe}i_b$$
$$v_e = i_e R_E = R_E(i_b + h_{fe}i_b)$$
$$v_b = h_{ie}i_b + v_e = i_b(h_{ie} + (1 + h_{fe})R_E)$$

The input impedance of the transistor is the base voltage divided by the base current:

$$R_{in} = \frac{v_b}{i_b} = h_{ie} + (1 + h_{fe})R_E \approx h_{ie} + h_{fe}R_E \qquad (4\text{-}14)$$

The voltage gain of the transistor is

$$A_{v(tr)} = \frac{v_{out}}{v_b} = \frac{v_e}{v_b} = \frac{i_b(R_E + h_{fe}R_E)}{i_b(h_{ie} + (1 + h_{fe})R_E)}$$

$$A_{v(tr)} = \frac{R_E(1 + h_{fe})}{h_{ie} + R_E(1 + h_{fe})} \qquad (4\text{-}15)$$

Equation 4-15 shows that the voltage gain of an emitter-follower must always be less than 1. It is also in phase with the input.

EXAMPLE 4-16

Find $A_{v(tr)}$, $A_{i(tr)}$, $A_{v(ck)}$, and $A_{i(ck)}$ for the circuit of Figure 4-15a if $R_S = 100\ \Omega$, $R_B = 200\ k\Omega$, $R_E = 1800\ \Omega$, $V_{CC} = 19.7\ V$, and $h_{fe} = 100$.

Solution

Because h_{ie} is not given and is needed in equations 4-14 and 4-15, the quiescent current must be found in order to find h_{ie}. From section 3-8 we have

$$V_{CC} = I_B R_B + V_{BE} + (h_{fe} + 1)I_B R_E$$

$$I_B \approx \frac{V_{CC} - V_{BE}}{R_B + h_{fe}R_E} = \frac{19\ V}{200\ k\Omega + 100 \times 1800\ \Omega} = 50\ \mu A$$

$$I_E \approx I_C = h_{fe}I_B = 5\ mA$$

$$h_{ie} = \frac{30\ h_{fe}}{I_E} = \frac{30 \times 100}{5\ mA} = 600\ \Omega$$

$$A_{v(tr)} = \frac{R_E(1 + h_{fe})}{h_{ie} + R_E(1 + h_{fe})} = \frac{181,800 \text{ }\Omega}{182,400 \text{ }\Omega} = 0.997$$

$$R_{in} = h_{ie} + (1 + h_{fe})R_E = 182,400 \text{ }\Omega$$

$$A_{i(tr)} = \frac{i_e}{i_b} = \frac{(h_{fe} + 1)i_b}{i_b} = 101 \qquad (3\text{-}10)$$

To find the circuit gains, the equivalent circuit of Figure 4-16 can be used. Because R_S is small compared to R_B and R_{in},

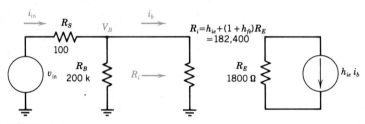

FIGURE 4-16
Circuit for example 4-16.

$$v_b \approx v_{in} \quad \text{so} \quad A_{v(ck)} = \frac{v_e}{v_{in}} = A_{v(tr)} \approx 1$$

$$A_{i(ck)} = \frac{i_e}{i_{in}} = \frac{i_e}{i_b} \times \frac{i_b}{i_{in}} = A_{i(tr)} \times \frac{i_b}{i_{in}}$$

In this circuit the current divides between R_B and R_i. By current division

$$i_b = \frac{R_B}{R_B + R_i} \times i_{in} = \frac{200 \text{ k}\Omega}{382,400 \text{ }\Omega} \times i_{in} = 0.523 \, i_{in}$$

$$A_{i(ck)} = A_{i(tr)} \times \frac{i_b}{i_{in}} = 101 \times 0.523 = 52.8$$

Equations 4-14 and 4-15 and the results of Example 4-16 show that the voltage gain of an emitter-follower is slightly less than 1, but it has a high input impedance. It is often used as a buffer amplifier because it will not load down the source. It also has a low output impedance (see section 4-7.1) so its output voltage will not fall when it is connected to a load (see Example 4-19).

4-7.1 The Output Impedance of the Emitter-follower

The output impedance of a two-port network, such as that of Figure 4-2, is defined as:

$$Z_{out} = \frac{v_2}{i_2}\bigg|_{v_1 = 0}$$

To find the output impedance of a circuit, a voltage can be applied to the output terminals with the input terminals short-circuited so the input voltage is 0 V. The ratio of this voltage to the resulting current is the output impedance.

Figure 4-17a shows a hypothetical voltage generator applied across the output (R_E) of an emitter-follower. The resulting current can be found by using Figure 4-17b, which is a transposed drawing of the emitter-follower equivalent circuit, Figure 4-15b.

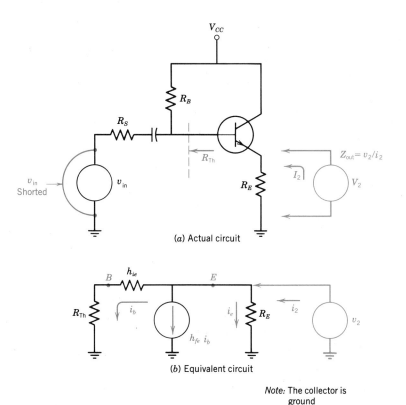

(a) Actual circuit

(b) Equivalent circuit

Note: The collector is ground

FIGURE 4-17
Finding the output impedance of an emitter-follower.

Figure 4-17b shows that the current, i_2, divides into 3 parts:

1. The emitter current $i_e = \dfrac{v_2}{R_E}$

2. The base current $i_b = \dfrac{v_2}{h_{ie} + R_{Th}}$

3. The current in the generator, $h_{fe}i_b$

Therefore:

$$i_2 = i_e + i_b + h_{fe}i_b = v_2 \left[\frac{1}{R_E} + \frac{1}{h_{ie} + R_{Th}} + \frac{h_{fe}}{h_{ie} + R_{Th}} \right]$$

By assuming the term i_b is negligible in comparison with $h_{fe}i_b$, we have

$$\frac{i_2}{v_2} = \frac{1}{R_{out}} \approx \frac{1}{R_E} + \frac{h_{fe}}{h_{ie} + R_{Th}}$$

Thus the output impedance of an emitter-follower is equal to the parallel combination of R_E and a resistor of $(h_{ie} + R_{Th})/h_{fe}$. This resistance is generally very small.

EXAMPLE 4-17

Find the output impedance of the emitter-follower for the circuit of Figure 4-16.

Solution

In Example 4-16 we found that $h_{ie} = 600\ \Omega$ for this circuit. The output impedance of the emitter-follower is, therefore, R_E in parallel with

$$\frac{h_{ie} + R_{Th}}{h_{fe}} = \frac{600\ \Omega + 100\ \Omega}{100} = 7\ \Omega$$

Again, R_{Th} is the Thevenin's resistance looking back from the base. In this circuit it is $100\ \Omega\ \|\ 200\ \text{k}\Omega \approx 100\ \Omega$.

$$Z_{out} \approx 1800\ \Omega\ \|\ 7\ \Omega \approx 7\ \Omega$$

EXAMPLE 4-18

What is the output impedance of the circuit of Figure 4-9? How does capacitively coupling a 1200 Ω resistor affect the gain of the circuit?

Solution

Figure 4-9 is a common-emitter circuit and its output impedance is R_C in parallel with $1/h_{oe}$. If $1/h_{oe}$ is ignored (assumed to be a large resistor in parallel with R_C), the output impedance is simply R_C or 1800 Ω for this circuit. The circuit is shown in Figure 4-18. It shows that the 1200 Ω resistor drops the gain

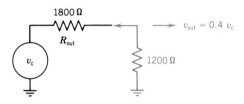

(a) Output circuit for Figure 4-8

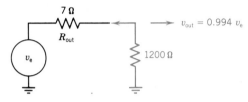

(b) Output circuit for Figure 4-14

FIGURE 4-18
Output circuits for a common-emitter amplifier and an emitter-follower.

to 40% [1200 Ω/(1800 Ω + 1200 Ω)] of its former value. Indeed, Example 4-12 showed that the addition of the 1200 Ω resistor dropped the gain of the circuit from 300 to 120.

EXAMPLE 4-19

How does capacitively coupling a 1200 Ω resistor across the 1800 load resistor of Figure 4-15 affect the gain of the circuit?

Solution

Example 4-17 showed that the output impedance of the emitter-follower circuit was 7 Ω. The output circuit is shown in Figure 4-18b. The gain will decrease by a factor of $1200/1207 = 0.994$, so the output will be practically unchanged.

Check: The addition of a capacitively coupled 1200 Ω resistor will not affect the dc biasing, but will reduce the ac value of R_E from 1800 Ω to 1800 Ω ∥

$1200\ \Omega = 720\ \Omega$. The gain of this circuit can be calculated from equation 4-15 as

$$A_{v(tr)} = \frac{R_E(1 + h_{fe})}{h_{ie} + R_E(1 + h_{fe})} = \frac{720 \times 101\ \Omega}{600\ \Omega + 720 \times 101\ \Omega} = \frac{72{,}720\ \Omega}{73{,}320\ \Omega} = 0.991$$

Examples 4-18 and 4-19 show that a 1200 Ω resistor will severely affect the gain of a common-emitter circuit with its high output impedance, but will have a negligible effect on an emitter-follower circuit, because of its low output impedance.

4-7.2 The Unbypassed Emitter Resistor

In some circuits an amplifier is used with an unbypassed emitter resistor, as shown in Figure 4-19. Fortunately, the input circuit for this type of amplifier is the same as for an emitter-follower, and this makes the transistor gain easy to calculate.

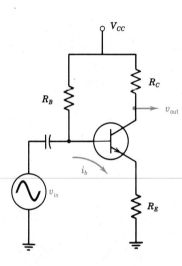

FIGURE 4-19
An amplifier with an unbypassed emitter resistor.

$$v_b = i_b\,[h_{ie} + (1 + h_{fe})\,R_E]$$
$$v_{out} = -i_c R_L = -h_{fe} i_b R_C$$
$$\frac{v_{out}}{v_{in}} = \frac{-h_{fe} i_b R_C}{i_b\,[h_{ie} + (1 + h_{fe})\,R_E]}$$

$$\boxed{A_{v(tr)} = \frac{-h_{fe} R_C}{h_{ie} + (1 + h_{fe})\,R_E}} \qquad (4\text{-}16)$$

Often the term $h_{fe}R_E$ dominates the denominator. Then equation 4-16 reduces:

$$A_{v(tr)} \approx \frac{-h_{fe}R_C}{h_{fe}R_E} = \frac{-R_C}{R_E}$$

so that the gain of this circuit depends only on the resistors, and not on the transistor parameters. This circuit is considered further in Chapter 8, "Feedback."

EXAMPLE 4-20

For the circuit of Figure 4-19, with $h_{fe} = 150$, $h_{ie} = 1000 \, \Omega$, $R_C = 2000 \, \Omega$, and $R_E = 400 \, \Omega$, find $A_{v(tr)}$
a. By using equation 4-16.
b. By inspection.

Solution
a. Using the equation

$$A_{v(tr)} = \frac{-h_{fe}R_C}{h_{ie} + (1 + h_{fe})R_E} = \frac{-150 \times 2000 \, \Omega}{1000 \, \Omega + 151 \times 400 \, \Omega} = \frac{-300,000}{61,400} = -4.89$$

b. By inspection:

$$A_{v(tr)} = \frac{-R_C}{R_E} = -5$$

In almost all cases the approximate equation is close enough to give a valid answer.

4-7.3 The Phase Splitter

The circuit of Figure 4-19 can be modified to take outputs from both the collector and emitter resistors. In this type of circuit the two outputs are 180° out of phase. Such a circuit is called a *phase splitter*. The voltage at the emitter is in phase with the source voltage and the voltage at the collector is 180° out of phase.

If $R_C = R_E$, the outputs are equal and out of phase. This is shown in Figure 4-20. The output impedances, however, are not equal. The output impedance of the emitter output is much less than the collector output. If these outputs must drive loads with high input impedances, the difference in

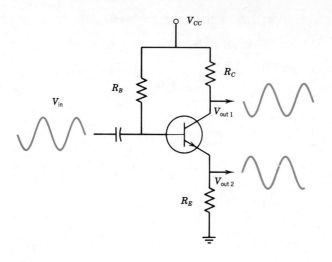

Note: If $R_E = R_C$, $V_{in} = V_{out\,1} = V_{out\,2}$

FIGURE 4-20
The phase splitter.

output impedance can be ignored. The phase splitter is used to drive circuits requiring differential inputs (see section 6-9.1), which generally have high input impedances.

4-8 SUMMARY

This chapter showed how to calculate the gains for common-emitter, common-base, and emitter-follower circuits. First, the equivalent circuits were developed, then the gain and input and output impedance equations were derived from them. The distinction between the gain of the transistor itself (v_c/v_b), and the gain of the circuit (v_{out}/v_{in}), was made to emphasize the effects of the circuit components, other than the transistor, on the circuit's performance. Finally, closely related circuits, such as the capacitive coupled load and the phase splitter, were discussed.

4-9 GLOSSARY

Equivalent circuit. A circuit consisting of resistors, capacitors, voltage sources, and current sources, used to represent an active device, such as a transistor.

Hybrid parameters. Four parameters whose values are used in the equivalent circuit for a transistor. They are called hybrid because they are a resistance, an admittance, a voltage ratio, and a current ratio.

Input impedance. The impedance seen looking into the input terminals of a network.

Output impedance. The impedance seen looking into the output terminals of a network.

Phase splitter. A circuit that produces two outputs that are 180° out of phase. The outputs are generally equal in magnitude.

R_{Th}. Thevenin's resistance seen looking outward from the base of a transistor.

Sieman (S). The new unit of admittance; formerly the mho.

Two-port. A circuit with four terminals and three or four components, often used as an equivalent circuit.

v_{Th}. Thevenin's voltage seen looking outward from the base of a transistor.

4-10 REFERENCES

ROBERT BOYLESTAD and LOUIS NASHELSKY, *Electronic Devices and Circuit Theory*, 4th Edition, Prentice-Hall, Englewood Cliffs, N. J., 1987.

J. F. PIERCE and T. J. PAULUS, *Applied Electronics*, Charles E. Merrill, Columbus, Ohio, 1972.

ROBERT L. RIDDLE and MARLIN P. RISTENBATT, *Transistor Physics and Circuits*, Prentice-Hall, Englewood Cliffs, N. J., 1958.

DONALD L. SCHILLING and CHARLES BELOVE, *Electronic Circuits, Discrete and Integrated*, 2nd Edition, McGraw-Hill, New York, 1979.

4-11 PROBLEMS

4-1 A voltmeter shows that the base voltage of a transistor is 35 mV.
  a. If the gain of the transistor is 150, what is its collector voltage?
  b. If the collector resistor is 2 kΩ, what is the ac collector current?

4-2 For the circuit of Figure P4-2, find $A_{v(tr)}$, $A_{v(ck)}$, $A_{i(tr)}$, and $A_{i(ck)}$.

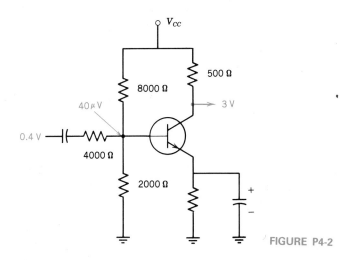

FIGURE P4-2

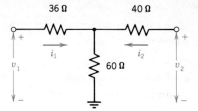

FIGURE P4-3

4-3 Find the h parameters for Figure P4-3.

4-4 If the circuit of Figure P4-3 is terminated with an 80 Ω resistor, and driven by a 20 V source, find i_1, i_2, and v_2. Use the h parameters found in problem 4-3.

4-5 For the **2N3904**, find all the h parameters for $I_C = 0.5, 1, 2,$ and 10 mA.

4-6 If we assume that $h_{ie} = Kh_{fe}/I_E \approx Kh_{fe}/I_C$, find K when $I_C = 0.5, 1, 2, 5,$ and 10 mA. Use the values found in problem 4-5.

4-7 A **2N3904** is operated with a collector current of 1 mA. Using the exact equations, find the voltage gain, the current gain, and the input and output impedances if $R_{Th} = 2$ kΩ and $R_C = 2$ kΩ. *Hint:* The results of problem 4-5 should help.

4-8 A transistor has the following parameters:

$$h_{fe} = 100$$
$$h_{ie} = 1000 \text{ Ω}$$
$$h_{re} = 1.5 \times 10^{-4}$$
$$h_{oe} = 2 \times 10^{-5}$$

Find the voltage gain, current gain, and the input and output impedances if $R_{Th} = 2$ kΩ and $R_C = 2$ kΩ.
  a. Use the exact equations.
  b. Use the approximiate equations.

4-9 For the circuit of Figure P4-9:

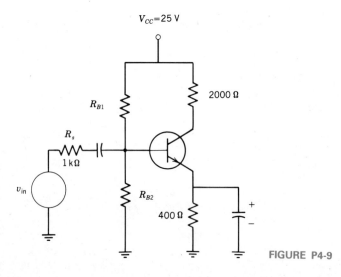

FIGURE P4-9

a. Find R_{B1} and R_{B2} so that a collector current of 4 mA flows. Use h_{fe} from the chart for the **2N3904**.
b. Find h_{ie}, h_{re}, and h_{oe} from the curves.
c. Find $A_{v(tr)}$, $A_{i(tr)}$, and R_i using the exact equations.
d. Repeat part (c) using the approximate equations.

4-10 For the circuit of Figure P4-10, find

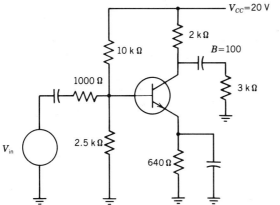

FIGURE P4-10

a. h_{ie}.
b. $A_{v(tr)}$.
c. $A_{v(ck)}$.
d. $A_{i(ck)}$.
e. $A_{i(tr)}$.

4-11 Repeat problem 4-10 for the circuit of Figure P4-11. Assume $h_{fe} = 120$.

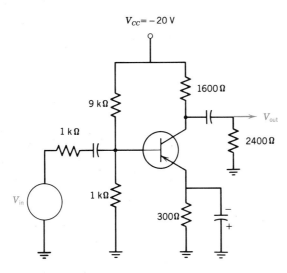

FIGURE P4-11

4-11 PROBLEMS **181**

4-12 Design a transistor circuit for $A_{v(tr)} = 200$ with $I_{CQ} = 3$ mA, $V_{CC} = 20$ V, and $h_{fe} = 150$.

4-13 Determine $A_{v(tr)}$ for the amplifier circuit of Figure P4-13.

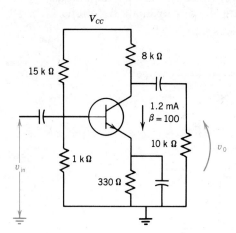

FIGURE P4-13

4-14 If the input of the amplifier of Figure P4-13 were connected to a transducer pickup with an output of 30 mV_{pp} and output impedance of 600 Ω, find $V_{o_{pp}}$.

4-15 For the circuit of Figure P4-15, find

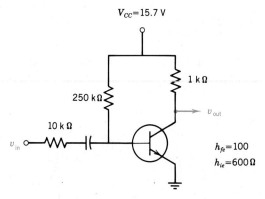

FIGURE P4-15

a. $A_{v(tr)}$.
b. $A_{v(ck)}$.
c. $A_{i(tr)}$.
d. $A_{i(ck)}$.

4-16 For the circuit of Figure 4-10 with $R_C = 1800$ Ω and $R_L = 1200$ Ω, and $R_E = 240$ Ω, change R_{B1} and R_{B2} so that $I_{CQ} = 6$ mA. Assume $h_{re} = 120$ and $V_{CC} = 20$ V.

a. Find the new values of R_{B1} and R_{B2}. Assume R_B is still ≈ 3000 Ω.

182 SMALL SIGNAL ANALYSIS AND ac GAIN

b. Draw the load lines on the characteristic curves.

c. Find the maximum ac swing.

4-17 A certain professor says, "Forget all about h_{fe}, h_{ie}, and even the collector resistor. The gain of a well-designed common-emitter amplifier is approximately 16 V_{CC}."

a. How close is he to the results of Example 4-6?

b. How would he justify this statement?

c. Is it true if a capacitively coupled load is added?

4-18 For the circuit of Figure 4-14, $R_S = 100\ \Omega$ and the following measurements were taken in the lab:

$$v_{in} = 312\ \text{mV}$$
$$v_b = 20\ \text{mV}$$
$$v_{out} = 5.28\ \text{V}$$

From these measurements, find h_{ib} and h_{fb}. How do they compare with the calculated values?

4-19 A transistor has an h_{fe} of 100 and an h_{ie} of 1000 Ω. Find h_{fb} and h_{ib}.

4-20 A transistor has the following common-base parameters:

$$h_{ib} = 50\ \Omega$$
$$h_{rb} = 3 \times 10^{-4}$$
$$h_{fb} = -0.985$$
$$h_{ob} = 2 \times 10^{-7}\ \text{S}$$

If R_E is 200 Ω and R_C is 1.5 kΩ, find the voltage and current gains of the circuit. Use

a. The exact equations.

b. The approximate equations.

4-21 For the circuit of Figure P4-21, with $R_L = \infty$ (open), find

$$A_{v(tr)}$$
$$A_{v(ck)}$$

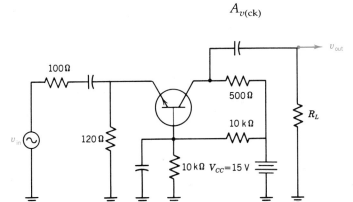

FIGURE P4-21

4-22 Repeat problem 4-21 if $R_L = 1500\ \Omega$.

4-23 For the circuit of Figure 4-16 (see Example 4-16), assume R_S is changed to 100 kΩ. Find
   a. $A_{v(tr)}$.
   b. $A_{v(ck)}$.
   c. R_o.
   Note that the assumption of Example 4-16, $v_b/v_{in} \approx 1$, is no longer valid.

4-24 In problem 4-23 find $A_{v(tr)}$ and $A_{v(ck)}$ if a 1200 Ω resistor is capacitively coupled across the 1800 Ω resistor.

4-25 Find the equivalent circuit of $A_{v(ck)}$ and R_o from problem 4-23. Show that adding a 1200 Ω resistor to it gives the same $A_{v(ck)}$ as found in problem 4-24.

4-26 Find the gain of the circuit of Figure 4-7 if the emitter capacitor becomes disconnected.
   a. Use exact formulas.
   b. Use approximations.

4-27 Repeat problem 4-24 for the circuit of Figure 4-9. Use the values given in Example 4-11. Assume $R_E = 200\ \Omega$.

4-28 For the circuit of Figure P4-28, $v_{in} = 0.5$ V. Find v_{out1} and v_{out2} if

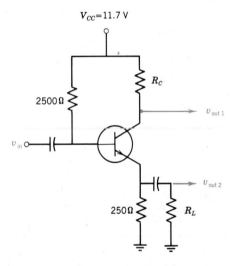

FIGURE P4-28

   a. $R_C = 0$, R_L is open.
   b. $R_C = 2$ kΩ, R_L is open.
   c. $R_C = 0$, $R_L = 1$ kΩ.
   d. $R_C = 2$ kΩ, $R_L = 1$ kΩ.

4-29 The input circuit of Figure P4-29 is connected to a transistor with $h_{fe} = 200$, $h_{ie} = 1$ kΩ. Find v_b if the transistor is connected in

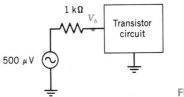

FIGURE P4-29

  a. Common-emitter.
  b. Common-base.
  c. Common-collector.
  Ignore the biasing resistors.

4-30 a. Sketch the circuit of a phase splitter that uses an emitter resistor of 1500 Ω.
  b. If the h_{fe} of the transistor is 120, $V = 25$ V, and the quiescent voltage is to be 6 V, design a bias circuit.
  c. Find the quiescent collector voltage.

After doing the problems, review the self-evaluation questions of section 4-2. If any of them still seem difficult, reread the appropriate sections of this chapter to find the answers.

CHAPTER FIVE

Field Effect Transistors

5-1 Instructional Objectives

This chapter introduces field effect transistors (FET)s and shows how to analyze circuits using them and how to design with them. After studying the chapter, the reader should be able to:

1. Understand the construction of JFETs and MOSFETs and explain why they have high input impedances.
2. Find I_D for a JFET using Shockley's equations.
3. Properly bias a JFET or MOSFET using fixed-bias or self-bias.
4. Use a computer program to determine the quiescent point for a self-bias circuit.
5. Analyze fixed and self-biased JFET amplifiers driving capacitively coupled loads.
6. Use a computer program to determine the optimum operating point for a JFET amplifier.
7. Analyze and design common-drain JFET circuits.
8. Design MOSFET bias circuits.
9. Find the gain of a MOSFET circuit.

5-2 Self-Evaluation Questions

Watch for the answers to the following questions as you read the chapter. They should help you to understand the material presented.

1. What is the difference between a JFET and a MOSFET?
2. Why can't a JFET have a positive V_{GS}? Is this also true for a MOSFET?
3. What is the pinch-off voltage? What is I_{DSS}?
4. Why can't a JFET be biased at its point of maximum g_m?
5. Why can V_{DD} be higher for a self-biased JFET than for a fixed-biased JFET?
6. How can a common-source amplifier, with a gain of less than 1, improve the gain of a circuit?
7. What is the difference between an enhancement type and a depletion type MOSFET?
8. How does VMOS construction differ from normal MOS construction? What are the advantages of a VMOS transistor?

5-3 INTRODUCTION

The field effect transistor (FET) has been increasingly used in recent years. It has some significant advantages over the bipolar junction transistor (BJT) such as a very high impedance (practically infinite), and a low noise figure, which is needed in circuits that amplify very small signals. FETs are often used in hi-fi FM receivers for this reason.

Field effect transistors can also be deposited in a much smaller area of silicon than a BJT, and be made to consume less power. These characteristics have led to their extensive use in digital circuits, especially large-scale integrated (LSI) circuits, such as memories, that contain thousands of transistors. As transistors, however, they have one major drawback compared to a BJT; the gain of an FET is much lower. Therefore, those applications requiring high voltage gain still use BJTs, but other circuits (such as digital circuits), where the gain is not as crucial, use FETs.

FETs are subdivided in two types: junction field effect transistors (JFETs), and metal oxide semiconductor FETs (MOSFETs). JFETs are used more often as amplifiers and will be discussed first. MOSFETs, and their derivative, complementary metal oxide semiconductor FETs (CMOS) are widely used in digital circuits.

To introduce FETs, consider the circuit of Figure 5-1, where a bar of lightly doped *n*-type semiconductor is attached to electrodes at either end. If a voltage is applied across the electrodes, current will flow through the semiconductor. The semiconductor is called an *n-channel*, the electrode where the positive potential is applied is called the *drain*, and the grounded electrode is named the *source*. It is the *source of the majority carriers of current* (electrons for an *n*-channel FET), and the drain is the electrode where the majority carriers leave the FET.

Figure 5-1 shows the principles of operation of an FET; a voltage is applied between the drain and the source (V_{DS}) that results in a current flow (I_{DS}) through the channel. Of course, *p*-channel FETs can also be built, and

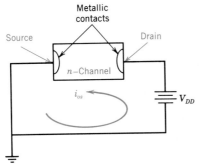

FIGURE 5-1
An *n*-channel device.

some FETs are symmetrical so that the source and drain can be interchanged. Because the mobility of electrons is greater than the mobility of holes, most FETs are *n*-channel. This also requires positive voltage supplies.

The circuit of Figure 5-1 is not an amplifier. It actually acts like a resistor because there is no element present to *control* the current flow. To make an amplifier such an element must be added.

5-4 JFET CONSTRUCTION

The junction FET (JFET) is produced by adding another area of semiconductor material, heavily doped and opposite to the channel doping, as shown in Figure 5-2. This semiconductor region is attached to a third electrode, called

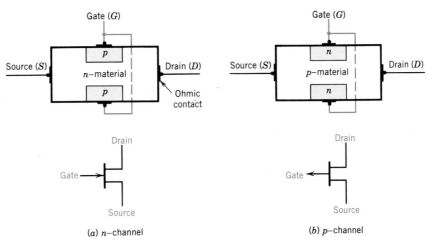

FIGURE 5-2
Structure of a JFET. (Robert Boylestad/Louis Nashelsky, *Electronic Devices & Circuit Theory*, 3rd ed., © 1982. Reprinted by permission of Prentice-Hall, Inc., Englewood Cliffs, New Jersey.

the *gate*, which controls the flow of current. The depth of the gate still leaves adequate room for the channel. Figure 5-2a shows the construction of an *n*-channel JFET and its symbol, and Figure 5-2b shows the *p*-channel FET. The direction of the arrow on the symbol is in toward the gate for an *n*-channel FET and out for a *p*-channel FET.

5-4.1 JFET Operation

In a JFET the gate must always be *reverse-biased*, or zero-biased at most, with respect to both the source and drain. This implies that the gate is always a reverse-biased diode, and consequently there is no significant gate current. The resulting high input impedance of the FET (almost ∞) simplifies bias calculations (see section 5-5).

To explain the operation of the JFET, assume an *n*-channel FET, where the gate and source are connected together and connected to ground. Figure 5-3a shows the situation when a small, positive voltage is applied to the drain. Notice that a depletion region starts to form around the junction between the gate region and the channel. The voltage drop across the channel is approximately uniform, so the portion of the channel closer to the drain is more positive than that near the source. Consequently the depletion region is *skewed*; it is larger near the drain where the reverse-bias is greater.

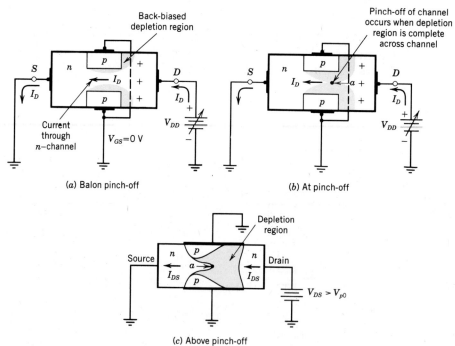

FIGURE 5-3
Operating regions of a JFET.

As the drain voltage increases, two things happen:

1. The current increases because of the higher drain-to-source voltage.
2. The depletion region increases.

Eventually the depletion region gets so large as to extend across the entire channel, as shown in Figure 5-3b. This is called *pinch-off*. The drain to source voltage at pinch-off is called V_P. The current at pinch-off depends on the voltage at point *a* in Figure 5-3b. Further increases in the drain voltage cause the pinch-off region to become larger, as shown in Figure 5-3c, but do not affect the voltage at point *a*, so the channel current remains essentially constant. This continues until the voltage becomes too large for the drain-to-gate junction. Then *avalanche breakdown* occurs, which may damage the FET.

Figure 5-4 shows the three regions of JFET operation. When the drain voltage, V_D, is less than pinch-off, the JFET acts as a resistor. The value of the resistance depends on the gate-to-source voltage. In some applications a JFET

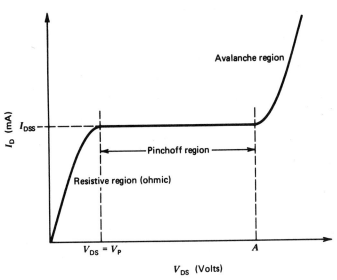

FIGURE 5-4
JFET characteristics as a function of V_{DS} ($V_{GS} = 0$). (From *Semiconductor Devices and Circuits* by Henry Zanger. Copyright © 1984 by John Wiley & Sons, Inc. Reprinted by permission of John Wiley & Sons, Inc.)

is used in this region as a voltage-controlled variable resistor. Between the pinch-off and the avalanche regions the current is essentially constant, but it increases rapidly if the avalanche region is exceeded. The JFET, when used as an amplifier, is meant to be operated in the pinch-off region.

5-4.2 The Common Drain Characteristics

The characteristics of the JFET are also affected by the gate-to-source voltage (V_{GS}). In the previous section it was assumed that both the gate and the source were connected to ground, or $V_{GS} = 0$. The gate cannot be made positive with respect to the source, for this would forward-bias the gate-to-source junction, allowing current to flow. The gate can be made negative with respect to the source, however. By making the gate negative, a depletion region is immediately set up, it requires less drain voltage to pinch-off the JFET, and the pinch-off current is less than when $V_{GS} = 0$.

A typical set of common drain characteristics for a **2N5459** JFET is shown in Figure 5-5a. It shows the drain current versus drain voltage for several negative values of V_{GS}. The highest value of drain current occurs at $V_{GS} = 0$. As stated before, when V_{GS} becomes negative, the pinch-off currents become smaller. When V_{GS} approximately equals the pinch-off voltage, $I_D \approx 0$. In Figure 5-5a, this occurs when the gate is about 5.8 V negative with respect to the source.

Figure 5-5b is the *common source* transfer characteristic for the same JFET. It is a plot of the drain current versus V_{GS}. Figure 5-5b is taken for $V_{DS} = 15$ V. After pinch-off, the drain current is almost constant so the changes in V_{DS}

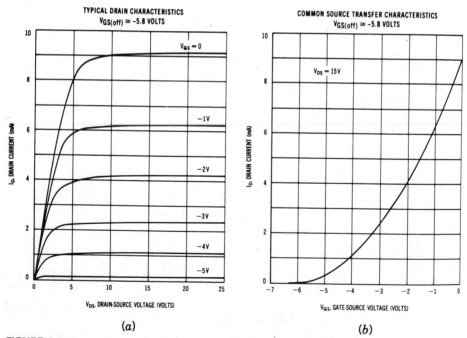

FIGURE 5-5
Typical characteristics for a **2N5459** JFET. (Copyright by Motorola, Inc. Used by permission.)

have a minimum effect on Figure 5-5b, and it can be used for almost any value of V_{DS} above pinch-off. One of the most important characteristics of a JFET is I_{DSS}. This is the drain-to-source current after pinch-off when $V_{GS} = 0$. Figure 5-5a shows that I_{DSS} is about 9 mA for a **2N5459**.

EXAMPLE 5-1

From Figure 5-5a, find the pinch-off voltage for the JFET.
a. Use the $V_{GS} = 0$ curve.
b. Use the $V_{GS} = -2$ V curve.
  How do these answers compare with the curves of Figure 5-5b?

Solution
Pinch-off occurs at the *knee* of the curve. For the $V_{GS} = 0$ curve, this is at about 6 V. For $V_{GS} = -2$ V, the knee of the curve is between 3.5 V and 4 V. If the 2 V bias is added to this, pinch-off becomes -5.5 V to -6 V.

Figures 5-5a and 5-5b state that pinch-off is at -5.8 V and the curve of Figure 5-5b starts to rise when $V_{GS} = -5.8$ V. In calculations in the following sections, however, more accurate results are obtained using a V_P of -6 V, and this is what will be used for the **2N5459**.

5-4.3 JFET Specifications
The specifications for three types of JFETs, the **2N5457, 2N5458,** and **2N5459,** are given in Figure 5-6. The maximum ratings show that the avalanche voltage across any junctions of these JFETs is 25 V.

EXAMPLE 5-2

If the gate is to be at -3 V for a **2N5459**, what is the maximum permissible drain voltage?

Solution
Because $V_{DG(max)} = 25$ V, the drain voltage should not exceed 22 V. This should also be the limit of the power supply for this circuit.

The electrical characteristics of Figure 5-6 show that there is considerable unit-to-unit variation between JFETs of the same type. This is similar to the large h_{FE} variations in a BJT.

From the *off* characteristics of Figure 5-6, we can see that the gate reverse current, I_{GSS}, is only 1 nA at 25°C. This is practically 0, and we will assume throughout this chapter that no gate current flows.

2N5457 2N5458 2N5459

CASE 29-05, STYLE 5
TO-92 (TO-226AA)

JFET
GENERAL PURPOSE

N-CHANNEL — DEPLETION

Refer to 2N4220 for graphs.

MAXIMUM RATINGS

Rating	Symbol	Value	Unit
Drain-Source Voltage	V_{DS}	25	Vdc
Drain-Gate Voltage	V_{DG}	25	Vdc
Reverse Gate-Source Voltage	V_{GSR}	−25	Vdc
Gate Current	I_G	10	mAdc
Total Device Dissipation @ T_A = 25°C Derate above 25°C	P_D	310 2.82	mW mW/°C
Junction Temperature Range	T_J	125	°C
Storage Channel Temperature Range	T_{stg}	−65 to +150	°C

ELECTRICAL CHARACTERISTICS (T_A = 25°C unless otherwise noted.)

Characteristic		Symbol	Min	Typ	Max	Unit
OFF CHARACTERISTICS						
Gate-Source Breakdown Voltage (I_G = −10 μAdc, V_{DS} = 0)		$V_{(BR)GSS}$	−25	—	—	Vdc
Gate Reverse Current (V_{GS} = −15 Vdc, V_{DS} = 0) (V_{GS} = −15 Vdc, V_{DS} = 0, T_A = 100°C)		I_{GSS}		— —	— —	−1.0 −200	nAdc
Gate Source Cutoff Voltage (V_{DS} = 15 Vdc, I_D = 10 nAdc)	2N5457 2N5458 2N5459	$V_{GS(off)}$	−0.5 −1.0 −2.0	— — —	−6.0 −7.0 −8.0	Vdc
Gate Source Voltage (V_{DS} = 15 Vdc, I_D = 100 μAdc) (V_{DS} = 15 Vdc, I_D = 200 μAdc) (V_{DS} = 15 Vdc, I_D = 400 μAdc)	2N5457 2N5458 2N5459	V_{GS}	— — —	−2.5 −3.5 −4.5	— — —	Vdc
ON CHARACTERISTICS						
Zero-Gate-Voltage Drain Current* (V_{DS} = 15 Vdc, V_{GS} = 0)	2N5457 2N5458 2N5459	I_{DSS}	1.0 2.0 4.0	3.0 6.0 9.0	5.0 9.0 16	mAdc
SMALL-SIGNAL CHARACTERISTICS						
Forward Transfer Admittance Common Source* (V_{DS} = 15 Vdc, V_{GS} = 0, f = 1.0 kHz)	2N5457 2N5458 2N5459	$|y_{fs}|$	1000 1500 2000	— — —	5000 5500 6000	μmhos
Output Admittance Common Source* (V_{DS} = 15 Vdc, V_{GS} = 0, f = 1.0 kHz)		$|y_{os}|$	—	10	50	μmhos
Input Capacitance (V_{DS} = 15 Vdc, V_{GS} = 0, f = 1.0 MHz)		C_{iss}	—	4.5	7.0	pF
Reverse Transfer Capacitance (V_{DS} = 15 Vdc, V_{GS} = 0, f = 1.0 MHz)		C_{rss}	—	1.5	3.0	pF

*Pulse Test: Pulse Width ≤ 630 ms; Duty Cycle ≤ 10%.

MOTOROLA SEMICONDUCTORS   SMALL-SIGNAL DEVICES

FIGURE 5-6
Specifications for the **2N5457, 2N5458** and **2N5459** JFETs. (Copyright by Motorola, Inc. Used by permission.)

The gate source cutoff voltage, $V_{GS(off)}$ is the pinch-off voltage. It is the voltage where essentially no current flows (10 nA in the specifications) and varies considerably from unit to unit. For a **2N5459**, it can be anywhere from -2 to -8 V.

The ON characteristics show I_{DSS} the pinch-off current when $V_{GS} = 0$. Since V_{GS} cannot be positive, this is the largest current that can flow in the JFET. One of the main differences between the **2N5457, 2N5458**, and **2N5459** is in I_{DSS}. Higher values of I_{DSS} generally result in a higher g_m (see section 5-6) and make the **2N5459** the preferred part.

EXAMPLE 5-3

For the JFET of Figure 5-5, what is I_{DSS}?

Solution
Figure 5-5 shows that the drain current I_D flattens out around 9 mA for $V_{GS} = 0$. At 10 V it is 9 mA for $V_{GS} = 0$, and at 20 V it is 9.2 mA. The specification sheet uses V_{DS} of 15 V, so I_{DSS} for this JFET is 9.1 mA.

The specifications state that a typical **2N5459** has an I_{DSS} of 9 mA. This matches the curves of Figure 5-5, so the figure can be assumed to be for a **2N5459**.

The small signal characteristics for the JFETs are discussed in section 5-6.

5-4.4 Shockley's Equation

After observing Figure 5-5a, the reader has probably noticed that the drain characteristics are *not uniform*; they are more closely spaced at more negative values of V_{GS}. W. Shockley of Bell Laboratories[1] developed an equation that predicts the drain current for any value of V_{GS}.

$$I_D = I_{DSS}\left(1 - \frac{V_{GS}}{V_P}\right)^2 \tag{5-1}$$

The common source transfer characteristic of Figure 5-5b has a parabolic shape that conforms to Shockley's equation.

---

[1] Shockley is also a co-inventor of the transistor.

EXAMPLE 5-4

For the JFET of Figure 5-5, find the drain current when $V_{GS} = -1$ V and $V_{GS} = -3$ V using Shockley's equation. Make the simplifying approximations that $I_{DSS} = 9$ mA and $V_P = -6$ V. (These are very close to the actual values.)

Solution
When $V_{GS} = -1$ V we have

$$I_D = I_{DSS}\left(1 - \frac{V_{GS}}{V_P}\right)^2 \quad \text{and} \quad \frac{V_{GS}}{V_P} = \frac{1}{6}$$

$$I_D = 9 \text{ mA}\left(\frac{5}{6}\right)^2 = 9 \text{ mA} \times \frac{25}{36} = 6.25 \text{ mA}$$

When $V_{GS} = -3$ V, $V_{GS}/V_P = 1/2$ and

$$I_D = 9 \text{ mA}\left(\frac{1}{2}\right)^2 = 2.25 \text{ mA}$$

Both of these results are exactly what the curves of Figure 5-5 show.[2]

5-4.5 Laboratory Investigation of a JFET

The simple circuit of Figure 5-7 was set up to test a **2N5459** JFET in the laboratory. As V_{DD} was varied, V_{DD} and V_D were measured and I_D was calculated from the measurements. The results are shown in Table 5-1. It can

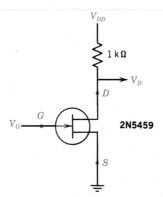

FIGURE 5-7
Laboratory circuit for obtaining the drain characteristics of a **2N5459** JFET.

---

[2]Our results show that using a V_P of -6 V instead of the -5.8 V on the spec sheet gives results that are closer to the values on curves of Figure 5-5. Therefore we will use a V_P of -6 V for the 2N5459.

Table 5-1
A Table of Drain Voltages and Current for a 2N5459 JFET.

V_{DD} (V)	$V_{GS} = 0$		$V_{GS} = -2$	
	V_D (V)	I_D (mA)	V_D (V)	I_P (mA)
20	9.50	10.50	17.75	2.25
19	8.55	10.45	16.76	2.24
18	7.55	10.45	15.76	2.24
17	6.56	10.44	14.80	2.20
16	5.58	10.42	13.80	2.20
15	4.59	10.41	12.84	2.16
14	3.84	10.16	11.77	2.23
13	3.13	9.73	10.79	2.21
12	2.61	9.39	9.81	2.19
11	2.21	8.79	8.81	2.19
10	1.89	8.11	7.85	2.15
9	1.62	7.38	6.84	2.16
8	1.38	6.62	5.82	2.18

be seen that I_{DSS} for this JFET is 10.45 mA and that the knee of the $V_{GS} = 0$ curve is at a V_{DS} between 3 and 4 V. When V_{GS} was set to -2 V, the drain current was constant at 2.16 mA. The drain currents at -3 V and -4 V were also measured. They were $I_D = 0.12$ mA at $V_{GS} = -3$, and $I_D = 0$ at $V_{GS} = -4$ V. This indicates that the pinch-off voltage for this particular JFET is between 3 and 4 V.

5-5 BIASING THE JFET

It is simpler to bias a JFET than to bias a BJT because the JFET gate does not conduct any current. In this section the analysis of JFET bias circuits and their design are considered. There are two types of bias circuits: the fixed-bias circuit where the bias voltage is obtained from a separate voltage source, and self-bias where the bias voltage is developed across a resistor between the source and ground. This is similar to biasing a BJT using a capacitively bypassed emitter resistor. It is important to realize that selecting the bias voltage, V_{GS}, determines the quiescent current, I_{DQ}, if the JFET is in pinch-off. For the **2N5459**, for example, if $V_{GS} = -1$ V, the curves of Figure 5-5 show that $I_D = 6.2$ mA. I_D can also be found from Shockley's equation if V_{GS} is specified, as shown in Example 5.4.

5-5.1 Analysis of the Fixed-bias Circuit

Figure 5-8 shows the fixed-bias JFET circuit. The bias voltage is provided by the biasing source, V_{GG}, and is applied to the gate via R_G. R_G is generally a large resistor (100 kΩ to 1 MΩ) to avoid loading the ac source. Its actual value is not too important because no current flows through it and the dc voltage at the gate is the same as V_{GG}. The ac signal enters via C_1, and the amplified output is picked off at C_2.

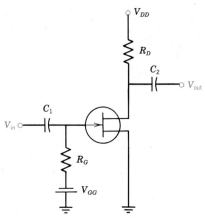

FIGURE 5-8
A fixed biased *n*-channel JFET.

EXAMPLE 5-5

For the circuit of Figure 5-8, assume the JFET is a **2N5459** whose characteristics are given in Figure 5-5. Find I_D and V_D if $V_{GS} = -2$ V, $V_{DD} = 20$ V, and $R_D = 2$ kΩ.

Solution

The precise way to handle this is to plot the load line on the characteristic curves, as shown in Figure 5-9. The load line is plotted in the same way as in Chapters 2, 3, and 4, by connecting the point $V_{DD} = 20$ V, $I_D = 0$ to the point $V_D = 0$, $I_D = V_D/R_D = 10$ mA. The quiescent point is at the intersection of this load line with the bias curve, $V_{GS} = -2$ V. From the figure we see that $V_{DQ} \approx 12$ V and $I_D \approx 4.2$ mA.

An approximate solution can be obtained by looking at the curves and stating that if pinch-off has occurred, then $I_D \approx 4.2$ mA. Therefore

$$V_D = V_{DD} - I_D R_D = 20 \text{ V} - 4.2 \text{ mA} \times 2 \text{ k}\Omega = 11.6 \text{ V}$$

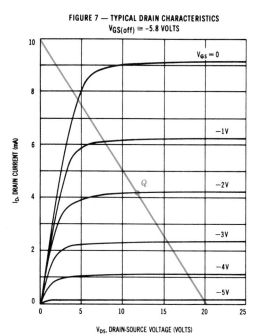

FIGURE 5-9
The load line for Example 5-5. (Copyright by Motorola, Inc. Used by permission.)

A third way to solve this problem is to use Shockley's equation. This gives a result of $I_D = 4$ mA and $V_{DS} = 12$ V.

5-5.2 Analysis of the Self-bias Circuit

The self-bias circuit for the JFET is shown in Figure 5-10. The gate is at ground potential, but the current flowing through the source resistor causes the source to be above ground by a voltage of $I_D R_S$. This has the same effect as applying a negative voltage of $I_D R_S$ to the gate.

There are three reasons for preferring self-bias to fixed-bias:

1. A self-bias circuit does not require a second power supply or voltage source.
2. A self-bias circuit stabilizes the quiescent point against unit-to-unit variations, just as it does for a BJT. If an FET in a self-bias circuit is replaced by another unit with a higher current, the source voltage, $I_D R_S$, will increase and result in a more negative bias, tending to reduce the current and hold the quiescent point stable.
3. The voltage of the drain power supply, V_{DD}, can now be as high as the maximum voltage rating (see Example 5-1).

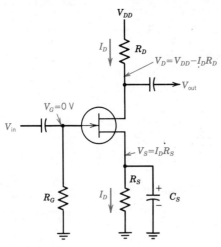

FIGURE 5-10
The self-biased circuit.

For these reasons the self-bias circuit is used most often. Of course, R_S must be bypassed by a large capacitor to short it out for the ac signals, or else the ac voltage developed at the source will reduce the gain of the circuit (see section 5-6.3).

To analyze a given self-bias circuit, the transfer characteristics for the JFET can be used. The source resistance line can be plotted on the transfer characteristics, and the point where it intersects the curve gives I_D and V_{GS}.

EXAMPLE 5-6

For the circuit of Figure 5-10, find V_D and I_D if the JFET is a **2N5459**, $V_{DD} = 25$ V, $R_D = 2000\ \Omega$, and $R_S = 400\ \Omega$.

Solution
Figure 5-11 shows the transfer characteristics with the resistance line plotted on it. For an R_S of 400 Ω, when $V_{GS} = -1$ V, $I_D = 2.5$ mA. This point (point A on the figure) and the origin determine the resistance line. The line intersects the curve at $V_{GS} = -1.8$ V, $I_D = 4.5$ mA. Therefore

$$V_D = V_{DD} - I_D R_D = 25\text{ V} = 4.5\text{ mA} \times 2000\ \Omega = 16\text{ V}$$

A second method of analyzing a self-bias circuit is to use Shockley's equation, $I_D = I_{DSS}[1 - (V_{GS}/V_P)]^2$ along with equation $V_{GS} = I_D R_S$. This

200 FIELD EFFECT TRANSISTORS

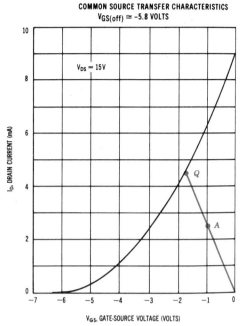

FIGURE 5-11
Plotting the bias line on the transfer curve for Example 5-6. (Copyright by Motorola, Inc. Used by permission.)

allows us to solve for I_D after eliminating V_{GS}. The equation, however, is quadratic, and the solution is best left to a computer program.

Figure 5-12 shows the computer program and results for the values of Example 5-6. The inputs are $V_P{}^3$, I_{DSS} (in mA) and R_S (in kΩ). The output 4.45 mA is the quiescent value of I_D in mA. Here it differs by about 1% from the values found on the graph.

5-5.3 Design of a JFET Biasing Circuit

If the engineer is faced with the problem of designing a bias circuit for a JFET, instead of analyzing a given circuit, there are several choices available, as there were in the design of BJT transistors, but the choices are more constrained, as we will see.

Perhaps the best way to design a bias circuit is to select a value of V_{DD} and draw a load line on the drain characteristics. The value of V_{DD} depends on the available power supplies, but can never be more than the maximum

---

[3]Although V_P is negative or 0, it is used in this program as a positive number.

```
10 Rem  This is a program to find the self bias of a JFET
20 print "type IDSS in mA and RS in K ohms."
30 input "Type VP,IDSS,RS"; V,I,R
40 B = (V+2*I*R)/(V*I)
50 A = (R^2)/(V^2)
60 Rem Find discriminant of the quadratic (c=1)
70 D = sqr((B^2)-(4*A))
80 J = (B-D)/(2*A)
90 Print "The quiescent value of ID is ";J;" mA"
100 End
```

(a) Program

6,9,0.4

(b) Input

```
type IDSS in mA and RS in K ohms.
The quiescent value of ID is  4.45114   mA
```

(c) Output

FIGURE 5-12
A program to find the quiescent current of a self-biased JFET and its results using the values of Example 5-6.

value of V_{DG} or V_{DS} (25 V for a **2N5457, 2N5458,** or **2N5459**). Of course, if fixed-bias is used, V_{DD} must be less than 25 V to allow for a negative voltage on the gate.

The load line starts at $I_D = 0$; $V_D = V_{DD}$ and generally goes through the knee of the $V_{GS} = 0$ curve or is somewhat shallower. The slope of the load line determines R_D (for a fixed-bias circuit) or $R_D + R_S$ (for a self-bias circuit). The operating region along the load line must stay in the pinch-off region, and the gate voltage swing can never be so high as to make V_{GS} positive. If the load line is too steep (too small a resistance), the voltage swing will be limited. In general, the higher the value of R_D, the higher the gain, provided the JFET is not capacitively coupled to a load. If there is a capacitive load, the optimum load line depends on the load and other factors (see section 5-6.2). Often this load line goes through the knee of the curve.

In many circuits the gain or the swing must be optimized. Using a V_{DD} of 20 V, three load lines are shown in Figure 5-13. Load line A goes through the knee of the $V_{GS} = 0$ curve, load line B is too steep, and load line C is shallower but gives a higher gain.

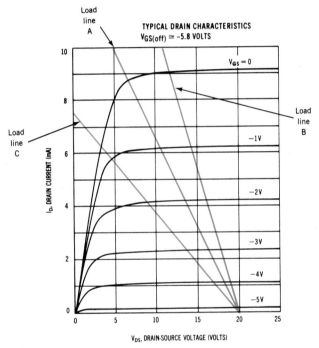

FIGURE 5-13
Load lines for a **2N5459** JFET. (Copyright by Motorola, Inc. Used by permission.)

EXAMPLE 5-7

If V_{GS} is to swing by ± 1 V, what will the corresponding swing of V_D be for each of the load lines in Figure 5-13?

Solution

The results are shown in Table 5-2. For load lines A and B, a quiescent V_{GS} of -1 V was selected. Load line C is shallower and would have little room to

Table 5-2
The Voltages on the Load Lines of Figure 5-13

	A	B	C
V_{DQ}	11	14	9
$V_{D(min)}$	7	11.5	5
$V_{D(max)}$	13.5	16	14

swing if $V_{GS} = -1$ V. Therefore a V_{GS} of -2 V was chosen. Table 5-2 shows that the output voltage swings ($V_{max} - V_{min}$) are 6.5 V for A, 4.5 V for line B (R is too small), and 9 V for point C.

The quiescent point for a JFET should be near the middle of the load line if maximum output swing is to be obtained. If only a small output swing can be useful, the quiescent point should be higher (closer to $V_{DS} = 0$) because this results in a higher quiescent drain current and a higher g_m (see section 5-6).

EXAMPLE 5-8

Given a 20 V power supply, design the bias circuit for a **2N5459** JFET so that $V_{GS} = -2$ V.
a. Use fixed-bias.
b. Use self-bias.

Solution

a. With $V_{DD} = 20$ V, the dc load line shown on Figure 5-14 seems like a reasonable choice. This load line corresponds to a load resistance of 2 kΩ (20 V/10 mA). The values are $V_{GS} = -2$ V and $R_D = 2$ kΩ. R_G can be anything from 100 kΩ to 1 MΩ.
b. For the self-bias circuit we can use the same 2 kΩ load line and the circuit of Figure 5-10. The load line intersects the $V_{GS} = -2$ V curve at $I_D = 4.2$ mA. Because the gate voltage is developed across the source resistor

$$R_S = \frac{V_{GS}}{I_D} = \frac{2\text{ V}}{4.2\text{ mA}} = 476\ \Omega$$

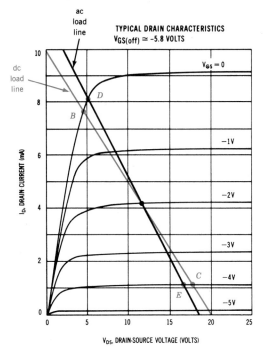

FIGURE 5-14
Load lines for Examples 5-8 and 5-9. (Copyright by Motorola, Inc. Used by permission.)

The slope of the load line for a self-bias circuit is $R_D + R_S$. Therefore

$$R_D = 2 \text{ k}\Omega - R_S = 1524 \text{ }\Omega$$

This completes the design.

EXAMPLE 5-9

For the circuit of Example 5-8, find the maximum swing and the ac gains of the **2N5459** JFET using the curves of Figure 5-14.

Solution
a. For the fixed-bias circuit the values can be taken directly from the curves. The circuit is biased at -2 V and the maximum input swing is therefore ± 2 V or from $V_{GS} = 0$ to $V_{GS} = -4$ V. From the dc load line we have $V_D = 5$ V at $V_{GS} = 0$ and $V_D = 18$ V at $V_{GS} = -4$ V, so the swing is 13 V.

The ac gain for this circuit is

$$|A_{v(tr)}| = \frac{v_D}{v_{GS}} = \frac{13\ \text{V}}{4\ \text{V}} = 3.25$$

b. For the self-bias circuit the ac load line must be plotted on Figure 5-14. Its slope is $-1/R_D$ but the quiescent point is still $V_D = 11.6$ V, $I_D = 4.2$ mA.

The point-slope formula can be used, but the simplest way to get a second point on the ac load line is to observe that $V_S = 2$ V, so when $I_D = 0$, $V_{DS} = V_{DD} - V_S = 18$ V. The ac load line is also plotted on Figure 5-14.

The swing along the ac load line is between point D, where $V_D = 5.5$ V, and point E, where $V_D = 16.5$ V. Thus the total output swing is 11 V, and the gain is

$$A_{v(tr)} = \frac{v_D}{v_S} = \frac{11}{4} = 2.75$$

Example 5-9 shows that the gain of a fixed-bias circuit is slightly higher than that of a self-bias circuit, but more importantly, it shows that the gain of a JFET is about 3. Compared to the gains we have seen for a BJT (100–300), this is very small and explains why BJTs are used in circuits requiring higher voltage gains. The current and power gains of a JFET amplifier are large, however. Due to its high input impedance, only a very small current enters the JFET.

5-6 SMALL SIGNAL ANALYSIS OF A JFET

As with the BJT, the ac analysis of a JFET amplifier requires an equivalent circuit where all dc or constant voltages are considered to be ground. The equivalent circuit for a JFET amplifier is shown in Figure 5-15. The input voltage is applied to the input resistor, R_G, and to the gate, which is essentially an infinite impedance. Thus the input impedance of a JFET amplifier is R_G (from 100 kΩ to 1 MΩ). The drain circuit consists of a current generator,

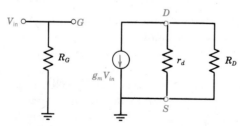

FIGURE 5-15
The equivalent circuit for a JFET.

$g_m v_{in}$, across an internal resistor, r_d. The output is taken across the drain resistor, R_D.

The term g_m^2 is the *transconductance* of the JFET. It is defined as

$$g_m = \frac{\partial i_D}{\partial v_{GS}} \approx \left.\frac{\Delta i_D}{\Delta v_{GS}}\right|_{V_D \text{ constant}} \tag{5-2}$$

It is a conductance that relates the output current (I_D for a JFET) to the input voltage.[4]

The resistor, r_d, is the internal resistance of the JFET. It is defined as

$$r_d = \left.\frac{\Delta v_D}{\Delta i_D}\right|_{V_{GS} \text{ constant}} \tag{5-3}$$

Since $g_m = \Delta i_D/\Delta v_{DS}$ it is the slope of the transfer characteristic (Figure 5-5b). An examination of Figure 5-5b reveals that the slope becomes steeper as V_{GS} becomes less negative and g_m is maximum when $V_{GS} = 0$. This is confirmed by the drain characteristics. The farther apart each increment of V_{GS} is, the larger the g_m. Unfortunately, a JFET amplifier cannot be biased at $V_{GS} = 0$ because any ac swing about that point would forward bias the gate-to-source voltage.

For a JFET, g_m can also be found using Shockley's equation.

$$I_D = I_{DSS}\left(1 - \frac{V_{GS}}{V_P}\right)^2$$

$$g_m = \frac{\partial I_D}{\partial V_{GS}} = \frac{-2I_{DSS}}{V_P}\left(1 - \frac{V_{GS}}{V_P}\right) \tag{5-4}$$

The term $-2I_{DSS}/V_P$ is often defined as g_{mo}, the g_m when $V_{GS} = 0$ V. Using this term, equation 5-4 becomes:

$$g_m = g_{mo}\left(1 - \frac{V_{GS}}{V_P}\right) \tag{5-5}$$

---

[4]The g_m concept was also used in pentode vacuum tubes; they function similarly to a JFET, but have higher values of g_m.

EXAMPLE 5-10

For the 2N5459 JFET, whose drain characteristics are given in Figure 5-5a, find g_{mo}.

Solution
Figure 5-5a shows that $I_{DSS} \approx 9$ mA and $V_p = -6$ V.

$$g_{mo} = \frac{-2I_{DSS}}{V_P} = \frac{-18 \text{ mA}}{-6 \text{ V}} = 3000 \times 10^{-6} \text{ S}$$

The specifications for g_{mo} (called $|y_{fs}|$ on the Motorola spec sheet, Figure 5-6) state that it is between 2000 and 6000 $\times 10^{-6}$ S.

EXAMPLE 5-11

From the drain characteristics of Figure 5-5, find g_m when $V_D = 15$ V, and $V_{GS} = -1$ V. Also find r_d when $V_{GS} = 0$. Verify the results using equation 5-5.

Solution
The data are obtained from the drain characteristics, as shown in Figure 5-16. The g_m about the point $= -1$ V, $V_D = 15$ V, can be found on the vertical blue line. We can see that at $V_{GS} = 0$, $I_D = 9.1$ mA and at $V_{GS} = -2$ V, $I_D = 4.2$ mA. Therefore

$$g_m = \left.\frac{\Delta I_D}{\Delta V_{GS}}\right|_{V_D \text{ constant } = 15 \text{ V}} = \frac{4.9 \text{ mA}}{2 \text{ V}} = 2450 \times 10^{-6} \text{ S}$$

Using equation 5-5 we have

$$g_m = g_{mo}\left(1 - \frac{V_{GS}}{V_P}\right) = 3000 \times 10^{-6}\left(1 - \frac{(-1)}{(-6)}\right)$$

$$g_m = 3000 \times 10^{-6} \times 0.83 = 2500 \times 10^{-6} \text{ S}$$

The two answers for g_m are very close.
The value of r_d can be found from the $V_{GS} = 0$ curve, as shown in Figure 5-16. We chose the points $V_D = 10$ V, $I_D = 9$ mA and $V_D = 25$ V, $I_D = 9 2$ mA, as shown on the figure.

$$r_d = \left.\frac{\Delta V_D}{\Delta I_D}\right|_{V_{GS} \text{ constant } = 0 \text{ V}} = \frac{15 \text{ V}}{12 \text{ mA}} = 75{,}000 \ \Omega$$

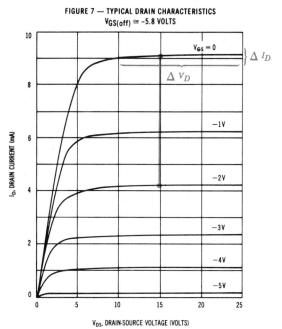

FIGURE 5-16
Curves for Example 5-11. (Copyright by Motorola, Inc. Used by permission.)

The spec sheet (Figure 5-6) calls the reciprocal of this parameter $|y_{os}|$ and gives a typical admittance value of 10 μS. This corresponds to an r_d of 1/10 μS = 100 kΩ, so the two answers are in reasonable agreement, especially considering that it is difficult to get highly accurate values from the curves.

5-6.1 The ac Gain of a JFET

The ac gain of a JFET transistor is v_{out}/v_{in}. For ac analysis, $\Delta I_D/\Delta V_{GS} = i_d/v_{in} = g_m$, and the equivalent circuit of Figure 5-15 shows that $v_{out} = -i_d \times R_D \| r_d$.

Example 5-11 shows that $r_d \approx 75{,}000$ Ω. In most cases, $r_d \gg R_D$, so it can be ignored, and $v_{out} = i_d R_D$. Therefore

$$A_v = \frac{v_{out}}{v_{in}} = \frac{-i_d R_D}{v_{in}}$$

$$\boxed{A_v = -g_m R_D} \tag{5-6}$$

The gain of a JFET can be calculated by the following steps:

1. Determine the quiescent value of V_{GS}.
2. Use equation 5-5 to find the g_m at its quiescent point.
3. Use equation 5-6 to find the gain.

EXAMPLE 5-12

Find the ac gain of the JFET circuit of Figure 5-17.

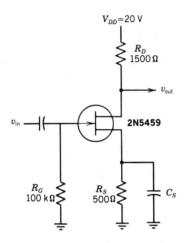

FIGURE 5-1
Circuit for Example 5-12.

Solution

This is a self-bias circuit. To find V_{GS}, we must find I_D. The computer program of Figure 5-12 gives a result of $I_D = 4$ mA.

$$V_{GS} = -4 \text{ mA} \times 500 \text{ } \Omega = -2 \text{ V}$$

$$g_m = g_{mo}\left(1 - \frac{V_{GS}}{V_P}\right)$$

For this JFET $g_{mo} = 3000 \times 10^{-6}$ S (see Example 5-10).

$$g_m = 3000 \times 10^{-6}\left(1 - \frac{-2 \text{ V}}{-6 \text{ V}}\right) = 2000 \times 10^{-6} \text{ S}$$

$$A_v = -g_m R_D = -2000 \times 10^{-6} \text{ S} \times 1500 \text{ } \Omega = -3$$

This circuit is very much like the circuit of Figure 5-14, whose gain was calculated in Examples 5-8 and 5-9, and the values developed here are within 10% of the values found in Example 5-9.

5-6.2 Design of a JFET Amplifier

Equation 5-6 indicates that the gain of a JFET increases as R_D increases. It is not as clear as it looks, however, because as R_D increases, I_D and V_{GS} decrease, and this causes g_m to decrease, so if R_D is doubled, for example, the gain will not double. Nonetheless, there is some net gain increase as R_D increases.

The simplest way to design for high gain, therefore, is to make R_D as large as possible, but as R_D becomes larger, the load line on the drain characteristics becomes flatter. Too large a value of R_D will reduce the swing if a capacitively coupled load is added, as explained later, and the value of R_D should also be well under r_d or the presence of r_d will reduce the gain.

EXAMPLE 5-13

Design an amplifier using a **2N5459** for high gain.
a. Use a fixed-bias circuit.
b. Use a self-bias circuit.

Solution

As in most design problems, several decisions must be made. Here V_{DD} and R_D must be chosen. For high gain, a high R_D and a high V_{DD} are essential. For the fixed-bias circuit an R_D of 10 kΩ and a V_{DD} of 20 V (to allow for a negative V_{GS}) are chosen. The load line appears in Figure 5-18. It shows that a reasonable value of V_{GS} is -4 V.

$$g_m = g_{mo}\left(1 - \frac{(-4)}{(-6)}\right) = 3000 \times 10^{-6} \text{ S } (1 - 0.667) = 1000 \times 10^{-6} \text{ S}$$

$$A_v = -g_m R_D = -1000 \times 10^{-6} \text{ S} \times 10{,}000 \text{ } \Omega = -10$$

For the self-bias circuit we can use $V_{DD} = 25$ V because the gate will be at ground. If R_D is to be 10 kΩ again, the load line has a slope of $-1/(R_D + R_S)$.

If V_{GS} is again to be -4 V, the drain characteristics show that $I_{DS} \approx$ 1 mA. Therefore $R_S = 4$ V/1 mA $= 4$ kΩ. The resulting load line goes through the point $V_{DS} = 0$, $I_{DS} = 25$ V/14 k$\Omega = 1.79$ mA and is also shown in Figure 5-18.

The quiescent point for both load lines is about the same ($V_{GS} = -4$ V, $I_D = 1$ mA, $V_{DS} = 9$ V). Therefore g_m, r_D and the gains are the same.

Another drawback of a high R_D is that it creates a high output impedance and the circuit gain will drop if a capacitively coupled load is driven by the circuit, as shown in Figure 5-19.

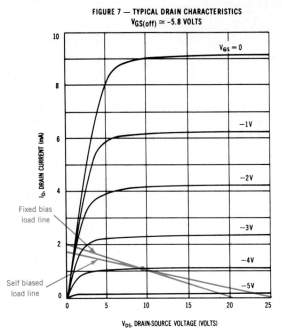

FIGURE 5-18
Load lines for Examples 5-13. (Copyright by Motorola, Inc. Used by permission.)

EXAMPLE 5-14

Repeat Example 5-13 if a 500 Ω capacitively coupled load resistor is added to the circuit, as shown in Figure 5-19.

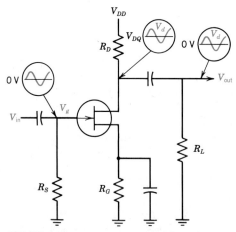

FIGURE 5-19
A JFET driving a comparatively coupled load.

212   FIELD EFFECT TRANSISTORS

Solution

With the same bias point, g_m will remain the same.

$$R'_L = R_L \| R_D = 500 \ \Omega \| 10{,}000 \ \Omega = 476 \ \Omega$$
$$A_v = -1000 \times 10^{-6} \ \text{S} \times 476 \ \Omega = -0.476$$

This gain is the same for both the fixed- and self-bias circuit because the values of R_D are the same.

Introducing the capacitively coupled load in Example 5-14 reduced the gain to less than 1. Once capacitively coupled loads are required, it is no longer true that the gain increases as R_D increases. In most cases it actually goes down as R_D increases.

A program to calculate the gain of a JFET when it must drive a capacitively coupled load is given in Figure 5-20a. The program requires the user to enter V_{DD}, I_{DSS}, V_P, and V_X. V_X is the voltage at the knee of the I_{DSS} curve. It is taken as 8 V for a **2N5459**. The program next asks the user to specify the quiescent point (V_{GS}, V_{DQ}) and R_L. Then the user can select fixed- or self-bias. The input format is shown in Figure 5-20b. This is the input format if the program is run on a VAX computer. It can also be run on an Apple or other

```
10 INPUT "Type VDD,VP,IDSS (in mA) and VX ";VD,VP,IR,VX
20 INPUT "Type the quiescent point,VDQ and VGS, and RL";VQ,VG,RL
30 INPUT "Fixed bias ? (Y/N)";A$
35 IS = IR/1000
40 RS = 0
50 ID = IS*(1-VG/VP)^2
60 RQ = (VD-VQ)/ID
70 IF A$ = "Y" THEN GO TO 90
80 RS = VG/ID
90 RD = RQ - RS
100 RP = RD*RL/(RD+RL)
110 VA = (IS-ID)*RP
120 VB = ID*RP
130 VC = (VQ - (VX*ID/IS))/(1 + VX/(IS*RP))
140 VE = VA
150 IF VB<VE THEN VE=VB
160 IF VC<VE THEN VE=VC
170 PRINT "RD IS   ";RD
180 PRINT "RS IS   ";RS
190 PRINT "THE MAXIMUM SWING IS ";VE
200 GM = 2*IS/VP*(1-VG/VP)
210 AV = GM*RP
220 PRINT "THE GAIN IS ";AV
230 END
```

(a) JFETQ Program

FIGURE 5-20
A program to calculate the gain of a JFET and its results.

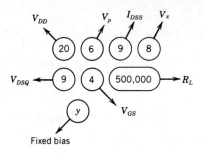

(b) Input format

```
20,6,9,8
9,4,500000
Y
```

Input file

```
RD IS    11000
RS IS    0
THE MAXIMUM SWING IS    7.49235
THE GAIN IS    10.7632
```

Output file

(c) Fixed bias input and results

```
25,6,9,8
9,4,500000
N
```

Input file

```
RD IS    12000
RS IS    4000
THE MAXIMUM SWING IS    7.53925
THE GAIN IS    11.7188
```

Output file

(d) Self biased input

```
20,6,9,8
9,4,500
Y
```

Input file

```
RD IS    11000
RS IS    0
THE MAXIMUM SWING IS    .478261
THE GAIN IS    .478261
```

Output file

(e) Fixed bias input with R_L = 500 ohms

FIGURE 5-20
(continued)

```
25,6,9,8
9,4,500
N
```

Input file

```
RD IS    12000
RS IS    4000
THE MAXIMUM SWING IS    .48
THE GAIN IS    .48
```

Output file

(f) Self biased input with R_L = 500 ohms

small microcomputer by typing in the input values. Note that the fixed-bias runs were made with $V_{DD} = 20$ V, to allow for a negative V_{GS}, while the self-bias runs were made with $V_{DD} = 25$ V.

The program gives output values for R_S, R_D, gain, and the maximum possible voltage swing for these values. Note that R_S is always 0 for a fixed-bias circuit.

Four computer runs were made for a **2N5459** operating at the quiescent point of Figure 5-18, which is essentially $V_{DSQ} = 9$ V, $V_{GS} = -4$ V. The first two runs used a very high value of R_L so that it was essentially not in the circuit. These are the values for Example 5-13. The results are shown in Figures 5-20c and 5-20d. Although the resistors came out to be 11 kΩ and 12 kΩ, because the drain characteristics do not fit Shockley's equation exactly, the gains were very close to those calculated in Example 5-13. Figures 5-20e and 5-20f show the input and outputs for the program run with R_L of 500 Ω with fixed- and self-bias. This circuit is the same as in Example 5-14 and the results are also very close.

[5]With a capacitively coupled load, the relationship between R_D, R_L, and g_m is complex and the optimum gain may no longer occur when R_D is very high. Indeed, it often occurs when R_D is quite low. A program to find the best value of R_D for a given V_{DD} and required output swing is given in Appendix C. It asks the user to specify the required output voltage swing, the load resistor, V_{DD}, and three parameters of the JFET, V_P, I_{DSS}, and V_X. It then asks the user to specify self- or fixed-bias by typing 1 or 0, respectively. It calculates the minimum value of V_{GS} consistent with the specified inputs and then finds the gains for various values of V_{GS} and R_D, starting at the highest possible value of V_{GS}.

EXAMPLE 5-15

Find the gain of a fixed-biased **2N5459** if the output swing must be $+2.5$ V, the capacitively coupled load resistor is 1000 Ω, and $V_{DD} = 20$ V.

Solution

The results of the program using these values is shown in problem 1 of Appendix C1. Using the given values of V_{DD}, R_L, and the swing, and putting in the parameters of the **2N5459**, $I_{DSS} = 9$ mA, $V_P = -6$ V, and $V_X = 8$ V, the program gives the best operating point at $I_D = 4.6225$ mA, $V_{GS} = -1.7$ V. At this point $R_D = 2197$ Ω and $A_v = 1.477$. The program shows that as R_D changes, the gain drops, so for these values this is the best point.

To check the gain

$$g_m = g_{mo}\left(1 - \frac{1.7}{6}\right) = 3000 \times 10^{-6}\,\text{S} \times 0.716 = 2150 \times 10^{-6}\,\text{S}$$

---

[5]The rest of this section may be omitted on first reading.

To find the gain R_P must first be found:

$$R_P = R_L \| R_D = \frac{1000\ \Omega \times 2197\ \Omega}{3197\ \Omega} = 687\ \Omega$$

$$A_v = g_m R_L = -2150 \times 10^{-6}\ S \times 687\ \Omega = -1.477$$

This checks the program exactly.

Two programs were run for a fixed-bias circuit. They are also shown in Appendix C1. Again, the user must specify the required voltage swing, R_L, and the transistor parameters. It gives the values of gain as the load resistor varies.

EXAMPLE 5-16

Find the best gain for a self-biased **2N5459** JFET if the swing is to be 2.5 V and $R_L = 1000\ \Omega$. Assume
a. $V_{DD} = 20$ V.
b. $V_{DD} = 25$ V.

Solution

Both of these problems were solved in Appendix C1 using the computer program of Appendix C. The input files, which contain the input numbers, and the resulting outputs are shown. Users that have BASIC, but cannot use input or output files, can simply type the numbers into the program.

Problem 2 in Appendix C1 uses a V_{DD} of 20 V. The problem is identical to Example 5-15, except that self-bias is used. The printout shows that the best gain is 1.37647 at a V_{GS} of -1.8 V. This is slightly less than the result for Example 5-15.

In Problem 3, V_{DD} was increased to 25 V. In self-bias circuits V_{DD} can equal the maximum rating of the JFET. Raising V_{DD} increased the maximum gain to 1.645. In general, raising V_{DD} allows the designer to increase the gain of the amplifier.

5-6.3 The Gain of a JFET with an Unbypassed Emitter Resistor

A JFET amplifier with an unbypassed source-resistor is shown in Figure 5-21a. The circuit is essentially the same as that of Figure 5-17, with the source bypass capacitor omitted. Omitting the capacitor does not affect the quiescent point or bias circuit because these depend only on dc values. The program of Figure 5-12 can still be used to calculate the bias point. As with the BJT, the

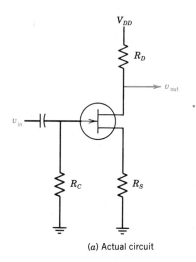

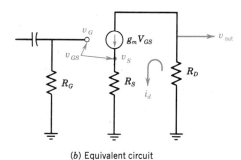

(b) Equivalent circuit

FIGURE 5-21
A JFET with an unbypassed service resistor.

gain of this circuit is reduced by the unbypassed source-resistor. Here, however, the gain reduction is not as severe. The gain can be calculated by using the equivalent circuit of Figure 5-21b. From it we see that

$$i_D = g_m v_{GS}$$
$$v_{GS} = v_G - i_D R_S = v_G - g_m V_{GS} R_S$$
$$v_G = v_{GS}(1 + g_m R_S)$$
$$v_{out} = -i_d R_D = -g_m v_{GS} R_D$$

and finally

$$\boxed{A_v = \frac{v_{out}}{v_G} = \frac{-g_m R_D}{1 + g_m R_S}} \quad (5\text{-}7)$$

EXAMPLE 5-17

Find the gain of Figure 5-21a if $R_D = 1500 \ \Omega$, $R_S = 500 \ \Omega$, and $g_m = 2000 \times 10^{-6}$ S.

Solution
Using equation 5-7 we have

$$A_v = \frac{-2000 \times 10^{-6} \text{ S} \times 1500 \ \Omega}{1 + 2000 \times 10^{-6} \text{ S} \times 500 \ \Omega} = -\frac{3}{2} = -1.5$$

The values given here are similar to the values in Example 5-12. Omitting the capacitor has reduced the gain of this circuit by one-half (from 3 to 1.5).

5-7 THE SOURCE-FOLLOWER

The source-follower is a JFET circuit where the output is taken at the source resistor and there is no drain resistor. It is shown in Figure 5-22 and is very similar to the emitter-follower circuit of a BJT. Its gain is less than 1, but it has

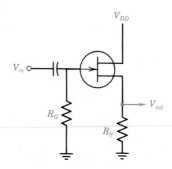

(a) Actual circuit

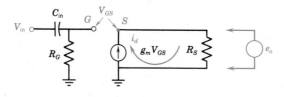

(b) Equivalent circuit

FIGURE 5-22
The source-follower.

a high input impedance and a low output impedance. Because of the JFET construction, the high input impedance is not needed, but the source follower is sometimes used because of its low output impedance.

5-7.1 The Gain of a Source-follower
The gain of a source-follower can be found from its equivalent circuit, Figure 5-22b, and is

$$A_v = \frac{g_m R_S}{1 + g_m R_S} \qquad (5\text{-}8)$$

Notice that it is positive (no phase inversion) and less than 1 (see problem 5-25).

The output impedance of the source follower can be found from Figure 5-22b by assuming $v_{in} = 0$ and applying the voltage e_o. As stated before, the output impedance of this circuit is the applied voltage divided by the current in the voltage generator with the input terminals short-circuited to ground. Under these conditions $V_G = 0$ and $V_S = e_o = -V_{GS}$. Thus the generator e_o causes the current e_o/R_S to flow in the resistor and the current $g_m e_o$ to flow in the current generator. The current flowing in response to e_o is

$$i_o = \frac{e_o}{R_S} + g_m e_o$$

$$Z_o = \frac{e_o}{i_o} = \frac{1}{\frac{1}{R_S} + g_m} \qquad (5\text{-}9)$$

To simplify equation 5-9, the output impedance of a common source circuit is the parallel combination of R_S and a resistor whose value is $1/g_m$.

5-7.2 Design of a Source-follower Circuit
Designing a source-follower generally involves only selecting R_S. The slope of the load line on the drain characteristics is $-1/R_S$. Thus R_S can be higher than it is for most common source amplifiers where the slope is $-1/(R_D + R_S)$. Choosing a very high value of R_S is not necessary, however, because the gain will always be less than 1.

EXAMPLE 5-18

Find the gain and output impedance of the source-follower of Figure 5-22 if the JFET is a **2N5459**, $V_{DD} = 20$ V, and $R_S = 2$ kΩ.

Solution

In order to find g_m, I_D must first be found. The JFET BIAS computer program (see Figure 5-12) can be used here because R_S is also the self-biasing resistor. It gives $I_D = 1.7$ mA. This value should be checked, if possible, from the curves; Figure 5-23 is a curve for this JFET with the proper load line. The quiescent point is shown at $V_{GS} = -3.4$ V, $I_D = 1.7$ mA. From equation 5-5 we have

$$g_m = g_{mo}\left(1 - \frac{V_{GS}}{V_P}\right)$$

where $g_{mo} = 3000 \times 10^{-6}$ (see Example 5-10).

$$g_m = 3000 \times 10^{-6}\left(1 - \frac{(-3.4 \text{ V})}{(-6.0 \text{ V})}\right) = 1300 \times 10^{-6} \text{ S}$$

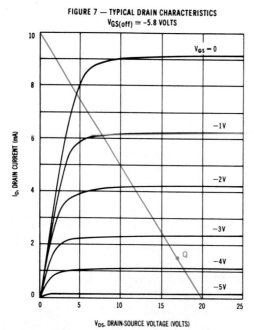

FIGURE 5-23
Load line and Q-point for a source-follower ($V_{DD} = 20$ V, $R_S = 2$ kΩ). (Copyright by Motorola, Inc. Used by permission.)

From equation 5-8

$$A_v = \frac{g_m R_S}{1 + g_m R_S} = \frac{1300 \times 10^{-6}\text{ S} \times 2000\text{ }\Omega}{1 + 1300 \times 10^{-6}\text{ S} \times 2000\text{ }\Omega} = \frac{2.6}{3.6} = 0.72$$

Notice the gain of a source-follower is also less than the gain of an emitter-follower.

To find the output impedance

$$Z_o = 2000\text{ }\Omega \parallel \frac{1}{g_m} = 2000\text{ }\Omega \parallel 769\text{ }\Omega = 555\text{ }\Omega$$

Because of its low output impedance, a source-follower can be used to improve the gain of a JFET circuit, as Example 5-19 demonstrates.

EXAMPLE 5-19

The circuit of Figure 5-24 is a two-stage JFET amplifier consisting of a common-source amplifier followed by a common-drain amplifier with a capacitively coupled 500 Ω load. Find the gain of this circuit.

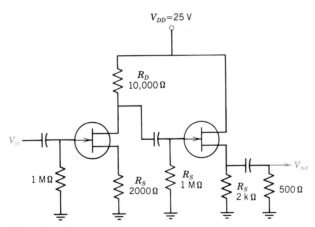

FIGURE 5-24
A two-stage amplifier using a common-source circuit.

Solution

The first stage is a common-source JFET, which is capacitively coupled to essentially an infinite load, so the effects of the load can be ignored. The quiescent drain current and g_m can be calculated as in Example 5-18.

For the common source stage

$$A_v = -g_m R_D = -1300 \times 10^{-6}\text{ S} \times 10{,}000\text{ }\Omega = -13$$

For the common drain circuit the 500 Ω, load resistor does not affect the bias or g_m, but it does affect the ac gain.

$$A_{v(\text{stage 2})} = \frac{g_m R'_S}{1 + g_m R'_S}$$

Here the effective value of R'_S is $R_S \parallel R_L$

$$R'_S = 2000\text{ }\Omega \parallel 500\text{ }\Omega = 400\text{ }\Omega$$

$$A_{v(\text{stage 2})} = \frac{1300 \times 10^{-6}\text{ S} \times 400\text{ }\Omega}{1 + 1300 \times 10^{-6}\text{ S} \times 400\text{ }\Omega} = \frac{0.52}{1.52} = 0.34$$

The gain of this circuit is

$$A_v = A_{v1} \times A_{v2} = -13 \times 0.34 = -4.42$$

Example 5-14 showed that if a common-source amplifier is directly coupled to a 500 Ω load, the gain is only -0.476. With the common drain amplifier added, the gain goes up to -4.42 because of the impedance transformation. The presence of the source-follower increased the gain by a factor of 9.

5-8 MOSFETS

The term MOSFET is a mnemonic for metal oxide semiconductor field effect transistor. A MOSFET is a field effect transistor with different construction and somewhat different characteristics than a JFET. MOSFETs find some use as amplifiers, but are primarily used in digital or switching circuits.

Figure 5-25 shows the construction and symbol for an *n*-channel MOSFET. Its construction starts with a lightly doped *p*-type substrate. Two *n*-doped regions are deposited in it and electrically connected to the source and drain. The current flows in the *n*-channel between the source and the drain and is controlled by the gate. One of the major differences between a JFET and a MOSFET is that the MOSFET gate is separated from the substrate by a very thin layer (10^{-4} in. or less) of *silicon dioxide* (SiO_2, the oxide in MOS) that *insulates* the gate from the substrate. Thus the high impedance of a MOSFET comes from the oxide insulator, instead of from a reverse-biased *p-n* junction, as in a JFET. This also means the gate can be made positive with respect to the source.

The symbol for an *n*-channel MOSFET is also shown in Figure 5-25.

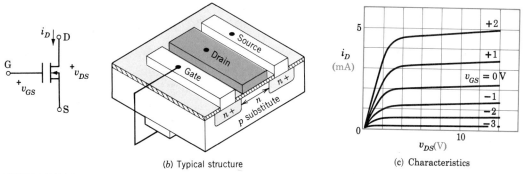

(b) Typical structure (c) Characteristics

FIGURE 5-25
An *n*-channel depletion- or enhancement-mode MOSFET. (*a,c*) (From *Electronics Circuits & Devices*, Third Edition, by Ralph J. Smith. Copyright © 1987 by John Wiley & Sons, Inc. Reprinted by permission of John Wiley & Sons, Inc.) (*b*) (From *Electronics Circuits & Devices* by Ralph J. Smith. Copyright © 1980 by John Wiley & Sons, Inc. Reprinted by permission of John Wiley & Sons, Inc.)

There are four connections in the symbol for the gate, drain, source, and substrate. Typically, however, the substrate is connected to the source so a MOSFET has only three terminals.

A *p*-channel MOSFET is just the reverse of the *n*-channel. It has an *n*-substrate, and the source and drain are connected to *p*-doped silicon. The symbol is the same, except that the substrate arrow is reversed and conventional current flows from source to drain. Historically, the first MOSFETs were *p*-channel, but because electrons have a greater mobility than holes, almost all modern MOSFETs are *n*-channel. The *p*-Channel MOSFET is rarely used, except in CMOS (complementary MOS).

5-8.1 Enhancement and Depletion Mode MOSFETs

MOSFETs are constructed in either enhancement or depletion configurations. Figure 5-26 shows diagrams of both types. These diagrams are not to scale; in particular the SiO$_2$ oxide insulator is very thin.

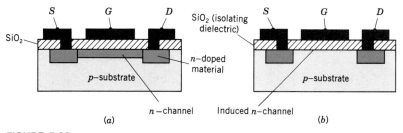

FIGURE 5-26
MOSFET construction: (*a*) depletion; (*b*) enhancement. (Robert Boylestad/Louis Nashelsky, *Electronic Devices & Circuit Theory*, 3rd ed., © 1982. Reprinted by permission of Prentice-Hall, Inc., Englewood Cliffs, New Jersey.

The difference between the MOSFET types can be seen by examining the channel beneath the gate. The depletion MOSFET has a built-in n-channel between the source and drain, whereas the enhancement most MOSFET does not. If the gate were left open, for example, and a voltage were applied between drain and source, current would flow through the channel in the depletion MOSFET. For an enhancement MOSFET there are two p-n junctions between drain and source and one of them is reverse-biased. Therefore no current will flow in an enhancement type MOSFET with an open gate.

5-8.2 Operation of a MOSFET

The operation of a MOSFET is controlled by the voltage on the gate. In a depletion type MOSFET if the gate voltage is positive with respect to the source and substrate, this positive gate voltage *attracts* electrons from the substrate into the channel. Consequently, the carrier density in the channel increases and the current rises. Conversely, a negative gate voltage repels electrons from the channel and the current falls.

For enhancement MOSFETs there is no initial channel. As the gate becomes more positive, however, it attracts the free electrons in the substrate until the concentration of electrons below the gate is so large that an effective n-channel is formed between the source and drain. Thus the gate must be enhanced (made positive) before current can flow.

As a general rule, depletion MOSFETs are used in linear amplifiers. Enhancement MOSFETs are widely used in digital circuits, especially as part of CMOS.

5-8.3 MOSFET Specifications and Curves

Figure 5-27 gives the specifications for two n-channel depletion-type MOSFETs, the **2N3796** and **2N3797**. The drain characteristics and transfer characteristics are given in Figure 5-28. The characteristics do not extend beyond the maximum values of V_{DS}, V_{GS}, and I_D given in the specifications.

These specifications and curves can be compared with the JFET specifications and curves. The following general observations result:

1. The g_{mo} of a MOSFET is somewhat less than the JFET (especially for the **2N3796**). Of course, the MOSFETs can be biased with V_{GS} positive, which would increase their g_m's. To operate at this point, however, means R_D must be small, reducing the gain.
2. The MOSFET drain resistance, r_d, is somewhat smaller than the JFET's. This can be seen from the more steeply sloped drain characteristic curves, and tends to reduce the gain.

For these reasons, MOSFETs are not as popular as JFETs for amplifying transistors. They are very widely used, however, in digital integrated circuits, because they can be fabricated on a very small silicon area.

2N3796 / 2N3797

CASE 22-03, STYLE 2
TO-18 (TO-206AA)

MOSFET
LOW-POWER AUDIO

N-CHANNEL — DEPLETION

MAXIMUM RATINGS

Rating	Symbol	Value	Unit
Drain-Source Voltage 2N3796 2N3797	V_{DS}	25 20	Vdc
Gate-Source Voltage	V_{GS}	±10	Vdc
Drain Current	I_D	20	mAdc
Total Device Dissipation @ T_A = 25°C Derate above 25°C	P_D	200 1.14	mW mW/°C
Junction Temperature Range	T_J	+175	°C
Storage Channel Temperature Range	T_{stg}	−65 to +175	°C

ELECTRICAL CHARACTERISTICS (T_A = 25°C unless otherwise noted.)

Characteristic		Symbol	Min	Typ	Max	Unit
OFF CHARACTERISTICS						
Drain-Source Breakdown Voltage (V_{GS} = −4.0 V, I_D = 5.0 μA) (V_{GS} = −7.0 V, I_D = 5.0 μA)	2N3796 2N3797	$V_{(BR)DSX}$	25 20	30 25	— —	Vdc
Gate Reverse Current(1) (V_{GS} = −10 V, V_{DS} = 0) (V_{GS} = −10 V, V_{DS} = 0, T_A = 150°C)		I_{GSS}	— —	— —	1.0 200	pAdc
Gate Source Cutoff Voltage (I_D = 0.5 μA, V_{DS} = 10 V) (I_D = 2.0 μA, V_{DS} = 10 V)	2N3796 2N3797	$V_{GS(off)}$	— —	−3.0 −5.0	−4.0 −7.0	Vdc
Drain-Gate Reverse Current(1) (V_{DG} = 10 V, I_S = 0)		I_{DGO}	—	—	1.0	pAdc
ON CHARACTERISTICS						
Zero-Gate-Voltage Drain Current (V_{DS} = 10 V, V_{GS} = 0)	2N3796 2N3797	I_{DSS}	0.5 2.0	1.5 2.9	3.0 6.0	mAdc
On-State Drain Current (V_{DS} = 10 V, V_{GS} = +3.5 V)	2N3796 2N3797	$I_{D(on)}$	7.0 9.0	8.3 14	14 18	mAdc
SMALL-SIGNAL CHARACTERISTICS						
Forward Transfer Admittance (V_{DS} = 10 V, V_{GS} = 0, f = 1.0 kHz)	2N3796 2N3797	$|y_{fs}|$	900 1500	1200 2300	1800 3000	μmhos
(V_{DS} = 10 V, V_{GS} = 0, f = 1.0 MHz)	2N3796 2N3797		900 1500	— —	— —	
Output Admittance (V_{DS} = 10 V, V_{GS} = 0, f = 1.0 kHz)	2N3796 2N3797	$|y_{os}|$	— —	12 27	25 60	μmhos
Input Capacitance (V_{DS} = 10 V, V_{GS} = 0, f = 1.0 MHz)	2N3796 2N3797	C_{iss}	— —	5.0 6.0	7.0 8.0	pF
Reverse Transfer Capacitance (V_{DS} = 10 V, V_{GS} = 0, f = 1.0 MHz)		C_{rss}	—	0.5	0.8	pF
FUNCTIONAL CHARACTERISTICS						
Noise Figure (V_{DS} = 10 V, V_{GS} = 0, f = 1.0 kHz, R_S = 3 megohms)		NF	—	3.8	—	dB

(1) This value of current includes both the FET leakage current as well as the leakage current associated with the test socket and fixture when measured under best attainable conditions.

MOTOROLA SEMICONDUCTORS   SMALL-SIGNAL DEVICES

FIGURE 5-27
Specifications for the *n*-channel depletion mode MOSFETS. (Copyright by Motorola, Inc. Used by permission.)

TYPICAL DRAIN CHARACTERISTICS

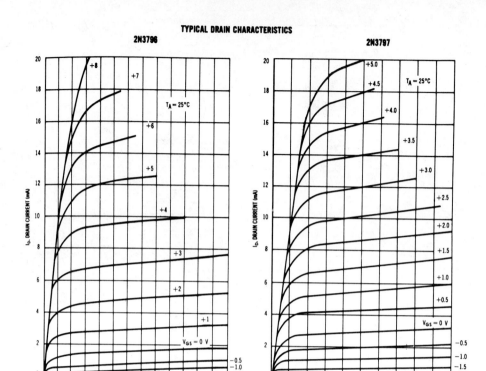

COMMON SOURCE TRANSFER CHARACTERISTICS

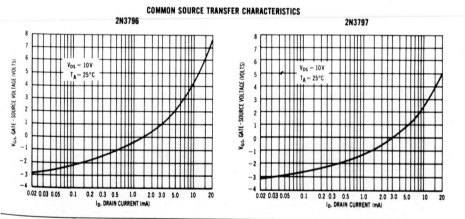

SMALL-SIGNAL DEVICES   MOTOROLA SEMICONDUCTORS

FIGURE 5-28
Drain and transfer characteristics for the **2N3796** and **2N3797** MOSFETS. (Copyright by Motorola, Inc. Used by permission.)

5-8.4 Biasing the MOSFET

Like JFETs, biasing circuits for MOSFETs are simple to design because of their high input impedance and very low gate current. Bias circuits are generally designed to use a single power supply. If V_{GS} is negative, a source resistance can be added, just as in a JFET. If V_{GS} is to be positive, a simple voltage divider circuit can be used to create the proper bias.

EXAMPLE 5-20

A **2N3796** is to be biased with $V_{DSQ} = 10$ V and $V_{DD} = 20$ V.
a. Find the bias circuit and R_D if V_{GS} is to be -1 V.
b. Repeat if V_{GS} is to be $+1$ V.

Solution

a. The load lines are plotted in Figure 5-29c. The $V_{GS} = -1$ V curve intersects the $V_{DS} = 10$ V line at point Q, where $I_D = 1$ mA. Because V_{GS} is negative, a source resistance is needed and $R_S = 1$ V/1 mA $= 1$ kΩ. To find R_D we can see that the load line intersects the I_D axis at 2.5 mA. Therefore

$$\frac{1}{R_D + R_S} = \frac{2.5 \text{ mA}}{20 \text{ V}}$$

$$R_D + R_S = 8 \text{ k}\Omega$$

$$R_D = 7 \text{ k}\Omega$$

The circuit is shown in Figure 5-29a. The dc voltages are shown in blue. As a check, the JFET BIAS program of Figure 5-12 was run with $R_S = 1$ kΩ, $I_{DSS} = 3$ mA, and $V_P = -3$ V. I_{DSS} was estimated from the $V_{GS} = 0$ curve and V_P was estimated from the lowest point on the transfer characteristics. The results of the computer run were $I_D = 1.145$ mA, reasonably close to the 1 mA obtained from the curve.

b. For $V_{GS} = +1$ V, the quiescent point is at Q2 on Figure 5-29c and the bias circuit can be a voltage divider circuit, as shown in Figure 5-29b. High resistors will not load down the driving circuit and a 20:1 voltage reduction is needed, so R_1 was set at 100 kΩ and R_2 at 1.9 MΩ.

To find R_D, note that the load line hits the I_D axis at 10.7 mA. Here $R_S = 0$ so the slope is simply $-1/R_D$.

$$R_D = \frac{20 \text{ V}}{10.7 \text{ mA}} = 1870 \text{ }\Omega$$

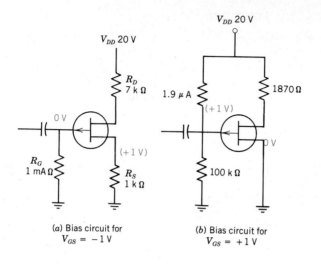
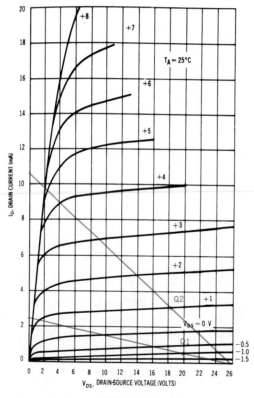

FIGURE 5-29
Bias circuits for a **2N3796** MOSFET. (Copyright by Motorola, Inc. Used by permission.)

228 FIELD EFFECT TRANSISTORS

5-8.5 AC Gain of a MOSFET

The gain of a MOSFET can be calculated using the same formulas as for a JFET. The gain is

$$\boxed{A_v = -g_m R_D} \quad (5\text{-}10)$$

As with the JFET

$$g_m = g_{mo}\left(1 - \frac{V_{GS}}{V_P}\right)$$

where

$$g_{mo} = \frac{-2I_{DSS}}{V_P}$$

For the **2N3797**, $I_{DSS} \approx 3$ mA and $V_P \approx -3$ V. Therefore

$$g_{mo} = \frac{-2I_{DSS}}{V_P} = 2000 \times 10^{-6}\ \text{S}$$

(The specification sheet value for the g_{mo} of a **2N3797** is 2300×10^{-6} S.)

EXAMPLE 5-21

Find the ac gain for the circuits of Figure 5-29.

Solution

a. For the circuit of Figure 5-29a

$$g_m = g_{mo}\left(1 - \frac{V_{GS}}{V_P}\right) = 2000 \times 10^{-6}\ \text{S} \times \left(1 - \frac{(-1)}{(-3)}\right) = 1333 \times 10^{-6}\ \text{S}$$

$$A_v = -g_m R_D = -1333 \times 10^{-6}\ \text{S} \times 7000\ \Omega = -9.3$$

b. For the circuit of Figure 5-29b that is biased at $V_{GS} = +1$ V

$$g_m = 2000 \times 10^{-6}\ \text{S} \times \left(1 - \frac{(+1)}{-3}\right) = 2666 \times 10^{-6}\ \text{S}$$

$$A_v = -2666 \times 10^{-6} \times 1870\ \Omega = -5.0$$

With the MOSFET biased positively, the g_m was higher than g_{mo}, but the higher value of R_D produced a greater gain despite the lower g_m. Again this is not necessarily true if a capacitively coupled load is added.

5-8.6 Enhancement Mode MOSFETs

The curves of a **2N4351** enhancement-type MOSFET are shown in Figure 5-30. No current flows when $V_{GS} = 0$, and the curves show that the gate must be enhanced (made positive) for any significant current. For the **2N4351**, no appreciable current flows until $V_{GS} = +4$ V.

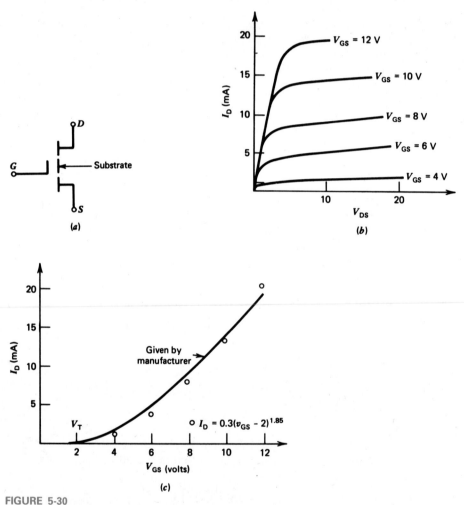

FIGURE 5-30
Characteristics of enhancement MOSFET (**2N4351**). (a) Symbol. (b) Drain characteristics (typical). (c) Transfer characteristics. (From *Semiconductor Devices and Circuits* by Henry Zanger. Copyright © 1984 by John Wiley & Sons, Inc. Reprinted by permission of John Wiley & Sons, Inc.)

This circuit can be easily biased by the voltage-divider method of Figure 5-29b. Its g_m and gains can be found from the curves or the manufacturer's specifications. Enhancement-mode MOSFETs are seldom used as amplifiers. They are used in digital circuits and in VMOS.

5-9 VMOS

Because of the narrow channel, directly under the gate, the power dissipation of a **2N3796** or **2N3797** MOSFET is severely limited. $V_{DD(max)}$ is 25 V for a **2N3796** and 20 V for a **2N3797** and $I_{D(max)} = 20$ mA. VMOS or "vertical" MOS allows much more current and power to be put through a MOSFET.

The cross section of a VMOS-FET is shown in Figure 5-31. The source is a ring of n-doped silicon and the p-area of the substrate surrounds the gate that penetrates it. Current flows vertically from drain to source under control of the gate.

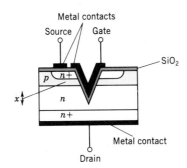

FIGURE 5-31
VMOS construction. (Robert Boylestad/Louis Nashelsky, *Electronic Devices & Circuit Theory*, 3rd ed., © 1982. Reprinted by permission of Prentice-Hall, Inc., Englewood Cliffs, New Jersey.)

VMOS construction allows much higher currents to flow through the FET. Figure 5-32 shows the common drain characteristics for a Motorola-type

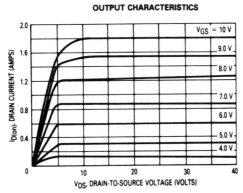

FIGURE 5-32
Output characteristics for a **2N6660** VMOS-FET. (Copyright by Motorola, Inc. Used by permission.)

2N6660 VMOS-FET. It is an *n*-channel enhancement-type VMOS. Several characteristics can be observed from these curves:

1. The current is in amperes instead of mA.
2. V_{DS} can go to 40 V instead of only 20 or 25 V.
3. The g_m of this FET is much greater than that of the conventional FET.

These features allow the VMOS-FET to absorb more current and power. They are not used for high gain, despite their high g_m, because reasonable values of resistance for Figure 5-32 are quite small, so the $g_m R_D$ product is also small.

5-10 SUMMARY

This chapter considered field-effect transistors, both JFETs and MOSFETs. The JFET was covered first, followed by the MOSFET. Their construction, biasing, and gain were examined. The common drain circuit for the JFET was also developed, and examples of its use were given.

The MOSFET section included discussions of *p*-channel and *n*-channel MOSFETs, enhancement and depletion modes of operation, and VMOS.

5-11 GLOSSARY

Channel. A path of doped silicon between the source and drain.

Common-drain amplifier. A JFET with the drain directly connected to V_{DD}. The output voltage is developed across R_S.

Common-source amplifier. A JFET with the ground connected to the source, possibly through a bypass capacitor.

Drain. The electrode where majority carriers leave the channel.

FET. Field effect transistor.

g_m. The symbol for transconductance.

g_{mo}. The transconductance when $V_{GS} = 0$.

Gate. The controlling electrode in an FET.

I_{DSS}. The current at pinch-off when $V_{GS} = 0$.

JFET. Junction FET.

MOSFET. Metal oxide semiconductor FET.

Pinch-off. The gate-to-source voltage where no current flows in an FET.

Silicon dioxide (SiO_2). The insulator used in a MOSFET.

Source. The electrode where majority carriers enter the channel.

Source-follower. A synonym for a common-drain amplifier.

Transconductance. A conductance that relates the input voltage to the output current.

V_{GS}. The gate to source voltage in an FET.

VMOS. Vertical MOS, a type of MOSFET used for high-power applications.

5-12 REFERENCES

Robert Boylestad and Louis Nashelsky, *Electronic Devices and Circuit Theory*, 4th Edition, Prentice-Hall, Englewood Cliffs, N.J., 1987.

Jacob Millman, *Microelectronics*, McGraw-Hill, New York, 1979.

Motorola Small Signal Transistor Data Handbook, Motorola, Inc., Phoenix, Arizona, 1983.

J.F. Pierce and T.J. Paulus, *Applied Electronics*, Charles E. Merrill, Columbus, Ohio, 1972.

Donald L. Schilling and Charles Belove, *Electronic Circuits, Discrete and Integrated*, 2nd Edition, McGraw-Hill, New York, 1979.

Henry Zanger, *Semiconductor Devices and Circuits*, John Wiley, New York, 1984.

5-13 PROBLEMS

5-1 From the curves of Figure P5-1, find pinch-off.

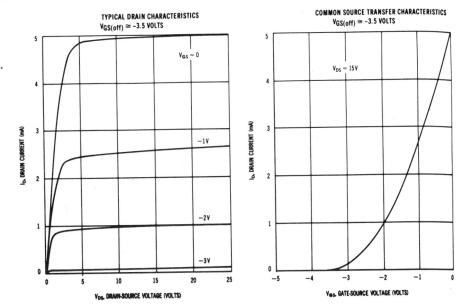

FIGURE P5-1
(Copyright by Motorola, Inc. Used by permission.)

    a. Use the $V_{GS} = 0$ curve.
    b. Use the drain characteristics.
    c. Use the transfer characteristics.

5-2 For a **2N5457** what is the reverse gate current at 25°C? At 100°C? How does this compare with the formulas developed in Chapter 1 for a reverse-biased diode?

5-3 From Figure 5-5a, plot the common-source transfer characteristic at $V_{DD} = 20$ V. How much does it differ from Figure 5-5b?

5-4 From the drain characteristics of Figure P5-1, plot the common source transfer characteristic at $V_{DD} = 20$ V. Also find I_{DSS}.

5-5 Use Shockley's equation on the JFET of Figure P5-1 to find I_D when $V_{GS} = -1$ V, -2 V, and -3 V.

5-6 Write a computer program for Shockley's equation. Use it to check the results of problem 5-5.

5-7 Using Table 5-1, and $V_{GS} = 0$, calculate I_D at $V_{GS} = -2$ V if V_P is assumed to be -3.8 V. How does this compare with the measured results?

5-8 For the JFET of Figure P5-1, design a fixed-bias circuit so that $V_{GS} = -1$ V. If $V_{DD} = 20$ V and $R_D = 4$ kΩ, find I_D and V_{DS}.

5-9 For the JFET of Figure P5-1, we have $V_{DD} = 25$ V, $R_D = 6.25$ kΩ. It must be biased so that $V_{DS} = 15$ V.
   a. Design a self-biased circuit to do this.
   b. Design a fixed-bias circuit to do this.

5-10 Repeat Example 5-6 if R_S is changed to 500 Ω.
   a. Use the transfer characteristics.
   b. Use the JFETBIAS computer program.

5-11 For the JFET of Figure P5-1, find the quiescent point if $V_{DD} = 20$ V, $R_D = 5000$ Ω, and $R_S = 1000$ Ω.
   a. Use the transfer characteristics.
   b. Use the JFETBIAS computer program.

5-12 For the JFET of Figure P5-1, use the load line that represents the conditions $V_{DD} = 24$ V, $R_D = 6$ kΩ.
   a. If fixed bias is used and V_{GS} is to be -1 V, find the gain if the swing is 1 V.
   b. Repeat for self-bias. Calculate R_S. The total resistance ($R_D + R_S$) is still to be 6 kΩ.

5-13 Given the n-channel JFET whose characteristics are $I_{DSS} = 6$ mA, $V_P = -4.5$ V. With $V_{DD} = 25$ V, it must be biased so that $V_{GSQ} = -1$, $V_{DSQ} = 10$ V. Find
   a. R_S.
   b. R_D.
   c. g_m.
   d. The gain of the stage.

5-14 For the JFET whose curves are given in Figure P5-1, find
   a. g_{mo}.
   b. g_m when $V_{GS} = -1$ V.
   c. g_m when $V_{GS} = -2$ V.

5-15 The circuit of Figure P5-15 uses the JFET whose characteristics are given in Figure P5-1. Find the gain of the circuit.

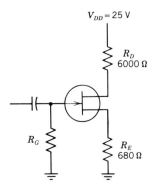

FIGURE P5-15

5-16 If the JFET of Figure P5-1 is biased at $V_{GS} = -2$ V, $V_{DS} = 10$ V, find its gain for both self- and fixed-bias. Use $V_{DD} = 22$ V for fixed-bias, and 25 V for self-bias.
   a. Use the curves.
   b. Use the JFETQ computer program in Figure 5-20.

5-17 For the JFET of Figure P5-1, find the best gain of the circuit if the output is capacitively coupled to a 1000 Ω load. Use $V_{DD} = 20$ V, $V_{DQ} = 10$ V, and $V_{GS} = -1$ V. Use the JFETQ program. Do this problem for both fixed- and self-bias.

5-18 For the JFET of Figure P5-1, find the best gain of the circuit if the output swing is to be 1 V, $V_{DD} = 20$ V, and the capacitively coupled load is 1000 Ω. Use the RSJFET program of Appendix C1. Do this problem for both self- and fixed-bias, and determine the optimum drain resistor for each case.

5-19 Find the best gain of the JFET of Figure P5-1 for
   a. Fixed-bias, $V_{DD} = 20$ V, $V_{swing} = 2$ V.
   b. Self-bias, $V_{DD} = 25$ V, $V_{swing} = 1$ V.
   Assume the JFET has the curves given in Figure P5-1.

5-20 Repeat problem 5-19 if a 1000 Ω, capacitively coupled resistor is added to the circuit.

5-21 Find the gain of the circuit of Figure 5-19 if $R_D = 2000$ Ω, $R_S = 400$ Ω, $R_L = 1000$ Ω, and $g_m = 4000 \times 10^{-6}$.

5-22 Repeat problem 5-21 if the source bypass capacitor is omitted.

5-23 The gain on the computer run of Figure 5-20d is 17% higher than predicted by Example 5-13. Explain the discrepancy by examining the output file.

5-24 For problem 5-12b, what would the gain be if the bypass capacitor were omitted?

5-25 Using the circuit of Figure 5-22b, derive equation 5-8.

5-26 Find the gain and output impedance of a source-follower using the JFET of Figure P5-1, if $V_{DD} = 20$ V, $R_S = 1$ kΩ.

5-27 For a **2N3797**, what is the maximum value of R_D if it is to be biased at $+1$ V, with $V_{DS} = +5$ V?

5-28 For Figure 5-29, find g_m from the curves about points Q_1 and Q_2. How do they compare with the values found in Example 5-21?

5-29 Find the gain of the circuits of Figure 5-29a and 5-29b by taking a ± 1 V swing around the quiescent points on their load lines, as shown in Figure 5-29c. How do they compare with the calculated results in Example 5-21?

5-30 A **2N3796** is designed so that $V_{DS} = 12$ V. Find I_D and g_m if $V_{GS} =$
   a. 0 V.
   b. -0.5 V.
   c. 1 V.
   d. 2 V.

5-31 For each point in problem 5-28 find the gain of the MOSFET. $V_{DD} = 22$ V and fixed-bias is used.

5-32 For the VMOS transistor of Figure 5-32, design a circuit to bias it at $V_D = 20$ V, $I_D = 0.8$ A, if V_{DD} is 40 V. Find the g_m and the gain of the circuit.

After doing the problems, go back to section 5-2 and review the self-evaluation questions. If any still seems difficult, review the appropriate sections to find the answers.

CHAPTER SIX

Multiple Transistor Circuits

6-1 Instructional Objectives

This chapter discusses several advanced transistor circuits. Most of these circuits use more than one transistor and are more complex than the single transistor circuits covered in the preceding chapters.

After reading the chapter, the student should be able to:

1. Find the gain of multiple-stage RC coupled circuits.
2. Find the gains of circuits containing emitter-followers and transistors with unbypassed emitter resistors.
3. Find the gain of multiple-stage FET amplifiers.
4. Analyze and design Darlington amplifiers.
5. Analyze and design direct-coupled amplifiers.
6. Find the operating range for a constant-current source.
7. Analyze and design comparator circuits.
8. Find the common-mode gain, difference-mode gain, and CMRR of a difference amplifier.
9. Convert gains and CMRRs from direct ratios to dBs and vice versa.
10. Analyze and design level-shifting cascode amplifiers.

6-2 Self-Evaluation Questions

Watch for the answers to the following questions as you read the chapter. They should help you to understand the material presented.

1. Can the gain of a multiple-stage transistor circuit be found by multiplying the individual stage gains together? Explain why or why not.
2. Why are capacitors necessary between stages of an amplifier?
3. Why is it more accurate to use approximate formulas to find circuit gains when the stages of the circuit are separated by an emitter-follower or have an unbypassed emitter-resistor?
4. Why are multistage JFET amplifiers easier to analyze than BJT amplifiers?
5. What is the advantage of direct-coupled amplifiers? What is their disadvantage?
6. Why can't the transistor in a constant current source saturate?
7. For a difference amplifier, why is a high CMRR desirable?
8. What are the advantages and disadvantages of balancing resistors?

6-3 INTRODUCTION

This chapter considers circuits that use more than one transistor. The use of several transistors provides greater gain and versatility.

The simplest concept of a multiple transistor circuit is shown in Figure 6-1, where each transistor and its accompanying resistors and capacitors are called a *stage*. The figure shows a three-stage amplifier connected in cascade. It is a good model for the RC coupled amplifiers discussed in section 6-4. It is not a good model for some of the circuits to be discussed later in this chapter, such as the difference amplifier, where the transistors are interconnected in a way that makes separating them into individual stages very difficult.

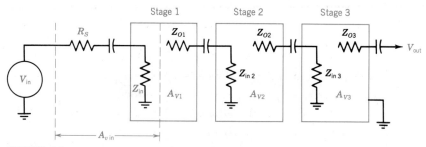

FIGURE 6-1
A three-stage amplifier.

If the gain of each stage of Figure 6-1 is A_{v1}, A_{v2}, and A_{v3}, then the gain of the total circuit will be their product.

$$A_v = A_{in} \times A_{v1} \times A_{v2} \times A_{v3} \tag{6-1}$$

where A_{in} is the reduction of the input voltage because it divides between the source resistance, R_s, and the input impedance of the first stage, Z_{in}.

$$A_{in} = \frac{Z_{in}}{Z_{in} + R_s}$$

If $Z_{in} \gg R_s$, then $A_{in} \approx 1$.

The input and output impedance of each stage may affect the gain of the adjacent stages. If the gains of the individual stages are measured or calculated independently and the stages are then connected together, we cannot expect equation 6-1 to be valid because of the interstage interaction. Or, to put it simply, each stage may *load down* the previous stage and affect its gain. Another problem to be considered is the possibility that one stage will *affect the bias* of an adjacent stage.

Although these facts may make it more difficult to analyze multiple-stage circuits, their operation can be determined by applying the principles studied in the preceding chapters.

6-3.1 Gain Calculations Using Input and Output Impedances

The effect of coupling between stages can be calculated if the input and output impedances of each stage are known. As shown in Figure 6-1, the output impedance of a stage can be considered as being connected to the input impedance of the next stage, so the unloaded gain of a stage is reduced by the factor

$$\frac{Z_{in2}}{Z_{out1} + Z_{in2}}$$

where Z_{out1} is the output impedance of the first stage and Z_{in2} is the input impedance of the next stage. If a stage has an inherently high input impedance, such as an FET or emitter-follower, Z_{in2} is often much larger than Z_{out1} and the reduction factor will be close to 1, so it can be ignored. For the RC coupled amplifiers of Section 6-4, Z_{out1} is generally larger than Z_{in} and the gain reduction is significant.

EXAMPLE 6-1

Find the gain of the three-stage amplifier of Figure 6-1 if R_s = 1 kΩ and for each stage A_v is 100, Z_{in} is 1 kΩ and Z_{out} is 4 kΩ.

Solution

The gain of each stage can be found and the entire gain calculated.

$$A_{in} = \frac{Z_{in}}{R_s + Z_{in}} = \frac{1 \text{ k}\Omega}{2 \text{ k}\Omega} = 0.5$$

After going through the first stage gain of 100, the gain is 50. The interstage coupling is 1 kΩ (Z_{in2}) in series with 4 kΩ (Z_{out1}) and reduces the gain to 10. After going through A_{v2} the gain is 100 × 10 = 1000. The interstage coupling between stages 2 and 3 reduces this gain to 200. Finally, A_{v3} is 100 so the overall gain is 20,000. Note that this is the gain with the output unloaded. Any load on stage 3 would be in series with its 4 kΩ output impedance and further reduce the gain. From this example, we can also see that a low output impedance would increase the gain of this circuit by increasing its coupling factors (see problem 6-1).

6-4 RC COUPLED CIRCUITS

Figure 6-1 is the ideal model for circuits cascaded together to increase their gain. If each stage of Figure 6-1 represents a transistor amplifier, the output of one stage is usually connected to the input of the next stage by a dc blocking capacitor that prevents the dc level of the driving stage from affecting the dc bias level of the succeeding stage. We have already seen one example of RC coupling in Figure 5-24, where a common-source JFET drove a common-drain JFET to improve the gain of a circuit driving a 500 Ω load resistor.

6-4.1 Cascaded Common-emitter Amplifiers

Bipolar junction transistor (BJT) amplifiers were discussed in Chapter 4. Two stages of common-emitter amplifiers can be connected together to produce a very high gain circuit.

Figure 6-2 is a redrawing of a single-stage, common-emitter amplifier, as

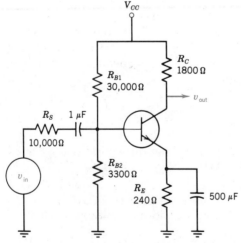

FIGURE 6-2
A single-stage common-emitter amplifier.

240 MULTIPLE TRANSISTOR CIRCUITS

described in Example 4-6 and section 4-5.3. Assuming $h_{fe} = 150$, the amplifier was found to have an h_{ie} of 900 Ω and the transistor gain was

$$A_{v(tr)} = \frac{-h_{fe}R_C}{h_{ie}} = \frac{-150 \times 1800\ \Omega}{900\ \Omega} = -300$$

Two of these stages can be cascaded for high gain as Example 6-2 shows.

EXAMPLE 6-2

If two identical stages of the amplifier of Figure 6-2 are connected together, as shown in Figure 6-3, find the overall gain of the circuit.

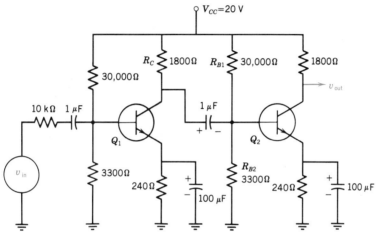

FIGURE 6-3
A two-stage common-emitter amplifier.

Solution

It is tempting to say that the gain of each transistor circuit is −300, and the overall transistor gain is $(-300)^2$ or 90,000, but this is wrong, because the second stage affects the first stage. Actually, the overall gain is the product of three gains, A_{in}, A_{v1}, and A_{v2}. A_{v2} is the gain of the second stage. If it is unloaded, A_{v2} is −300 as before.

A_{v1} is the gain of the first stage. Here, however, R_C is capacitively connected to the bias resistors and h_{ie} of the second stage. This gives stage 1 an effective load resistance of R'_L where

$$R'_L = R_C \parallel R_{B1} \parallel R_{B2} \parallel h_{ie2} = 1800\ \Omega \parallel 30{,}000\ \Omega \parallel 3300\ \Omega \parallel 900\ \Omega$$

$$R'_L = 500\ \Omega$$

$$A_{v1} = \frac{-h_{fe}R'_L}{h_{ie}} = \frac{-150 \times 500\ \Omega}{900\ \Omega} = -83.3$$

Finally, A_{in} is the portion of the input voltage that reaches the base of the first transistor. The input circuit is a voltage divider of 900 Ω ∥ 3300 Ω ∥ 30,000 Ω = 690 Ω in series with a 10 kΩ resistor. Therefore

$$A_{in} = \frac{690 \ \Omega}{10{,}690 \ \Omega} = 0.0645$$

$$A_v = A_{in} \times A_{v1} \times A_{v2} = 0.0645 \times -83.3 \times -300 = 1612$$

EXAMPLE 6-3

For the circuit of Figure 6-3, find the output impedance of stage 1 and the input impedance of stage 2. Use these values to check the results of Example 6-2.

Solution

The output impedance of stage 1 (with h_{oe} ignored) is simply R_{C1} or 1800 Ω. The input impedance of stage 2 is $R_B \parallel h_{ie2}$ = 30,000 Ω ∥ 3,000 Ω ∥ 900 Ω = 692.3 Ω. The gain of stage 1 is then A_{v1} times the coupling factor, or

$$-300 \times \frac{692.3 \ \Omega}{1800 \ \Omega + 692.3 \ \Omega} = -83.3$$

This checks the gain as found in Example 6-2.

6-4.2 The Cascaded Common-emitter Circuit in the Laboratory

The circuit of Figure 6-3 was set up in the laboratory. The major problem with setting up such a high-gain circuit is its tendency to oscillate and produce an output at its own resonant frequency. This destroys its capabilities as an amplifier. By keeping a CRO probe on the output, we knew the circuit was constructed correctly when oscillations started.

Defeating spurious oscillations is as much an art as a science. We managed to stop oscillations by keeping the wires short, adding small capacitors at appropriate places, and capacitively coupling a 680 Ω resistor across the output. This reduced the gain, but also reduced the probability of oscillation.

With the circuit working, the following voltages were measured:

$$v_{in} = 3.1 \ \text{mv}$$

$$v_{C1} = 25 \text{ mv}$$
$$v_{C2} = 1.65 \text{ v}$$

EXAMPLE 6-4

If the transistor parameters of Figure 6-3 are taken as $h_{fe} = 150$, $h_{ie} = 900 \ \Omega$, calculate the gain of the laboratory circuit. How does it compare with the measured results?

Solution
For stage 1

$$A_v = A_{in} \times A_{v1}$$
$$= 0.0645 \times 83.3 \quad \text{(see Example 6-1)}$$
$$= 5.37$$

The measured gain is 25 mV/3.1 mV = 8.

At this low level there is about a 33% discrepancy. We could not measure the voltage at the base of Q_1 because it was too small, and because inserting the voltmeter probe at this point caused the circuit to break into oscillations.

For stage 2

$$R'_L = 680 \ \Omega \ \| \ 1800 \ \Omega = 494$$

$$A_{v2} = \frac{-h_{fe}R'_L}{h_{ie}} = \frac{-150 \times 494 \ \Omega}{900 \ \Omega} = -82.3$$

The measured gain is 1.65 V/25 mV = 66

Here the measured gain is slightly less than the calculated gain, but the discrepancies are within reason.

6-4.3 Circuits with Unbypassed Emitter-Resistors

Circuits that contain transistors with unbypassed emitter-resistors can often be analyzed quickly by making certain simplifying assumptions. One effect of an unbypassed emitter-resistor is to increase the input impedance of the stage, which means that it does not load down the previous stage as heavily.

In addition, the gain of a circuit with an unbypassed emitter-resistor can often be approximated quite accurately. The gain of a BJT emitter-follower, for example, is assumed to be 1. This assumption is usually 99% accurate.

EXAMPLE 6-5

Find the transistor gain of the circuit of Figure 6-3 if the emitter capacitor of the second stage is removed.
a. Use approximate analysis.
b. Use precise analysis.

Solution

In this problem, the reduction in gain due to the input resistor A_{in} is ignored because A_{in} is dependent on the input resistor. If it is still 10 kΩ, the gains calculated here can be multiplied by 0.0645 to give the overall circuit gain.

a. For approximate analysis, we assume the gain of stage 2 is R_C/R_E (see section 4-7.2).

$$\frac{-R_C}{R_E} = \frac{-1800\ \Omega}{240\ \Omega} = -7.5$$

Since the input impedance of stage 2 is high, it will not load down stage 1. Therefore, the gain of stage 1 is 300. The total transistor gain is

$$A_{v(tr)} = A_{v1} \times A_{v2} = -300 \times -7.5 = 2250$$

b. Using exact analysis, we have

$$A_{v2} = \frac{-h_{fe}R_C}{h_{ie} + (1 + h_{fe})R_E} = \frac{-150 \times 1800\ \Omega}{900\ \Omega + (151 \times 240\ \Omega)} = -7.27$$

To find the gain of the first stage, the loading effect of the second stage must be determined.

$$R_{in} = h_{ie} + (h_{fe} + 1)R_E = 37{,}140\ \Omega$$

Here R'_L is the parallel combination of R_C, the biasing resistors, and R_{in}.

$$R'_L = 37{,}140\ \Omega\ ||\ 30{,}000\ \Omega\ ||\ 3300\ \Omega\ ||\ 1800\ \Omega = 1088\ \Omega$$

$$A_{v1} = \frac{-h_{fe}R'_L}{h_{ie}} = \frac{-150 \times 1088}{900} = -181$$

$$A_{v(tr)} = -181 \times -7.27 = 1316$$

This analysis shows that there is about a 40% discrepancy between the approximate and precise analysis. This discrepancy is primarily because the approximate analysis ignored the loading effect of the biasing resistors. When extreme accuracy is not required, engineers often use the quicker approximate analysis.

6-4.4 The Common-emitter–Common-collector Amplifier

The gain of a common-emitter stage can be improved by following it with a common-collector or emitter-follower stage. This is the common-emitter–common-collector (CE–CC), or common-emitter–emitter-follower circuit. It is useful when driving low impedance capacitively coupled loads.

EXAMPLE 6-6

If the circuit of Figure 6-2 must drive a capacitively coupled 600 Ω load, find its gain and output impedance.

Solution

If $1/h_{oc}$ is ignored, the output impedance is simply R_C or 1800 Ω. The gain is $-h_{fe}R'_L/h_{ie}$ where R'_L is 1800 Ω ∥ 600 Ω = 450 Ω.

$$A_v = \frac{-150 \times 450 \text{ Ω}}{900 \text{ Ω}} = -75$$

Check: The gain of this circuit without the 600 Ω resistor is 300. The 600 Ω resistor forms a voltage divider with the output impedance, R_o. Therefore $A_v = -300 \times 600\text{Ω}/2400\text{Ω} = -75$.

Example 6-7 shows the effect of adding an emitter-follower to the circuit.

EXAMPLE 6-7

For the circuit of Figure 6-4, find R_{B3} if the quiescent emitter voltage must be 9 V. Also find the gain of the circuit and its output impedance. Assume the transistors have an h_{fe} of 150. For simplicity, also assume that $h_{fe} + 1 = h_{fe}$.

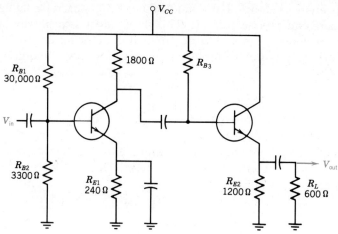

FIGURE 6-4
A common-emitter common-collector amplifier.

Solution
To find R_o, I_E and I_B must be found first.

$$I_E = \frac{V_E}{R_E} = \frac{9 \text{ V}}{1200 \text{ }\Omega} = 7.5 \text{ mA}$$

$$I_B = \frac{I_E}{h_{fe}} = \frac{7.5 \text{ mA}}{150} = 50 \text{ mA}$$

$$R_{B3} = \frac{V_{CC} - (V_E + V_{BE})}{I_B} = \frac{20 \text{ V} - 9.7 \text{ V}}{50 \text{ mA}} = 226 \text{ k}\Omega$$

$$h_{ie} = \frac{30 \, h_{FE}}{I_E} = \frac{30 \times 150}{7.5 \text{ mA}} = 600 \text{ }\Omega$$

For an emitter-follower

$$A_v = \frac{h_{fe} R'_E}{h_{ie} + h_{fe} R'_E} \quad \text{where } R'_E = 600 \text{ }\Omega \parallel 1200 \text{ }\Omega = 400 \text{ }\Omega$$

$$A_v = \frac{150 \times 400 \text{ }\Omega}{600 + 150 \times 400 \text{ }\Omega} = \frac{60{,}000}{60{,}600} = 0.99$$

The input impedance of the emitter-follower is $h_{ie} + h_{fe} R'_E = 60{,}600 \text{ }\Omega$.
For the first stage $h_{ie} = 900 \text{ }\Omega$ (see section 6-4.1).

$$R'_L = 1800 \text{ }\Omega \parallel 226{,}000 \text{ }\Omega \parallel 60{,}600 \text{ }\Omega = 1735 \text{ }\Omega$$

$$A_{v1} = \frac{-h_{fe} R'_L}{h_{ie2}} = \frac{-150 \times 1735 \text{ }\Omega}{900 \text{ }\Omega} = -289$$

The total gain is $A_{v1} \times A_{v2} = -289 \times 0.99 = -286$. This is quite close to the approximate gain of 300 for the common-emitter stage times 1 for the common-collector stage or 300. The impedance transformation caused by the emitter-follower raised the gain from -75 in Example 6-6 to -289 in Example 6-7.

The output impedance of this circuit is

$$R_E \| \frac{h_{ie} + R_{Th}}{h_{fe}}$$

Here R_{Th} is 1.8 kΩ $\|$ 226 kΩ (R_{B3}) = 1.786 kΩ so

$$R_o = 1200\ \Omega \| \frac{600\ \Omega + 1786\ \Omega}{150} = 1200\ \Omega \| 15.9\ \Omega = 15.7\ \Omega$$

6-4.5 The Multistage JFET Amplifier

Multistage JFET amplifiers can be analyzed very simply because their high input impedance means there is little interaction between the stages. Therefore the total gain is simply the product of the gains of the individual stages. Also A_{in} is usually 1 because of the high input impedance of the first stage.

EXAMPLE 6-8

Find the gain of the two-stage JFET amplifier of Figure 6-5.

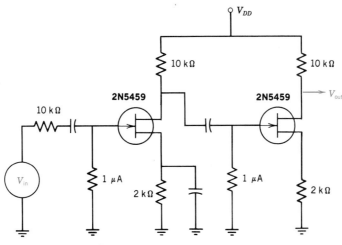

FIGURE 6-5
A two-stage JFET amplifier.

Solution

Again the gain is

$$A_{v(t)} = A_{in} \times A_{v1} \times A_{v2}$$

A_{in} is 1 MΩ in series with 10 kΩ, so that:

$$A_{in} = \frac{1 \text{ M}\Omega}{1.01 \text{ M}\Omega} = 0.99 \approx 1$$

The gain of each of the JFET stages is the same because there is no interaction due to their high input impedances. The JFET BIAS computer program for a **2N5459** ($V_P = -6$ V, $I_{DSS} = 9$ mA) gives a quiescent I_D of 1.7 mA. Therefore $V_{GS} = -2$ k$\Omega \times 1.7$ mA $= -3.4$ V.

$$g_m = g_{mo}\left(1 - \frac{V_{GS}}{V_P}\right)$$

$$g_m = 3000 \times 10^{-6} \text{ S}\left(1 - \frac{[-3.4]}{[-6]}\right) = 1300 \times 10^{-6} \text{ S}$$

$$A_v = -1300 \times 10^{-6} \text{ S} \times 10{,}000 \, \Omega = -13$$

Because both circuits have the same gain and $A_{in} = 1$

$$A_{v(total)} = A_{in} \times A_{v1} \times A_{v2} = 1 \times -13 \times -13 = 169$$

This example shows that cascading JFET amplifiers can result in a high gain. The gain and output swing would be smaller, however, if the second stage had to drive a capacitively coupled load.

6-5 DIRECT-COUPLED AMPLIFIERS

Direct-coupled amplifiers are designed to amplify variations in a dc voltage. Their frequency response must go down to dc (0 Hz). This means that they cannot include capacitors and results in unbypassed emitter-resistors and relatively low gain. In direct-coupled circuits, the bias of a circuit is determined by the dc level of the preceding circuit, so that the circuits are highly interdependent.

EXAMPLE 6-9

The circuit of Figure 6-6 is a two-stage, direct-coupled amplifier. Assume h_{FE} for each transistor is 100. Find the gain of the amplifier.

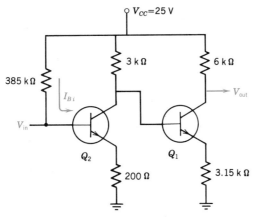

FIGURE 6-6
A two-stage direct-coupled amplifier.

Solution
The first step in an exact analysis is to find the dc voltages throughout the circuit. This allows us to determine h_{ie} and to establish the dc operating points. I_{B1} can be found first.

$$I_{B1} = \frac{V_{CC} - V_{BE}}{R_B + h_{fe}R_E} = \frac{24.3 \text{ V}}{385{,}000 \, \Omega + 20{,}000 \, \Omega} = \frac{24.3 \text{ V}}{405{,}000 \, \Omega} = 60 \, \mu\text{A}$$

$$I_{C1} = h_{FE}I_{B1} = 100 \times 60 \, \mu\text{A} = 6 \text{ mA}$$

$$h_{ie1} = \frac{30 \, h_{FE}}{I_E} = \frac{30 \times 100}{6 \text{ mA}} = 500 \, \Omega$$

$$V_{C1} = V_{CC} - I_{C1}R_{C1} = 25 \text{ V} - (6 \text{ mA} \times 3 \text{ k}\Omega) = 7 \text{ V}$$

V_{C1} is also the dc voltage on the base of Q_2. The emitter voltage of Q_2 must be $V_{C1} - V_{BE2} = 6.3$ V. Now it follows that:

$$I_{E2} = \frac{V_{E2}}{R_{E2}} = \frac{6.3 \text{ V}}{3.15 \text{ k}\Omega} = 2 \text{ mA} \approx I_{C2}$$

$$h_{ie2} = \frac{30 \, h_{fe}}{I_{E2}} = \frac{3000}{2 \text{ mA}} = 1500 \, \Omega$$

and

$$V_{C2} = V_{CC} - I_2R_{C2} = 25 \text{ V} - (2 \text{ mA} \times 6 \text{ k}\Omega) = 13 \text{ V}$$

This allows for about a 6 V swing of the output voltage (from 13 V to 7 V) and completes the dc analysis.

For the ac analysis, both stages are amplifiers with unbypassed emitter-resistors. The gain of such a stage was developed in Chapter 4.

$$A_{v(tr)} = \frac{-h_{fe}R_L}{h_{ie} + (1 + h_{fe})R_E} \approx \frac{-h_{fe}R_L}{h_{ie} + h_{FE}R_E}$$

Here R_L is $3{,}000 \parallel Z_{in2}$

$$Z_{in2} = h_{ie2} + h_{FE}R_E \approx 1500 \, \Omega + (100 \times 3.15 \text{ k}\Omega) = 316{,}500 \, \Omega$$

Therefore, because of the large unbypassed emitter-resistor in stage 2, Z_{in2} is so large it can be ignored.

$$A_{v1} = \frac{-h_{fe}R_L}{h_{ie} + h_{fe}R_{E1}} = \frac{-100 \times 3000 \, \Omega}{500 + (100 \times 200 \, \Omega)} = -14.6$$

The same formula can be used for A_{v2}.

$$A_{v2} = \frac{-h_{fe}R_{L2}}{h_{ie2} + h_{fe}R_{E2}} = \frac{-100 \times 6000}{1500 \, \Omega + (100 \times 3150)} = \frac{-600{,}000}{316{,}500} = -1.9$$

The overall gain is the product of the stage gains.

$$A_{vT} = A_{v1} \times A_{v2} = -14.6 \times (-1.9) = 27.7$$

Note that approximate calculations are extremely accurate for this circuit because of the high input impedance of stage 2. The gain of each stage is approximately $-R_C/R_E$. For stage 1 this is $-3 \text{ k}\Omega/200 \, \Omega = -15$. For stage 2 this is $-6 \text{ k}\Omega/3.15 \text{ k}\Omega = -1.9$, and the total approximate gain is

$$A_{v1} = -15 \times (-1.9) = 28.5$$

6-5.1 Design of a Direct-coupled Amplifier

A major problem in designing direct-coupled amplifiers is to get the quiescent point of each stage correct, because the dc voltage on one stage becomes the biasing voltage for the succeeding stage. The circuit of Figure 6-6 was deliberately designed for a low dc level on the collector of stage 1 (7 V). This allowed us to hold down the value of R_{E2}. R_{E2} is an unbypassed emitter, and if the bias voltage is high, this resistor must have a high value to limit the current. A high R_E, however, reduces the gain of the circuit.

Operational amplifiers (op-amps, see Chapter 12) are a form of direct-coupled amplifier, because they have a dc gain. In most cases, designers will use op-amps instead of designing direct-coupled amplifiers, because op-amps are readily available, easy to use, and eliminate the problem of having to design a circuit.

6-6 DARLINGTONS

A Darlington or *Darlington pair* is the name given to a pair of transistors connected as shown in Figure 6-7. It can be seen that the collectors of both transistors are tied together and that the emitter current of Q_1 becomes the base current of Q_2. The Darlington pair is sometimes considered as a single transistor with the collector, base, and emitter, as shown in Figure 6-7.

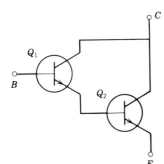

FIGURE 6-7
The basic Darlington pair.

Darlingtons provide a *very high current gain*, approximately equal to the product of the gains of two transistors ($h_{fe1} \times h_{fe2}$). They are often packaged as a single, three-terminal unit instead of being constructed from two discrete transistors, and their overall current gain, $h_{fe1} \times h_{fe2}$, is specified by the manufacturer. They are usually used in an emitter-follower configuration, so their input impedance is very high and their voltage gain is less than 1. They function as a super emitter-follower. In practice, low-power Darlingtons are used in difference amplifiers (see section 6-7) and optical couplers (see Chapter 1). Some of them have a minimum h_{fe} of 20,000. Higher power Darlingtons are used in power supplies, but they have smaller h_{fe}s (1000 minimum is typical).

6-6.1 The Darlington Circuit

The basic Darlington circuit is shown in Figure 6-8a and an equivalent circuit is shown in Figure 6-8b. The current i_{b1}, which enters the base of Q_1, is amplified to produce an emitter current of $h_{fe1}i_{b1}$. This becomes the base current in Q_2, and is further amplified to produce an output current of $h_{fe2}h_{fe1}i_{b1}$.

Two observations should be made here. The current in transistor Q_2 is usually much higher than the current in Q_1, and this leads to the following conclusions:

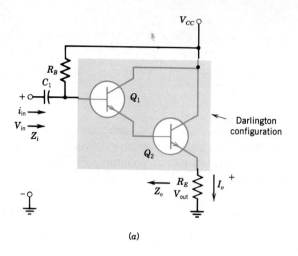

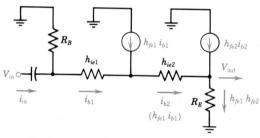

FIGURE 6-8
The Darlington circuit.

1. h_{ie2} is usually much smaller than h_{ie1} because of the current difference in the transistors.
2. h_{fe2} is also usually less than h_{fe1} (the h_{fe} curves tend to fall off at the high current; see Appendix A), but the reduction in h_{fe} is not nearly as pronounced as the reduction in h_{ie}.

EXAMPLE 6-10

If the values for Figure 6-8a are $V_{CC} = 25$ V, $R_E = 100\ \Omega$, $h_{fe1} = 100$, $h_{fe2} = 50$, and the quiescent value of V_{out} is to be 15 V, find
a. R_B.
b. R_{in}.
c. A_v.
d. A_i.

252   MULTIPLE TRANSISTOR CIRCUITS

Solution

a. This problem can be approached by working backwards from the emitter, because the conditions at the emitter are known.

Since $R_E = 100\ \Omega$ and $V_E = 15\ \text{V}$

$$I_E = \frac{15\ \text{V}}{100\ \Omega} = 150\ \text{mA}$$

$$I_{b2} = h_{fe1}I_{b1} = \frac{I_E}{h_{fe2}} = \frac{150\ \text{mA}}{50} = 3\ \text{mA}$$

$$I_{b1} = \frac{I_{b2}}{h_{fe1}} = \frac{3\ \text{mA}}{100} = 30\ \mu\text{A}$$

The dc voltage at R_B is R_E plus the base-to-emitter voltage drops across the two transistors. If these are both taken as 0.7 V, we have

$$R_B = \frac{V_{CC} - (V_E + V_{BE1} + V_{BE2})}{I_{B1}} = \frac{25\ \text{V} - 16.4\ \text{V}}{30\ \mu\text{A}} = \frac{8.6\ \text{V}}{30\ \mu\text{A}} = 287{,}000\ \Omega$$

b. To find R_i and A_v, the values of h_{ie} must be calculated first.

$$h_{ie2} = \frac{30 \times h_{fe2}}{I_{E2}} = \frac{30 \times 50}{150} = 10\ \Omega$$

$$h_{ie1} = \frac{30 \times h_{fe1}}{I_{E1}} = \frac{30 \times 100}{3} = 1000\ \Omega$$

The input voltage is the sum of the output voltage and the voltages across h_{ie2} and h_{ie1}.

$$V_{out} = h_{fe2}h_{fe1}i_{b1}R_E = 500{,}000\ i_{b1}$$

The voltage across h_{ie2} is

$$i_{b2}h_{ie2} = h_{fe1}h_{ie2}i_{b1} = 1000\ i_{b1}$$

The voltage across h_{ie1} is

$$h_{ie1}i_{b1} = 1000\ i_{b1}$$

$$V_{in} = 502{,}000\ i_{b1}$$

$$R_i = \frac{V_{in}}{i_{b1}} = \frac{502{,}000\ i_{b1}}{i_{b1}} = 502{,}000\ \Omega$$

c. The voltage gain is $A_v = \dfrac{V_{out}}{V_{in}} = \dfrac{500{,}000\, i_{b1}}{502{,}000\, i_{b1}} = 0.996$.

d. The current gain of the circuit is

$$A_i = \dfrac{i_{out}}{i_{in}} = \dfrac{i_b}{i_{in}} \times \dfrac{i_{out}}{i_b}$$

The term i_b/i_{in} is determined by current division between the input impedance R_i and the biasing resistor R_B.

$$\dfrac{i_b}{i_{in}} = \dfrac{R_b}{R_b + R_{in}} = \dfrac{287{,}000}{789{,}000} = 0.36$$

$$A_{i(ck)} = 0.36 \times 5000 = 1800$$

6-6.2 Power Considerations

Because of the high currents, the power ratings of the transistors must be considered when constructing Darlingtons. The power in the transistor is the $V_{CE}I_C$ product and cannot exceed the maximum power dissipation of the transistor.

EXAMPLE 6-11

Can the power transistor in Example 6-10 be a **2N3904**?

Solution

Example 6-10 shows that the current in the second transistor is 150 mA. Because V_{CC} is 25 V and the emitter voltage is 15 V, the voltage across the transistor is 10 V. The transistor therefore absorbs

$$P_D = 10\text{ V} \times 150\text{ mA} = 1.5\text{ watts}$$

The specifications on the **2N3904** (see Appendix A) show that it can dissipate 350 mW at 25°C (no heat sink)[1] or 1 W (with a heat sink). In neither case can it absorb 1.5 W, so a **2N3904** cannot be used as the second stage in this circuit.

---

[1] A discussion of heat sinking is given in section 9-5.

6-6.3 Exact Equations for the Darlington

Boylestad and Nashelsky (see References) give the following exact equations for the current gain and input impedance of a Darlington pair:

$$A_i = \frac{h_{fe1}h_{fe2}}{1 + h_{oe1}h_{fe2}R_E}$$

$$Z_i = \frac{h_{fe1}h_{fe2}\,R_E}{1 + h_{oe1}h_{fe2}\,R_E}$$

These equations take into account the fact that h_{oe1} may not be negligible in comparison with the input impedance of the second stage of the Darlington. If the term $h_{oe}h_{fe}R_E$ is < 0.1; however, the formulas reduce to the simple approximations.

$$A_i = h_{fe1}h_{fe2} \qquad (6\text{-}2)$$

$$R_i = h_{fe1}h_{fe2}R_E \qquad (6\text{-}3)$$

In most practical cases, the term $h_{oe}h_{fe}R_E$ is small because the Darlington's output current is large, and if R_E were also large, the emitter voltage would be too high to be practical. This limits the value of R_E and makes $h_{oe}h_{fe}R_E$ small enough so that it can usually be ignored. Therefore, the approximate equations 6-2 and 6-3 are often used.

6-6.4 Power Darlingtons

Darlingtons often come with both transistors in a single package. Figure 6-9 shows the schematic and case for the **2N6383** manufactured by RCA. This Darlington is rated at 10 A and 40 W, but the 40 W is at a case temperature of 25°C. It requires excellent heat sinking to keep the case at that temperature. The TO-3 case for this Darlington is designed for heat sinking. This Darlington has a typical current gain of 1000 at an output current of 5 A.

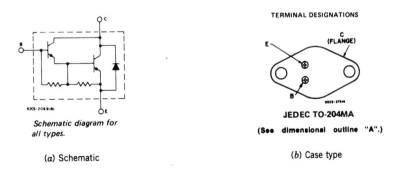

(a) Schematic

(b) Case type

FIGURE 6-9
The RCA **2N6383** power Darlington (GE/RCA Solid State Division.)

2N6034, 2N6035, 2N6036 PNP
2N6037, 2N6038, 2N6039 NPN

PLASTIC DARLINGTON COMPLEMENTARY SILICON POWER TRANSISTORS

... designed for general-purpose amplifier and low-speed switching applications.

- High DC Current Gain —
  h_{FE} = 2000 (Typ) @ I_C = 2.0 Adc
- Collector-Emitter Sustaining Voltage — @ 100 mAdc
  $V_{CEO(sus)}$ = 40 Vdc (Min) — 2N6034, 2N6037
  = 60 Vdc (Min) — 2N6035, 2N6038
  = 80 Vdc (Min) — 2N6036, 2N6039
- Forward Biased Second Breakdown Current Capability
  $I_{S/b}$ = 1.5 Adc @ 25 Vdc
- Monolithic Construction with Built-In Base-Emitter Resistors to Limit Leakage Multiplication
- Space-Saving High Performance-to-Cost Ratio TO-126 Plastic Package

DARLINGTON 4-AMPERE COMPLEMENTARY SILICON POWER TRANSISTORS
40, 60, 80 VOLTS
40 WATTS

*MAXIMUM RATINGS

Rating	Symbol	2N6034 2N6037	2N6035 2N6038	2N6036 2N6039	Unit
Collector-Emitter Voltage	V_{CEO}	40	60	80	Vdc
Collector-Base Voltage	V_{CB}	40	60	80	Vdc
Emitter-Base Voltage	V_{EB}		5.0		Vdc
Collector Current — Continuous Peak	I_C		4.0 8.0		Adc
Base Current	I_B		100		mAdc
Total Power Dissipation @ T_C = 25°C Derate above 25°C	P_D		40 0.32		Watts W/°C
Total Power Dissipation @ T_A = 25°C Derate above 25°C	P_D		1.5 0.012		Watts W/°C
Operating and Storage Junction Temperature Range	T_J, T_{stg}		−65 to +150		°C

THERMAL CHARACTERISTICS

Characteristic	Symbol	Max	Unit
Thermal Resistance, Junction to Case	θ_{JC}	3.12	°C/W
Thermal Resistance, Junction to Ambient	θ_{JA}	83.3	°C/W

*Indicates JEDEC Registered Data.

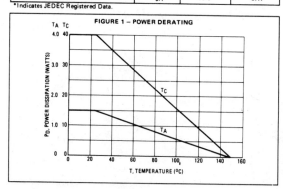

FIGURE 1 — POWER DERATING

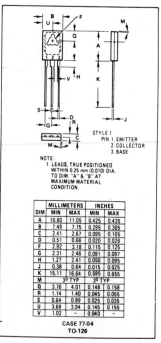

STYLE 1
PIN 1. EMITTER
2. COLLECTOR
3. BASE

NOTE:
1. LEADS, TRUE POSITIONED WITHIN 0.25 mm (0.010) DIA. TO DIM. "A" & "B" AT MAXIMUM MATERIAL CONDITION.

DIM	MILLIMETERS		INCHES	
	MIN	MAX	MIN	MAX
A	10.80	11.05	0.425	0.435
B	7.49	7.75	0.295	0.305
C	2.41	2.67	0.095	0.105
D	0.51	0.66	0.020	0.026
F	2.92	3.18	0.115	0.125
G	2.31	2.46	0.091	0.097
H	1.27	2.41	0.050	0.095
J	0.38	0.64	0.015	0.025
K	15.11	16.64	0.595	0.655
M	3° TYP		3° TYP	
Q	3.76	4.01	0.148	0.158
R	1.14	1.40	0.045	0.055
S	0.64	0.89	0.025	0.035
U	3.68	3.94	0.145	0.155
V	1.02	—	0.040	—

CASE 77-04
TO-126

FIGURE 6-10
Specifications for the Motorola **2N6034-39** Darlingtons. (Copyright by Motorola, Inc. Used by permission.)

The Motorola **2N6034–2N6039**s are Darlingtons available either in *npn* or *pnp* configurations. Their specifications are given in Figure 6-10. The curves of h_{fe} versus I_C are given in Figure 6-11. It can be seen that h_{fe} is very dependent on I_C, and at 25°C it reaches a maximum of almost 4000 (at $I_C = 0.7$ A) for the *pnp* Darlingtons, and almost 3000 for the *npn* Darlingtons.

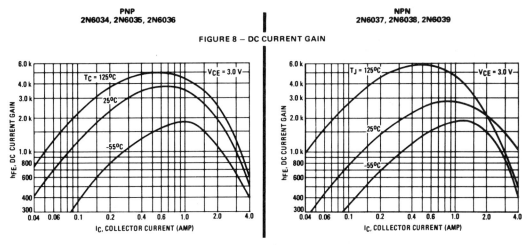

FIGURE 6-11
dc current gain vs. I_C for the **2N6034-39** power Darlingtons. (Copyright by Motorola, Inc. Used by permission.)

Manufacturers like Motorola and RCA have produced a variety of Darlingtons for most common applications. Darlington packages with smaller power dissipations typically have higher current gains.

6-7 CONSTANT-CURRENT SOURCES

A constant-current source is a circuit that delivers a constant current into a resistor or load, regardless of changes in the load. Figure 6-12 shows a simple

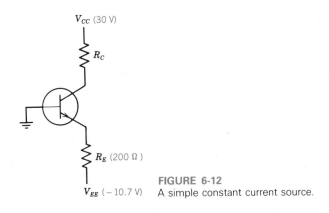

FIGURE 6-12 A simple constant current source.

one-transistor constant-current source. An examination of the circuit shows that the emitter current is fixed.

$$I_E = \frac{V_{BE} - V_{EE}}{R_E} \qquad (6\text{-}4)$$

As long as the value of the collector resistor is low enough so that the transistor *does not saturate*,[2] $I_E \approx I_C$ and the *collector current will remain constant*. The transistor can be driven into saturation, however, by raising the value of R_C. With a constant current flowing, as R_C increases the collector voltage decreases until it goes below the base voltage, which causes the transistor to saturate. Then the collector current will fall rapidly and the circuit will no longer function as a constant-current source.

EXAMPLE 6-12

For the circuit of Figure 6-12, with the values given in blue ($V_{CC} = 30$ V, $V_{EE} = -10.7$ V, $R_E = 200\ \Omega$), find
a. The range of resistance values where it will function as a constant current source.
b. V_C and I_C if $R_C = 500\ \Omega$.
c. V_C and I_C if $R_C = 1000\ \Omega$.

Solution
a. If the circuit operates as a constant-current source, the constant current will be

$$I_E = \frac{V_{BE} - V_{EE}}{R_E} = \frac{-0.7\ \text{V} - (-10.7\ \text{V})}{200\ \Omega} = \frac{10\ \text{V}}{200\ \Omega} = 50\ \text{mA}$$

The circuit will function as a constant-current source as long as the transistor stays out of saturation. Therefore the drop across R_C must be less than 30 V to keep the base-to-collector junction reverse-biased (positive), so that the transistor does not saturate.

$$R_C I_C \le 30\ \text{V} \qquad R_C \le \frac{30\ \text{V}}{50\ \text{mA}} = 600\ \Omega$$

The current in R_C will be constant at 50 mA for all values of R_C between 0 and 600 Ω.

---

[2]Saturation was discussed in section 2-10.

b. If $R_C = 500\ \Omega$, the resistance is in the range where the circuit will function as a constant current source. Therefore

$$I_C = 50\ \text{mA}$$

and

$$V_C = V_{CC} - I_C R_C$$
$$V_C = 30\ \text{V} - 50\ \text{mA} \times 500\ \Omega$$
$$V_C = 5\ \text{V}$$

c. If $R_C = 1000\ \Omega$, the resistance is too high for the circuit to function as a constant current source. Here $V_C = V_e + V_{ce(\text{sat})}$. If $V_{ce(\text{sat})}$ is taken as 0.1 V and the emitter is at -0.7 V, then $V_C = -0.6$ V.

$$I_C = \frac{30.6\ \text{V}}{1\ \text{k}\Omega} = 30.6\ \text{mA}$$

The reader can see that the collector current dropped from 50 mA to 30.6 mA because the resistance is too high. Also, because the transistor is saturated, it is operating at a forced β (see section 2-10). Here, $I_E = 50$ mA, and $I_C = 30.6$ mA, so the base current must be the difference, or 19.4 mA.

$$h_{FE(\text{forced})} = \frac{I_C}{I_B} = \frac{30.6\ \text{mA}}{19.4\ \text{mA}} = 1.58$$

Thus the transistor in this circuit is operating at a very low forced β.

The basic characteristics of a constant-current source are that it has a fixed base voltage, V_B, driving current into a fixed emitter voltage, V_{EE}. Sometimes Zener diodes (see problem 6-18) or voltage dividers are used to set the base voltage of a constant-current source.

6-8 COMPARATORS

A comparator is a circuit that is meant to *compare* two different voltages and produce a *digital output* that essentially can *be only one of two levels*, V_1 or V_2. It often has one input, V, and a reference voltage, V_{ref}. If $V > V_{\text{ref}}$, then $V_{\text{out}} = V_1$. If $V < V_{\text{ref}}$, then $V_{\text{out}} = V_2$. The output voltage should always be either V_1 or V_2.

The basic comparator circuit is shown in Figure 6-13. The reference voltage in this circuit is 6 V, as established by the source connecting it to base

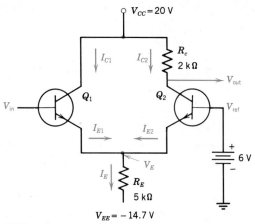

FIGURE 6-13
The basic comparator circuit.

of Q_2. The key to understanding the comparator is to realize that the voltage at the emitters, V_E, is $V_{ref} - V_{BE}$, or $V_{in} - V_{BE}$, whichever is greater. If V_{in} is 4 V, for example, and $V_{BE} = 0.7$ V, then current flows in Q_2, V_E is 5.3 V, and the base–emitter junction of transistor Q_1 is reverse-biased by 1.3 V.

As V_{in} increases, no current flows in Q_1 until V_{in} reaches about 5.8 V. At this point, the base–emitter junction of Q_1 is forward-biased by 0.5 V and a trickle of current flows. As V_{in} increases further, more current flows in Q_1. When V_{in} becomes greater than 6 V, Q_1 starts to conduct most, if not all, of the current in R_E and Q_2 quickly becomes cut off. When V_{in} is 7 V, for example, $V_E = 6.3$ V and Q_2 is completely cut off.

EXAMPLE 6-13

For the circuit of Figure 6-13, find V_E and V_{out} when V_{in} is
a. 4 V.
b. 6 V.
c. 8 V.

Solution

a. When $V_{in} = 4$ V, $V_{in} - V_{BE} = 3.3$ V, but $V_{ref} - V_{BE} = 5.3$ V. Since $V_{ref} - V_{BE}$ is greater, $V_E = 5.3$ V and Q_1 is reverse-biased, so no current flows in it ($I_{C1} = 0$). For Q_2

$$I_E = \frac{V_E - V_{EE}}{R_E} = \frac{5.3 \text{ V} - (-14.7) \text{ V}}{5 \text{ k}\Omega} = 4 \text{ mA}$$

$$I_{C2} \approx I_E = 4 \text{ mA}$$

$$V_{out} = V_{CC} - I_{C2}R_C = 20 \text{ V} - (4 \text{ mA} \times 2 \text{ k}\Omega) = 12 \text{ V}$$

b. When $V_{in} = 6$ V $= V_{ref}$, V_E still equals 5.3 V and therefore I_E still equals 4 mA. Now, however, the best assumption is that the current divides equally between Q_1 and Q_2 (because their base and emitter voltages are the same). Therefore

$$I_{C2} = \frac{I_E}{2} = 2 \text{ mA}$$

$$V_{out} = V_{CC} - I_{C2}R_C = 20 \text{ V} - (2 \text{ mA} \times 2 \text{ k}\Omega) = 16 \text{ V}$$

c. When $V_{in} = 8$ V, $V_E = 7.3$ V. Transistor Q_2 is reverse-biased, so $I_{C2} = 0$ and $V_{out} = 20$ V. The emitter current has increased

$$I_E = \frac{V_E - V_{EE}}{R_E} = \frac{7.3 \text{ V} - (-14.7) \text{ V}}{5 \text{ k}\Omega} = 4.4 \text{ mA}$$

but this current flows only through Q_1 and has no effect on V_{out}.

Figure 6-14 is a curve of V_{out} versus V_{in} for the circuit of Figure 6-13. As long as V_{in} is less than about 5.8 V, transistor Q_2 conducts all the current that flows in R_E and $V_{out} = 12$ V. When V_{in} is 6.2 V or greater, transistor Q_1 conducts all the current, cuts off Q_2, and $V_{out} = 20$ V (V_{CC}).

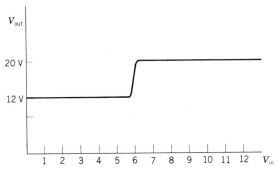

FIGURE 6-14
Curve of V_{in} vs. V_{out} for the comparator of Figure 6-13.

Note that V_{out} changes only when $V_{in} \approx 6$ V. Otherwise, if $V_{in} < 6$ V, $V_{out} = 12$ V, and if $V_{in} > 6$ V, $V_{out} = 20$ V. Therefore V_{out} assumes one of two voltages (12 V or 20 V), depending on whether V_{in} is greater or less than V_{ref}.

6-8.1 Commercial Comparators

Commercial comparators contain much more circuitry than the rudimentary comparator discusssed in the previous section. They are still essentially three-terminal devices, however, with two inputs, one of which could be considered as the reference, and one output whose value denotes which input is larger.

Commercial comparators have two significant advantages over the rudimentary comparator:

1. Their cutoff is much sharper; it occurs over a much smaller range of input voltages.
2. Their outputs are often 0 V and 5 V. This makes them compatible with TTL and other digital circuits.

Figure 6-15 shows the symbol and transfer characteristics of a typical commercial comparator. The transfer characteristics show that the output is 3 V if the difference between the two input voltages is less than -2 mV, and 0 if the difference between the two input voltages is greater than 2 mV. Thus the comparator output will be one of these voltages if there is at least a 2 mV difference between the input voltages.

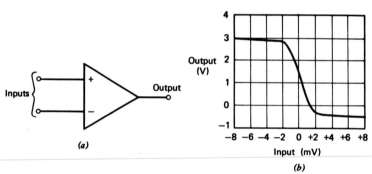

FIGURE 6-15
An analog voltage comparator. (From *Integrated Circuits in Digital Electronics* by Arpad Barna and Dan I. Porat. Copyright © 1973 by John Wiley & Sons, Inc. Reprinted by permission of John Wiley & Sons, Inc.)

Figure 6-16 shows the transfer characteristics of the **LM106**, a popular comparator manufactured by Texas Instruments Inc., and others. The transfer characteristics show that the output at 25°C is either 0 V or 5.5 V and the excursion occurs over a 0.2 mV range, so this comparator reacts to a very small difference between the input voltages. If the voltage of the plus input terminal is at least 0.2 mV greater than the voltage at the minus input terminal, the output voltage will be 5.5 V; and if the voltage at the input terminals are equal, or the minus input voltage is greater, the output voltage will be 0 V.

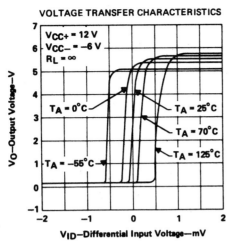

FIGURE 6-16
Transistor characteristics of an **LM106** comparator. (Courtesy of Texas Instruments, Inc.)

6-9 DIFFERENCE AMPLIFIERS

A *difference amplifier* is an amplifier with two input terminals, like a comparator. It is designed to amplify voltages that are applied *differentially* to the two terminals while rejecting any voltage that is applied to both inputs simultaneously.

The situation is illustrated by the transformer circuit of Figure 6-17. Any voltage applied by v_d results in a voltage difference between outputs a and b. Any voltage applied by v_{cm} results in an identical change of voltage at both outputs a and b. This is called the *common-mode* voltage. In many cases v_{cm} is much larger than the difference voltage, v_d, but v_d is the voltage of interest. Thus, a difference amplifier that amplifies v_d, but not v_{cm}, is required.

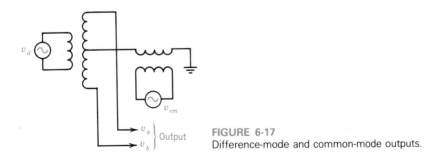

FIGURE 6-17
Difference-mode and common-mode outputs.

In practical circuits, difference amplifiers are used in medical electronics, where very small signals are generated by the brain or the heart and picked off the patient's body by electrodes. They are often found in the presence of much higher common-mode signals. Another use for difference amplifiers is in transmission lines, where the signal is applied to a long line that may also be picking up external noise, possibly 60 Hz from a power source. The 60 Hz is a common-mode signal, while the information is contained in the difference-mode input signal. Difference amplifiers are also used as the inputs to operational amplifiers (see Chapter 12).

6-9.1 The Basic Difference Amplifier

The basic difference amplifier is shown in Figure 6-18. It looks much like the comparator circuit of Figure 6-13, except that there are two output voltages, v_{out1} and v_{out2}. These two voltages should be equal and out of phase. R_{B1} and R_{B2} serve as biasing resistors for this circuit, and V_{BB} is the biasing source. Often, V_{BB} is ground in these circuits. As with comparators, positive and negative supply voltages (V_{CC} and V_{EE}) are often used.

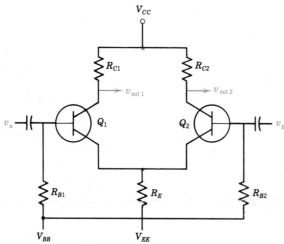

FIGURE 6-18
The basic difference amplifier.

The circuit of Figure 6-18 operates as follows:

1. In common-mode, the inputs v_a and v_b both go up or down by identical amounts. This causes a current increase in R_E that raises the emitter voltage to compensate for the input voltage rise. In common-mode operation, the circuit acts approximately like an amplifier with an unbypassed emitter-resistor and has a small gain.
2. In difference-mode, the *differential voltage* is applied at v_a and v_b. Assume v_a goes up and v_b goes down. The emitter voltage and current remain constant, but Q_1 gets most of the current and Q_2 gets very little, which

causes a large voltage difference between v_{out1} and v_{out2}. The circuit acts like an ordinary amplifier, and produces a large voltage gain.

In order to function well, the circuit of Figure 6-18 must be symmetrical. Therefore, $R_{C1} = R_{C2}$ and $R_{B1} = R_{B2}$. The h_{fe}s of the two transistors should also be as close as possible. One can purchase a matched pair of transistors for this purpose.

6-9.2 Biasing the Difference Amplifier

Figure 6-19 shows a biasing circuit for a difference amplifier. It is drawn from the standpoint of one of the transistors. The current that flows in R_E is twice the collector current because it is the sum of the currents in both transistors. In the single transistor equivalent circuit of Figure 6-19, R_E is doubled (it is twice the actual value of the resistor), to take this into account.

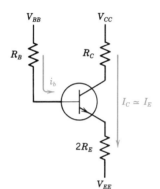

FIGURE 6-19
The bias circuit for a difference amplifier.

In Figure 6-19, if the voltages across R_B, the base-to-emitter junction, and R_E are considered, we have

$$i_b = \frac{V_{BB} - V_{BE} - V_{EE}}{[R_B + (1 + h_{FE})2R_E]} \tag{6-5}$$

In many circuits, $R_B \ll 2 h_{FE} R_E$ and this simplifies equation 6-5 considerably.

EXAMPLE 6-14

In Figure 6-18, assume $V_{BB} = 0$ V (ground), $R_{B1} = R_{B2} = R_{C1} = R_{C2} = R_E = 1$ kΩ, $h_{FE1} = h_{FE2} = 100$, $V_{CC} = 20$ V and $V_{EE} = 20$ V. Find I_b, I_{C1} and the quiescent output voltage at the collector of Q_1.
a. Use equation 6-5.
b. Use reasonable approximations.

Solution

a. Equation 6-5 gives

$$i_b = \frac{-0.7 \text{ V} - (-20 \text{ V})}{1 \text{ k}\Omega + 101 \text{ (2 k}\Omega)} = \frac{19.3 \text{ V}}{203{,}000} = 95 \text{ μA}$$

$$i_c = h_{FE}i_b = 9.5 \text{ mA}$$

$$V_{CQ1} = V_{CC} - I_{C1}R_{C1} = 20 \text{ V} - (9.5 \text{ mA} \times 1 \text{ k}\Omega) = 10.5 \text{ V}$$

b. Using approximations, we ignore the small voltage drop across R_B and assume the base of the transistors is also at ground. This places the emitter at -0.7 V, and the voltage across R_E is 19.3 V.

$$I_E = \frac{19.3 \text{ V}}{1 \text{ k}\Omega} = 19.3 \text{ mA}$$

The collector current in Q_1 is half this, or 9.65 mA, and

$$V_{C1} = 20 \text{ V} - 9.65 \text{ mA} \times 10 \text{ k}\Omega = 10.35 \text{ V}$$

Not only has the approximate method given results very close (within 2%) to the exact equation, but it also gives the reader a better feel for what is happening in the circuit.

6-9.3 AC Analysis of the Difference Amplifier

There are two voltage gains to be considered for a difference amplifier; the common-mode gain, A_c, and the difference mode gain, A_d. The ratio of these two gains, A_d/A_c, is called the *common-mode rejection ratio* (CMRR) and is a figure of merit for a difference amplifier. Good difference amplifiers have high CMRRs.

The common mode gain, A_c, is

$$A_c = \frac{\Delta v_{out}}{\Delta v_{cm}}$$

where Δv_{out} is the voltage at one of the collectors, and Δv_{cm} is the common-mode input voltage applied to the base of both transistors simultaneously. Ideally, A_c should be very small or even 0, because the amplifier should not transmit common-mode signals.

The difference mode gain is

$$A_d = \frac{\Delta v_{out}}{\Delta v_d}$$

where Δv_d is the total difference signal. It is ideally obtained by applying a signal $\Delta v_d/2$ to the base of one transistor and a signal $-\Delta v_d/2$ to the base of the other transistor.

EXAMPLE 6-15

A difference amplifier has $A_c = 0.1$, $A_d = 100$. Find Δv_{out} in response to a 5 V common-mode change and a 0.1 V difference-mode change. Also find the CMRR.

Solution
For the common-mode

$$\Delta v_{out} = A_c \times \Delta v_{cm} = 0.1 \times 5 \text{ V} = 0.5 \text{ V}$$

For the difference mode

$$\Delta v_{out} = A_d \times \Delta v_d = 100 \times 0.1 \text{ V} = 10 \text{ V}$$

$$\text{CMRR} = \frac{A_d}{A_c} = \frac{100}{0.1} = 1000$$

The difference voltage has a much larger effect on the output, although it is much smaller.

The equations for A_c and A_d for the basic difference amplifier circuit of Figure 6-18 are derived in Appendix D. They are

$$A_c = \frac{h_{fe}R_C}{h_{ie} + R_{Th} + 2h_{fe}R_E} \approx \frac{R_C}{2R_E} \quad \text{if} \quad (2h_{fe}R_E \gg R_{Th} + h_{ie}) \quad (6\text{-}6)$$

$$A_d = \frac{h_{fe}R_C}{2(R_{Th} + h_{ie})} \quad (6\text{-}7)$$

where R_{Th} is the Thevenin's resistance looking from the base towards the source.

From equations 6-6 and 6-7, the CMRR can be found.

$$\text{CMRR} = \frac{A_d}{A_c} \approx \frac{h_{fe}R_C}{2(R_{Th} + h_{ie})} \bigg/ \frac{R_C}{2R_E} \approx \frac{h_{fe}R_E}{(R_{Th} + h_{ie})}$$

EXAMPLE 6-16

For the circuit of Figure 6-18, assume $V_{CC} = 20$ V, $V_{EE} = -18.7$ V, $R_{C1} = R_{C2} = 1$ kΩ, $R_E = 3$ kΩ, $h_{fe1} = h_{fe2} = 150$, and the source is connected directly to the base so that $R_{Th} = 0$. Find A_c, A_d, and the CMRR.

Solution

To use equations 6-6 and 6-7, h_{ie} of the transistors must be found. The current in each must be found first. With the given values, we can assume $V_B = 0$ V and $V_E = -0.7$ V (see Example 6-14).

$$I_E = \frac{V_E - V_{EE}}{R_E} = \frac{-0.7\text{ V} - (-18.7\text{ V})}{3\text{ k}\Omega} = 6\text{ mA}$$

This means that 3 mA flow in each transistor.

$$h_{ie} = \frac{30\, h_{fe}}{I_E} = \frac{30 \times 150}{3} = 1500\ \Omega$$

$$A_c = \frac{h_{fe} R_C}{R_{Th} + h_{ie} + 2 h_{fe} R_E} = \frac{100 \times 1000\ \Omega}{1500\ \Omega + 2 \times 100 \times 3000\ \Omega}$$

$$= \frac{10^5\ \Omega}{601{,}500\ \Omega} = 0.166$$

The common-mode gain of less than 1 means that the circuit attenuates the common-mode signal.

$$A_d = \frac{h_{fe} R_C}{2(R_{Th} + h_{ie})} = \frac{10^5\ \Omega}{3000\ \Omega} = 33.3$$

$$\text{CMRR} = \frac{A_d}{A_c} = \frac{33.3}{0.166} = 200$$

Check:

$$\text{CMRR} = \frac{h_{fe} R_E}{R_{Th} + h_{fe}} = \frac{100 \times 3000\ \Omega}{1500\ \Omega} = 200$$

6-9.4 The Difference Amplifier in the Laboratory

Figure 6-20 is a difference amplifier that was set up in the laboratory. To attempt to achieve a reasonable symmetry, the resistors all had a 1% tolerance. The transistors, however, were randomly selected **2N3904**s (the first ones in the transistor bin). Nevertheless, the results were very close to the predicted values, despite the transistor mismatch.

The measured dc values for Figure 6-20 were

$$V_{C1} = 12.36\text{ V}$$

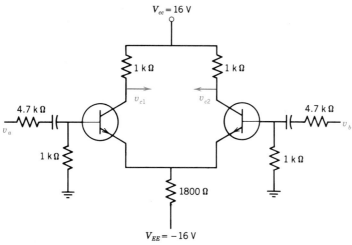

FIGURE 6-20
A laboratory difference amplifier.

$$V_{C2} = 11.16 \text{ V}$$
$$V_E = -0.7 \text{ V}$$

For the common-mode test, a sine wave was applied to both input terminals, v_a and v_b. The ac results were

$$v_{in} = v_a \text{ and } v_b = 10 \text{ V}$$
$$v_{c1} = 0.42 \text{ V}$$
$$v_{c2} = 0.55 \text{ V}$$

For the difference mode test, the sine wave was applied to v_a while v_b was grounded. Appendix D shows that this has the same effect as applying a true differential voltage, $v_{in/2}$ at v_a, and $-v_{in/2}$ at v_b. The ac results were

$$v_{in} = 0.252 \text{ V}$$
$$v_{c1} = 1.902 \text{ V}$$
$$v_{c2} = 1.916 \text{ V}$$

EXAMPLE 6-17

Based on the given values, find A_c, A_d, and CMRR for the circuit of Figure 6-20.

Solution
The common-mode gains for the two transistors are

$$A_c = \frac{v_{out}}{v_{in}} = 0.042 \text{ for } Q_1 \quad \text{and} \quad 0.055 \text{ for } Q_2$$

The average common mode gain is 0.0485.
The difference mode gain is

$$A_d = \frac{v_{out}}{v_{in}} = \frac{1.91 \text{ V}}{0.252 \text{ V}} = 7.54$$

$$\text{CMRR} = \frac{A_d}{A_c} = \frac{7.54}{0.0485} = 155.5$$

EXAMPLE 6-18

Calculate A_c, A_d, and the CMRR for the circuit of Figure 6-20. How do they compare with the measured values?

Solution
For the difference amplifier itself, the gains can be found from equations 6-6 and 6-7, but first h_{fe} and h_{ie} must be found. The transistors are randomly selected **2N3904s**, so an h_{fe} of 150 is assumed. To find h_{ie} the emitter current of the transistor must be found. The average dc collector voltage is 11.76 V (average of the 12.36 and 11.16 collector voltages), and therefore

$$I_C = \frac{V_{CC} - V_C}{R_C} = \frac{16 \text{ V} - 11.24 \text{ V}}{1 \text{ k}\Omega} = 4.26 \text{ mA}$$

Check:

$$I_E = \frac{V_E - V_{EE}}{R_E} = \frac{15.3 \text{ V}}{1800 \text{ }\Omega} = 8.5 \text{ mA}$$

This is the sum of the current in both transistors and should be twice the current calculated before.

$$h_{ie} = \frac{30 h_{fe}}{I_E} = \frac{30 \times 150}{4.26} = 1056 \text{ }\Omega$$

R_{Th}, the Thevenin's resistance at the base, is the parallel combination of 4.7 kΩ and 1 kΩ. In this circuit

$$R_{Th} = 4 \text{ k}\Omega \parallel 1 \text{ k}\Omega = 825 \text{ }\Omega$$

Now that h_{fe}, h_{ie}, and R_{Th} have been found, equations 6-6 and 6-7 can be used.

$$A_c = \frac{h_{fe}R_C}{R_{Th} + h_{ie} + 2h_{fe}R_E} = \frac{150 \times 1 \text{ k}\Omega}{825 \text{ }\Omega + 1056 \text{ }\Omega + (2 \times 150 \times 1800)\Omega}$$

$$A_c = \frac{150{,}000 \text{ }\Omega}{541{,}891 \text{ }\Omega} = 0.277$$

(Note the approximate formula $A_c = R_C/2R_E = 1 \text{ k}\Omega/3600 \text{ }\Omega = 0.277$ is highly accurate here.)

$$A_d = \frac{h_{fe}R_C}{2(R_S + h_{ie})} = \frac{150 \times 1000 \text{ }\Omega}{2(825 + 1056)\Omega} = \frac{150{,}000 \text{ }\Omega}{2(1881)\Omega} = 39.9$$

So far the gains do not look much like the answers obtained in Example 6-15, but these gains are only for the difference amplifier itself. For the entire circuit, v_a and v_b must also be Thevenized. The Thevenized input voltage is v_{oc}, where the 1 kΩ and 4.7 kΩ resistors form a voltage divider. Thus $v_{base} = v_{in}/5.7$ and both amplifier gains must be divided by 5.7 to get the actual circuit gains.

$$A_c = \frac{0.277}{5.7} = 0.0486$$

$$A_d = \frac{39.9}{5.7} = 7.0$$

Now the common-mode gain found here matches the calculated common-mode gain found in Example 6-15 almost exactly. The difference mode gains were 7 versus 7.54, or a discrepancy of 7%. Of course, the common-mode gain does not depend on h_{fe} (if the approximate formula is used), so it is highly accurate.

The measured CMRR is

$$\text{CMRR} = \frac{39.9}{0.277} = \frac{7}{0.0485} = 144$$

as compared to a CMRR of 155 calculated in Example 6-17.

6-9.5 Balancing Resistors

Balancing resistors are resistors or a potentiometer placed in the emitter circuit of a difference amplifier in order to balance the currents in each half of the circuit. Thus, balancing resistors compensate for gain differences between the transistors and optimize the circuit performance. A difference amplifier with balancing resistors is shown in Figure 6-21. The balancing resistors are R_1 and R_2. Usually their values are quite close. Often a potentiometer is placed between the emitter and point A. This takes the place of the balancing resistors and allows the user to adjust the potentiometer to equalize the currents in each transistor.

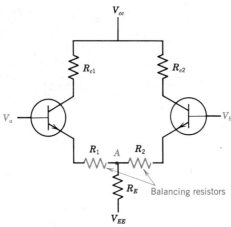

FIGURE 6-21
A difference amplifier with balancing resistors.

The disadvantage of balancing resistors is that they reduce the difference mode gain without changing the common-mode gain, and thus reduce the CMRR. The formula for the difference mode gain of an amplifier with balancing resistors, also derived in Appendix D, is

$$A_d = \frac{h_{fe}R_C}{2(R_{Th} + h_{ie} + h_{fe}R_B)} \tag{6-8}$$

where R_B is the average value of the balancing resistors.

In the laboratory circuit of Figure 6-20 a 100 Ω potentiometer was placed in the emitter circuit and the dc voltage of each output then balanced at 11.8 V. The ac results were

Common mode

$$v_{in} = 15.65 \text{ V}$$

$$v_{out} = 1.0 \text{ V}$$

Difference mode

$$v_{in} = 0.73 \text{ V}$$
$$v_{out} = 1.0 \text{ V}$$

Therefore

$$A_c = \frac{0.77}{15.65} = 0.0492$$

and

$$A_d = \frac{1.0}{0.73} = 1.37$$

The common mode gain did not change, but the difference mode gain decreased significantly.

EXAMPLE 6-19

How does the difference mode gain, calculated by equation 6-8, compare with the measured difference mode gain?

Solution

Because a 100 Ω potentiometer is used, it is reasonable to assume that the resistance in each leg of the circuit is 50 Ω. Therefore

$$A_d = \frac{h_{fe}R_C}{2(R_S + h_{ie} + h_{fe}R_1)} = \frac{150 \times 1000}{2[825 + 1056 + (150 \times 50)]} = \frac{150{,}000 \text{ Ω}}{18{,}762 \text{ Ω}} = 8$$

To get A_d for the circuit, we Thevenize the input by dividing by 5.7.

$$A_{d(ck)} = \frac{8}{5.7} = 1.40$$

This is within 2% of the predicted value.

6-9.6 Decibels

Commercial difference amplifiers and some other circuits often specify parameters such as gain and CMRR in terms of *decibels* (dBs). The decibel was originally defined as the smallest power change in an amplifier perceptible to the human ear, and is a logarithmic function, because large signals require a greater power change than small signals to cause a perceptible change in volume. Mathematically, decibels are a comparison between two power levels and are defined as

$$dB = 10 \log_{10} \frac{P_2}{P_1} \qquad (6\text{-}9)$$

Equation 6-9 is often used to express the power gain of a circuit or system in decibels, where P_2 is the output power and P_1 is the input power. Often, however, parameters that depend on voltages, such as gains or CMRRs, are measured in dBs. Since power across a resistor, for example, is proportional to voltage squared, equation 6-9 becomes

$$dB = 10 \log_{10}\left(\frac{V_2}{V_1}\right)^2 = 20 \log_{10}\left(\frac{V_2}{V_1}\right) \qquad (6\text{-}10)$$

EXAMPLE 6-20

An amplifier has a voltage output 100 times its input. How many dBs does it gain?

Solution
From equation 6-10 we have

$$dB = 20 \log_{10} 100 = 40 \text{ dB}$$

because $\log_{10} 100 = 2$.

EXAMPLE 6-21

A difference amplifier has a CMRR of 80 dB. What is its ratio of A_d/A_c?

Solution

$$db = 80 = 20 \log_{10} \text{CMRR}$$
$$4 = \log_{10} \text{CMRR}$$
$$\text{CMRR} = 10{,}000$$

6-9.7 Commercial Difference Amplifiers

Difference amplifiers have been packaged in integrated circuits by various manufacturers. In general, commercial difference amplifiers are much more elaborate and have better CMRRs than the generic difference amplifiers of Figure 6-18 or Figure 6-20.

Commercial difference amplifiers make two improvements over the generic circuits.

1. They generally use Darlington pairs on the inputs.
2. They all use constant current sources in place of R_E.

The use of a constant current source theoretically makes $R_E = \infty$ and reduces A_c to 0. Actually, a very small A_c persists, but the CMRR is quite high. Commercial difference amplifiers have CMRRs from 80 to 120 dBs (10^4 to 10^6).

Figure 6-22 gives the specifications and circuit diagram on the **CA3000**, a difference amplifier manufactured by RCA. The Darlington pairs on the inputs (Q_1 and Q_2, Q_3 and Q_4) are clearly shown, as is the constant current generator (Q_5). The **CA3000** has a CMRR of 98 dB (typ) $\approx$ 100 dB $\approx 10^5$.

EXAMPLE 6-22

Using the circuit for the **CA3000**, find the dc voltages if pins 1, 2, and 6 are tied to ground, $V_{CC} = +6$ V, $V_{EE} = -6$ V, and pins 4 and 5 are left open.

Solution

The first part of this problem requires that the current, I_{C5}, in the constant current source be found. I_1 must be found, then the base voltage of Q_5 and then I_E can be found.

Assuming the voltage drop across each diode is 0.7 V

$$I_1 = \frac{4.6 \text{ V}}{10 \text{ k}\Omega} = 0.46 \text{ mA}$$

$$V_{B(Q5)} = -I_1 R_1 = -0.46 \text{ mA} \times 5 \text{ k}\Omega = -2.3 \text{ V}$$

$$V_{E(Q5)} = V_B - V_{BE} = -3.0 \text{ V}$$

$$I_E = \frac{3.0 \text{ V}}{R_9 + R_{10}} = \frac{3 \text{ V}}{3 \text{ k}\Omega} = 1 \text{ mA}$$

Assuming the current divides evenly, $V_{c2} = V_{c3} = V_{CC} - I_C R_1 = 6$ V $-$ (0.5 mA $\times$ 8 kΩ) = +2 V.

The voltage at the collector of Q_5 is $V_{C(Q5)} = V_{in} -$ (2 base-to-emitter drops) $- (I_{C1} \times R_4) = 0 - 1.4$ V $-$ (0.5 mA $\times$ 50 Ω) ≈ -1.4 V.

This shows that the constant current source is still unsaturated because $V_{C(Q5)} > V_{B(Q5)}$.

Operational amplifiers (see Chapter 12) have characteristics similar to that of difference amplifiers and are often used for this purpose in modern electronics.

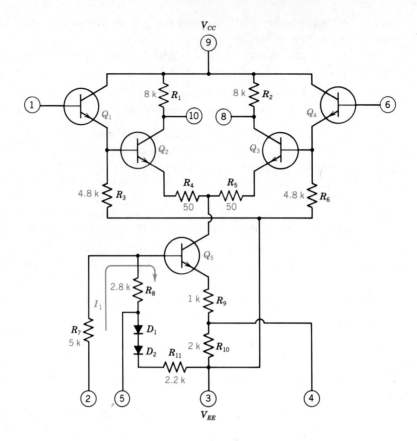

DC Amplifier

Features:

- Designed for use in Communication, Telemetry, Instrumentation, and Data-Processing Equipment
- Balanced differential-amplifier configuration with controlled constant-current source to provide outstanding versatility
- Built-in temperature stability for operation from $-55°$ C to $+125°$ C
- Companion Application Note, ICAN 5030 "Applications of RCA CA3000 Integrated Circuit DC Amplifier" covers characteristics of different operating modes, frequency considerations. 10 MH_Z narrow band tuned amplifier design, crystal oscillator design, and many other application aids
- Input impedance—195 kΩ typ.
- Voltage gain—37 dB typ.
- Common-mode rejection ratio—98 dB typ.
- Input offset voltage—1.4 mV typ.
- Push-pull input and output
- Frequency capability—DC to 30 MH_Z (with external C and R)
- Wide AGC range—90 dB typ.

Applications

- Schmitt trigger
- RC-coupled feedback amplifier
- Mixer
- Comparator
- Modulator
- Crystal oscillator
- Sense amplifier

FIGURE 6-22
The **CA3000** difference amplifier. (Adapted from *Integrated Circuits for Linear Applications Databook*. GE/RCA Solid State Division.)

6-10 THE CASCODE AMPLIFIER

There are two different types of amplifiers, called *cascode amplifiers,* described in the literature. One type is used as a level shifter, the other type is used to improve the high-frequency gain of a stage.[3] The level shifter is discussed here. The high-frequency type is discussed in Chapter 7 on frequency response.

The basic level-shifting cascode amplifier is shown in Figure 6-23. The inputs are an ac input (v_i) riding on a dc value (V_1). The output is an ac voltage, v_l, also riding on a dc output, V_L. The cascode circuit consists essentially of an emitter-follower stage (Q_1) and a constant-current source (Q_2). In this type of amplifier, $v_i = v_L$, but $V_1 \neq V_L$. Thus, this amplifier does not produce any gain ($A_V \approx 1$) but allows for level shifting. V_L is often designed to be 0 V or ground. One use of the cascode circuit is to take a transistor collector voltage that has a dc level and shift it so the dc level of the output is ground. Of course, this has already been done with an ac load resistor and a blocking capacitor, but the cascode amplifier does not load down the circuit or restrict the output swing in the same way that a capacitively coupled load does.

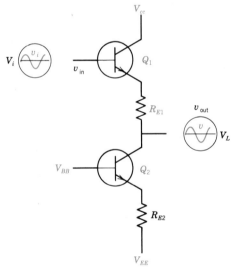

FIGURE 6-23
The basic cascode amplifier.

6-10.1 Design of a Cascode Amplifier

Figure 6-24 shows a transistor amplifier (Q_3) driving a cascode amplifier. In this circuit, the designer has many options in selecting the dc output level and the values of the resistors to suit the needs of the circuit.

---

[3]The level-shifter is described in Schilling and Belove. The high-frequency type is described in Pierce and Paulus, and Boylestad and Nashelsky (see References).

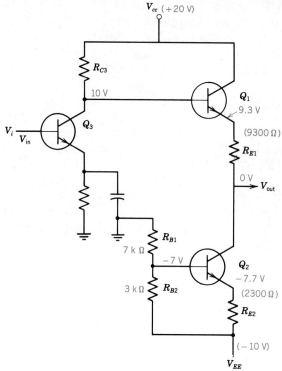

FIGURE 6-24
A transistor amplifier driving a cascode amplifier.

EXAMPLE 6-23

Design a cascode amplifier for the circuit of Figure 6-24 with

$$V_{CC} = +20 \text{ V}$$
$$V_{EE} = -10 \text{ V}$$

Assume the dc level of the input, at the collector of Q_1, is 10 V and the required dc output level is 0 V. Also assume that the ac output voltage can swing by ± 5 V around the quiescent point. Find R_{B1}, R_{B2}, R_{E1}, and R_{E2}.

Solution
The first choice to be made here is the value of the current in the constant current generator. Let us choose a value of 1 mA and then calculate the resistors needed to produce this current. A second choice that must be made

278   MULTIPLE TRANSISTOR CIRCUITS

is the value of the base voltage at Q_2. The design requires the dc output voltage V_{out} to be 0. Thus, the output voltage will vary from $+5$ V to -5 V. This voltage is also on the collector of Q_2. The base voltage of Q_2 must be less than -5 V in order to keep Q_2 unsaturated. For the design, a base voltage of -7 V is therefore selected.

If R_{B1} and R_{B2} are not too large, the current in them will be large, compared to the base current, I_{B2}. Then R_{B1} and R_{B2} form a voltage divider. Reasonable choices for R_{B1} and R_{B2} to produce a base voltage of -7 V with $V_{EE} = -10$ V, are $R_{B1} = 7$ kΩ and $R_{B2} = 3$ kΩ (see problem 6-34).

Now that the voltage and currents have been determined, the values of R_{E1} and R_{E2} can be found. $V_{E2} = -7.7$ V and since I_{E2} has been chosen as 1 mA

$$R_{E2} = \frac{2.3 \text{ V}}{1 \text{ mA}} = 2300 \text{ }\Omega$$

The voltage at the emitter of Q_1 is 9.3 V. Because the output voltage is to be 0 V (dc) and the current in R_{E1} is 1 mA

$$R_{E1} = \frac{9.3 \text{ V}}{1 \text{ mA}} = 9300 \text{ }\Omega$$

The values are shown in blue in Figure 6-24.

Analysis of a cascode amplifier is simpler than the design. Once the current of the constant-current source is determined, the dc levels throughout the circuit can be found (see problem 6-35).

6-11 SUMMARY

Various multiple-transistor special purpose circuits were discussed in this chapter. It began with the multiple-transistor amplifier, used when the gain of a single transistor is insufficient. Both multistage BJT and FET amplifiers were considered. Then two special purpose circuits, Darlingtons and constant-current sources, were considered. This led to an explanation of two commonly used circuits, comparators and difference amplifiers. Because these circuits are widely used, their discussion was quite detailed. Finally, an examination of the cascode amplifier was included to complete the chapter.

6-12 GLOSSARY

Balancing resistors. Resistors placed in the legs of a difference amplifier to balance the performance of the legs.

Blocking capacitor. A capacitor used to isolate the dc level of one stage from the dc or biasing level of the next stage of a circuit.

Cascode amplifier. An amplifier consisting of an emitter-follower and a constant-current source. It is designed to shift the dc level of an ac signal.

Common mode gain (A_c). The ratio of the change of output voltage to the change of common mode input voltage in a differential amplifier.

Common mode rejection ratio (CMRR). The ratio of A_d to A_c and a measure of the merits of a difference amplifier.

Common mode voltage. A voltage that causes both inputs to an amplifier to change equally.

Comparator. A circuit that compares two input voltages and produces 1 of 2 output voltages, depending upon which input voltage is larger.

Constant-current source. A source that delivers a constant current to a load, regardless of changes in the load.

Darlington. A two-transistor configuration for high current gain (see Figure 6-8 or 6-9).

Decibel (dB). A logarithmic measure of gain (see equations 6-9 and 6-10).

Difference amplifier. An amplifier designed to amplify difference voltages while rejecting common-mode voltages.

Difference mode voltage (A_d). The ratio of the change of output voltage to the change of difference mode input voltage in a differential amplifier.

Direct-coupled amplifier. An amplifier that uses no capacitors, or can amplify changes in a dc input level.

6-13 REFERENCES

ROBERT BOYLESTAD and LOUIS NASHELSKY, *Electronic Devices and Circuit Theory*, 4th Edition, Prentice-Hall, Englewood Cliffs, N. J., 1987.

JACOB MILLMAN, *Microelectronics*, McGraw-Hill, New York, 1979.

J.F. PIERCE and T.J. PAULUS, *Applied Electronics*, Charles E. Merrill, Columbus, Ohio, 1972.

DONALD L. SCHILLING and CHARLES BELOVE, *Electronic Circuits, Discrete and Integrated*, 2nd Edition, McGraw-Hill, New York, 1979.

HENRY ZANGER, *Semiconductor Devices and Circuits*, Wiley, New York, 1984.

6-14 PROBLEMS

6-1 Repeat Example 6-1 if Z_{out} of each stage is 1 kΩ, instead of 4 kΩ.

6-2 Find the gain of the circuit of Figure P6-2. If $V_{in} = 1$ mV, find V_{C1} and V_{C2}.

6-3 Find the gain of the circuit of Figure P6-2 if capacitor C_2 is removed.

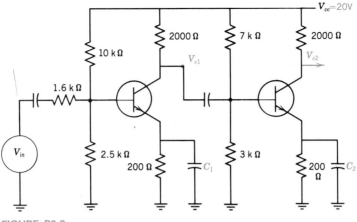

FIGURE P6-2

6-4 Find the exact gain for the circuit of Figure P6-4, taking h_{re} and h_{oe} into account.

6-5 For Figure P6-4, find the gain of the circuit if $h_{re} = h_{oe} = 0$.

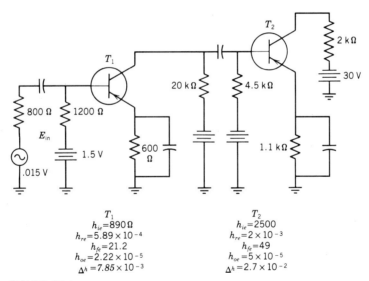

T_1
$h_{ie} = 890\,\Omega$
$h_{re} = 5.89 \times 10^{-4}$
$h_{fe} = 21.2$
$h_{oe} = 2.22 \times 10^{-5}$
$\Delta h = 7.85 \times 10^{-3}$

T_2
$h_{ie} = 2500$
$h_{re} = 2 \times 10^{-3}$
$h_{fe} = 49$
$h_{oe} = 5 \times 10^{-5}$
$\Delta h = 2.7 \times 10^{-2}$

FIGURE P6-4

6-6 Find the gain, the input impedance, and the output impedance of the circuit of Figure P6-6. Assume h_{fe} for all transistors is 100.

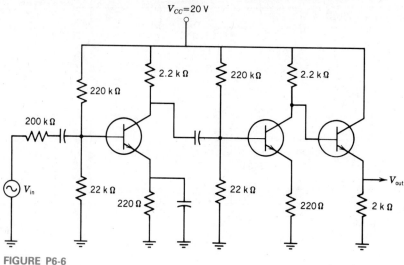

FIGURE P6-6

6-7 For the circuit of Figure P6-7, assume $h_{FE} = 120$ for both transistors. Find the dc levels throughout the circuit, h_{ie} for both the *npn* and *pnp* transistors, and the gain of the circuit.

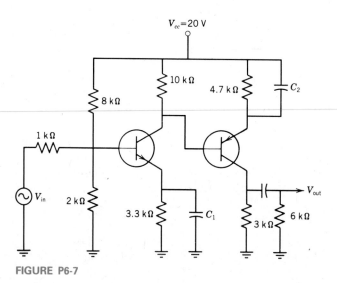

FIGURE P6-7

6-8 Repeat Problem 6-7 if capacitor C_1 is open-circuited.

6-9 Repeat Example 6-8 if the output of stage 2 is capacitively coupled to a 1 kΩ load.

6-10 Find the gain of the circuit of Figure P6-10.

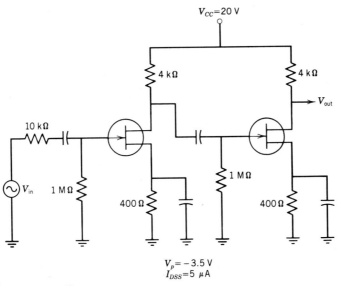

FIGURE P6-10

6-11 Repeat Problem 6-10 if a capacitively coupled 500 Ω resistor is added to the output.

6-12 The circuit of Figure P6-7 can become a direct coupled amplifier if all the capacitors are removed. Find its gain. Assume the 6 kΩ resistor is also removed.

6-13 For the circuit of Figure P6-13, assume both transistors have an h_{FE} of 100.

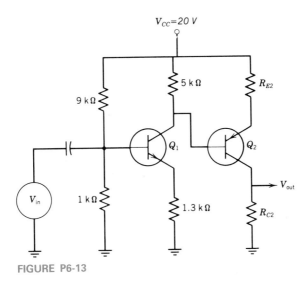

FIGURE P6-13

6-14 PROBLEMS

a. Find R_{E2} so that the current in Q_2 is 1 mA.
b. Find R_{C2} so that V_{C2} is halfway between V_{B2} and ground.
c. Find the gain of the dc coupled circuit.

6-14 Using exact formulas, find the current gain and input impedance of a Darlington similar to Figure 6-8, if $h_{fe1} = 120$, $h_{fe2} = 40$, $R_E = 50\ \Omega$, and $V_{CC} = 20$ V. Also find R_B. Assume V_{out} is 8 V.

6-15 Repeat problem 6-14 using the approximate formulas.

6-16 a. Find the gain of a **2N6034** Darlington at a collector current of 0.3 A at 25°C.
b. Design a circuit to work at these values with $V_{EQ} = 10$ V and $V_{CC} = 20$ V. Find the emitter resistor and the bias resistor.

6-17 Repeat Example 6-10 if R_E is changed to 300 Ω.

6-18 For the constant current of Figure P6-18, find the constant current. Over what range of R_C will it remain constant?

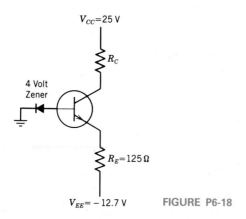

FIGURE P6-18

6-19 For the circuit of Figure P6-19, find the current in each transistor and the voltage at each collector if
a. $V_{in} = .7$ V
b. $V_{in} = 4.7$ V
c. $V_{in} = 6.7$ V
Fill in the table in the figure.

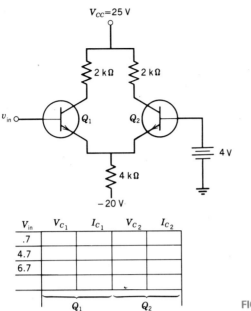

V_{in}	V_{C_1}	I_{C_1}	V_{C_2}	I_{C_2}
.7				
4.7				
6.7				
	Q_1		Q_2	

FIGURE P6-19

6-20 For the circuit of Figure P6-20, sketch the output as v_{in} goes from 0 to 10 V.

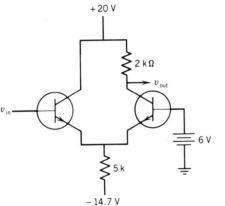

FIGURE P6-20

6-21 Using the circuit of Figure P6-21, design a comparator circuit that will provide an output swing of 10 V. Use $V_{CC} = +15$ V, $V_{EE} = -15$ V, $V_R = 3$ V. Find the collector voltage of Q_1 if
  a. $v_i = 0$ V
  b. $v_i = 10$ V

6-14 PROBLEMS 285

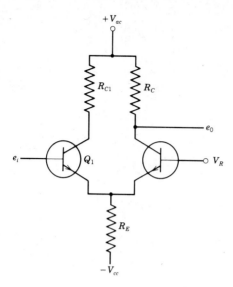

FIGURE P6-21

6-22 Some stereo sets have rows of light-emitting diodes (LEDs). The higher the volume, the more LEDs light up. Conceptually design this part of the stereo set.

6-23 Rework Example 6-14 if V_{BB} is 5 V, and $h_{FE1} = h_{FE2} = 150$.

6-24 In response to a 2 V common-mode input voltage, the output voltage of a difference amplifier changes by 0.1 V. In response to a 0.05 V difference voltage, its output changes by 2 V. Find A_c, A_d, and its CMRR.

6-25 A difference amplifier has $R_{C1} = R_{C2} = R_E = 2000\ \Omega$. If the transistors have an h_{fe} of 150 and an h_{ie} of 1000 Ω, find A_c, A_d, and its CMRR. Assume R_{Th} is 500 Ω.

6-26 For the difference amplifier of Figure P6-26, find the difference-mode gain, the common-mode gain and the CMRR. Assume $h_{fe} = 100$ and $h_{ie} = 30 h_{fe}/I_e$. Assume the average dc level of the sine wave is 0.

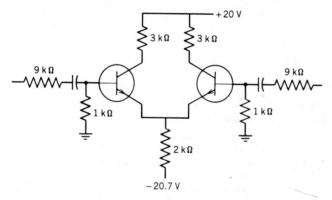

FIGURE P6-26

6-27 Repeat problem 6-26 if a 50 Ω potentiometer is used as a balancing resistor (assume it places 25 Ω in each leg of the difference amplifier).

6-28 A circuit has a gain of 25,000. How many dBs is this?

6-29 A difference amplifier has a CMRR of 75 dB. Find the ratio of A_c to A_d.

6-30 For problems 6-26 and 6-27, express the CMRR in dBs.

6-31 Repeat Example 6-22 if pin 5 is connected to -3 V and pin 4 is connected to -4 V. Is Q_5 saturated under these conditions?

6-32 In Figure 6-23, if $R_{E1} = 7$ kΩ, $R_{E2} = 5$ kΩ, $V_{BB} = -5$ V, $V_{EE} = -10.7$ V, and $V_{CC} = +10$ V, and the dc level of the input is $+5$ V, what is the dc level of the output?

6-33 Design a circuit similar to Figure 6-23 with $V_{CC} = +20$ V and $V_{CC} = -20$ V so that an input signal with a $+10$ V dc level is changed to an output signal with a ground level.

6-34 In Figure 6-24, if $h_{fe} = 100$ for Q_2, find the precise currents in R_{B1} and R_{B2}, and the voltage on the base of Q_2.

6-35 Find the gain of the cascode amplifier of Figure 6-24 if $R_{B1} = R_{B2} = 5$ kΩ, $R_{C3} = 2$ kΩ, $V_{CC} = +20$ V, $V_{EE} = -15.7$ V, $R_{E2} = 3$ kΩ, and $R_{E1} = 5$ kΩ. Assume all transistors have an h_{fe} of 100 and the dc level at the collector of Q_3 is 10 V. What is the dc level of the output?

After attempting the problems, return to section 6-2 and reread the self-evaluation questions. If any of them seem unclear, review the appropriate sections of the chapter to find the answers.

CHAPTER SEVEN

The Frequency Response of Amplifiers

7-1 Instructional Objectives

The gain of the amplifiers studied in the previous chapters tends to fall off at both high and low frequencies. The reasons for these frequency response limitations are explained in this chapter. After reading it, the student should be able to:

1. Construct the Bode plot for an amplifier.
2. Find f_L for the high-pass circuit and for circuits using blocking capacitors.
3. Calculate the low-frequency response of an amplifier due to the emitter bypass capacitor.
4. Find f_H for the low pass circuit.
5. Find f_β and $C_{b'e}$ for a transistor.
6. Calculate f_H for a BJT amplifier.
7. Find the frequency response of JFET and multistage circuits.
8. Use percentage tilt and rise time to determine the high- and low-frequency response of a circuit.
9. Compensate a scope probe.
10. Determine the resulting rise time when two circuits are cascaded.

7-2 Self-Evaluation Questions

Watch for the answers to the following questions as you read the chapter. They should help you to understand the material presented.

1. What capacitors cause the low-frequency response? What capacitors cause the high-frequency response?
2. Why are blocking capacitors usually less important than emitter-capacitors in determining f_L for an amplifier?
3. Why is the collector-to-base capacity especially important in determining f_H?
4. What is the collector pole? Why is it usually ignored in high-frequency BJT calculations?
5. Explain why the gain–bandwidth product of an amplifier tends to remain constant.
6. Why do stage interactions affect the frequency response of multistage circuits?
7. How are bandwidth and rise time of a square-wave output related?
8. What is the advantage of attenuating scope probes? Why must they be perfectly compensated?

7-3 CAPACITIVE EFFECTS

In the previous chapters capacitors have been considered to be ideal; they were perfect short circuits for ac signals and open circuits for dc. The actual effects of the capacitors in an amplifier is to make the gain of the amplifier dependent on the frequency of the applied ac signal.

A plot of the gain of an amplifier versus frequency is called a *frequency response curve*. A typical frequency response curve for the single-stage amplifiers discussed in Chapters 4 and 5 is given in Figure 7-1. It can be divided into three regions:

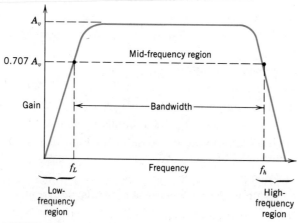

FIGURE 7-1
A frequency-response curve for a typical amplifier.

1. The *low-frequency* region. In the low-frequency region the gain drops off because of the coupling capacitors and the emitter capacitors. The gain decreases (the effects get worse) as the frequency decreases.
2. The *mid-frequency* region. In the mid-frequency range the effects of the capacitors are ignored and they are assumed to function ideally. The gain in this region, typically from 500 Hz to 100,000 Hz, is constant and is the gain of the amplifier as calculated in Chapters 4 and 5.
3. The *high-frequency* region. In the high-frequency region the gain drops off due to the unavoidable stray capacitance between the wires and the electrodes of the transistor. The gain decreases as the frequency increases.

In Figure 7-1, the maximum, mid-frequency gain is A_v. The two frequencies where the gain is $0.707A_v$ are called the *lower-half-power* and *upper-half-power* frequencies (f_L and f_H) respectively and occur at the intersection of the dashed line and the blue frequency response curve on the figure. They are also called the *3dB points* because the gain is 3 decibels[1] below maximum at these frequencies (see problem 7-1).

EXAMPLE 7-1

Show that the power output of the amplifier at f_L and f_H is half the maximum power output.

Solution

Assume an input voltage of constant magnitude V_{in}. At mid-frequency the output voltage is A_vV_{in} into a load resistor, R. Thus the power of the ac output is $[A_vV_{in}]^2/R$. At the half-power frequencies the output voltage is $0.707A_vV_{in}$ and the output power is $[0.707A_vV_{in}]^2/R = 0.5[A_vV_{in}]^2/R$ or half the power at mid-frequency.

7-3.1 Bode Plots

A *Bode plot* is a simplified, straight-line drawing that *approximates* the frequency response curve of an amplifier. A Bode plot for the frequency response curve of Figure 7-1 is shown in Figure 7-2. It is assumed to be a straight line at maximum gain for all frequencies between f_L and f_H. Beyond these frequencies, the gain, plotted on a decibel scale, drops off at the rate of 6 dB per octave (an octave means a doubling of the frequency) or 20 dB per decade (a decade means the frequency changes by a factor of 10).

In Figure 7-2 the actual frequency response curve is shown in blue and

---

[1] Decibels were introduced in section 6-9.6.

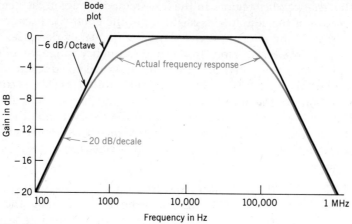

FIGURE 7-2
Bode plot and actual frequency response of an amplifier.

the Bode plot is shown in black. The maximum difference between the Bode approximation and the actual gain occurs at f_L and f_H, where the actual gain is 3 dB down, but the Bode plot still shows maximum gain. At all other frequencies the difference is less than 3 dB. Figure 7-2 is a logarithmic plot for an amplifier with f_L assumed to be 1000 Hz and f_H assumed to be 100,000 Hz for simplicity. The frequency axis is logarithmic, as it is in most frequency plots. The gain axis is in decibels, which are a logarithmic measure of the gain.

7-3.2 Phase Shift

It will be shown in section 7-4 that the calculations of the frequency response of an amplifier involve complex numbers. These complex numbers, of the form $R + jX_C$, produce both a *magnitude* and a *phase angle*. The magnitude is the "gain" of the circuit caused by the capacitor. (This gain is usually less than 1 and is often called an *attenuation*.) The phase angle indicates that a *phase shift* has occurred, so the output sine wave is not in phase with the input.

For circuits whose frequency response is given in Figure 7-1 or 7-2, there is no phase shift in the mid-frequency range. At very low frequencies the phase shift is 90°, leading, and at high frequencies the phase shift is 90°, lagging. At f_L and f_H the phase shift is 45°. Precise calculations of both gain and phase shift are given in sections 7-4 and 7-5.

7-3.3 Poles and Zeros[2]

Figures 7-1 and 7-2 show that the gain of an amplifier is a function of frequency. In general the gain equation is

---

[2]Readers who are unfamiliar with poles and zeros may omit this section.

$$A_v(f) = A_{vmax}\frac{(s-z_1)(s-z_2)\cdots(s-z_n)}{(s-p_1)(s-p_2)\cdots(s-p_n)} \quad (7\text{-}1)$$

where s is the complex frequency and z and p are the complex plane locations of the zeros and poles of the function. From equation 7-1 we can see that the gain is zero for any value of s that equals one of the z values, and the gain is infinite for any value of s that equals one of the p values.

Normally we are only interested in the response at actual frequencies, but they occupy the *imaginary axis* on the complex plane, where $s = jw$ or $s = jZ\pi f$. For an amplifier, all of the poles and most of the zeros are complex numbers with a real component, so s for actual frequencies will not equal them.

Another way of looking at thte situation is to say that the gain of an amplifier is equal to a constant A_{vmax} times the vector distances to the zeros divided by the vector distances to the poles. This attribute will be discussed in the following sections.

Bode plots also rely on the poles and zeros of a function. They turn downward by 6 dB per octave at each pole and go upward at 6 dB per octave at each zero. This is illustrated in section 7-4.3, on the effects of the emitter bypass capacitor.

7-4 THE LOW-FREQUENCY RESPONSE OF A BJT AMPLIFIER

As Figure 7-1 shows, the gain of an amplifier falls off at low frequencies. This decrease in gain is due to the effects of the blocking capacitors and the emitter bypass capacitors on the performance of the circuit.

7-4.1 The Blocking Capacitor at Low Frequencies

The low-frequency effect of a blocking capacitor on an amplifier is the same as its effect on a resistor network. Figure 7-3 shows the basic resistor network.

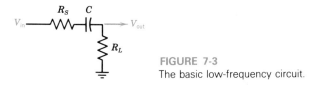

FIGURE 7-3
The basic low-frequency circuit.

R_S is the source resistance and R_L is the load. The basic equation is:

$$V_{out} = V_{in}\frac{R_L}{R_L + R_S + jX_C} \quad (7\text{-}2)$$

As the frequency increases, X_C decreases. At high frequencies X_C becomes negligible compared to the resistors, and equation 7-2 reduces to

$$V_{out} = V_{in} \frac{R_L}{R_L + R_S}$$

This is the same as the gain of the circuit with the capacitor short-circuited.

We shall call the gain of the circuit at mid frequencies, where the capacitor has a negligible effect, A_{vm}. In this circuit $A_{vm} = R_L/(R_L + R_S)$. At lower frequencies

$$V_{out} = V_{in} \frac{R_L}{R_L + R_S + X_C}$$

$$= V_{in} \frac{R_L}{R_L + R_S} \left[\frac{1}{1 + \frac{X_C}{R_L + R_S}} \right]$$

$$= A_{vm} V_{in} \frac{1}{1 + \frac{X_C}{R_C + R_S}}$$

But

$$X_C = \frac{1}{jwC} = \frac{1}{j2\pi fC}$$

$$\frac{V_{out}}{V_{in}} = A_v(f) = A_{vm} \left[\frac{1}{1 + \frac{1}{j2\pi f(C(R_S + R_L))}} \right] \qquad (7\text{-}3)$$

For this circuit we define the lower-half-power frequency as

$$f_L = \frac{1}{2\pi C(R_S + R_L)} \qquad (7\text{-}4)$$

Then equation 7-3 reduces to

$$A_v(f) = A_{vm} \frac{1}{1 + \frac{f_L}{jf}} = A_{vm} \frac{1}{1 - \frac{jf_L}{f}}$$

Note that if $f \gg f_L$, the term jf_L/f becomes small compared to 1 and $A_{vm} \approx A_v$. If $f \ll$ than f_L, the term jf_L/f dominates the denominator and

$$A_v(f) \approx A_{vm} \frac{1}{\frac{-jf_L}{f}} = A_{vm} \frac{jf}{f_L}$$

This is the gain in the extreme low frequency range of Figure 7-1 or 7-2. It is proportional to the frequency f. The j factor indicates a 90° phase shift between the input and the output.

Equation 7-3 is of the form

$$A_v(f) = A_{vm} \frac{1}{1 + \frac{jf_1}{f_2}} \tag{7-5}$$

The term

$$\frac{1}{1 + \frac{jf_1}{f_2}}$$

often occurs when calculating frequency responses. The value of the gain and phase shift for it can be found by calculations involving complex numbers, or by the BASIC computer program given in Figure 7-4.

```
5   REM  This program calculates the gain and phase shift of an
6   REM  amplifer. For low frequencies, F1 is the lower half power
7   REM  frequency and F2 is the frequency of interest. For high
8   REM  frequencies, F2 is the upper half power frequency and F1
9   REM  is the frequency of interest.
10  INPUT F1,F2
20  K = F1/F2
30  D = SQR(1 + K^2)
40  A =   ATN(K)*180/3.1416
50  V = 1/D
60  PRINT   "The gain is ", V
65  PRINT   "The phase shift is ", A, "degrees."
70  END
```

FIGURE 7-4
A BASIC computer program, FREQ. BAS, for calculating frequency responses. *Note:* ATN is this computer's code for arctangent.

EXAMPLE 7-2

In Figure 7-3, $R_S = 3000\ \Omega$ and $R_L = 1000\ \Omega$. Find the value of the blocking capacitor if f_L is to be 100 Hz. Also find the gain and phase shift of the circuit at 50 Hz.

Solution
From equation 7-3 we have:

$$f_L = \frac{1}{2\pi C(R_L + R_S)}$$

or

$$C = \frac{1}{2\pi f(R_L + R_S)}$$

$$C = \frac{1}{2\pi \times 100\ \text{Hz} \times 4000} = 0.398\ \mu\text{F}$$

To find the gain at 50 Hz, the computer program of Figure 7-4 can be used with $f_1 = f_L = 100$ Hz and $f_2 = 50$ Hz. It gives a gain of 0.447 and a phase shift of 63.4°. Here

$$A_{vm} = \frac{R_L v_{in}}{R_L + R_S} = \frac{1000\ \Omega \times v_{in}}{1000\ \Omega + 3000\ \Omega} = 0.25\ v_{in}$$

and

$$v_{out} = v_{in} \times 0.25 \times 0.447 = 0.112\ v_{in}$$

so the gain (actually attenuation) of this circuit at 50 Hz is 0.112. Note that the gain can also be calculated approximately as

$$A = A_{vm} \times \frac{f}{f_L} = 0.25 \times \frac{1}{2} = 0.125\ v_{in}$$

Even at 50 Hz, which is as high as half of f_L, it is close to the actual gain.

EXAMPLE 7-3

For the circuit of Figure 7-5, which is Figure 4-9 repeated here for convenience, find f_L for the input circuit if $C = 10\ \mu\text{F}$.

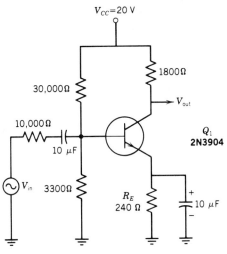

FIGURE 7-5
Circuit for Example 7-3.

Solution
In this circuit, $R_S = 10\ k\Omega$ and $R_L = 30{,}000\ \|\ 3300\ \|\ h_{ie}$. Assuming $h_{ie} = 900\ \Omega$ as before, $R_L = 691\ \Omega$.

$$f_L = \frac{1}{2\pi \times (R_L + R_S)C} = \frac{1}{2\pi \times 10{,}691 \times 10^{-5}} = 1.49\ \text{Hz}$$

Thus, with a 10 μF blocking capacitor, this circuit has a very low f_L.

7-4.2 Pole and Zero Consideration of the Low-frequency Circuit

The low-frequency circuit of Figure 7-3 can also be calculated using pole and zero considerations. The equation for the circuit is

$$v_{out} = v_{in} \frac{R_L}{R_L + R_S + \frac{1}{sC}}$$

$$\frac{v_{out}}{v_{in}} = A_v(f) = \frac{R_L}{R_L + R_S} \times \frac{1}{1 + \frac{1}{(R_L + R_S)sC}}$$

$$A_v(f) = A_{vm} \frac{s}{s + \frac{1}{(R_L + R_S)C}} \qquad (7\text{-}6)$$

Equation 7-6 shows that there is a zero at $s = 0$ Hz and a pole at $s = -1/[C(R_L + R_s)]$. The pole–zero plot for this circuit is shown in Figure 7-6. Note that the frequencies are radian frequencies and must be divided by 2π to obtain the frequency in Hz.

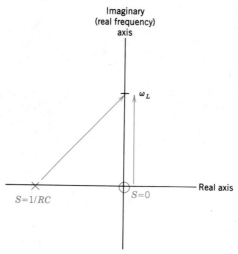

FIGURE 7-6
The low-frequency poles and zeros.

In section 7-3.3 we stated that the magnitude of the gain at any frequency is the gain at mid-frequency, A_{vm}, times the distance from the zero divided by the distance from the pole. At 0 Hz the frequency is right on the zero and the gain is zero, as we would expect. High frequencies are high up on the imaginary axis and the distance from the zero is approximately equal to the distance from the pole. Therefore the ratio is 1, and the gain is A_{vm}, again as we would expect.

EXAMPLE 7-4

Find the gain of the circuit at f_L using pole–zero considerations.

Solution
At f_L, f equals $1/2\pi RC$ and $w = 2\pi f = 1/RC$. The situation at f_L is shown in Figure 7-6. The operating point is on the imaginery (or real frequency) axis at jw_L. Let us call the distance, in radians, to the zero D_Z. Here

$$D_Z = 1/RC$$

We shall also call the distance to the pole D_P. Because the pole is on the real axis at $-1/RC$ and forms an isosceles right triangle with the frequency axis

$$D_P = \sqrt{2}/RC$$

and

$$A_v(f) = A_{vm}\frac{D_Z}{D_P} = A_{vm}/\sqrt{2} = 0.707 A_{vm}$$

Again the gain at f_L agrees with the gain found previously. The phase angle can also be found as the difference between the vector from the zero and the vector from the pole. At f_L, this angle is 45°.[3]

7-4.3 The Emitter Bypass Capacitor

The previous section has shown that it is easy to obtain a low f_L due to the blocking capacitor. Unfortunately, the effects of the emitter bypass capacitor are more complex and more severe.

The gain equation for the emitter bypass resistor is derived in Appendix E. It is

$$A_{v(ck)} = \frac{-h_{fe}R_C\left[s + \frac{1}{R_E C_E}\right]}{[R_{Th} + h_{ie}]\left[\frac{s + (1 + h_{fe})R_E + R_{Th} + h_{ie}}{R_E C_E [R_{Th} + h_{ie}]}\right]} \times \frac{V_{Th}}{V_{in}} \quad (7\text{-}7)$$

Equation 7-7 has a pole at

$$f = \frac{(1 + h_{fe})R_E + R_{Th} + h_{ie}}{2\pi R_E C_E (R_{Th} + h_{ie})}$$

and a zero at $f = 1/(2\pi R_E C_E)$. In practical circuits the pole is at a much higher frequency than the zero, and usually at a *much higher frequency than the pole due to the blocking capacitor*. Therefore the pole frequency due to the emitter capacitor is taken as f_L.

A typical Bode plot for an amplifier at low frequencies is given in Figure 7-7. At high frequencies the gain is $A_{(vm)}$, the mid-frequency gain. Below f_L

---

[3] A thorough discussion of poles and zeros is given in Millman (see References).

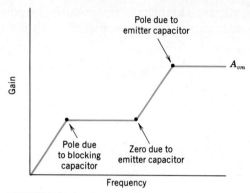

FIGURE 7-7
A typical Bode plot for the low-frequency response of an amplifier.

the gain starts to fall off at 6 dB per octave until the frequency of the zero is reached. Then the gain steadies and remains so until the pole due to the blocking capacitor takes effect. Remember, however, that the Bode plot is an idealized approximation of the actual frequency response.

EXAMPLE 7-5

How does equation 7-7 behave
a. At high frequencies?
b. At low frequencies?

Solution
a. At high frequencies s is much greater than either the zero or the pole. By ignoring the zero and the pole, equation 7-7 reduces to:

$$A_v = \frac{-h_{fe}R_C s}{[R_{Th} + h_{ie}]s} \times \frac{V_{Th}}{V_{in}} = \frac{-h_{fe}R_C}{R_{Th} + h_{ie}} \times \frac{V_{Th}}{V_{in}}$$

But this is exactly the mid-frequency gain formula found in Chapter 4 (see section 4-5.2). As we would expect, at high frequencies the capacitors have no effect and the gain is the mid-frequency gain.

b. At low frequencies, s is small compared to the magnitude of both the pole and the zero. If s is taken as 0, equation 7-7 reduces to

$$A_v = \frac{-h_{fe}R_C}{R_{Th} + h_{ie} + (1 + h_{fe}R_E)} \approx \frac{-R_C}{R_E}$$

300 THE FREQUENCY RESPONSE OF AMPLIFIERS

But this is the gain of a transistor circuit with an unbypassed emitter resistor (see section 4-7.2), or a circuit without an emitter capacitor. Again this is exactly what we would expect because capacitors become an open circuit at low freqencies.

EXAMPLE 7-6

If the emitter bypass capacitor in Figure 7-5 is 10 µF, find the pole and zero frequencies.

Solution
For the circuit of Figure 7-5

$$R_{Th} = 30{,}000\ \Omega\ \|\ 3{,}300\ \Omega\ \|\ 10{,}000\ \Omega = 2292\ \Omega \qquad \text{(see Example 4-7)}$$

Assuming the transistor has an h_{fe} of 150, and an h_{ie} of 900 Ω we have

$$f = \frac{(1 + h_{fe})R_E + R_{Th} + h_{ie}}{2\pi R_E C_E (R_{Th} + h_{ie})}$$

$$= \frac{151 \times 240\ \Omega + 2292\ \Omega + 900\ \Omega}{2\pi \times 240\ \Omega \times 10^{-5} F (3192\ \Omega)} = \frac{39432\ \Omega}{48.1} = 820\ \text{Hz}$$

The zero is at $f = \dfrac{1}{2\pi R_E C_E} = \dfrac{10^5}{6.28 \times 240} = 66.3\ \text{Hz}$

Thus the pole is well above the zero.

EXAMPLE 7-7

Find the gain and phase shift of the circuit of Figure 7-5 at a frequency of 500 Hz.

Solution
From equation 7-7 we have

$$A_v = \frac{-V_{Th}}{V_{in}} \times \frac{h_{fe}R_C}{[R_{Th} + h_{ie}]} \times \frac{s + \dfrac{1}{R_E C_E}}{s + \dfrac{(1 + h_{fe})R_E + R_{Th} + h_{ie}}{R_E C_E (R_{Th} + h_{ie})}}$$

The first term is

$$\frac{V_{Th}}{V_{in}} = \frac{3300\ \Omega \parallel 30{,}000\ \Omega}{10{,}000\ \Omega + 3300\ \Omega \parallel 30{,}000\ \Omega}$$

$$= \frac{2973\ \Omega}{12973\ \Omega} = 0.23$$

The second term is

$$\frac{-h_{fe}R_C}{R_{Th} + h_{ie}} = \frac{-150 \times 1800\ \Omega}{3192\ \Omega} = -84.6$$

The third term is

$$\frac{s + \dfrac{1}{R_E C_E}}{s + \dfrac{(1 + h_{fe})R_E + R_{Th} + h_{ie}}{R_E C_E (R_{Th} + h_{ie})}}$$

If s is replaced by $j2\pi f$ and both the numerator and denominator are divided by $j2\pi$, we have

$$\frac{500 - j\dfrac{1}{2\pi R_E C_E}}{500 - j\dfrac{(1 + h_{fe})R_E + R_{Th} + h_{ie}}{2\pi R_E C_E (R_{Th} + h_{ie})}}$$

But the expressions in parentheses have already been calculated in Example 7-6, so this factor reduces to

$$\frac{500 - j66}{500 - j820} = \frac{504.3 < -7.5°}{960.4 < -58.6°} = 0.525 < 51.1°$$

Thus the total circuit gain is

$$A_v = -0.23 \times -84.6 \times 0.525 < 51.1° = 10.2 < -128.9°$$

The gain in dB is

$$A_{v(dB)} = 20 \log_{10} 10.2 \approx 20\ \text{dB}$$

7-4.4 A Frequency Response Program

Equation 7-7 is of the form

$$A_v = K \frac{S + Z1}{S + P1} \qquad (7-8)$$

where $Z1$ is the zero, $P1$ is the pole, and K is the constant.

$$K = \frac{V_{Th}}{V_{in}} \times \frac{-h_{fe}R_C}{R_{Th} + h_{ie}} \quad \text{or} \quad A_{vm}$$

The program of Figure 7-8 requests the user to input $P1$, $Z1$ and K. It then calculates the frequency response of the circuit for a range of frequencies from 10 Hz to 5000 Hz. The gain is calculated in both magnitude and dBs. The dBs are normalized to be 0 at mid-frequency. Therefore all dB calculations are negative.

```
5   PRINT " INPUT THE POLE AND ZERO IN HERTZ"
10  INPUT P1,Z1
12  PRINT "ENTER MIDFREQUENCY GAIN OR VALUE"
15  INPUT AV
20  DATA 10,20,50,70,100,200,500,700,1000,2000,5000
30  PRINT "    FREQ          GAIN         GAIN(DB)         ANGLE "
40  PRINT
50  FOR N = 1 TO 11
60  READ F
70  M1 = SQR (F^2 + Z1^2)
80  A1 = ATN (F/Z1)
90  M2 = SQR (F^2 + P1^2)
100 A2 = ATN (F/P1)
110 K = M1/M2
120 PRINT  F, AV*K, 20*LOG10(K),  180*(A1-A2)/3.1416
130 NEXT N
140 END
```

FIGURE 7-8
A BASIC program, FREQ Z, B4S, for Equation 7-8.

EXAMPLE 7-8

Use the program of Figure 7-8 to find the frequency response of the circuit of Examples 7-6 and 7-7.

Solution

For this circuit, $P1 = 820$ Hz, $Z1 = 66.3$ Hz, and $K = 0.23 \times 84.6 = 19.5$. Figure 7-9 shows the input and output files for the given values. The values at 500 Hz are in exact agreement with those obtained in Example 7-8. The computer values can also be plotted to give a curve similar to Figure 7-7.

```
                    820,66.3
                    19.5
```

(*a*) Input file

```
   INPUT THE POLE AND ZERO IN HERTZ
   ENTER MIDFREQUENCY GAIN OR VALUE
        FREQ           GAIN           GAIN(DB)          ANGLE

         10           1.59436         -21.749          7.87853
         20           1.64633         -21.4704         15.3892
         50           1.97108         -19.9066         33.5323
         70           2.28447         -18.625          41.6756
        100           2.83225         -16.7581         49.5025
        200           4.86792         -12.0538         57.9525
        500          10.2407           -5.5941         51.0735
        700          12.7173           -3.71281        44.1033
       1000          15.1118           -2.21435        35.5585
       2000          18.0523            -.67003        20.3949
       5000          19.2446            -.114502        8.55388
```

(*b*) Output file

FIGURE 7-9
A computer printout for Example 7-8.

7-4.5 The Low-frequency Circuit in the Laboratory

The circuit of Figure 7-5 was set up in the laboratory using 1% resistors and a randomly selected transistor. The results are shown in Figure 7-10. From it we see that the maximum gain of the circuit is 25.5 dB. It is 3 dB down at 22.5 dB, or about 960 Hz, compared to the 820 Hz calculated in Example 7-6. The general shape of the curve is similar to the Bode plot of Figure 7-7, and the gain at 500 Hz is 20.5 dB, compared to 20 dB calculated in Example 7-7. Thus the results obtained in the laboratory are quite close to the theoretical results of Examples 7-6 and 7-7.

If the mid-frequency gain of 25.2 dB is subtracted from the dB values in Figure 7-10, the differences between the curve and the printout of Figure 7-9 is less than 1 dB throughout the entire range.

7-5 THE HIGH-FREQUENCY RESPONSE OF A BJT AMPLIFIER

Figure 7-1 shows that the gain of an amplifier also decreases at high frequencies. This gain reduction is caused by small unavoidable capacitance around

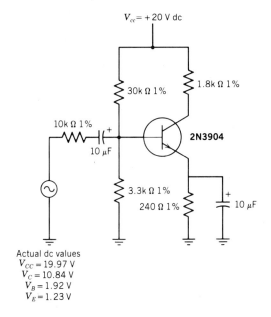

Actual dc values
$V_{CC} = 19.97$ V
$V_C = 10.84$ V
$V_B = 1.92$ V
$V_E = 1.23$ V

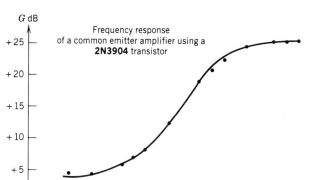

FIGURE 7-10
Frequency response of a common-emitter amplifier using a **2N3904** transistor.

the circuit, such as the capacity between the wires and the ground plane and each other, and the interelectrode capacitance of the transistor (capacity between base and emitter, base and collector, etc.). These capacitances are called *stray capacity* and are too small to be significant at low frequencies, but become important at high frequencies. The gain of a circuit is further reduced by parameters, such as h_{fe}, that decrease at high frequencies.

7-5.1 The Basic High-frequency Circuit

Figure 7-11 shows the basic high-frequency circuit. Both the Thevenin's and Norton's equivalent circuit are shown because the Norton's equivalent circuit is often used in this case. This circuit is sometimes called the *low-pass* circuit because it passes low frequencies ($V_{out} \approx V_{in}$ at low frequencies) while attenuating high-frequency signals.

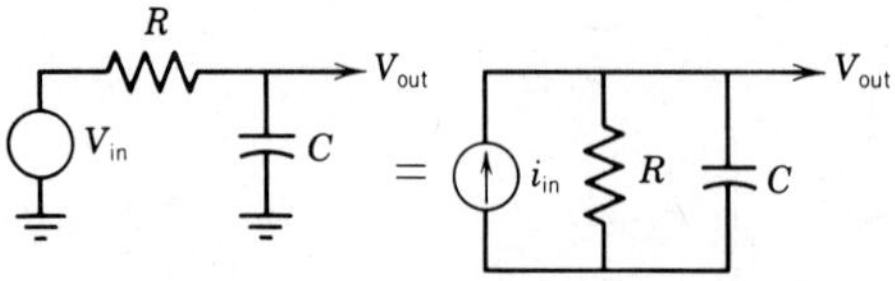

(a) Thevenin   (b) Norton

FIGURE 7-11
The basic high-frequency circuit.

The basic equation for the circuit of Figure 7-11 is

$$v_{out} = \frac{\frac{1}{sC}}{R + \frac{1}{sC}} v_{in} = \frac{1}{s\,CR + 1} v_{in} \qquad (7\text{-}9)$$

If s is replaced by $j2\pi f$, equation 7-9 becomes

$$v_{out} = \frac{1}{1 + j2\pi RCf} v_{in}$$

We define $f_H = 1/2\pi RC$. Then equation 7-9 becomes

$$v_{out} = \frac{1}{1 + \dfrac{jf}{f_H}} v_{in} \qquad (7\text{-}10)$$

It can be seen from equation 7-10 that $v_{out} \approx v_{in}$ if $f \ll f_H$. If $f = f_H$, $v_{out} = 0.707\, v_{in}$ with a 45° phase shift. If $f \gg f_H$, $v_{out} = -jf_H/f \times v_{in}$. In this region, v_{out} falls by 6 dB per octave as f increases.

EXAMPLE 7-9

A current generator is feeding a 1 kΩ resistor. The wiring has a stray capacity of 10 pF. Find f_H.

Solution

$$f_H = \frac{1}{2\pi RC} = \frac{1}{2\pi \times 1000\,\Omega \times 10^{-11}} = \frac{10^8}{2\pi} = 15.9 \text{ MHz}$$

EXAMPLE 7-10

Find the attenuation of the circuit of Example 7-9 at 10 MHz.

Solution

Because 10 MHz is less than f_H, the attenuation factor must be greater than 0.707. This problem can be solved using complex arithmetic, but it can also be solved by the computer program of Figure 7-4. Running the program with $f_1 = 10$ MHz and $f_2 = 15.9$ MHz gives a gain of 0.8465 and a phase angle of $-32.2°$.

7-5.2 Pole–Zero Considerations[4]

Equation 7-9 for the high-frequency or low-pass circuit can be rewritten as

$$v_{out} = \frac{v_{in}}{RC} \times \frac{1}{\left(s + \frac{1}{RC}\right)} \tag{7-11}$$

Equation 7-11 shows that the high-frequency circuit has a pole at $s = -1/RC$ and no zero. Generally $1/RC$ is a high frequency. At 0 Hz the distance to the pole is $1/RC$, and equation 7-11 reduces to $v_{in} = v_{out}$. At $f = f_H$, the distance to the pole is $\sqrt{2}/RC$ so that $v_{out} = 0.707\, v_{in}$ (see problem 7-10). At infinite frequencies the distance to the pole is infinite and v_{out} is 0.

7-6 THE BJT AT HIGH FREQUENCIES

The analysis of the BJT (bipolar junction transistor) at high frequencies is more complex than the simple RC circuit discussed in section 7-5. It depends on the capacities around the circuit, and on the variation of h_{fe} with frequency.

7-6.1 The Transistor Capacitors

Figure 7-12 shows the transistor with the capacitors that affect the frequency response shown in blue. Capacitors that affected the low-frequency response, such as blocking capacitors and the emitter capacitor, are short circuits at high

---

[4]This section may be omitted on first reading.

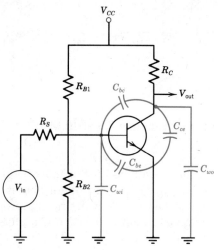

FIGURE 7-12
The capacitors associated with a transistor that affect the high-frequency response.

frequencies and can be ignored. Therefore, in Figure 7-12, the blocking capacitors are not shown and the emitter is shown connected to ground because it is effectively short-circuited by the emitter capacitor.

The capacitors shown in Figure 7-12 are

1. C_{wi} This is the stray wiring capacity on the input side of the transistor. It is the capacity between the wire to the base and ground.
2. C_{be} This is the base-to-emitter interelectrode capacity. It is the capacity between the base and emitter junctions of the transistor and is in parallel with C_{wi}.
3. C_{bc} The collector-to-base interelectrode capacity. This is the capacity between the base and collector junctions of the transistor.
4. C_{ce} The collector-to-emitter interelectrode capacity. This is the capacity between the collector and emitter of the transistor.
5. C_{wo} The stray wiring capacity on the output side of the transistor. It is in parallel with C_{ce}.

Low-power transistors, such as the **2N3904**, usually have small leads and are connected by small wires. As a result, the capacitances listed before are also very small (usually less than 5 pF). This makes it almost impossible to measure them, and their values are often arrived at by an "educated guess."

The interelectrode capacitances for the **2N3904** transistors are given in the manufacturer's specifications (see Appendix A). The curves are for C_{ib} and C_{ob} versus the reverse-bias voltage. These capacitors are partially due to the charges on either side of the depletion region of the junction and become smaller as the reverse voltage increases because the depletion region becomes wider. C_{ib} is the input capacitor and is the same as C_{be}. C_{ob} is the same as C_{bc}.

When the transistor is connected as an amplifier, the base is forward-biased so C_{ib} is the capacity at about a reverse voltage of 0 V or $C_{ib} = C_{be} \approx 5$ pf. The collector-to-base junction is reverse-biased, however. If the reverse bias is assumed to be 5 V (the collector is 5 V positive with respect to the base), the curve gives C_{ob} or C_{bc} as 2 pF. This is a reasonable value of C_{bc} for this type of transistor.

7-6.2 The Hybrid-pi Equivalent Circuit

The *hybrid-pi* is the name given to the equivalent circuit used for the high-frequency analysis of BJT amplifiers. It is shown in Figure 7-13. The base, collector, and emitter terminals of the transistor are shown by circles. In this model the *base junction* is at b', but the *actual base contact* is at b. In the hybrid-pi model, the base resistance is divided into two components, $r_{b'b}$ and $r_{b'e}$, where $r_{b'b}$ is called the *base-spreading resistance* and is the ohmic resistance between the base contact and the base junction, and $r_{b'e}$ is the ohmic resistance of the forward-biased junction.

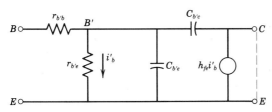

FIGURE 7-13
The hybrid-pi equivalent circuit of a transistor.

The input resistance of the transistor is still h_{ie}. Here

$$h_{ie} = r_{b'b} + r_{b'e}$$

In most transistors, $r_{b'b}$ is very small, typically between 10 and 50 Ω, and is very difficult to measure. In these cases, $r_{b'b}$ is often ignored, the base and b' are considered as the same point, and $r_{b'e}$ equals h_{ie}.

The capacitor $C_{b'e}$ is the *diffusion capacitor*. It occurs because an increase of v_{be} causes more charges to move across the junction. At high frequencies, however, the voltage v_{be} reverses itself very rapidly, before all the charges have been able to respond to the change of voltage. This phenomenon increases as the frequency increases and is represented by the *diffusion capacitor*, $C_{b'e}$. $C_{b'e}$ is typically between 50 and 200 pF, and is much larger than the stray and interelectrode capacities.

7-6.3 The Gain–Bandwidth Product and $C_{b'e}$

In the hybrid-pi circuit of Figure 7-13, the collector can be shorted to the emitter as shown by the blue dashed line in the figure. The current that flows in the short circuit is $h_{fe}i'_b$ or Bi'_b. For high frequency circuits, most literature uses B as the *short-circuit-current gain*.

Figure 7-14 shows the variation of h_{fe} or β with frequency. The curves

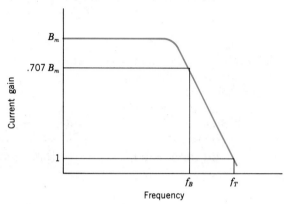

FIGURE 7-14
The variation of gain with frequency.

look very much like the high-frequency response of an *RC* circuit, and the equation is

$$\beta(f) = \frac{i_c}{i'_b} = \frac{\beta}{1 + j\dfrac{f}{f_\beta}} \tag{7-12}$$

There are three frequencies of interest:

1. The mid-frequency. In this range β is maximum and is equal to the h_{fe} of the transistor.
2. f_β. This is the frequency where β is down by 3 dB, or is 0.707 of its mid-frequency value.
3. f_T. This is a very high frequency where the short-circuit-current gain is down to 1. f_T is defined as the *gain–bandwidth product* partially because it is the frequency or bandwidth where the current gain is 1. Transistor manufacturers generally specify f_T. For the **2N3904** the specifications show a minimum f_T of 300 MHz.

From equation 7-12 a relationship between f_T and f_β can be found. At f_T, β = 1 and the frequency is much higher than f_β. Therefore the denominator of equation 7-12, $1 + jf/f_\beta$, becomes f/f_β. Substituting 1 for β(f) and f_T for f, equation 7-12 becomes

$$f_T = \beta f_\beta \tag{7-13}$$

310 THE FREQUENCY RESPONSE OF AMPLIFIERS

EXAMPLE 7-11

A transistor has an h_{fe} of 150 and an f_T of 300 Mhz. Find its f_β.

Solution
From equation 7-13 we have

$$300 \text{ MHz} = 150 f_\beta$$

and

$$f_\beta = 2 \text{ MHz}.$$

After f_β has been found, $C_{b'e}$ can be found. In Figure 7-13 with the collector short-circuited to the emitter, $C_{b'c}$ is in parallel with $C_{b'e}$. The upper-half-power frequency for the transistor input circuit is calculated just as though it were an *RC* circuit (see section 7-5). Using the equations developed there, we find

$$f_B = \frac{1}{2\pi r_{b'e}(C_{b'e} + C_{b'c})} \qquad (7\text{-}14)$$

$C_{b'c}$ is typically small compared to $C_{b'e}$ and is sometimes neglected.

EXAMPLE 7-12

A transistor has $r_{b'e} = 900 \;\Omega$ and $f_B = 2$ MHz. Find $C_{b'e}$. Neglect $C_{b'c}$.

Solution
Equation 7-14 can be transposed as

$$C_{b'e} = \frac{1}{2\pi r_{b'e} f_\beta} = \frac{1}{2\pi \times 900 \;\Omega \times 2 \times 10^6 \text{ MHz}} = 88.5 \text{ pF}$$

7-6.4 The Miller Capacitance

The collector-to-base capacity, C_{bc}, consists of the interelectrode capacity, $C_{b'c}$ and the wiring and terminal between the base and collector leads. For a small transistor like the **2N3904**, C_{bc} is typically about 4 pF. This small capacitance is, however, crucial to the determination of the high frequency response of a transistor circuit. Figure 7-15 shows why the base-to-collector capacitance is so important. Consider first the base-to-emitter capacitor. The voltage at the

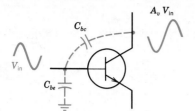

FIGURE 7-15
The Miller effect.

base is v_{in} and the emitter voltage is ground for high frequency signals. Therefore, the current that flows in C_{be} is v_{in}/X_C. Now looking at the base-to-collector junction, Figure 7-15 shows that the base voltage is v_{in}, but the voltage on the collector is $-A_v v_{in}$. Consequently, the voltage *across* C_{bc} is $(1 + A_v)v_{in}$ and a much larger current flows in this capacitor. To account for this larger capacitance in equivalent circuits, C_{bc} is multiplied by the factor $(1 + A_v)$ to give an equivalent capacity to ground. This is called the *Miller capacitor*, C_M, and

$$C_M = C_{bc}(1 + A_v) \tag{7-15}$$

EXAMPLE 7-13

A transistor has the following parameters: $C_{be} = 5$ pF, $C_{b'e} = 100$ pF, $C_{bc} = 3$ pF, and the gain of the transistor is 100. Find the equivalent input capacitance.

Solution
The equivalent input capacity is the sum of all the capacities seen at the input:

$$C_{eq} = C_{be} + C_{b'e} + (1 + A_v)C_{bc}$$
$$= 5 \text{ pF} + 100 \text{ pF} + (101) \times 3 \text{ pF}$$
$$C_{eq} = 408 \text{ pF}$$

Although C_{bc} is the smallest capacitance, it makes the largest contribution to C_{eq} because of the Miller effect.

Many authors calculate the transistor gain, A_v, as $-g_m R_L$, where g_m is the *transconductance* of the transistor as defined in Chapter 5. If a load resistor, R_L, is placed at the collector, then

$$g_m = \frac{\Delta i_{out}}{\Delta v_{in}}$$

In Figure 7-13 $i_{out} = h_{fe}i_{b'}$, and $v_{in} = r_{b'e}i_{b'}$. Therefore, $g_m = h_{fe}/r_{b'e}$ and $A_v = -g_m R_L = -h_{fe}R_L/r_{b'e}$. When $r_{b'b}$ is small, $r_{b'e} \approx h_{ie}$, and $A_v = -h_{fe}R_L/h_{ie}$, the same equation used in previous chapters.

7-6.5 Calculating the High-frequency Response

The high-frequency equivalent circuit at the input of a BJT amplifier generally determines its high-frequency response and can be used to find f_H. This circuit is shown in Figure 7-16, where R_{Th} is the Thevenin's equivalent resistance looking outward from the base terminal, $C_{b'e}$ is the diffusion capacitance, and C_M is the Miller capacitance. In this figure stray capacitances are ignored because they are generally small compared to the Miller and diffusion capacitances.

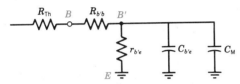

FIGURE 7-16
The high-frequency input circuit for a BJT.

If the input terminals of Figure 7-16 are shorted, the circuit becomes an RC circuit and f_H can be found from it. The steps needed to find f_H are as follows:

1. Find f_T, h_{fe} and $r_{b'b}$ from the manufacturer's specifications. If $r_{b'b}$ is not given, assume a small value (20 to 50 Ω). Assuming that $r_{b'b} = 0$ Ω is sufficiently accurate in most cases and simplifies the calculations.
2. Find h_{ie}, either from the manufacturer's specifications or from the equation $h_{ie} = 30\, h_{fe}/I_E$.
3. With h_{ie} and $r_{b'b}$ determined, $r_{b'e}$ can be found.
4. The resistance of the circuit is now $r_{b'e}$ in parallel with $R_{Th} + r_{b'b}$. Note that the high-frequency response cannot be taken if a generator is connected directly to the base of the transistor, because R_{Th} is zero under this condition and effectively shorts out $r_{b'e}$.
5. Find f_B and $C_{b'e}$ as shown in Examples 7-11 and 7-12.
6. Find g_m, and use it to calculate A_v and the Miller capacitance.
7. Once all the resistances and capacitances have been found, the value of f_H can be calculated.

EXAMPLE 7-14

The transistor of Figure 7-5 has the following parameters: $h_{fe} = 150$, $h_{ie} = 920$ Ω, $r_{b'b} = 20$ Ω, and $f_T = 300$ MHz. Assume $C_{bc} = 4$ pF. Find f_H for this circuit.

Solution
Each of the resistors and capacitors in Figure 7-16 must be calculated before f_H can be found.

$$r'_{be} = h_{ie} - r_{b'b} = 920\ \Omega - 20\ \Omega = 900\ \Omega$$

$$f_\beta = \frac{f_T}{h_{fe}} = \frac{300\ \text{MHz}}{150} = 2\ \text{MHz}$$

$$C_{b'e} = 88.5\ \text{pF} \quad \text{(see Example 7-12)}$$

$$R_{Th} = 10{,}000\ \Omega\ \|\ 30{,}000\ \Omega\ \|\ 3{,}300\ \Omega = 2292\ \Omega$$

$$R_{Th} + r_{b'b} = 2312\ \Omega$$

The resistance at the base is then $(R_{Th} + r_{b'b})\ \|\ r_{b'e}$ or

$$2312\ \Omega\ \|\ 900\ \Omega = 648$$

$$g_m = \frac{h_{fe}}{r_{b'e}} = \frac{150}{900} = 0.16667\ \text{S}$$

$$A_{v(tr)} = g_m R_C = 0.16667\ \text{S} \times 1800\ \Omega = -300$$

The Miller capacity is

$$C_M = (1 + A_v) C_{bc} = 301 \times 4\ \text{pF} = 1204\ \text{pF}$$

The total capacity is $C_M + C_{b'e} = 1293\ \text{pF}$.
Finally:

$$f_H = \frac{1}{2\pi RC} = \frac{1}{2\pi \times 1293\ \text{pF} \times 648} = 190\ \text{kHz}$$

7-6.6 The Collector Pole
In the basic transistor circuit of Figure 7-5, a second pole exists that also affects the high frequency response. This pole is at the collector and occurs because the stray and interelectrode capacitance, as seen at the collector, is in parallel with the collector resistor, R_C. This capacitance is the sum of the collector-to-base capacitance, C_{bc}, the collector-to-emitter capacitance, C_{ce}, and any wiring capacity.

Two important points should be made about the collector pole. The first is that the Miller effect on C_{bc} is negligible when viewed from the collector (see problem 7-18). The second, and more important point is that the collector pole is usually at a much higher frequency than the pole at the base. Therefore, the pole at the collector does not significantly affect f_H for the circuit. At frequencies above the collector pole frequency, however, the gain of the circuit drops off at 12 dB per octave due to the combined effects of the pole at the base and the pole at the collector.

EXAMPLE 7-15

For the circuit of Figure 7-5, find the frequency of the pole at the collector if the total collector capacitance is assumed to be 10 pF.

Solution
This circuit is basically 10 pF in parallel with 1800 Ω. Therefore:

$$f_H = \frac{1}{2\pi RC} = \frac{1}{2\pi \times 1800 \text{ Ω} \times 10 \text{ pF}} = \frac{10^9}{36\pi} = 8.84 \text{ MHz}$$

The pole at the base is at 190 kHz, and is clearly the dominant pole in determining f_H for this circuit.

7-6.7 The High-frequency Circuit in the Laboratory

The circuit of Figure 7-5 was set up in the lab and a frequency response curve was taken. The results are shown in Figure 7-17.

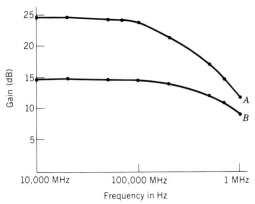

FIGURE 7-17
Laboratory results for the frequency response of the circuit of Figure 7-5.

The A curve is for the circuit as shown, which has a theoretical transistor gain of 300. The curve is down by 3 dB at 170 kHz, compared to the 190 kHz calculated in Example 7-14. This is quite close, but accurate results are very difficult to achieve because C_{bc}, a crucial parameter, is almost impossible to measure and must be approximated.

The lower or B curve in Figure 7-17 is for the circuit with a 680 Ω load resistor placed in parallel with the 1800 Ω collector resistor. This lowers the transistor gain to about 83, and the circuit gain in dB, as shown on the figure is correspondingly lower. But f_H for the circuit has increased significantly. An examination of the lower curve shows that the gain is not down 3 dB until 550 kHz.

7-6.8 The Gain–Bandwidth Product

The *bandwidth* of an amplifier is defined as the range of frequencies between f_L and f_H and was shown in Figure 7-1. But in most circuits f_H is much larger than f_L, so the bandwidth of an amplifier is often considered to be the same as f_H. Generally an increase in the gain of an amplifier is accompanied by a decrease in bandwidth, so that the *gain–bandwidth product*, the product of the mid-frequency gain times f_H, tends to remain constant. For a BJT amplifier, the gain is $g_m R_L$, but the Miller capacitance is also proportional to $g_m R_L$, and this capacitance is predominantly responsible for the limitation on f_H. Therefore any increase in gain due to an increase in g_m, or, more likely, R_L also increases the Miller capacitance and reduces the bandwidth.

EXAMPLE 7-16

What is the gain–bandwidth product for each of the circuits whose curves are shown in Figure 7-17?

Solution

The circuit of curve A, with the 1800 Ω resistor as the load, had a gain of about 25 dB or 17.5. From the curves its f_H is 170 kHz, so the product is

$$\text{gain–bandwidth} = 17.5 \times 170 \text{ kHz} = 3 \text{ MHz}$$

When the 680 Ω resistor was placed in parallel with the collector resistor, the gain dropped to about 15 dB or 5.5, but f_H went up to 550 kHz. The product for this circuit is

$$5.5 \times 550 \text{ kHz} = 3 \text{ MHz}$$

So for the laboratory circuit, the gain–bandwidth product remained constant despite the decrease in gain caused by the addition of the 680 Ω resistor.

7-7 THE FREQUENCY RESPONSE OF JFETS

The characteristics and gains of JFET amplifiers were discussed in Chapter 5. Because of the high impedance at the gate of a JFET, which practically isolates it from the output circuit, it is easier to determine its frequency response.

7-7.1 The Low-frequency Response of a JFET

In Chapter 5, we discussed both fixed- and self-biased JFETs. The low-frequency response of a fixed-bias JFET depends entirely on the gate resistor and the blocking capacitor and is very easy to calculate.

EXAMPLE 7-17

Find the lower-half-power frequency, f_L for the fixed-bias JFET circuit of Figure 7-18.

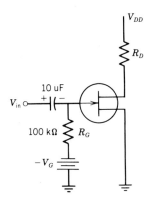

FIGURE 7-18
The fixed-bias JFET for Example 7-17.

Solution
Because f_L depends only on R_G and C, we have by equation 7-4

$$f_L = \frac{1}{2\pi R_G C} = \frac{1}{2\pi \times 10^5 \, \Omega \times 10^{-5} \, F} = \frac{1}{2\pi} = 0.159 \text{ Hz}$$

A fixed-bias circuit requires an additional power supply to bias the gate, but f_L can be made very low, as Example 7-17 demonstrates.

A self-biased JFET is shown in Figure 7-19, which is a repeat of Figure 5-17. Its low-frequency response depends on both the blocking capacitor and the source bypass capacitor, C_S. In most cases the source bypass capacitor determines the frequency response, and the effect of the blocking capacitor can be ignored.

The low-frequency effects of the source bypass capacitor can be determined from equation 5-7, developed in Chapter 5 for a JFET with an unbypassed source resistor:

$$A_v = \frac{-g_m R_D}{1 + g_m R_S}$$

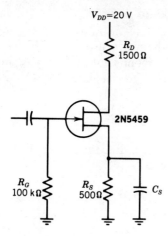

FIGURE 7-19
The self-biased JFET.

At low frequencies, where the source capacitor is important, R_S can be replaced by the parallel combination of R_S and C_S:

$$Z_S = R_S \| C_S = \frac{R_S \times \dfrac{1}{sC_S}}{R_S + \dfrac{1}{sC_S}}$$

With a little algebra this equation becomes

$$A_{v(f)} = \frac{A_{vm}\left(s + \dfrac{1}{R_S C_S}\right)}{\left(s + \dfrac{1 + g_m R_S}{R_S C_S}\right)} \qquad (7\text{-}16)$$

Where A_{vm} is the mid-frequency gain of the JFET, $-g_m R_D$. Equation 7-16 shows that the low-frequency response has a zero at $f = 1/(2\pi R_S C_S)$ and a pole at $f = (1 + g_m R_S)/2\pi R_S C_S$. This equation is of the same form as equation 7-7 for the low-frequency response of a BJT amplifier with an emitter bypass capacitor, and the frequency response curve is similar to that of Figure 7-7 or Figure 7-10.

EXAMPLE 7-18

Find the low-frequency response of the circuit of Figure 7-19. Use $g_m = .002$ S, as found in Example 5-12.

Solution
Using equation 7-16 we find

$$A_{vm} = g_m R_D = 0.002 \text{ S} \times 1500 \text{ }\Omega = 3$$

$$Z1 = \frac{1}{2\pi R_s C_s} = \frac{1}{2\pi \times 500 \text{ }\Omega \times 5 \text{ }\mu\text{F}} = \frac{10^3}{5 \text{ s}} = 63.7 \text{ Hz}$$

$$P1 = \frac{1 + g_m R_S}{2\pi R_s C_s} = \frac{1 + 0.002 \text{ S} \times 500}{2\pi \times 500 \text{ }\Omega \times 5 \text{ }\mu\text{F}} = \frac{2 \times 10^3}{5 \text{ s}} = 127.4 \text{ Hz}$$

Note that the pole and zero are much closer in an FET amplifier than in a BJT amplifier.

Equation 7-16 is in the form of equation 7-8. Therefore the program of Figure 7-8 can be used to calculate the frequency response curve. The results are shown in Figure 7-20.

```
127.4,63.7
3.0
```

(a) Input file

```
INPUT THE POLE AND ZERO IN HERTZ
ENTER MIDFREQUENCY GAIN OR VALUE
     FREQ             GAIN            GAIN(DB)            ANGLE

      10             1.51372          -5.94154           4.43369
      20             1.55317          -5.71802           8.50898
      50             1.77508          -4.55804           16.7011
      70             1.95327          -3.72719           18.9111
     100             2.1962           -2.70897           19.3734
     200             2.65549          -1.05952           14.8304
     500             2.93061          -.203259            7.0344
     700             2.96371          -.105709            5.11532
    1000             2.98198          -.523367E-01        3.61555
    2000             2.99545          -.131834E-01        1.82056
    5000             2.99927          -.211359E-02         .729674
```

(b) Output file

FIGURE 7-20
Printout for Example 7-18.

A JFET using the circuit of Figure 7-19 was tested in the laboratory. We found that the results agreed almost exactly with the printout of Figure 7-20.

7-7.2 The High-frequency Response of a JFET

Figure 7-21a shows the JFET amplifier with the stray capacitors, which are important at high frequencies, in blue. These capacitors are:

C_{Gg} The gate-to-ground capacitance, including all wiring and stray capacitance. (Note that here we are using G for gate and g for ground.)

C_{GD} The gate-to-drain capacitance.

C_{Dg} The drain-to-ground capacitance, including the stray wiring capacitance on the output.

The equivalent circuit is shown in Figure 7-21b. The Miller effect is the same as for the BJT and

$$C_M = C_{GD}(1 + A_v) = C_{GD}(1 + g_m R_D)$$

For the JFET, the Miller capacitance is generally smaller than for the BJT because the gain of the transistor is smaller.

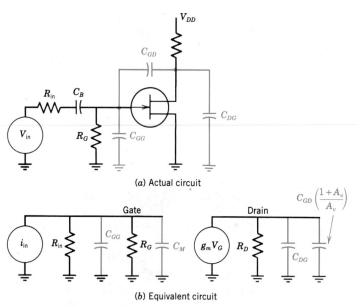

FIGURE 7-21
The high-frequency JFET circuit.

An examination of Figure 7-21 shows that, like the BJT, there are poles on the input side and the output side. Here, however, the dominant pole depends on the resistor and capacitor values for the particular circuit.

EXAMPLE 7-19

For the circuit of Figure 7-22, $C_{Gg} = 6$ Pf, $C_{GD} = 4$ pF, $C_{Dg} = 1$ pF, and the g_m of the JFET is 1000 µS. Find the frequency response of the circuit if R_i equals
a. 100 Ω.
b. 10,000 Ω.
c. 1,000 Ω.

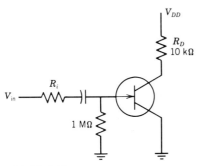

FIGURE 7-22
Circuit for Example 7-19.

Solution

For each case the resistance and capacitance on the output side is approximately the same:

$$R_D = 10,000 \text{ }\Omega$$

$$C_{output} \approx C_{DG} + C_{Gg} = 4 \text{ pF} + 1 \text{ pF} = 5 \text{ pF}$$

The output pole is therefore at

$$f_H = \frac{1}{2\pi RC} = \frac{1}{2\pi \times 5 \text{ pF} \times 10,000 \text{ }\Omega} = 3.18 \text{ MHz}$$

On the input side the gain is

$$A_{v(tr)} = g_m R_D = 0.001 \text{ S} \times 10,000 \text{ }\Omega = 10$$

and

$$C_M = (1 + A_v) C_{GD} = 11 \times 4 \text{ pF} = 44 \text{ pF}$$

The input capacitance is $C_M + C_{Gg} = 44$ pF + 6 pF = 50 pF. When $R_i = 100$ Ω, the input resistance is 100 Ω ∥ 1 MΩ ≈ 100 Ω and

$$f_H = \frac{1}{2\pi RC} = \frac{1}{2\pi \times 100 \text{ }\Omega \times 50 \text{ pF}} = 31.8 \text{ MHz}$$

Here the output pole clearly dominates and $f_H = 3.18$ MHz. When $R_i = 10$ kΩ, the resistance is 10 kΩ ∥ 1 MΩ ≈ 10 kΩ.

$$f_H = \frac{1}{2\pi RC} = \frac{1}{2\pi \times 10{,}000 \text{ Ω} \times 50 \text{ pF}} = 318 \text{ kHz}$$

Here the input pole clearly dominates and $f_H = 318$ kHz.

When $R_i = 1$ kΩ, both the input and output poles are at 3.18 MHz. Neither pole dominates, but the response at any frequency can always be found by multiplying the attenuation factors due to each pole together. At 3.18 MHz, for example, the gain is half of the mid-frequency gain (0.707) × (0.707), and is falling off at 12 dB per octave.

7-8 THE FREQUENCY RESPONSE OF MULTISTAGE AMPLIFIERS

Multistage amplifiers were introduced in Chapter 6 and the methods of calculating their mid-frequency gains were discussed. As we saw in Chapter 6, a *cascaded amplifier* is an amplifier having several stages; it is generally used to increase the gain of the circuit. Each additional stage, however, has its own frequency limitations, and the bandwidth of a multistage amplifier is always smaller than the bandwidth of any individual stage in the circuit.

A typical cascaded amplifier has several poles and zeros that affect its bandwidth, and the general expression for the gain as a function of frequency is

$$A_{v(f)} = A_{v(\text{mid-frequency})} \times \frac{\left(1 + \frac{jf_{Z1L}}{f}\right)\left(1 + \frac{jf_{Z2L}}{f}\right)\cdots}{\left(1 + \frac{jf_{P1L}}{f}\right)\left(1 + \frac{jf_{P2L}}{f}\right)\cdots \times \left(1 + \frac{jf}{f_{P1H}}\right)\left(1 + \frac{jf}{f_{P2H}}\right)} \qquad (7\text{-}17)$$

where $f_{Z1L}, f_{Z2L}, \ldots$ are the zeros of the frequency expression (due to the low-frequency response), $f_{P1L}, f_{P2L}, \ldots$ are the poles of the low-frequency response and $f_{P1H}, f_{P2H}, \ldots$ are the poles due to the high-frequency response. We will call each of the terms in equation 7-7 an *attenuation factor* because the main effect of each term is to attenuate the gain.

If all the poles and zeros are considered, calculating the frequency response by equation 7-17 can become lengthy and tedious, and if it must often be done we suggest using a computer. As a first approximation, which is usually valid in most cases, two simplifications are usually made:

1. The zeros are ignored. This means they are assumed to be at a much lower frequency than the corresponding pole, which is usually true (see Example 7-8).

2. If one of the poles is at a much higher frequency than the others (on the low-frequency end) or a much lower frequency than the others (on the high-frequency end), this pole is considered to be the *dominant pole*. It effectively determines f_L or f_H for the circuit and the other poles can be ignored.

EXAMPLE 7-20

A circuit has a gain of 1000 and high-frequency poles at 800 kHz and 1000 kHz. Find its gain at 800 kHz.

Solution
Using equation 7-17 the expression for this circuit becomes

$$A(f) = 1000 \left(\frac{1}{1 + \dfrac{j800 \text{ kHz}}{800 \text{ kHz}}} \right) \left(\frac{1}{1 + \dfrac{j800 \text{ kHz}}{1000 \text{ kHz}}} \right)$$

The first attenuation factor is 0.707 or 3 dB. The second attenuation factor can be found either by calculation or by using the program of Figure 7-4, and is 0.781. Therefore

$$A(800 \text{ kHz}) = 1000 \times 0.707 \times 0.781 = 553$$

At 800 kHz this circuit's gain is 55% of its mid-frequency gain.

7-8.1 Analysis of a Two-stage BJT Circuit

In a multistage amplifier the poles and zeros of each stage must be calculated with the stages connected together, because the coupling between them must be considered. A straightforward approach to determining the frequency response of an amplifier is developed in Example 7-21.

EXAMPLE 7-21

For the circuit of Figure 7-23, find
a. The overall gain.
b. The gain at 1000 Hz.
c. The gain at 1 MHz.

For each transistor, assume $h_{fe} = 100$, $f_T = 250$ MHz, $C_{be} = 8$ pF, $C_{ce} = 5$ pF, and $C_{cb} = 4$ pF. Ignore $r_{b'b}$.

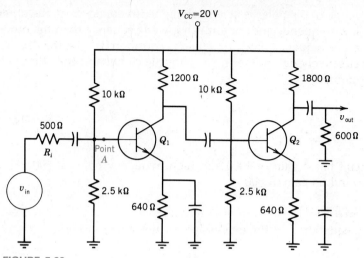

FIGURE 7-23
A two-stage BJT amplifier for Example 7-21. Note: All capacitors are 10 μF.

Solution

Because h_{ie} is not given, it will be necessary to find it. To do so, we must first find the dc value of I_E using the methods developed in Chapter 3.

$$V_{BB} = \frac{2.5 \text{ k}\Omega}{12.5 \text{ k}\Omega} \times 20 \text{ V} = 4 \text{ V}$$

$$R_B = 10 \text{ k}\Omega \parallel 2.5 \text{ k}\Omega = 2 \text{ k}\Omega$$

$$V_{BB} = I_B R_B + h_{fe} I_B R_E + V_{BE} \quad \text{(assume } h_{fe} + 1 = h_{fe}\text{)}$$

$$4 \text{ V} = 2000 \, I_B + 64{,}000 \, I_B + 0.7 \text{ V}$$

$$3.3 \text{ V} = 66{,}000 \, I_B$$

$$I_B = 50 \text{ μA}, \quad I_C = h_{fe} I_B = 5 \text{ mA}$$

$$h_{ie} = \frac{30 \, h_{fe}}{I_C} = \frac{3000}{5} = 600$$

Fortunately, the bias circuits for both transistors are identical so the calculations apply to both.

To find the gain of each transistor, the equivalent collector resistor must be found.

$$R_{L1} = R_{C1} \parallel R_{B2} \parallel h_{ie2} = 1200 \, \Omega \parallel 2000 \, \Omega \parallel 600 \, \Omega = 333 \, \Omega$$

$$R_{L2} = 1800 \, \Omega \parallel 600 \, \Omega = 400 \, \Omega$$

$$A_{v(tr)1} = \frac{-h_{fe}R_{L1}}{h_{ie}} = \frac{-100 \times 333 \, \Omega}{600} = -55.5$$

$$A_{v(tr)2} = \frac{-h_{fe}R_{L2}}{h_{ie}} = \frac{-100 \times 400 \, \Omega}{600} = -66.7$$

To find the circuit gain, equation 4-11 can be used.

$$A_{v(ck)} = \frac{v_{Th}}{v_{in}} \times \frac{-h_{fe}R_L}{(h_{ie} + R_{Th})}$$

R_{Th} for circuit 1 is $500 \, \Omega \, \| \, 2500 \, \Omega \, \| \, 10 \, k\Omega = 400 \, \Omega$

Remembering that the Thevenin's circuit is looking back from point A

$$\frac{V_{Th}}{v_{in}} = \frac{R_B}{R_B + R_i} = \frac{2000 \, \Omega}{2500 \, \Omega} = 0.8$$

$$A_{v(ck)} = 0.8 \times \frac{-100 \times 333 \, \Omega}{1000 \, \Omega} = -26.64$$

The overall circuit gain is

$$A_{v(ck)} = A_{v(ck1)} \times A_{v(ck2)} = -26.64 \times -66.7 = 1777$$

The low-frequency response can now be calculated. From equation 7-7 the pole of the low-frequency response is at

$$f_P = \frac{h_{fe}R_E + (R_{Th} + h_{ie})}{2\pi \times (h_{ie} + R_{Th}) X R_E X C_E} = \frac{100 \times 640 \, \Omega + 1000 \, \Omega}{2\pi \times 1000 \, \Omega \times 640 \, \Omega \times 10 \, \mu F}$$

$$= \frac{65000}{2\pi \times 6.4} = 1616 \, Hz$$

The zero of the frequency response is at

$$f_Z = \frac{1}{2\pi R_E C_E} = \frac{10^5}{2\pi \times 640} = 26 \, Hz$$

This frequency is so far below the pole frequency that it can be ignored. The pole frequencies due to the blocking capacitor are also very low compared to 1616 Hz, and can also be ignored.

R_{Th} for transistor 2 is

$$R_B \, \| \, R_{C1} = 2000 \, \Omega \, \| \, 1200 \, \Omega = 750 \, \Omega$$

Now the pole frequency for transistor 2 can be calculated by equation 7-7.

$$f_P = \frac{100 \times 640\,\Omega + 750\,\Omega + 600\,\Omega}{2\pi \times 1350\,\Omega \times 640\,\Omega \times 10\,\mu F} = \frac{65{,}350}{54.2} = 1200\text{ Hz}$$

To find the gain at 1000 Hz we can use

$$A_v(1000\text{ Hz}) = A_{vm} \frac{1}{\left(1 + \frac{jf_{P1L}}{f}\right) \times \left(1 + \frac{jf_{P2L}}{f}\right)}$$

$$= A_{vm} \frac{1}{\left(1 + \frac{j1616}{1000}\right)\left(1 + \frac{j1206}{1000}\right)}$$

$$= 1777 \times 0.526 \times 0.638 = 596$$

The two attenuation factors, 0.526 and 0.638, were found using the computer program of Figure 7-4. The gain of this circuit at 1000 Hz is about one-third of its mid-frequency value, due primarily to the emitter bypass capacitors. To improve the low-frequency response the emitter bypass capacitors would have to be larger than 10 μF. Very large capacitors are often used to achieve a low f_L.

The high-frequency response is calculated for the first stage as follows:

$$f_T = 250\text{ MHz} \qquad f_B = \frac{f_T}{h_{fe}} = \frac{250\text{ MHz}}{100} = 2.5\text{ MHz}$$

$$C_{bc} + C_{b'e} = \frac{1}{2\pi \times h_{ie} \times f_B} = \frac{1}{2\pi \times 600\,\Omega \times 2.5 \times 10^6}$$

$$= \frac{10^{-6}}{3000\pi} = 106\text{ pF}$$

$$C_{b'e} = 102\text{ pF}$$

$$C_M = (1 + A_{v(tr)})C_{bc} = 56.5 \times 4\text{ pF} = 226\text{ pF}$$

$$C_{total} = C_{be} + C_{b'e} + C_M = 8\text{ pF} + 102\text{ pF} + 226\text{ pF} = 336\text{ pF}$$

The equivalent resistance is

$$R_{EQ} = R_B \| R_i \| h_{ie} = 500\,\Omega \| 2000\,\Omega \| 600\,\Omega = 240\,\Omega$$

$$F_{(P1H)} = \frac{1}{2\pi R_{EQ}C_T} = \frac{10^{12}}{2\pi \times 240\,\Omega \times 336\text{ pF}} = 1.97\text{ MHz}$$

For the interstage between transistors Q1 and Q2

$$C_M = (1 + A_{v(tr)2}) C_{bc} = 67.7 \times 4 \text{ pF} = 271 \text{ pF}$$

$$C_{total} = C_{be} + C_M + C_{ce} + C_{b'e}$$
$$= 8 \text{ pF} + 271 \text{ pF} + 5 \text{ pF} + 102 \text{ pF} = 386 \text{ pF}$$

$$R_{EQ} = h_{ie2} \parallel R_B \parallel R_{C1} = 600 \text{ } \Omega \parallel 2000 \text{ } \Omega \parallel 1200 \text{ } \Omega = 333 \text{ } \Omega$$

$$f_{(P2H)} = \frac{1}{2\pi \times 333 \text{ } \Omega \times 386 \text{ pF}} = 1.24 \text{ MHz}$$

At 1 MHz the high-frequency response is

$$A_{v(1MHz)} = A_{vm} \times \frac{1}{\left(1 + \frac{jf}{f_{P1H}}\right)\left(1 + \frac{jf}{f_{P2H}}\right)}$$

$$= 1777 \times \frac{1}{\left(1 + \frac{j1 \text{ MHz}}{1.97 \text{ MHz}}\right)} \frac{1}{\left(1 + \frac{j1 \text{ MHz}}{1.24 \text{ MHz}}\right)}$$

$$A_{v(1MHz)} = 1777 \times 0.891 \times 0.718 = 1232$$

The gain of this entire circuit is down about 3 dB at 1 MHz.

The frequency response for other two-stage amplifiers, such as two-cascaded JFET stages or a BJT driving an emitter-follower, can be calculated using methods similar to those of Example 7-21. Unfortunately, such calculations are also long and tedious, so examples will not be presented here. For circuits that incorporate low gain stages, such as an emitter-follower, the frequency response of the low gain stage is usually much broader than that of the amplifier, so often only the amplifier limits the frequency response.

7-9 TRANSIENT RESPONSE OF AMPLIFIERS

In the previous sections we have considered the frequency response of amplifiers when sine waves were applied to their inputs. In this section we will consider the response of the high-pass circuit, the low-pass circuit, and amplifiers in general when a square wave is applied to their inputs.

If a square wave is applied to the input of an amplifier, the output is a *distorted* square wave. The distortion is caused by the limited frequency response of the amplifier, and an analysis of this distortion can indicate the upper- and lower-half-power frequencies of the amplifier.

Before considering the response of an amplifier, the response of the high-pass and low-pass RC circuits to a square wave will be considered. The equations developed will be directly applicable to amplifiers.

7-9.1 The Response of the High-pass Circuit

The high-pass or low-frequency RC circuit was discussed in section 7-4. When a square wave is applied to this circuit, as shown in Figure 7-24, the output is a *tilted* square wave. The *percentage tilt*, *P*, is defined as

$$\% P = \frac{\Delta V}{V/2} \times 100$$

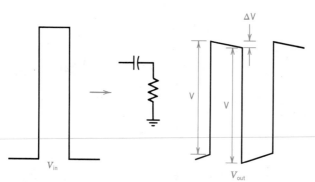

FIGURE 7-24
A square wave applied to a high-pass circuit.

Millman and Taub (see References) have shown that, if *P* is reasonably small, the lower-half-power frequency can be found from the percentage tilt by the equation

$$f_L = \frac{P}{\pi T} \qquad (7\text{-}18)$$

where *T* is the period of the applied square wave.

328 THE FREQUENCY RESPONSE OF AMPLIFIERS

EXAMPLE 7-22

A square wave is applied to a high-pass circuit, and the resulting waveform is shown in Figure 7-25. If the sweep time is 1 ms per cm (assume each box is 1 cm), find f_L.

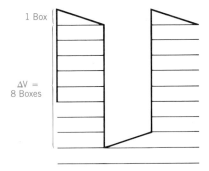

FIGURE 7-25
Oscilloscope trace for Example 7-22.

Solution
Figure 7-25 shows that the abrupt change of voltage is 8 boxes and ΔV is one box. Therefore

$$P = \frac{\Delta V}{V/2} = \frac{1}{4} = 0.25 \quad \text{or} \quad 25\%$$

The figure also shows that the period of the wave, T, is 6 boxes or 6 ms. By equation 7-18

$$f_L = \frac{0.25}{6\pi \times 10^{-3}} = \frac{250}{6\pi} = 13.3 \text{ Hz}$$

7-9.2 Response of the Low-pass Circuit to a Square Wave

Figure 7-26 shows the response of a low-pass or high-frequency circuit to a square wave input. Because of the high-frequency limitations of the circuit, the output is rounded as shown in Figure 7-26b.

If the total voltage swing for the waveform of Figure 7-26b is ΔV, the *rise time* (t_r) is defined as the time it takes for the output waveform to go between 10% of ΔV and 90% of ΔV. The output waveform must be observed on an oscilloscope so that the rise time can be measured. Some oscilloscopes have

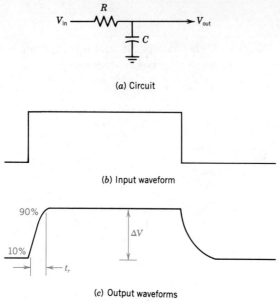

FIGURE 7-26
Response of a low-pass circuit to a square wave.

their graticules marked at the 10 and 90% points, to facilitate rise time measurements. Rise time and its measurements are shown in Figure 7-27.

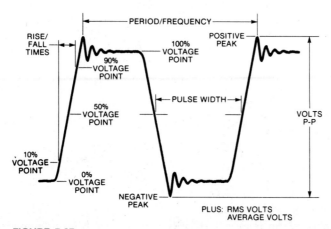

FIGURE 7-27
Rise time and its measurement. (Courtesy of Hewlett Packard Co.)

EXAMPLE 7-23

If a square wave is applied to an amplifier, its output swings from 5 V to 15 V. At what voltages would the rise time be measured?

Solution

The total ΔV is 15 V − 5 V = 10 V. The 10 and 90% points are therefore at 1 V and 9 V, but this is *above the base* of 5 V. The 10% point is at 6 V and the 90% point is at 14 V. The rise time is the time the waveform takes to transverse these voltages.

It has been shown[5] that the rise time for the RC circuit is

$$t_r = 2.2\, RC \tag{7-19}$$

But in section 7-5.1, the high-frequency 3 dB point was defined as

$$f_H = \frac{1}{2\pi RC} \tag{7-20}$$

Combining equations 7-19 and 7-20, we obtain

$$t_r = \frac{2.2}{2\pi f_H} = \frac{0.35}{f_H} \tag{7-21}$$

Equation 7-21 shows that there is a simple relationship between the rise time of an amplifier and its upper-half-power frequency. Perhaps more importantly, it shows that *the higher f_H is, the more square the output will appear* because high f_H values lead to short rise times.

EXAMPLE 7-24

An RC circuit has $R = 10\,\Omega$ and $C = 100$ pF. Find its rise time and frequency response.

Solution
By equation 7-19

$$t_r = 2.2\, RC = 2.2 \times 10^4\,\Omega \times 10^{-10}\,F = 2.2\,\mu s.$$

$$f_H = \frac{0.35}{t_r} = \frac{0.35}{2.2\,\mu s} = 159\text{ kHz}$$

---

[5]Consult Chapter 2 of Millman and Taub (see References).

7-9.3 Amplifying Square Waves

If a square wave is put into an amplifier, the same sort of distortion can be found at the output as if it were entered into an RC circuit. No matter how complex the amplifier is, a fairly accurate estimate of its effective f_L can be obtained by observing the slope of the output, and f_H can be approximated from the rise time. An amplifier with a high f_H will amplify both sine waves and square waves well, but an amplifier with a low f_H will also distort square waves.

Many engineers use square wave testing to determine the frequency response of an amplifier, by driving the amplifier with a small square wave and observing the output. The reader is cautioned to be sure the output waveform is not clipping on either end; this will distort the measurements. One way of determining that the output is not clipping is to raise the amplitude of the input square wave slightly; if the output also increases, on both the high and low levels, the amplifier is not clipping.

The output of an amplifier, when viewed on an oscilloscope, will exhibit both tilt and rise time distortions. Unless the amplifier has a very narrow frequency response, however, the sweep speed required to find the rise time will be much higher than the sweep required to find the tilt. Therefore the two types of distortion can be kept separated and will not interfere with each other (see problem 7-37).

If a square wave is applied to an amplifier, its frequency must be low enough so that it presents a reasonable output picture. If the frequency is too

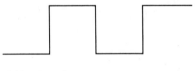

(a) Input waveform

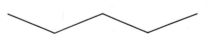

(b) "Integrated output"—the period of the applied square wave is too short

(c) More reasonable output—the period of the applied wave is long enough

FIGURE 7-28
Outputs when a square wave is applied to an amplifier.

large, the amplifier will "integrate" it, as shown in Figure 7-28b, and the results will be meaningless.

As a rule of thumb, we suggest the minimum period of the applied square wave should be

$$T = \frac{2}{f_H}$$

This gives a reasonable picture, as shown in Figure 7-28c. The equation also shows that circuits with low bandwidths require long periods for square wave testing.

EXAMPLE 7-25

A square wave, with a period of 1 ms, is fed into an amplifier. The output is observed to have a 20% tilt and a rise time 2 μs. Find f_H and f_L for the amplifier.

Solution
Equation 7-18 can be used to find f_L.

$$f_L = \frac{P}{\pi T} = \frac{0.2}{\pi \times 10^{-3}} = \frac{200}{\pi} = 63.5 \text{ Hz}$$

Equation 7-21 can be used to find f_H.

$$f_H = \frac{0.35}{2 \text{ μs}} = 175 \text{ kHz}$$

7-10 OSCILLOSCOPE PROBE COMPENSATION

The input signal to a *cathode-ray oscilloscope* (CRO), which is the waveform to be observed, is picked up by the oscilloscope probe, and then sent to the vertical amplifier within the CRO. The ideal scope probe would have no effect on the circuit under observation when it is connected; it would have infinite resistance and no capacitance. Unfortunately, real scope probes are made up of coaxial cables to prevent 60 Hz and other noise from distorting the signal, and a coaxial cable introduces considerable capacitance into the circuit.

To reduce the amount of capacitance coupled into a circuit under test, most scope probes have an *attenuator* in the probe tip. This attenuator reduces the signal available to the vertical amplifier, usually by a 10 to 1 ratio, and therefore reduces the *sensitivity* of the CRO to small signals, but it also reduces the probe capacitance.

To operate optimally, and not distort the input waveform, these probes must be *compensated*, which means that the capacitance within the probe tip must be properly adjusted. On modern scope probes a small screwdriver adjustment is provided for this purpose. The reasons for probe compensation are explained by an extension of the low-pass circuit, and are discussed in this section.

7-10.1 The Capacitively Loaded Attenuator

The capacitively loaded attenuator is a resistive voltage divider that must drive a capacitive load. The circuit is shown in Figure 7-29a, where R_1 and R_2 form the voltage divider and C_1 is the capacitance of the load. Usually C_1 consists of unavoidable stray and cable capacitance.

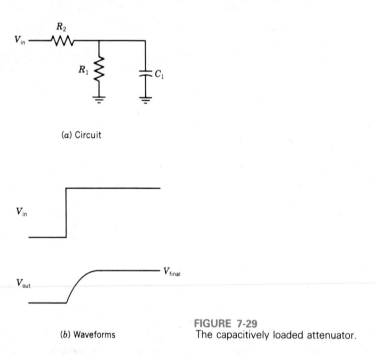

FIGURE 7-29
The capacitively loaded attenuator.

The response of this circuit to a step voltage is shown in Figure 7-29b. It starts at 0 V and rises to V_{final}, where V_{final} is the normal output of the attenuator:

$$V_{final} = \frac{R_1}{R_1 + R_2} V_{in}$$

The *rise time* of this circuit is determined by the Thevenin's impedance looking back into it:

$$t_r = 2.2 \times R_1 \| R_2 \times C_1$$

334   THE FREQUENCY RESPONSE OF AMPLIFIERS

If we are being very precise, it takes the output infinite time to reach V_{final}, because it approaches it asymptotically. As a practical rule, however, the output is considered to reach V_{final} after five times constants, where the *time constant* for this circuit is its RC product.

EXAMPLE 7-26

The capacitively loaded attenuator of Figure 7-29 has $R_1 = 4 \text{ k}\Omega$, $R_2 = 2 \text{ k}\Omega$, and $C_1 = 100 \text{ pF}$. Find the response of this circuit to the waveform of Figure 7-30a.

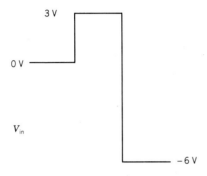

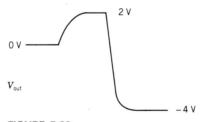

FIGURE 7-30
Waveforms for Example 7-26.

Solution
The response is shown in Figure 7-30b. The attenuator causes the output to be 2/3 of the input so the output rises to 2 V when the input is at 3 V and falls to -4 V when the input goes to -6 V.

The time constant of this circuit is

$$RC = R_1 \| R_2 \times C_1 = 1500 \text{ }\Omega \times 100 \text{ pF} = 0.15 \text{ }\mu\text{s}$$

Since the circuit requires five constants (0.75 µs) to reach its final voltage, the width of the pulse in Figure 7-29 should not be shorter than 0.75 µs.

7-10.2 The Compensated Attenuator

The circuit of Figure 7-31 is a *compensated attenuator*. It differs from the capacitively loaded attenuator because a second capacitor, C_2, has been placed across R_2.

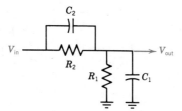

FIGURE 7-31
The compensated attenuator.

The response of the compensated attenuator to step changes in the input voltage, V_{in}, is interesting and important. It can be summarized as follows:

1. At any time greater than five time constants after the step, the circuit acts as a resistive voltage divider, and *the resistors alone determine the output voltage*. At this time the capacitors are charged up to the voltages across their respective resistors.
2. At the time of the step the voltage across each capacitor *jumps instantly*. The *change* of voltage across each capacitor is given by

$$\Delta V_{C1} = \Delta V_{in} \times \frac{C_2}{C_1 + C_2} \qquad (7\text{-}22)$$

$$\Delta V_{C2} = \Delta V_{in} \times \frac{C_1}{C_1 + C_2}$$

where ΔV_{in} is the magnitude of the step or instantaneous change of the input, and ΔV_{C1} or ΔV_{C2} is the instantaneous change or *jump* of the voltage across C_1 or C_2. The output voltage immediately following the step is the voltage across the capacitor *before the step* occurred *plus the jump voltage* (ΔV_{C1} or ΔV_{C2}).

The response of the circuit is best explained by an example.

EXAMPLE 7-27

If the values of the components in Figure 7-31 are $R_1 = 4\ \text{k}\Omega$, $C_1 = 100\ \text{pF}$, $R_2 = 2\ \text{k}\Omega$, $C_2 = 150\ \text{pF}$, find the response of the circuit to the waveform of Figure 7-32a.

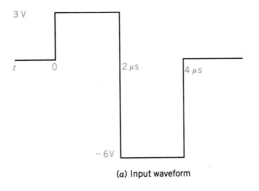

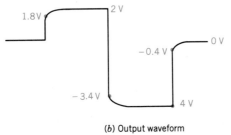

FIGURE 7-32
Waveforms for Example 7-27.

Solution

The output waveform is shown in Figure 7-32b. At $t = 0$, the input jumps by 3 V. The output, which is across C_1 and R_1, jumps by

$$\Delta V_{C1} = \frac{C_2}{C_1 + C_2} \Delta V_{in} = \frac{150 \text{ pF}}{250 \text{ pF}} \times 3 \text{ V} = 1.8 \text{ V}$$

After the jump it rises toward its final voltage, 2 V, with a time constant equal to the RC product of the circuit. In this circuit the RC product is the parallel combination of the resistors and the parallel combination of the capacitors:

$$t = 2000 \text{ }\Omega \parallel 4000 \text{ }\Omega \times (150 \text{ pF} + 100 \text{ pF}) = 1200 \text{ }\Omega \times 250 \text{ pF} = 300 \text{ ns}.$$

At $t = 2$ μs, more than five time constants have elapsed and the output is at 2 V. Then the input jumps by -9 V (from $+3$ V to -6 V), and the output voltage jumps by 0.6×-9 V $= -5.4$ V or from 2 V to -3.4 V.

At $t = 4$ μs, the input jumps by $+6$ V and the output jumps by 3.6 V, or from -4 V to -0.4 V.

The compensated attenuator has three types of compensation: undercompensation, overcompensation, and perfect compensation. These are shown in Figure 7-33.

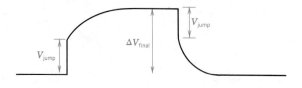

(a) Undercompensation

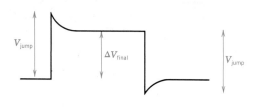

(b) Overcompensation

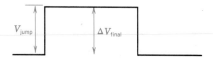

(c) Perfect compensation

FIGURE 7-33
The three types of compensation.

The *undercompensated* case is shown in Figure 7-33a. The jump voltage, V_{jump}, is not as large as the differences between the final voltages, ΔV_{final}. Consequently, the output jumps only part of the way when the input changes. It then charges toward the final voltage at the time constant of the circuit. Undercompensation occurs when $R_1 C_1 > R_2 C_2$, as in Example 7-27.

The *overcompensated* case is shown in Figure 7-33b. Here the jump voltage is larger than the difference between the final voltages. The output jumps too far and must come back to the final voltage. For overcompensation $R_1 C_1 < R_2 C_2$.

Perfect compensation occurs when $R_1C_1 = R_2C_2$. With perfect compensation, the jump voltage equals ΔV_{final}, and the output appears to jump to its final voltage instantly, as shown in Figure 7-33c.

7-10.3 The Oscilloscope Probe

Whenever a measuring instrument is placed in a circuit, it *changes* or *distorts* that circuit. Good instrument design holds the distortion to a minimum, and in most cases the effect of the test instrument is negligible. *Attenuated oscilloscope probes* are usually used because they reduce the effect of the probe on the circuit under test.

Most modern oscilloscope probes attenuate the signal being observed by a 10:1 ratio in order to reduce the capacitance introduced into the circuit when the probe is applied, and also to increase the impedance of the probe. A typical probe is shown in Figure 7-34. It consists of four parts: the *probe tip*, which is applied to the circuit; the *barrel*, which contains the attenuating resistor and capacitor; the *coaxial cable*, which connects the probe to the amplifier within the oscilloscope; and the *amplifier* itself. The coaxial cable is a cable with a center conductor for the signal, inside a shield or mesh of wire that is connected to ground. The shield isolates the signal from external electrical noise.

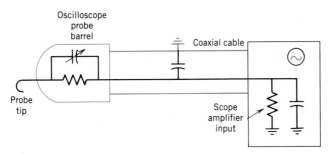

FIGURE 7-34
The oscilloscope probe.

The probe and amplifier form a compensated attenuator circuit similar to the circuit of Figure 7-31. The voltage source is the signal being observed and the output is the signal at the input terminals of the amplifier; this is the waveform that will be displayed on the screen. R_2 and C_2 in Figure 7-31 are the resistor and capacitor in the barrel of the probe. R_1 is the input resistance of the amplifier, and C_1 is the input capacitance of the amplifier plus the capacitance of coaxial cable between the signal lead and its shield, which is connected to ground.

As in the circuit of Figure 7-31, perfect compensation provides the best output waveform with the least distortion. To achieve perfect compensation, the capacitor in the barrel is adjustable. Generally a small screwdriver adjustment is provided in the barrel.

Whenever a probe is connected to an amplifier, or moved from one amplifier to another, it should be *compensated*. To compensate a probe, a square wave is applied to the probe tip (many oscilloscopes provide a *calibrated output* for this purpose) and the output is observed on the screen. The capacitor in the barrel is then adjusted until perfect compensation is attained, and the screen looks like Figure 7-33c instead of Figure 7-33a or 7-33b.

Looking into the circuit from the probe tip, it can be seen that the resistors are in series, which increases the resistance, and the capacitors are in series, which decreases the capacitance at the tip. These are the reasons attenuating probes are used.

EXAMPLE 7-28

The circuit of Figure 7-34 is to be a 10:1 attenuator. If the input impedance of the scope amplifier is 10 pF and 1 M Ω, and the capacitance of the coaxial cable is 40 pF, find the resistance and capacitance in the barrel. Also find the impedance at the probe tip.

Solution

Given that the resistance of the amplifier is 1 M Ω, the resistance in the barrel required to make the circuit a 10:1 attenuator is 9 M Ω. The capacitance in the barrel should be chosen for perfect compensation. In section 7-10.2 we stated that, for perfect compensation, $R_1 C_1 = R_2 C_2$. Here C_1 is the amplifier capacitance in parallel with the coaxial cable.

$$1 \text{ M} \Omega \times 50 \text{ pF} = 9 \text{ M} \Omega \times C_2$$

$$C_2 = 5.45 \text{ pF}$$

The input resistance seen at the probe tip is 10 M Ω. The capacitance is the 5.45 pF barrel capacitance in series with the 50 pF of the cable and amplifier, or 5 pF. Thus the probe tip has reduced the capacitance seen by the circuit by a factor of 10.

7-10.4 Practical Scope Probes

Oscilloscope manufacturers produce a variety of probes for various purposes. The Tektronix Corporation markets 10:1 probes (listed as 10× in their catalog) with about 22 pF per meter of cable in addition to the amplifier and stray capacitance. A 2 meter probe has an input impedance of 11 pF and 10 MΩ.

If very small signals are to be viewed, the 10× attenuation may not be tolerable and a probe that does not attenuate the signal (a 1× probe) must be used. Of course the capacitance of the circuit cannot be reduced by this probe. A 1× probe similar to the 10× probe has a capacitance of 54 pF and an impedance of 1 M Ω.

7-10.5 Special Purpose Probes

The 1× and 10× probes discussed in section 7-10.5 are called *passive* probes because they contain only passive elements (resistors and capacitors). Oscilloscope manufacturers also provide a variety of probes for special purposes, such as *current probes* that have a collar that clips around a wire to measure current in that wire by transformer action, *active probes* that have active elements in the probe to reduce capacitance and are used for high-frequency measurements, *differential probes*, and *high voltage* probes. The reader should consult the manufacturers' catalogs for further information on the use and availability of special-purpose probes.

7-11 THE FREQUENCY RESPONSE LIMITATIONS OF OSCILLOSCOPES

In order to measure high frequencies or fast rise times, one must have a very good oscilloscope. Scope amplifiers are specified by their frequency response; inexpensive amplifiers might have a frequency response of 10 MHz, whereas more costly amplifiers can have responses of 60 or 100 MHz or even higher. To measure high frequencies, the frequency response of the scope should be higher than the frequency to be measured.

A good scope is also required if fast rise times are to be measured. The *rise time* of a scope amplifier is given approximately by equation 7-21:

$$t_r = \frac{0.35}{BW}$$

where BW is the specified bandwidth (f_H) of the amplifier.

Millman and Taub (see References) state that if two amplifiers, with rise times t_{r1} and t_{r2}, are connected together, the rise time of the resulting waveform is

$$t_r = 1.05 \sqrt{t_{r1}^2 + t_{r2}^2} \qquad (7\text{-}23)$$

This is the situation when a probe is connected to observe the rise time of a square wave. It becomes t_{r1} in equation 7-23 and the rise time of the oscilloscope becomes t_{r2}, and the rise time displayed on the screen will be t_r. To observe rise times accurately, the rise time of the scope must be shorter than the rise time under observation, which implies a high bandwidth for the scope amplifier.

Figure 7-35 gives the measurement error for rise times depending on the ratio of the scope rise time to the waveform rise time. From it we can see that if the scope rise time is 7 times faster than the observed rise time, the distortion is less than 1%, but if the waveform rise time is as fast as the scope rise time, the distortion will be 40%.

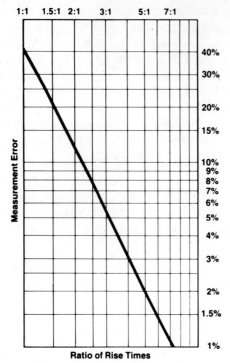

FIGURE 7-35
Measurement error vs. rise time ratios for an oscilloscope. (Courtesy of Tektronix, Inc.)

EXAMPLE 7-29

A waveform has a rise time of 7 ns. How will it appear if observed
a. On a 100 MHz scope?
b. On a 10 MHz scope?

Solution
a. On a 100 MHz scope the rise time is

$$t_r = \frac{0.35}{BW} = \frac{0.35}{10^8} = 3.5 \text{ ns}$$

Thus the ratio of the waveform rise time to the scope rise time is 2:1, and Figure 7-35 shows that there will be a 12% distortion. So the 7 ns rise time will appear as 8.4 ns when viewed on the scope.

b. The rise time of a 10 MHz scope is 35 ns. In Figure 7-35 this is a ratio of 1:5 and is off the chart, so equation 7-23 must be used:

$$t_r = 1.05 \sqrt{(35)^2 + (7)^2} = 1.05 \sqrt{1225 + 49} = 37.5 \text{ ns}$$

The displayed rise time will be 37.5 ns, and almost all of it will be due to the slow rise time of the scope rather than to the waveform under observation. This example shows that it is useless to try to measure a fast waveform with a slow oscilloscope amplifier.

7-12 SUMMARY

This chapter considered the frequency response of the amplifiers introduced in earlier chapters. The high- and low-frequency limitations of RC circuits were explored, and these concepts were extended to BJT and FET amplifiers. The frequency response of multistage amplifiers was also considered. In the latter sections of this chapter, square wave testing of amplifiers and the oscilloscope were discussed. Scope probe compensation, rise time measurements, and the bandwidth limitations of an oscilloscope were explained.

7-13 GLOSSARY

Bode plot. A straight-line logarithmic approximation to the frequency response of an amplifier.

$C_{b'e}$ (the diffusion capacity). The equivalent capacitive effect caused by the inability of charges to move rapidly enough at high frequencies.

Capacitively loaded attenuator. An attenuator circuit where the output is loaded with a capacitor.

Coaxial cable. Cable with a center conductor surrounded by a shield, which is usually connected to ground.

Compensated attenuator. An attenuator with capacitors in parallel with both resistors.

Compensated scope probes. Scope probes that display a square wave in response to a square wave input (without overshoot or undershoot).

f_T. The frequency where the short-circuit current gain of a BJT is 1. It is also called the gain–bandwidth product for an amplifier.

Gain–bandwidth product. The product of the gain times of the bandwidth (f_H) for an amplifier (also see f_T).

Lower-half-power frequency (f_L). The low frequency where the gain is 3 dB down from the mid-frequency gain.

Miller capacitance (C_M). The equivalent capacity between the base and collector of a transistor, usually $(1 + A_v) C_{bc}$.

Percentage tilt. A measure of the tilt of a square wave after being put through a circuit. It indicates the low-frequency response of the circuit.

Pole. A complex frequency where the gain is infinite.

Rise time. The time it takes for a square wave output to go from 10 to 90% of its change in voltage.

Stray capacity. Capacity of wires and electrodes to ground.

Time constant. The RC product of a circuit.

Upper-half-power frequency (f_H). The high frequency where the gain is 3 dB down from the mid-frequency gain.

V_{final}. The voltage (dc level) a circuit will reach five time constants after a step change in the input voltage.

Zero. A complex frequency where the gain is zero.

7-14 REFERENCES

ROBERT BOYLESTAD and LOUIS NASHELSKY, *Electronic Devices and Circuit Theory*, 4th Edition, Prentice-Hall, Englewood Cliffs, N. J., 1987.

Hewlitt Packard, *Measurement Computation Systems*, Palo Alto, CA, 1985.

JACOB MILLMAN, *Microelectronics*, 2nd Edition, McGraw-Hill, New York, 1987.

JACOB MILLMAN and HERBERT TAUB, *Pulse, Digital, and Switching Waveforms*, McGraw-Hill, New York, 1965.

J. F. PIERCE and T. J. PAULUS, *Applied Electronics*, Charles E. Merrill, Columbus, Ohio, 1972.

DONALD L. SCHILLING and CHARLES BELOVE, *Electronic Circuits, Discrete and Integrated*, 2nd Edition, McGraw-Hill, New York, 1979.

TEK Products Catalog, Tektronix, Inc., Beaverton, OR, 1985.

HENRY ZANGER, *Semiconductor Devices and Circuits*, John Wiley, New York, 1984.

7-15 PROBLEMS

7-1 Show that the gain of an amplifier at its upper- and lower-half-power frequencies is 3 dB below its mid-frequency gain.

7-2 The gain of an amplifier is of the form $A_v = K/f$ where K is a constant. Show that the gain drops off by a 6 dB every time f doubles (6 dB per octave) and drops by 20 dB whenever f increases by a factor of 10 (10 dB per decade).

7-3 In Example 7-2, f_L was 100 Hz and the operating frequency was 100 Hz. How many dB below mid-frequency is this
  a. Using the actual calculations.
  b. Using the Bode plot.

7-4 In Figure 7-3 $R_L = 2$ kΩ, $R_s = 1$ kΩ, and $C = 2$ μf. Find f_L and the gain and phase shift at 10 Hz.

7-5 In Examples 7-6 and 7-7 the pole due to the blocking capacitor was ignored. Explain why.

7-6 In Figure 7-5, the emitter capacitor must be changed so that f_L is 100 Hz. Find the new value of the emitter capacitor. What is the frequency of the zero under these conditions?

7-7 For the circuit of Figure P7-7, the transistor parameters are $h_{fe} = 80$ and $h_{ie} = 1600$ Ω. Find
  a. The mid-frequency gain of the circuit and the transistor.
  b. The frequency of the pole and zero due to emitter-capacitor.

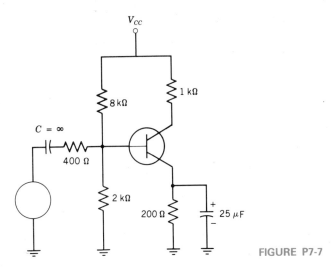

FIGURE P7-7

c. The frequency of the pole due to the blocking capacitor.
d. The gain of the circuit at 100 Hz.

7-8 In Figure 7-10 find the dB drop
   a. In the octave between 100 Hz and 200 Hz.
   b. In the octave between 200 Hz and 400 Hz.
   c. In the decade between 100 Hz and 1000 Hz.

7-9 A 1000 Ω resistor is paralleled by a 50 pF capacitor. Find
   a. f_H
   b. The attenuation at 1 MHz.
   c. The attenuation at 5 MHz.

7-10 Draw the pole–zero diagram for the high-frequency circuit. From it show that the gain at f_H is 0.707 of the mid-frequency gain.

7-11 A transistor has the following parameters: f_T = 200 MHz, h_{fe} = 120, h_{ie} = 1000 Ω, $r_{b'b}$ = 0, R_L = 1500 Ω, R_{Th} = 900 Ω, C_{be} = 5 pF, and C_{bc} = 3 pF. Find
   a. f_β
   b. $C_{b'e}$
   c. The Miller capacitance
   d. f_H

7-12 The circuit of Figure P7-12 has the following parameters: h_{fe} = 100, C_{bc} = 10 pF, and f_T = 200 MHz. Assume $r_{b'b}$ = 0. Find
   a. h_{ie}
   b. g_m
   c. A_v for the transistor and the circuit

7-15 PROBLEMS   345

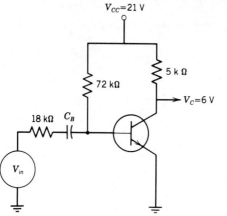

FIGURE P7-12

    d. $C_{b'e}$
    e. f_H
    f. The circuit voltage gain at 100 KHz

7-13 Find the gain and upper-half-power frequency for the circuit of Figure P7-13. Assume $C_{b'c} = 8$ pF, $C_{b'e} = 100$ pF, and $r_{b'b} = 0$.

7-14 Find f_H due to the pole at the collector of Figure P7-13 if the total collector-to-ground capacitance is 10 pF.

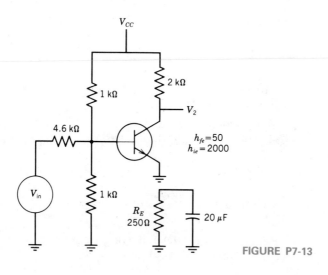

FIGURE P7-13

7-15 Find the gain and f_H for the circuit of Figure P7-15.

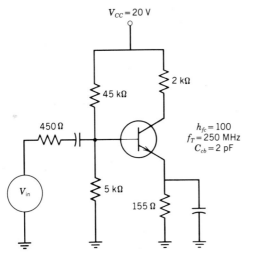

FIGURE P7-15

7-16 Design an amplifier for a 1 MHz bandwidth using a **2N3904** transistor. Design it so that $I_C = 2$ mA. Let $V_{CC} = 20$ V. Assume that the source impedance is 500 Ω and that $C_{b'c}$ is 4 pF. Find the values of all the resistors in your circuit.

7-17 Repeat problem 7-16 for a circuit with a 2 MHz bandwidth. What is the gain–bandwidth product for each circuit?

7-18 Show that the Miller effect, when viewed from the collector, causes C_{bc} to be multiplied by the factor $(1 + A_v)/A_v$. What is this factor if $A_v = 150$?

7-19 For the laboratory circuit discussed in section 7-6.7, find the theoretical gain and bandwidth if the 1800 Ω collector resistor is paralleled by a 680 Ω load resistor.

7-20 In Example 5-13 of Chapter 5, a JFET circuit was designed. The circuit had $R_D = 10$ kΩ, $R_S = 4$ kΩ, and $g_m = 1000$ μS. If R_S is bypassed by a 10 μF capacitor, find the low frequency pole and zero for this circuit.

7-21 Generate a frequency response curve for problem 7-20 using the program of Figure 7-8.

7-22 In Example 7-20, find the required value of the bypass capacitor if f_L is to be 80 Hz.

7-23 A JFET amplifier has $R_D = 5000$ Ω, $R_G = 100$ kΩ, $g_m = 1000$ μS, and $C_{Gg} = C_{GD} = C_{Dg} = 5$ pF. Find the input and output poles if
  a. $R_i = 100$ Ω.
  b. $R_i = 10$ kΩ.

7-24 For the circuit of Figure 7-22, assume R_D is capacitively coupled to a 2 kΩ load resistor. Using the parameters given in Example 7-19, find
  a. The gain.

b. The input pole.
c. The output pole.
d. The gain–bandwidth product.

Do this problem for $R_i = 100\ \Omega$ and for $R_i = 10\ k\Omega$. How does the gain–bandwidth product compare with that of Example 7-19?

7-25 In Example 7-19, when R_i was 1 kΩ, both the input and output pole were at 3.18 MHz. Find the frequency at which the gain is down by 3 dB from its mid-frequency gain.

7-26 A two-stage circuit has low frequency poles at 100 Hz and 200 Hz. Find the low frequency attenuation at 100 Hz and 200 Hz. Ignore all other poles and zeros.

7-27 Figure P7-27 is Figure 6-3, repeated here for the reader's convenience. Find its low frequency poles and zeros and the gain of the circuit at 200 Hz. The reader can use the values found in Example 6-2 to help with this problem.

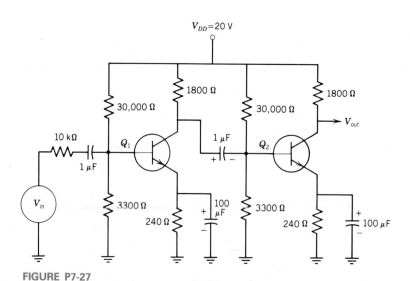

FIGURE P7-27

7-28 Find the high frequency poles for the circuit of Figure P7-27. Assume $f_T = 300$ MHz and $C_{be} = C_{bc} = C_{ce} = 4$ pF.

7-29 For the circuit of Figure P7-29 find
a. I_{EQ}
b. h_{ie}
c. The voltage gain of the transistor.
d. The voltage gain of the circuit.
e. The low frequency pole and zero.
f. The upper-half-power frequency if $C_{bc} = 5$ pF and $f_T = 200$ MHz. Ignore the other interelectrode capacitances and $r_{b'b}$.

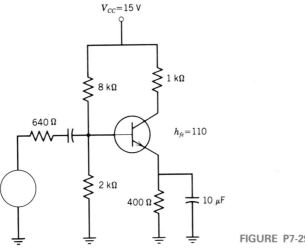

FIGURE P7-29

7-30 If the circuit of Figure P7-29 is capacitively coupled to another identical circuit, find
  a. The overall mid-frequency gain.
  b. The low frequency poles and zeros.
  c. The high frequency poles.
  d. The circuit gain at 100 Hz.
  e. The circuit gain at 1 MHz.

7-31 If the JFET circuit of Figure 7-19 is capacitively coupled to another identical circuit, find its low frequency poles and zeros and its gain at 100 Hz.

7-32 For the circuit of Figure 7-22, find the high frequency poles if it is coupled to another identical circuit and $R_i = 1\ k\Omega$.

7-33 A square wave is applied to a high-pass circuit. The output is shown in Figure P7-33. Find

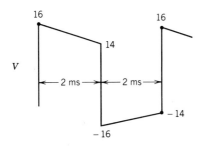

FIGURE P7-33

a. Percentage tilt.
b. The lower-half-power frequency of the circuit.
c. The capacitance of the circuit if R is 50,000 Ω.
d. The waveform of the input.

7-34 An amplifier has an f_L of 100 Hz. What is the percentage tilt of its output if a 1 kHz square wave is applied to its inputs?

7-35 A low-pass circuit consists of a 20 kΩ resistor in parallel with a 100 pF capacitor. What is the time constant of this circuit? What is its rise time?

7-36 A square wave is put into an amplifier that has a gain of 200 and a bandwidth of 500 kHz.
a. What is the rise time of the resulting output waveform?
b. If the maximum output swing of the amplifier is 8 V, how large can the input be?
c. What is the minimum period of the input for reasonable representation at the output?

7-37 In Example 7-25, what scope speeds would probably be used to observe
a. The tilt.
b. The rise time.

7-38 A capacitively loaded attenuator, similar to that in Figure 7-29, has $R_1 = 8$ kΩ, $R_2 = $ to 2 kΩ, and is loaded by a 200 pF capacitor.
a. Sketch its output to a ± 5 V square wave.
b. Find the time constant and rise time of the output.
c. What is the minimum period of the input square wave to allow five time constants so the output can reach V_{final}?

7-39 A relatively slow square wave is applied to the circuit of Figure P7-39. The square wave is 16 V peak to peak.
a. Sketch the output, indicating any x time constants involved.
b. Repeat part a if the 2 kΩ resistor is shunted by a 300 pF capacitor.
c. What is the minimum half period the square wave may have that will allow all voltages to reach their final values?

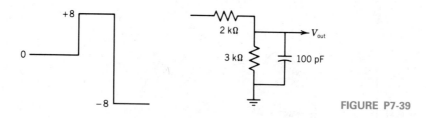

FIGURE P7-39

7-40 Sketch the response of a low-pass circuit with an upper-half-power frequency of 500 kHz to a single 1 μs pulse of 10 V.

7-41 An oscilloscope amplifier has an impedance of 500 kΩ and a capacitance of 15 pF. If the cable capacity is 45 pF, design a scope probe for
   a. 2:1 attenuation.
   b. 10:1 attenuation.
   What is the impedance at the probe tip in each case?

After attempting the problems, go back to section 7-2 and reread the questions. If any of them seem difficult, review the appropriate sections of the text to find the answers.

CHAPTER EIGHT

Feedback

8-1 Instructional Objectives

This chapter introduces the topic of feedback, where a portion of the output signal is sent back to the input.
After reading this chapter, the student should be able to:

1. Determine whether a feedback circuit has series or shunt feedback.
2. Find the gain of a feedback amplifier.
3. Find the input impedance of a feedback amplifier.
4. Find the output impedance of a series feedback circuit.
5. Calculate the change of gain of a feedback circuit due to a change in the gain of the basic amplifier.
6. Analyze feedback circuits without using the β concept.
7. Determine the frequency response of a feedback amplifier.

8-2 Self-Evaluation Questions

Watch for the answers to the following questions as you read the chapter. They should help you to understand the material being presented.

1. In section 8-3.1, what is the difference between gains A_1 and A_2?
2. How does feedback affect the input impedance of a series feedback amplifier? Of a shunt feedback amplifier?
3. What are the advantages and disadvantages of negative feedback? Of positive feedback? Which type of feedback is most often used in amplifiers?

4. How is series feedback most commonly injected into a circuit?
5. When shunt feedback is introduced into a circuit, by what factor is the current gain reduced? By what factor is the voltage gain reduced?
6. Does feedback affect the gain–bandwidth product of an amplifier?

8-3 BASIC FEEDBACK CONCEPTS

A *feedback circuit* is an electronic or mechanical circuit in which a *portion* of the output is *fed back* to the input. Electronic circuits are generally amplifiers, and a fraction of the output voltage or current is either added to or subtracted from the input. Feedback has a significant effect on the performance of these circuits.

Feedback is a complex subject, and students often have difficulty trying to understand it. The literature describes series and shunt feedback and subdivides both of these into voltage-driven and current-driven feedback. In sections 8-3 to 8-7 we will discuss only series and shunt feedback, and our methods will not adhere strictly to rigorous feedback theory, but the answers obtained in the examples are correct in all cases. We hope to gain in clarity and simplicity what we lose in rigor, render the subject understandable to the student encountering it for the first time, and help the engineer who is looking for an answer. Rigorous feedback circuit analysis is discussed in section 8-8.

In a *series feedback* circuit, a *voltage* is fed back that is placed in *series* with the *source voltage*. Figures 8-1 through 8-6 are examples of series feedback circuits. Series feedback circuits are characterized by the fact that *all the current provided by the source enters the amplifier*, both before and after the feedback network is connected.

In a *shunt feedback* circuit, the feedback element is placed *in parallel* with the amplifier. Thus the feedback element injects or drains off some of the input current. This type of amplifier is discussed in section 8-5.

8-3.1 The Series Feedback Circuit

The basic series feedback circuit is shown in Figure 8-1. It consists of

1. The source voltage generator, v_s. For simplicity, the input resistance of v_s is assumed to be 0.
2. The amplifier.
3. The β (feedback) network. This is usually a passive network, like a voltage divider, that divides the output voltage. β is the *ratio of the voltage fed back to the output voltage*.

We may assume that the β network is *unilateral;* that is, there is transmission from the output to the input, but no transmission from the input (v_s) to the output. This assumption is justified for almost all feedback circuits.

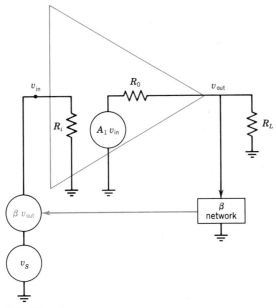

FIGURE 8-1
The basic series feedback amplifier.

4. The load on the amplifier. This load consists of the β network and any additional load resistor, R_L, connected to the output.
5. The feedback voltage generator, $\beta\, v_{out}$. This generator is in series with the source voltage.

EXAMPLE 8-1

In the circuit of Figure 8-2, the amplifier inverts the input signal. Find v_s if $v_{out} = -20$ V.

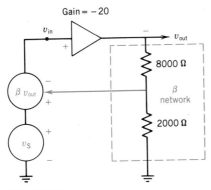

FIGURE 8-2
Circuit for Example 8-1.

8-3 BASIC FEEDBACK CONCEPTS

Solution
1. The β network is a 2000 Ω resistor in series with an 8000 Ω resistor. They form a voltage divider. The ratio of voltage fed back to the v_{out} is β. Here β = 0.2.
2. The voltage fed back is βv_{out} = 0.2 × 20 V = 4 V. Notice that this voltage opposes the source voltage.
3. Because the gain of the amplifier is −20 and v_{out} = −20 V, V_{in} must be +1 V.
4. V_s must equal V_{in} plus the voltage being fed back.

$$v_s = v_{in} + \beta v_{out} = 1\text{ V} + 4\text{ V} = 5\text{ V}$$

8-3.2 The Series Feedback Equations

The circuit of Figure 8-1 can be used to develop the equations for the series feedback circuit. The basic amplifier is enclosed in the triangular box. It consists of

1. The input impedance of the amplifier, R_i.
2. A voltage generator, whose voltage is the amplified input voltage, $A_1 v_{in}$.
3. An output impedance, R_o.

The gain of the amplifier with the load impedance (if any) and the β network in place is A_2 where

$$A_2 = \frac{v_{out}}{v_{in}}$$

Note the difference between A_1 and A_2. A_1 is the gain of the amplifier without any external load and without the β network.

The gain of the circuit with feedback is A_f. Assuming the feedback voltage generator opposes the source voltage, as shown in Figure 8-2, we have

$$A_f = \frac{v_{out}}{v_s} = \frac{A_2 v_{in}}{v_s}$$

But

$$V_{in} = V_s - \beta v_{out} = v_s - \beta A_2 v_{in}$$

or

$$v_s = v_{in}(1 + \beta A_2)$$

Therefore

$$A_f = \frac{A_2 v_{in}}{v_s} = \frac{A_2}{1 + \beta A_2} \quad (8\text{-}1)$$

Equation 8-1 gives the relationship between the gain without feedback (but with the feedback network in place), A_2, and the gain with feedback, A_f.

The input impedance of a feedback circuit is also important. With feedback, a circuit, I_f, flows in the input circuit. The input impedance with feedback is defined as

$$R_{if} = \frac{v_s}{i_f}$$

From the circuit we find

$$v_s - \beta v_{out} = I_f R_i$$
$$v_s = I_f R_i + \beta v_{out} = i_f R_i + \beta A_2 v_{in}$$

But $V_{in} = I_f R_i$. Therefore

$$v_s = i_f R_i + A_2 \beta i_f R_i$$

$$\frac{v_s}{i_f} = R_{if} = R_i (1 + \beta A_2) \quad (8\text{-}2)$$

The output impedance of the amplifier, R_o, is also affected by feedback. In the basic series feedback amplifier it is

$$R_{of} = \frac{R'_o}{1 + A_2 \beta} \quad (8\text{-}3)$$

where R'_o is the parallel combination of the amplifier's output impedance and the impedance of the β network. Equation 8-3 is derived in Appendix F1.

Equations 8-2 and 8-3 show that negative feedback, which causes βA to be a positive value (see section 8-3.3), increases the input impedance and decreases the output impedance. These are both desirable because the increased input impedance reduces the current the source must provide, and the smaller output impedance means the output is less dependent on the load.

EXAMPLE 8-2

For the circuit of Figure 8-3, the transistor has $h_{fe} = 100$ and $h_{ie} = 1000\ \Omega$.

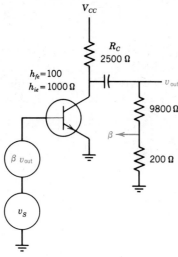

FIGURE 8-3
Circuit for Example 8-2.

a. Find A_1, A_2, R_i, and R_o without feedback.
b. Repeat part a for the circuit with feedback.
c. If $v_s = 0.1$ V, find v_{in} and v_{out}.
d. If a 500 Ω load resistor is placed in parallel with the output, find the gain.

Solution
a. This type of transistor circuit was studied in Chapter 4 (see equations 4-8, 4-10, and so forth). The transistor gain is

$$A_{v(tr)} = \frac{-h_{fe}R_C}{h_{ie}}$$

A_1 is the gain without the β network in place.

$$A_1 = \frac{-h_{fe}R_C}{h_{ie}} = \frac{-100 \times 2500\ \Omega}{1000\ \Omega} = -250$$

A_2 is the actual circuit gain with the β network in place.

$$R'_L = R_C \parallel R_\beta, \quad \text{where } R_\beta \text{ is the resistance of the β network.}$$
$$R'_L = 2500\ \Omega \parallel 10{,}000\ \Omega = 2000\ \Omega$$

358 FEEDBACK

$$A_2 = \frac{-h_{fe}R'_L}{h_{ie}} = \frac{100 \times 2000 \: \Omega}{1000} = 200$$

$$R_i = h_{ie} = 1000 \: \Omega$$

$$R'_o = R'_L \text{ (neglecting } h_{oe}) = 2000 \: \Omega$$

b. The β network is a voltage divider of 200 Ω and 9800 Ω. Therefore β = 0.02. With feedback we have

$$A_f = \frac{A_2}{1 + \beta A_2} = \frac{200}{1 + 4} = -40$$

$$R_{if} = R_i(1 + \beta A_2) = 1000 \: \Omega \times 5 = 5000 \: \Omega$$

$$R_{of} = \frac{R'_o}{1 + \beta A_2} = \frac{2000 \: \Omega}{1 + 0.02 \times 200} = \frac{2000 \: \Omega}{5} = 400 \: \Omega$$

c. If $V_s = 0.1$ V, then

$$v_{out} = A_f v_s = 40 \times 0.1 \text{ V} = 4 \text{ V}$$

The voltage fed back is $\beta v_{out} = 0.02 \times 4$ V $= 0.08$ V. The input voltage at the base of the transistor is

$$v_{in} = 0.1 \text{ V} - 0.08 \text{ V} = 0.02 \text{ V}$$

$$i_f = \frac{0.02 \text{ V}}{1000} = 20 \text{ mA}$$

$$R_{if} = \frac{v_s}{i_f} = \frac{0.1 \text{ V}}{20 \text{ mA}} = 5000 \: \Omega$$

This agrees with the answer found in part b.

d. If a 500 Ω resistor is placed across the load, the gain can be found either by recalculating A_2 or by using the output impedance.

By recalculating A_2, we have

$$R'_L = 2500 \: \Omega \: \| \: 10{,}000 \: \Omega \: \| \: 500 \: \Omega = 400 \: \Omega$$

$$A'_2 = \frac{-h_{fe}R'_{L2}}{h_{ie}} = -40$$

$$\beta A'_2 = 0.02 \times 40 = 0.8$$

$$A_{f2} = \frac{-A'_2}{1 + \beta A'_2} = \frac{-40}{1.8} = -22.2$$

Alternately, the circuit can be considered as having a gain of 40 and an output impedance of $R_{of} = 400 \: \Omega$.

When a 500 Ω load resistor is added to this network:

$$A_{f2} = A_f \frac{R_L}{R_L + R'_{of}} = -40 \times \frac{500 \, \Omega}{900 \, \Omega} = -22.2$$

8-3.3 Positive and Negative Feedback

If the polarity of the voltage of the feedback generator aids the source voltage, the feedback is positive; if it *opposes* the source voltage, as in Examples 8-1 and 8-2, the feedback is *negative*. For positive feedback, the term βA_2 in equation 8-1 is a negative number and the positive feedback increases the gain.

If βA is -0.5 in equation 8-1, $A_f = 2A$, so this value doubles the gain. If $\beta A = -1$, the denominator is 0 and the gain is infinite. Unfortunately, this really means the circuit is *oscillating* and is *useless* as an amplifier. Because high gains are usually easy to obtain, and because positive feedback destabilizes a circuit and increases its tendency to oscillate, positive feedback is very rarely used in an amplifier. It is used in the design of oscillators (see Chapter 11).

Negative feedback occurs when βA is positive and the voltage is fed back in opposition to the applied voltage. It reduces the gain of the amplifier. In exchange for the gain reduction, negative feedback amplifiers provide greater stability and increase the bandwidth of the circuit by raising f_H and lowering f_L. Stability is discussed in section 8-3.4 and the frequency response of feedback circuits is discussed in section 8-7.

8-3.4 Stability

Stability is the ability of an amplifier to function as designed, despite changes in its parameters that cause changes in its gain. The stability of transistor amplifiers was previously discussed in section 3-9. The stability of a feedback amplifier can be found by differentiating equation 8-1.

$$A_f = \frac{A_2}{1 + \beta A_2}$$

$$\frac{dA_f}{dA_2} = \frac{1 + \beta A_2 - \beta A_2}{(1 + \beta A_2)^2} = \frac{1}{(1 + \beta A_2)^2} = \frac{A_2}{(1 + \beta A_2)} \cdot \frac{1}{A_2} \cdot \frac{1}{(1 + \beta A_2)}$$

$$\frac{dA_f}{dA_2} = \frac{A_f}{A_2 (1 + \beta A_2)} \tag{8-1}$$

Therefore

$$\frac{dA_f}{A_f} = \frac{1}{1 + \beta A_2} \cdot \frac{dA_2}{A_2} \tag{8-4}$$

Equation 8-4 shows that if the gain A_2 changes by a factor of dA_2/A_2, the change of gain in the feedback amplifier, dA_f/A_f, will be reduced by the factor $(1 + \beta A_2)$.

EXAMPLE 8-3

A negative feedback amplifier has $\beta A = 9$. If A_2 changes by 10% (due to a change of h_{fe}, h_{ie}, or for any other reason), what is the percentage change in the gain of the feedback amplifier?

Solution
Here $dA_2/A_2 = 10\% = 0.1$.
By equation 8-4

$$\frac{dA_f}{A_f} = \frac{1}{1 + \beta A} \frac{dA}{A} = \frac{1}{10} \times 0.1 = 1\%$$

In Example 8-3 a 10% change in the intrinsic gain of the amplifier produced a 1% gain change in the feedback circuit. Therefore, the gain of the feedback amplifier is much more stable than its gain without feedback.

If there is a large amount of negative feedback $\beta A \gg 1$. In this case, equation 8-1 reduces to

$$A_f = \frac{A_2}{1 + \beta A_2} \approx \frac{A_2}{\beta A_2} = \frac{1}{\beta} \tag{8-5}$$

When equation 8-5 is valid, the gain of the circuit depends *entirely on the feedback network* and *not* on the parameters of the amplifier (h_{fe}, h_{ie}, and so on), provided only that these parameters are stable enough to maintain a high value of βA. It also means that the gains of these amplifiers is independent of external factors, such as temperature variations. Engineers often approximate the gain of feedback amplifiers by $1/\beta$ (see problem 8-9).

8-4 ACTUAL VOLTAGE FEEDBACK CIRCUITS

The conceptual discussion in the previous section assumed a feedback voltage generator that does not actually exist in most circuits. Feedback is usually injected by other means. In this section, several circuits that use voltage feedback will be analyzed.

The gain of most feedback circuits can be found by following these steps:

1. Find the gain of the amplifier without feedback. This is the gain of the amplifier with β assumed to be 0.

2. Find β.
3. Use equation 8-1 to find the gain of the amplifier.

8-4.1 The JFET Feedback Circuit

The JFET feedback circuit of Figure 8-4 is the simplest and most easily analyzed feedback circuit. The feedback voltage is developed across the β network (R_A and R_B) and is connected directly in series with the source.

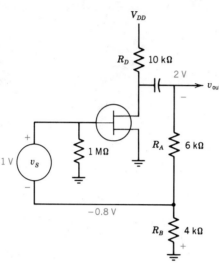

FIGURE 8-4
A JFET feedback circuit.

EXAMPLE 8-4

a. Find the gain of the feedback circuit of Figure 8-4 if the JFET has $g_m = 0.002$ S
b. Find the voltage at gate if $v_s = 1$ V.

Solution

a. The gain of the JFET circuit without feedback is

$$A_2 = -g_m (R_D \| R_\beta)$$

where R_β is the resistance of the feedback network.

$$R_\beta = 4\text{ k}\Omega + 6\text{ k}\Omega = 10\text{ k}\Omega$$
$$A_2 = 0.002 \times 5000\text{ }\Omega = -10$$

In this circuit $\beta = 0.4$ because of the voltage divider action of R_A and R_B and

$$A_f = \frac{-A_2}{1 + \beta A_2} = \frac{-10}{5} = -2$$

b. If $v_s = 1$ V, the voltages around the circuit are shown in blue of Figure 8-4.

$$v_{out} = A_f v_s = -2 \text{ V}$$

The voltage at the gate is

$$v_g = v_s - \beta v_{out} = 1 \text{ V} - 0.8 \text{ V} = 0.2 \text{ V}$$

Check: The gain of the circuit, from gate to drain is

$$\frac{v_D}{v_G} = \frac{-2}{0.2} = -10 = A_2$$

8-4.2 The Unbypassed Emitter-resistor

The circuit of Figure 8-4 is seldom used because neither end of the source voltage, v_s, is connected to ground. Most voltage sources have one end directly connected to ground, so this circuit will not work in these cases.

In most voltage feedback circuits, the feedback voltage is introduced across an unbypassed emitter- or source-resistor. For negative feedback, this voltage must be in phase with the input voltage so it opposes the input voltage, thereby reducing the input current and the circuit gain.

An amplifier with an unbypassed emitter-resistor, as shown in Figure 8-5, can be considered as a voltage feedback circuit. This circuit has already

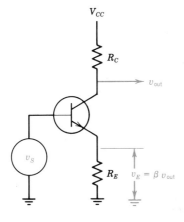

FIGURE 8-5
A transistor amplifier with an unbypassed emitter-resistor (viewed from a feedback perspective).

been discussed in section 4-7.2 without reference to feedback, and we can arrive at the same equations using feedback.

In Figure 8-5 the voltage across R_E is the feedback voltage, v_r, and also equals $\beta\, v_{out}$. Notice that this voltage is in phase with the source voltage, and opposes it, causing the input current to be small. This, therefore, is a negative feedback circuit.

To use the feedback equations developed in section 8-3, the gain of the circuit without feedback must be found first. This gain can be found by setting β equal to 0, which effectively grounds the emitter. Then we have

$$A_2 = \frac{-h_{fe}R_C}{h_{ie}}$$

β is the fraction of the output voltage fed back to the input. In the circuit of Figure 8-5

$$v_E = (1 + h_{fe})i_b R_E$$

$$v_{out} = h_{fe} i_b R_C$$

$$\beta = \frac{v_E}{v_{out}} = \frac{(1 + h_{fe})R_E}{h_{fe}R_C}$$

$$A_f = \frac{A_2}{1 + \beta A_2} = \frac{\dfrac{-h_{fe}R_C}{h_{ie}}}{1 + \dfrac{(1 + h_{fe})R_E}{h_{fe}R_C} \cdot \dfrac{h_{fe}R_C}{h_{ie}}} = \frac{-h_{fe}R_C}{h_{ie} + (1 + h_{fe})R_E}$$

This is the same as equation 4-16, developed in Chapter 4 for this circuit.

EXAMPLE 8-5

Find the input impedance for the circuit of Figure 8-5 using the feedback formulas.

Solution

The input impedance, R_i, without feedback, or when the emitter is grounded, is h_{ie}. With feedback

$$R_{if} = (1 + \beta A)R_i$$

Here

$$\beta A_2 = \frac{(1 + h_{fe})R_E}{h_{ie}}$$

$$R_{if} = \left[1 + \frac{(1 + h_{fe})R_E}{h_{ie}}\right] h_{ie} = h_{ie} + (1 + h_{fe})R_E$$

This is the same input impedance as found in Chapter 4.

8-4.3 Multistage Feedback Amplifiers

Amplifiers are often built with a much higher gain than the circuit requires. Then feedback is introduced to lower the gain and stabilize the amplifier. As we saw in Chapter 6, multistage amplifiers are often used to provide high gain.

Figure 8-6 is a multistage amplifier with feedback. The output is con-

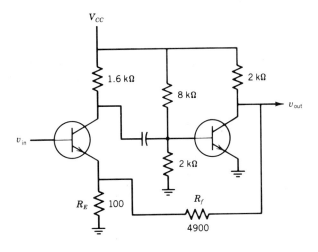

For each transistor, $h_{fe} = 100$, $h_{ie} = 800$.

FIGURE 8-6
A multistage amplifier with feedback. *Note:* For each transistor $h_{fe} = 100$, $h_{ie} = 800$.

nected through R_f to the unbypassed emitter resistor. This output is in phase with the input and this circuit is a series voltage feedback circuit with negative feedback.

EXAMPLE 8-6

For the circuit of Figure 8-6, find its gain and the input impedance.

Solution

As discussed in section 8-4, the first step is to find the gain of the circuit without the feedback. The load resistance at the collector of Q_1 is the parallel combination of the resistors there and h_{ie2}.

$$R_{L1} = 1600\ \Omega\ ||\ 8\ k\Omega\ ||\ 2\ k\Omega\ ||\ 800\ \Omega = 400\ \Omega$$

The feedback resistor, R_E, acts as an unbypassed emitter-resistor in the first stage. Therefore its gain, calculated exactly, is

$$A_{Q1} = \frac{-h_{fe}R_{L1}}{h_{ie} + (1 + h_{fe})R_E} = \frac{-100 \times 400\ \Omega}{800\ \Omega + (101)\,100\ \Omega} = \frac{-40{,}000\ \Omega}{10{,}900\ \Omega} = -3.67$$

The load resistor for the second stage is

$$R_{L2} = 2\ k\Omega\ ||\ 5\ k\Omega = 1428\ \Omega$$

$$A_{Q2} = \frac{-h_{fe}R_{L2}}{h_{ie}} = \frac{-100 \times 1428\ \Omega}{800\ \Omega} = -178.5$$

$$A_T = A_{Q1} \times A_{Q2} = -3.67 \times -178.5 = 655.1$$

β for this circuit is the feedback voltage, the emitter voltage at Q_1, divided by V_{out} at the collect of Q_2. The resistors form a voltage divider so that

$$\beta = \frac{R_E}{R_E + R_{f2}} = \frac{100\ \Omega}{100\ \Omega + 4900\ \Omega} = 0.02$$

$$\beta A = 655.1 \times 0.02 = 13.1$$

From equation 8-1 we have

$$A_f = \frac{A}{1 + \beta A} = \frac{655.1}{14.1} = 46.46$$

The input impedance to this circuit without feedback, but with the 100 Ω emitter-resistor in place, is

$$R_i = h_{ie} + (1 + h_{fe})R_E = 10{,}900\ \Omega$$

With feedback

$$R_{if} = (1 + \beta A)R_i = 14.1 \times 10{,}900\ \Omega = 153{,}690\ \Omega$$

There are many variations on multiple stage circuits with feedback, but the circuit of Figure 8-6 is probably the most commonly used circuit.

8-5 CURRENT FEEDBACK

Current or shunt feedback occurs when a resistor is placed *between* the input and the output of an amplifier. The effect of this resistor is to *shunt* a portion of the input current to the output, reducing the input current into the amplifier and, therefore, the gain.

The situation is shown in Figure 8-7 with the feedback resistor shown in blue. For shunt feedback, it is clearer to use current generators than voltage generators. The input current is supplied by the current generator i_{in}, and the amplifier is considered to be a current generator, with a current gain, h_{fe}, that amplifies the current at its input, I_1.

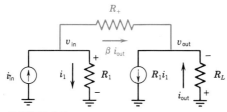

FIGURE 8-7
The basic shunt or current feedback circuit.

As with voltage feedback, the feedback resistor, R_f, must be included. For $\beta = 0$, this is done by connecting R_f between the output and ground.

The concept of current feedback is often complex and confusing to the student. We hope to clarify it by offering precise definitions of the terms.

A_i is the current gain of the circuit with the feedback resistor in the circuit, but connected to ground so that $\beta = 0$.

$$A_i = \frac{i_{out}}{i_{in}}$$

where I_{out} is the current flowing in the output load resistor (R_L in Figure 8-7) and I_{in} is the input current to the circuit. If the circuit is actually driven by a voltage generator, its Norton's equivalent current generator should be used.

β is the ratio of the current in the feedback resistor to I_{out}. We can assume that V_{in} is very low or approximately 0, so that R_f is effectively in parallel with R_L. For most circuits, this approximation is very accurate.

The circuit without feedback ($\beta = 0$) is considered first.

EXAMPLE 8-7

Find the voltage gain and the current gain of the circuit of Figure 8-7 if the feedback resistor is disconnected from the input and connected to ground.

Solution
For the voltage gain without feedback, we have

$$i_{in} = i_1$$
$$v_{in} = i_{in} R_i$$
$$v_{out} = h_{fe} i_{in} (R_L \| R_f)$$

$$A_v = \frac{v_{out}}{v_{in}} = \frac{h_{fe} i_{in}(R_L \| R_f)}{i_{in} R_i} = \frac{h_{fe}(R_L \| R_f)}{R_i} \quad (8\text{-}6)$$

The current gain of the circuit A_i is defined as the current in R_L, i_{out}, divided by the input current. Without feedback (with R_f connected to ground), the current produced by the current generator $h_{fe} i_1$ divides between R_L and R_f, and by current division

$$i_{out} = h_{fe} i_1 \times \frac{R_f}{R_L + R_f}$$

but, without feedback, $i_1 = i_{in}$

$$A_i = \frac{i_{out}}{i_{in}} = \frac{h_{fe} R_f}{R_L + R_f}$$

These equations are basically the same as for any BJT amplifier.

When the feedback resistor is connected to the input, a portion, β, of the output current is drawn off from the input, as shown in blue in Figure 8-7.

In practical circuits the input voltage, v_{in}, is small compared to the output voltage. If v_{in} is taken as 0 V, then R_f is in parallel with R_L. The current in R_f is found by simple current division to be

$$i_f = \frac{R_L}{R_f} \times i_{out}$$

Therefore

$$\beta = \frac{R_L}{R_f} \quad (8\text{-}8)$$

for this circuit.

8-5.1 The Current Feedback Equations

Using the circuit of Figure 8-7, the equations for the current gain and input impedance of the shunt feedback circuit with feedback can be found. With the feedback resistor R_f in place

$$i_{in} = i_1 + i_f = i_1 + \beta i_{out} \tag{8-9}$$

But $I_{out} = A_i I_1$, where A_i is the current gain without feedback. Therefore, equation 8-9 becomes

$$i_{in} = i_1 + \beta A_i i_1$$

The current gain with feedback is

$$A_f = \frac{i_{out}}{i_{in}} = \frac{A_i i_1}{i_1 + \beta A_i i_1}$$

$$\boxed{A_f = \frac{A_i}{1 + \beta A_i}} \tag{8-10}$$

The current gain equation with feedback looks very much like the voltage gain equation found in section 8-3.

To find the input impedance with feedback, R_{if}, we have

$$R_{if} = \frac{v_{in}}{i_{in}}$$

but $v_{in} = i_1 R_i$ and $i_{in} = i_1 + \beta i_{out} = i_1 + \beta A_i i_1$. Therefore:

$$\boxed{R_{if} = \frac{R_i}{1 + \beta A_i}} \tag{8-10}$$

Thus the input impedance of a current feedback circuit is *reduced* by the factor $(1 + \beta A_i)$. This smaller input impedance helps justify the assumption made earlier that v_{in} is small compared to v_{out}. This does increase the load on the source, but this circuit is starting to exhibit one of the characteristics of an op-amp ($v_{in} \approx 0$).

If the current source is replaced by a voltage source with an input impedance, R_S, it can be shown (see Appendix F2) that the voltage gain of the circuit, defined as v_{out}/v_S, decreases by the factor $1 + \beta A_i$ when feedback is introduced. Thus the factor that reduces the current gain also reduces the voltage gain.

8-5.2 The Single Stage Shunt Feedback Circuit

The simplest example of shunt or current feedback is the single-stage circuit of Figure 8-8. The feedback resistor, R_f, provides the feedback path and shunts some of the input current directly to the output.

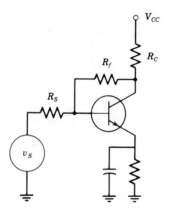

FIGURE 8-8
The basic single-stage current feedback circuit.

EXAMPLE 8-8

The circuit of Figure 8-9 is driven by a voltage source with a 150 Ω source impedance. The biasing resistors are 9 kΩ and 1 kΩ ($R_{BB} = 900$ Ω), the

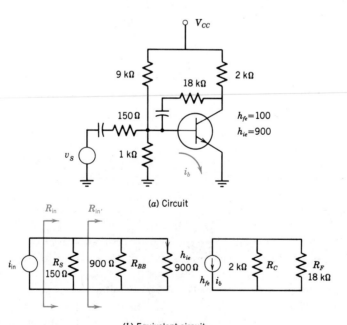

FIGURE 8-9
Circuit for Example 8-8.

370 FEEDBACK

collector resistor is 2 kΩ, and $R_f = 18$ kΩ. Find the current and voltage gains of the circuit and its input impedance:
a. Without feedback.
b. With feedback.

Solution
a. Without feedback. The equivalent circuit is shown in Figure 8-9b. The input impedance is

$$R_i = h_{ie} \| R_{BB} \| R_S = 900\ \Omega \| 900\ \Omega \| 150\ \Omega = 112.5\ \Omega$$

The current gain i_{out}/i_{in} can be found from Figure 8-9b:

$$i_b = \frac{i_{in}}{8}$$

$$h_{fe}i_b = \frac{100\ i_{in}}{8} = 12.5\ i_{in}$$

We define the output current as the current in the 2 kΩ load resistor.

$$i_{out} = h_{fe}i_b \times \frac{R_f}{R_f + R_C} = 12.5\ i_{in} \times \frac{18,000\ \Omega}{20,000\ \Omega} = 11.25\ i_{in}$$

$$A_i = \frac{i_{out}}{i_{in}} = 11.25$$

The voltage gain of this circuit is

$$A_v = \frac{V_{base}}{V_{in}} \times \frac{h_{fe}R_{EQ}}{h_{ie}} = \frac{450\ \Omega}{600\ \Omega} \times 100 \times \frac{(2000\ \Omega \| 18,000\ \Omega)}{900\ \Omega}$$

$$A_v = \frac{450\ \Omega}{600\ \Omega} \times 100 \times \frac{1800\ \Omega}{900\ \Omega} = 150$$

b. With feedback. With feedback we have

$$\beta = \frac{R_L}{R_f} = \frac{2000\ \Omega}{18,000\ \Omega} = 0.111$$

$$\beta A_i = 0.111 \times 11.25 = 1.25$$

$$1 + \beta A_i = 2.25\ [1]$$

---

[1] Some other authors have used A_i at the collector of the transistor. For them, $A_i = 12.5$ here, but by this definition, β also changes to 0.1, so the β A_i product is still 1.25 (see problem 8-18).

$$A_{if} = \frac{A_i}{1 + \beta A_i} = \frac{11.25}{2.25} = 5$$

$$A_{vf} = \frac{A_v}{1 + \beta A_i} = \frac{150}{2.25} = 66.7$$

$$R_{if} = \frac{R_i}{1 + \beta A_i} = \frac{112.5 \, \Omega}{2.25} = 50 \, \Omega$$

The results of this example can be checked by applying a voltage at the source and finding the resulting voltages and currents throughout the circuit.

EXAMPLE 8-9

In the circuit of Figure 8-9, find the voltages and currents throughout the circuit if $v_{in} = 0.3$ V.

Solution

The first problem is to determine how much current flows when $v_s = 0.3$ V. Figure 8-9a shows that the resistance seen by V_s is R_s in parallel with R'_{in}. We know that R_{in} with feedback is 50 Ω. Therefore

$$R_S \parallel R'_{in} = 50 \, \Omega$$

$$R'_{in} = \frac{150 \, \Omega \times 50 \, \Omega}{(150 \, \Omega - 50 \, \Omega)} = 75 \, \Omega$$

This shows the total resistance seen by v_s is 225 Ω and that

$$v_{base} = \frac{R'_{in} \times v_S}{R_S + R'_{in}} = \frac{75 \, \Omega}{225 \, \Omega} \times 0.3 \text{ V} = 0.1 \text{ V}$$

$$i_{base} = \frac{v_{base}}{h_{ie}} = 0.1 \text{ V}/900 \, \Omega = 0.111 \text{ mA}$$

$$h_{fe} i_b = 11.1 \text{ mA}$$

$$i_{out} = \frac{R_f}{R_C + R_f} \times h_{fe} i_b = \frac{9}{10} \times 11.1 \text{ mA} = 10 \text{ mA}$$

$$v_{out} = i_{out} R_C = 10 \text{ mA} \times 2000 = 20 \text{ V}$$

$$A_v = \frac{v_{out}}{v_{in}} = \frac{20 \text{ V}}{0.3 \text{ V}} = 66.7$$

To find the current gain, the voltage generator must be Nortonized. Then $i_{in} = i_{RS} + i_{R_{BB}} + i_b + i_f$.

The voltage across the equivalent input circuit is 0.1 V and has already been found as 0.111 mA.

$$i_{R_{BB}} = 0.111 \text{ mA}$$

$$i_S = \frac{v_S}{R_S} = \frac{0.1 \text{ V}}{150 \text{ }\Omega} = 0.666 \text{ mA}$$

$$i_f = \frac{20 \text{ V}}{18{,}000 \text{ }\Omega} = 1.111 \text{ mA}$$

$$i_{in} = 0.666 \text{ mA} + 0.111 \text{ mA} + 0.111 \text{ mA} + 1.111 \text{ mA} = 2 \text{ mA}$$

$$A_i = \frac{i_{out}}{i_{in}} = \frac{10 \text{ mA}}{2 \text{ mA}} = 5$$

$$R'_{if} = \frac{v_{in}}{i_{in}} = \frac{0.1 \text{ V}}{2 \text{ mA}} = 50 \text{ }\Omega$$

The values of A_v, A_i, and R_{if} check with Example 8-8. The currents and

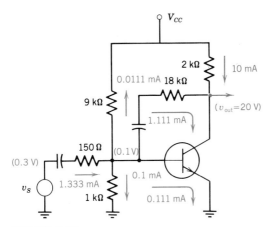

FIGURE 8-10
The currents and voltages as calculated in Example 8-9.

voltages throughout the circuit are shown in Figure 8-10. Note that in this problem, $v_{out} = 20$ V, and $v_{base} = 0.1$ V.

The 0.1 V at the base is indeed small and justifies the assumption that R_f is in parallel with R_C.

8-5 CURRENT FEEDBACK

8-5.3 An Alternative Method of Shunt Feedback Analysis[2]

An interesting variation on the method of analyzing shunt feedback circuits exists. The circuit can be analyzed with an ideal current generator applied to its input terminals. This leads to a higher value of A_i. Now, however, if a voltage source in series with a resistor, R_S, is applied to the circuit, the ratio of the voltage gain with feedback to the voltage gain without feedback is given by

$$\frac{A_{vf}}{A_v} = \frac{A_i}{A_{if}} \times \frac{R_S + R_i}{R_S + R_{if}} = \frac{1}{1 + \beta A_i} \times \frac{R_S + R_i}{R_S + R_{if}} \qquad (8\text{-}12)$$

Equation 8-12 also shows that if R_S is zero, the voltage gains with and without feedback are the same.

EXAMPLE 8-10

Analyze the circuit of Figure 8-9 by the methods of this section.

Solution

In the equivalent circuit of Figure 8-9b, the 150 Ω resistor associated with the source is no longer in the circuit. If an ideal current generator is applied to the input, the two 900 Ω resistors, then

$$A_i = 45$$
$$\beta = 0.111 \quad \text{(as before)}$$
$$\beta A_i = 5$$
$$1 + \beta A_i = 6$$
$$A_{if} = \frac{A_i}{1 + \beta A_i} = \frac{45}{6} = 7.5$$
$$R_i = 900\ \Omega\ ||\ 900\ \Omega = 450\ \Omega$$
$$R_{if} = \frac{450\ \Omega}{6} = 75\ \Omega$$

Now, if a voltage source with a 150 Ω impedance is applied to the circuit, the ratio of the voltage gains with and without feedback is

---

[2]This section may be omitted at first reading.

$$\frac{A_{vf}}{A_v} = \frac{1}{1 + \beta A} \times \frac{R_S + R_i}{R_S + R_{if}} = \frac{1}{6} \times \frac{600 \ \Omega}{225 \ \Omega} = \frac{1}{2.25}$$

This checks with the results of Example 8-8.

8-5.4 Multistage Shunt Feedback Amplifiers

Multistage feedback amplifiers can be analyzed by the methods used in the previous sections. Figure 8-11 shows a multistage amplifier whose voltage

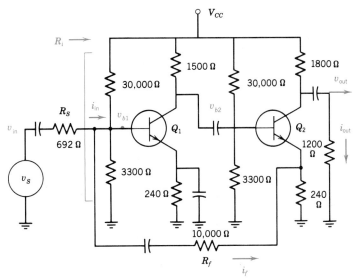

FIGURE 8-11
A multistage shunt feedback amplifier.

gain is to be found. It is a two-stage amplifier with a feedback resistor, R_f, and the feedback path is from the base of $Q1$ to the emitter of $Q2$. In any analysis, one must first determine whether voltage (series) or current (shunt) feedback is being used. Because R_f shunts current from the input to the output, this is a shunt feedback amplifier.

Although we are interested in finding the voltage gain, this shunt feedback amplifier must be analyzed by the same steps used to analyze the single-stage amplifier of section 8-5.2, namely:

1. Determine where the input and output voltages and currents are. The choices we have made are shown in blue on Figure 8-11.
2. Find A_v without feedback.
3. Find A_i without feedback.

4. Find β and βA_i.
5. Calculate the voltage gain with feedback.
6. Check the answers. Feedback circuits are very forgiving of errors, so the check should be very close (within 1%) to the answers found previously.

EXAMPLE 8-11

Find the voltage gain of the circuit of Figure 8-11. For both transistors, assume $h_{fe} = 150$, and $h_{ie} = 900\ \Omega$. Reasonable engineering assumptions to simplify the arithmetic will be allowed.

Solution

The solution proceeds in accordance with the steps listed previously. The chosen input and output points are shown in blue. For step 2, finding the voltage gain without feedback, R_f is assumed to be connected to ground. Then we have

$$R'_i = 30{,}000\ \Omega\ ||\ 3300\ \Omega\ ||\ h_{ie}$$

To simplify, we assume $30{,}000\ \Omega\ ||\ 3300\ \Omega = 3{,}000\ \Omega$

$$R'_i = 3000\ \Omega\ ||\ 900\ \Omega = 692\ \Omega$$

Because $R'_i = R_S$, $v_{b1} = v_{S/2}$

$$A_{v(\text{trl})} = \frac{-h_{fe} R'_L}{h_{ie}}$$

R'_L is the parallel combination of $1500\ \Omega$, $3300\ \Omega$, $30{,}000\ \Omega$, and the input impedance of Q_2. Because R_{E2} is unbypassed, this impedance is

$$R_{i2} = h_{ie} + (h_{fe} + 1)R_E = 900\ \Omega + 151 \times 240\ \Omega = 37{,}140\ \Omega$$

This impedance is in parallel with $1500\ \Omega\ ||\ 3000\ \Omega = 1000\ \Omega$ and will be ignored for simplicity.

$$A_{v(\text{trl})} = \frac{-h_{fe} R'_L}{h_{ie}} = \frac{-150 \times 1000\ \Omega}{900\ \Omega} = -166.7$$

Transistor 2 has an unbypasssed emitter-resistor. Its voltage gain is approximately R'_L/R_E. Here $R'_L = 1800\ \Omega\ ||\ 1200\ \Omega = 720\ \Omega$ and $R_E = 240\ \Omega$. Strictly speaking, R'_E should be used where R'_E is R_E ($240\ \Omega$) in parallel with R_f ($10{,}000\ \Omega$). Because $R_f \gg R_E$, it can be ignored to simplify the mathematics.

Therefore

$$A_{v2} = \frac{-R_L'}{R_E} = \frac{-720\ \Omega}{240\ \Omega} = -3$$

$$A_v = \frac{v_b}{v_S} \times A_{v1} \times A_{v2} = 0.5 \times -166.7 \times -3 = 250$$

In step 3 the current gain, A_i, without feedback must be found. The voltage generator v_S must be replaced by a current generator, i_{in}, in parallel with R_S as shown in Figure 8-12.

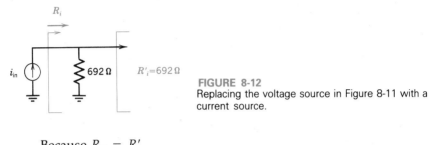

FIGURE 8-12
Replacing the voltage source in Figure 8-11 with a current source.

Because $R_S = R_i'$

$$i_1 = \frac{i_{in}}{2}$$

This feeds h_{ie} in parallel with R_{BB}.

$$i_{b1} = \frac{3000\ \Omega}{3900\ \Omega} \times i_1 = \frac{3000}{3900} \times \frac{i_{in}}{2}$$

$$i_{C1} = h_{fe}\ i_{b1} = 150\ i_{b1}$$

The resistance at the collector of Q_7 is 37,140 Ω in parallel with 1000 Ω. The current flowing in the base of Q_2 is, by current division

$$i_{b2} = \frac{1000\ \Omega}{38,140\ \Omega} i_{C1}$$

$$i_{c2} = h_{fe}\ i_{b2} = 150\ i_{b2}$$

The collector current, i_{C2} divides between the 1800 Ω resistor and the 1200 Ω resistor. By current division

$$i_{out} = \frac{1800\ \Omega}{1200\ \Omega + 1800\ \Omega} i_{c2} = 0.6\ i_{c2}$$

Finally

$$A_i = 0.5 \times \frac{3000\ \Omega}{3900\ \Omega} \times 150 \times \frac{1000\ \Omega}{38{,}140\ \Omega} \times 150 \times 0.6$$

$$A_i = 136.1$$

The next step is to find β. This is the ratio of the output current, flowing in the 1200 Ω load resistor, to the current flowing in the feedback resistor, R_f. If i_{out} is in the 1200 Ω load resistor, then 0.667 i_{out} must be flowing in the 1800 Ω collector resistor and the current in the emitter is 1.667 i_{out}. This current divides between R_f and R_E.

$$i_f = \frac{240\ \Omega}{10{,}240\ \Omega} \approx \frac{240\ \Omega}{10{,}000\ \Omega} = 0.024\ I_E$$

$$\beta = \frac{i_f}{i_{out}} = 0.024 \times 1.667 = 0.04$$

$$\beta A_i = 0.04 \times 136 = 5.44$$

Now the voltage gain with feedback can be found.

$$A_{vf} = \frac{A_v}{1 + \beta A_i} = \frac{250}{6.44} = 38.82$$

This problem can be checked as shown in Example 8-9.

$$R_{if} = \frac{R_i}{1 + \beta A_i}$$

R_i without feedback, but using a current generator as an input is

692 Ω || 692 Ω = 346 Ω

$$R_{if} = \frac{346\ \Omega}{6.44} = 53.72\ \Omega$$

$$R'_{if} = \frac{R_S R_{if}}{R_S - R_{if}} = \frac{53.72\ \Omega \times 692\ \Omega}{638.28\ \Omega} = 58.24\ \Omega$$

With feedback

$$v_{b1} = \frac{R'_{if}}{R_S + R'_{if}} \times v_{in} = \frac{58.24\ \Omega}{750.24\ \Omega} \times v_{in} = 0.07763\ v_{in}$$

$$v_{out} = 0.07763\ v_{in} \times A_{v1} \times A_{v2} = 0.07763\ v_{in} \times 166.7 \times 3$$

$$v_{out} = 38.823\ v_{in}$$

$$A_v = \frac{v_{out}}{V_{in}} = 38.823$$

Despite some assumptions, the two methods check to about 0.01%. This is acceptable accuracy. For a further check see problem 8-20.

8-6 A NEW LOOK AT FEEDBACK

Through the years, a mystique has arisen about feedback; that feedback circuits can be analyzed only by deciding whether the circuit contains series or shunt feedback, and then calculating A and β. We maintain that feedback circuits can be analyzed in a straightforward manner, using the circuit equations developed in previous chapters.

Perhaps the reason feedback circuits have not been analyzed in this manner by other authors is because most engineers want to start at the beginning, at the source voltage. This is difficult, however, when feedback is used. The trick is to find the *proper starting point*. For BJT circuits, the proper starting point is usually the *current* in the *base of the first transistor*. Two examples are presented in this section to demonstrate the procedure.

EXAMPLE 8-12

Find the voltage gain and input impedance of the two-stage series feedback circuit of Figure 8-6, by the methods described previously.

Solution

In accordance with the foregoing discussion, the base current in Q1 is taken as the starting point. This current is unknown, but will be called i_{b1}. It then follows that

$$i_{c1} = h_{fe}\ i_{b1} = 100\ I_{b1}$$

The collector current divides between R_C (1.6 kΩ), R_{BB} (1.6 kΩ), and h_{ie2} (800 Ω). By current division

$$i_{b2} = i_{c1/2} = 50\ i_{b1}$$
$$i_{C2} = 100\ i_{b2} = 5000\ i_{b1}$$

If the effect of the current flowing in R_f due to the emitter current of transistor Q1 is ignored (it is shown to be negligible in Appendix F3), the output load resistor is

$$R_{C2} \parallel (R_f + R_E) = 2000 \ \Omega \parallel 5000 \ \Omega = 1428.57 \ \Omega$$

$$v_{out} = 1428.57 \ \Omega \times 5000 \ I_{b1} = 7{,}142{,}850 \ I_{b1}$$

The current in R_f is

$$i_f = v_{out}/(R_f + R_E) = \frac{7{,}142{,}850 \ i_{b1}}{5000 \ \Omega} = 1428.57 \ i_{b1}$$

The current in R_E is

$$i_{RE} = i_f + (1 + h_{fe})i_{b1} = 1428.57 \ i_{b1} + 101 \ i_{b1} = 1529.57 \ i_{b1}$$

The currents are shown in blue in Figure 8-13. Once the currents have been found, as a function of i_{b1}, the voltages can be calculated.

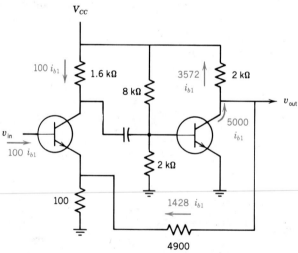

FIGURE 8-13
Circuit for Example 8-12. *Note*: For each transistor $h_{ie} = 100$, $h_{ie} = 800$.

The voltage at the emitter is

$$v_{RE} = i_{RE} \times R_E = 152{,}957 \ i_{b1}$$

The input voltage to the circuit is

$$v_{in} = h_{ie} \ i_{b1} + v_{RE} = 800 \ i_{b1} + 152{,}957 \ i_{b1} = 153{,}757 \ i_{b1}$$

Finally, the voltage gain of the circuit is

$$A_v = \frac{v_{out}}{v_{in}} = \frac{7{,}142{,}850\, i_{b1}}{153{,}757\, i_{b1}} = 46.455 \quad (46.46)$$

$$R_i = \frac{v_{in}}{i_{in}} = \frac{153{,}757\, i_{b1}}{i_{b1}} = 153{,}757\ \Omega \quad (153{,}690\ \Omega)$$

The numbers in parentheses are the answers calculated in Example 8-6 for this circuit using A and β. The discrepancy is less than 0.1%.

EXAMPLE 8-13

Find the voltage gain and input impedance of the circuit of Figure 8-9.

Solution

Figure 8-9 is redrawn for convenience as Figure 8-14. Again, the current in the emitter is called i_{b1}.

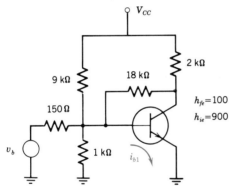

FIGURE 8-14
Circuit for Example 8-13.

$$i_{c1} = 100\, i_{b1}$$

If the voltage at the base is ignored

$$v_{out} = 18\ k\Omega\ \|\ 2\ k\Omega \times 100\, i_{b1} = 180{,}000\, i_{b1}$$

The voltage at the base is

$$v_b = h_{ie}\, i_{b1} = 900\, i_{b1}$$

This is indeed small, compared to $180{,}000\, i_{b1}$.
The current in R_f is

$$i_{Rf} = \frac{v_{out}}{R_f} = \frac{180{,}000\, i_{b1}}{18{,}000} = 10\, i_{b1}$$

The biasing resistor is $R_{BB} = 900\,\Omega$

$$i_{bias} = \frac{V_{bias}}{R_{BB}} = \frac{900\, i_{b1}}{900} = i_{b1}$$

Thus the total current flowing into the combination of R_f, h_{ie} and R_{BB} is $12\, i_{b1}$.

$$R'_{if} = \frac{v_b}{i_{total}} = \frac{900\, i_{b1}}{12\, i_{b1}} = 75\,\Omega$$

The voltage at the source is

$$v_S = v_b + i_{total}\, R_S = 900\, i_{b1} + 12\, i_{b1} \times 150\,\Omega$$

$$v_S = 2700\, i_{b1}$$

Finally

$$A_v = \frac{v_{out}}{v_S} = \frac{180{,}000\, i_{b1}}{2700\, i_{b1}} = 66.7$$

Both answers agree exactly with the results found in Example 8-8.

In this section we have analyzed both a series and a shunt feedback circuit without using A or β. In both cases, the results agreed with the results obtained previously.

8-7 FEEDBACK AND FREQUENCY RESPONSE

One of the advantages of negative feedback is that it improves the frequency response of a circuit. The lower gain of the circuit with feedback results in a lower f_L and a higher f_H, thus increasing the bandwidth of the feedback amplifier.

Figure 8-15 shows a typical frequency response curve for an amplifier with and without feedback, where BW is the bandwidth of the amplifier without feedback, and BW_f is the bandwidth of the amplifier with feedback.

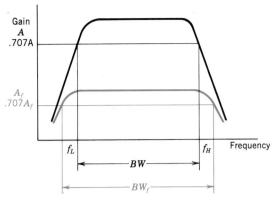

FIGURE 8-15
The effect of feedback on gain and bandwidth.

EXAMPLE 8-14

A feedback amplifier has a gain of 100 and $\beta = 0.04$. Find the gain of the feedback amplifier at f_H and f_L (the 3 dB points of the amplifier without feedback).

Solution
At mid-frequency, $A_v = 100$ and

$$A_{vf} = \frac{A_v}{1 + \beta A_v} = \frac{100}{5} = 20$$

At high and low frequencies, both the gain reduction and phase shift must be considered. When A_v is considered as a function of frequency

$$A_v(f) = \frac{A_v}{1 + \dfrac{jf}{f_H}}$$

where A_v is the mid-frequency gain. When $f = f_H$

$$A_v(f) = \frac{A_v}{1 + j1} = 0.707\, A_v \angle -45° = 50 - j50$$

The gain with feedback at this frequency is

$$A_{vf}(f) = \frac{70.7 \angle -45°}{1 + 0.04\,(50 - j50)}$$

8-7 FEEDBACK AND FREQUENCY RESPONSE

$$A_{vf}(f) = \frac{70.7 \angle -45°}{1 + 2 - j2} = \frac{70.7 \angle -45°}{3 - j2} = \frac{70.7 \angle -45°}{3.6 \angle -33.42°}$$

$$A_{vf}(f) = 19.64 \angle -11.58°$$

The gain of the amplifier with feedback has been reduced from 20 to 19.64. This is only 0.158 dB for the amplifier with feedback, compared to a gain reduction of 3 dB for the amplifier without feedback.

A simple relationship for the gain of a feedback amplifier can be established. If $A_{vf}(f)$ is the gain of the amplifier with feedback, as a function of frequency:

$$A_{vf}(f) = \frac{A_v(f)}{1 + \beta A_v(f)} = \frac{\dfrac{A_v}{\left(1 + \dfrac{jf}{f_H}\right)}}{1 + \dfrac{\beta A_v}{\left(1 + \dfrac{jf}{f_H}\right)}} \tag{8-13}$$

$$A_{vf}(f) = \frac{A_v}{1 + \beta A_v + \dfrac{jf}{f_H}}$$

$$A_{vf}(f) = \frac{\dfrac{A_v}{(1 + \beta A_v)}}{1 + \dfrac{jf}{(1 + \beta A_v) f_H}}$$

$$A_{vf}(f) = \frac{A_{vf}}{1 + \dfrac{jf}{f'_H}} \tag{8-14}$$

where

$$f'_H = f_H (1 + \beta A_v) \tag{8-15}$$

Thus the gain of the amplifier with feedback as a function of frequency is the mid-frequency gain of the feedback amplifier reduced by the factor $1 + jf/f'_H$, where f'_H is an upper-half-power frequency that is $(1 + \beta A_v)$ times the upper-half-power frequency of the amplifier without feedback. Therefore, although feedback reduces the gain of an amplifier by the factor $(1 + \beta A_v)$, it also increases the bandwidth by the same factor. Thus the *gain–bandwidth product* of an amplifier is *not* affected by feedback.

EXAMPLE 8-15

The feedback amplifier of Example 8-14 ($A_v = 100$, $\beta = 0.04$) has an f_H of 100 kHz.

a. At what frequency is the gain of the amplifier with feedback down by 3 dB?
b. Check the answer by finding the gain of the amplifier with and without feedback at the frequency found in part a.

Solution

a. The simplest way to find the upper-half-power frequency is to use equation 8-14.

$$f'_H = f_H (1 + \beta A) = 100 \text{ kHz} \times 5 = 500 \text{ kHz}$$

b. This problem can be checked by substituting a frequency of 500 kHz into equation 8-12.

$$A_{vf}(f) = \frac{\dfrac{A_v}{\left(1 + \dfrac{jf}{f_H}\right)}}{1 + \dfrac{\beta A_v}{\left(1 + \dfrac{jf}{f_H}\right)}} = \frac{\dfrac{100}{1 + \dfrac{j500 \text{ kHz}}{100 \text{ kHz}}}}{1 + \dfrac{0.04 \times 100}{1 + \dfrac{j500 \text{ kHz}}{100 \text{ kHz}}}}$$

The numerator of this fraction is

$$\frac{100}{1 + j5} = \frac{100}{5.1 \angle -78.7°} = 19.6 \angle -78.7°$$

The denominator of this fraction is

$$1 + \frac{0.04 \times 100}{1 + \dfrac{j500 \text{ kHz}}{100 \text{ kHz}}} = 1 + \frac{4}{1 + j5}$$

$$= 1 + \frac{4}{5.1 \angle 78°}$$

$$= 1 + 0.784 \angle -78°$$

$$1 + 0.154 + j0.769$$

$$= 1.154 + j0.769$$

$$= 1.387 \angle -33.68°$$

Therefore

$$A_{vf}(f) = \frac{19.6 \angle -78.7°}{1.387 \angle -33.68°}$$

$$= 14.13 \angle -45.02°$$

This is almost exactly 3 dB down from the mid-frequency gain of the amplifier with feedback, which is 20. Note that the phase shift at f_H for the feedback amplifier is also 45°.

8-7.1 The Feedback Circuit in the Laboratory

The circuit of Figure 8-16 was set up in the laboratory. The 47 pF capacitor between the collector and base was placed there to lower f_H so the effects of feedback could be observed.

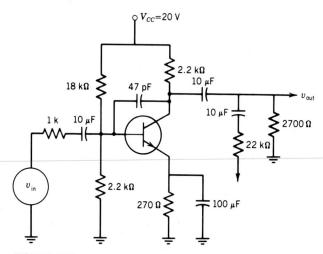

FIGURE 8-16
The feedback circuit in the laboratory.

The voltage gain was measured and the results are given in Table 8-1, and are plotted in Figure 8-17.

The measured mid-frequency gain of the amplifier without feedback was 73, and 18.1 with feedback. These compare to calculated values of 77 and 17.1, respectively (see problem 8-19). The value of f_H is 37 kHz, as shown in Figure 8-17, and f'_H, the upper-half-power frequency with feedback, was 123 kHz.

Table 8-1
The Gain of the Amplifier of Figure 8-16 as a Function of Frequency

Frequency (Hz)	A_v	A_{vf}
20	10	7.7
40	20	12.5
70	32.5	15.1
100	42	16.5
200	59	17.6
400	68.5	18
700	72	18.1
1000	72.5	18.1
2000	73	18.1
4000	73	18.1
7000	73	18
10,000	72.5	18
20,000	68	17.6
40,000	50	17
70,000	34	15.6
100,000	19.5	14.1
200,000		8.9
400,000		3.5

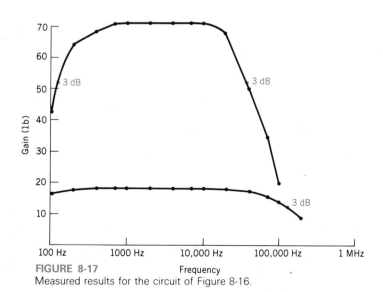

FIGURE 8-17
Measured results for the circuit of Figure 8-16.

8-8 A MORE RIGOROUS APPROACH TO FEEDBACK CIRCUIT ANALYSIS

In the previous sections a simplified, "first-cut" approach to feedback circuits has been used. It can be summarized as follows:

1. Determine whether series or shunt feedback is being used, primarily by determining whether any of the input current is being drawn off by the feedback path.
2. If series feedback is being used find β as a function of the output voltage.
3. If shunt feedback is being used, Nortonize the input and find β as a function of the output current.

Although this approach gives the current result for two of the most important parameters of a feedback circuit, gain and input impedance, in all cases, rigorous theory objects to the approach on several grounds as will be explained in the following paragraphs.

Rigorous theory concentrates on the *cause* of the feedback as well as the type of feedback (series or shunt). The cause of the feedback is either the output voltage or the output current. Thus rigorous theory divides feedback circuits into four types:

1. Voltage-series feedback. The voltage of the voltage generator in series with the input voltage depends on the output voltage of the circuit.
2. Current-series feedback. The voltage of the voltage generator in series with the input depends on the output current of the circuit.
3. Voltage-shunt feedback. The amount of current drained from the input (see Figure 8-7) depends on the output voltage of the circuit.
4. Current-shunt feedback. The amount of current drained from the input depends on the output current.

Rigorous theory also subdivides amplifiers into four types: voltage, transconductance, transresistance, and current, consistent with the divisions listed in the previous paragraph. Table 8-2 gives the characteristics of each type of amplifier.

In actual amplifiers, the type of amplifier and whether the feedback is voltage or current driven may not be readily apparent to the inexperienced user. This is true because the load voltage and load current are generally interrelated and dependent on each other. Possibly the best guide in making the distinction is to use the dictum that *the feedback factor, β, should be independent of the load resistance.*

8-8.1 Rigorous Theory and the Circuits in the Previous Sections

Some of the circuits analyzed in the previous sections conform to rigorous theory while others do not. In the next few paragraphs we shall make the distinctions, with appropriate comments.

Table 8-2
Feedback Amplifier Analysis

Characteristic \ Topology	(1) Voltage series	(2) Current series	(3) Current shunt	(4) Voltage shunt
Feedback signal X_f	Voltage	Voltage	Current	Current
Sampled signal X_o	Voltage	Current	Current	Voltage
Input circuit: set[a]	$V_o = 0$	$I_o = 0$	$I_o = 0$	$V_o = 0$
Output circuit: set[a]	$I_i = 0$	$I_i = 0$	$V_i = 0$	$V_i = 0$
Signal source	Thévenin	Thévenin	Norton	Norton
$\beta = X_f/X_o$	V_f/V_o	V_f/I_o	I_f/I_o	I_f/V_o
$A = X_o/X_i$	$A_v = V_o/V_i$	$G_M = I_o/V_i$	$A_I = I_o/I_i$	$R_M = V_o/I_i$
$D = 1 + \beta A$	$1 + \beta A_V$	$1 + \beta G_M$	$1 + \beta A_I$	$1 + \beta R_M$
A_f	A_v/D	G_m/D	A_I/D	R_M/D
R_{if}	$R_i D$	$R_i D$	R_i/D	R_i/D
R_{of}	$\dfrac{R_o}{1 + \beta A_v}$	$R_o(1 + \beta G_m)$	$R_o(1 + \beta A_i)$	$\dfrac{R_o}{1 + \beta R_m}$
$R'_{of} = R_{of} \| R_L$	$\dfrac{R'_o}{D}$	$R'_o \dfrac{1 + \beta G_m}{D}$	$R'_o \dfrac{1 + \beta A_i}{D}$	$\dfrac{R'_o}{D}$

[a]This procedure gives the basic amplifier circuit without feedback but taking the loading of β, R_L, and R_s into account. (From *Microelectronics*, Jacob Millman. Published by McGraw-Hill, Inc., 1979.)

The circuits of Figure 8-4 and 8-6 are examples of voltage-driven series feedback circuits. Their analyses in Examples 8-4 and 8-6 conform to rigorous theory.

The circuit of Figure 8-5 is a current-driven series feedback circuit. It does not conform to rigorous theory because the β used in section 8-4.2 is a function of the output resistance, R_C, and because the circuit is analyzed as though it were a voltage-driven circuit. Rigorous theory arrives at the same conclusions presented here through a more laborious procedure.

The circuits of Figures 8-8 and 8-9 are examples of voltage-driven shunt feedback. Though the methods used here do not conform to rigorous theory, the results are correct.

The circuit of Figure 8-11 is an example of a current-driven shunt feedback circuit. The methods used here do conform to rigorous theory.

8-8.2 Output Impedance of Feedback Amplifiers

The gain and input impedances of a feedback amplifier depend only on whether series or shunt feedback is being used. The output impedance, however, depends on whether the feedback is voltage or current driven. Table 8-3 shows this dependency. Generally the terms *increasing* and *decreasing* mean the impedance is multiplied or divided by the factor $1 + \beta A$.

Table 8-3
Effect of negative feedback on amplifier characteristics

	Type of feedback			
	Voltage-series	Current-series	Current-shunt	Voltage-shunt
Reference	Figure 12-9a	Figure 12-9b	Figure 12-9c	Figure 12-9d
R_{of}	Decreases	Increases	Increases	Decreases
R_{if}	Increases	Increases	Decreases	Decreases
Improves characteristics of	Voltage amplifier	Transconductance amplifier	Current amplifier	Transresistance amplifier
Desensitizes	A_{vf}	G_{Mf}	A_{If}	R_{Mf}
Bandwidth	Increases	Increases	Increases	Increases
Nonlinear distortion	Decreases	Decreases	Decreases	Decreases

(From *Microelectronics*, Jacob Millman. Published by McGraw-Hill, Inc., 1979.)

Unfortunately, the output impedances cannot be calculated without some thought, despite the help given by Table 8-3. The output impedance of the circuit of Figure 8-6, for example, is unaffected by feedback.

The fact that this is an introductory book, with space limitations, precludes a further and deeper discussion of rigorous feedback theory. Those interested in pursuing it further should consult the references at the end of the chapter.

8-9 SUMMARY

This chapter introduced feedback circuit. The concepts of positive and negative feedback and the advantages and disadvantages of both were explained, and the basic feedback equations were developed.

Feedback circuits were divided into two types: voltage or series feedback, and current or shunt feedback. Methods of calculating voltage gains, current gains, and input impedances for both types of circuits were presented and demonstrated for several common types of feedback circuits. The effects of feedback on the frequency response of amplifiers was also considered.

8-10 GLOSSARY

A_f. The gain of an amplifier with feedback.

$A_f(f)$. The gain of an amplifier with feedback as a function of frequency.

β. The percent of the output voltage or current that is sent back to the input.

β network. The network that determines the percent of the output fed back to the input.

Current feedback. See Shunt feedback.

Negative feedback. A circuit in which the feedback signal decreases the input signal or the feedback voltage opposes the source voltage.

Positive feedback. A circuit in which the feedback signal increases the input signal or the feedback voltage aids the source voltage.

R_{if}. The input impedance of an amplifier with feedback.

Series feedback. Feedback where the feedback voltage is placed in series with the input voltage generator.

Shunt feedback. A circuit in which a portion of the output current is fed back to the input. This circuit is usually characterized by a resistor that shunts a portion of the input current to the output.

Stability. The ability of an amplifier to maintain a constant gain despite variations in its parameters.

Voltage feedback. See Series feedback.

8-11 REFERENCES

ROBERT BOYLESTAD and LOUIS NASHELSKY, *Electronic Devices and Circuit Theory*, 4th Edition, Prentice-Hall, Inc., Englewood Cliffs, N. J., 1987.

JACOB MILLMAN, *Microelectronics*, 2nd Edition, McGraw-Hill Co., New York, 1987.

J. F. PIERCE and T. J. PAULUS, *Applied Electronics*, Charles E. Merrill Co., Columbus, Ohio, 1972.

RODNEY B. FABER, *Essentials of Solid State Electronics*, John Wiley, New York, 1985.

8-12 PROBLEMS

8-1 Repeat Example 8-1 if the gain of the amplifier is changed to 100. Find v_S and v_{in} if v_{out} is -20 V.

8-2 Repeat Example 8-1 if the gain of the amplifier is 50 and the bottom resistor of the β network is changed from 2000 Ω to 1000 Ω. Again assume $v_{out} = -20$ V.

8-3 An amplifier has a gain of 50 and an input impedance, R_i, of 1000 Ω. Find its gain and input impedance if feedback is added to the amplifier and $\beta = 0.05$.

8-4 An amplifier has a gain of 100, but the circuit requires a gain of only 25. If feedback is to be added to this circuit to reduce its gain, find β.

8-5 The amplifier of problem 8-4 deteriorates so that its gain drops to 80. What is the gain of the amplifier with feedback? What is the % change in gain for the amplifier with and without feedback?

8-6 If the circuit of Example 8-3 has an equivalent biasing resistor, R_{BB}, of 3000 Ω across its inputs, find its input impedance with and without feedback.

8-7 A **2N3904** transistor has $h_{fe} = 150$ and $h_{ie} = 900$ Ω. If it is used in an amplifier with $R_C = 1800$ Ω and connected to a voltage divider feedback network consisting of a 500 Ω resistor and an 8500 Ω resistor, find:
  a. A_1
  b. A_2
  c. A_f
  d. R_{if}
  e. R_{of}

8-8 Repeat problem 8-7 if a 600 Ω load resistor is connected to the circuit.

8-9 A high-gain amplifier has negative feedback added to it with $\beta = 0.1$ Find the gain of the amplifier with feedback.

8-10 If the circuit of Figure 8-4, with $g_m = .002$ S, is capacitively coupled to a 5000 Ω load resistor, find:
  a. The gain without feedback.
  b. The gain with feedback.
  In each case what is the percent change between these gains and the gains found in Example 8-4?

8-11 If the circuit of Figure 8-5 has $h_{fe} = 100$, $h_{ie} = 1000$ Ω, $R_E = 100$ Ω, and $R_C = 1000$ Ω, find A_2, β, and the gain with feedback.

8-12 Using the methods of section 8-4.2, find the gain of a JFET amplifier with an unbypassed source-resistor.

8-13 Using the methods of section 8-4.2, derive the equation for the gain of an emitter-follower.

8-14 Repeat Example 8-6 if the transistor parameters are $h_{fe} = 150$ and $h_{ie} = 1000$ Ω. How much does the gain of the circuit, with feedback, change due to the change of parameters?

8-15 For the circuit of Figure P8-15, assume h_{fe} and h_{ie} of the transistors are reasonable. Find

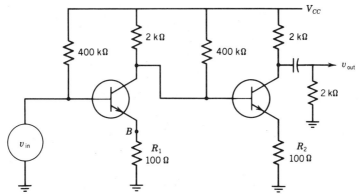

FIGURE P8-15

a. The gain of the circuit as shown.
b. The bottom end of the 2 kΩ output resistor is removed from ground and tied to point B. The input voltage is adjusted so that v_{out} is 1 V.
   (1) Is this voltage or current feedback?
   (2) Find v_{in}.
   (3) Find the voltage across R_1 due to feedback only.
   (4) Find the difference voltage $v_{in} - v_{R1}$.
   (5) The difference voltage of part 4 multiplied by the open loop gain should equal the output voltage. Show that it does.

8-16 For the circuit of Figure P8-16, find R_f so that $\beta = 0.01$. Then find the voltage gain of the circuit.

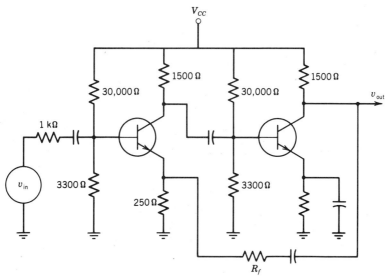

FIGURE P8-16
For each transistor $h_{fe} = 120$, $h_{ie} = 1000$.

8-17 For the circuit of Figure P8-17, find the following:
   a. A_i and A_v of the circuit without feedback, $R_{in}i$ and R'_{in}.
   b. With the 23 kΩ resistor between collector and base find A_i of the circuit, R_{in}, R'_{in}, and A_v of the circuit.

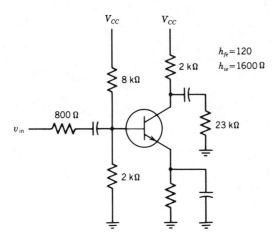

FIGURE P8-17
$h_{fe} = 120$, $h_{ie} = 1600$.

8-18 For the circuit of Figure 8-9, if A_i is defined as the collector current divided by the input current, find the new values of β and A_i. Show that the product of β and A_i does not change.

8-19 For the circuit of Figure 8-16, the output voltage and current are taken across the 2700 Ω resistor. Assume $h_{fe} = 150$. Find A_v and A_i if
   a. The bottom of the 22,000 Ω resistor is connected to ground.
   b. The bottom of the 22,000 Ω resistor is connected to the base of the transistor.

8-20 If the source voltage in Figure 8-11 is adjusted so the voltage at the base of Q1 is 0.09 V, find the current and voltage in each component of the circuit. Show that A_i and A_v agree with the values calculated in Example 8-10.

8-21 Find the voltage and current gains of the circuit of Figure 8-11 using the methods of section 8-6.

8-22 Repeat problem 8-19 using the methods of section 8-6.

8-23 A circuit that uses shunt feedback has $A_v = 100$, $A_i = 60$, $\beta = 0.05$, $f_L = 200$ Hz, and $f_H = 100$ kHz. Find
   a. The values of f_L and f_H for the circuit with feedback.
   b. The gain of the circuit without feedback at 200 kHz.
   c. The gain of the circuit with feedback at 200 kHz.

8-24 Analyze the circuit of Figure 8-11 using the methods of section 8-5.3 and Equation 8-12. Show that the ratio of A_v to A_{vf} is 6.44.

8-25 For the circuit of Figure P8-25

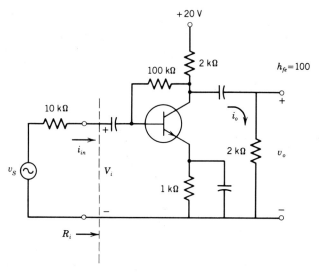

FIGURE P8-25

a. Determine v_o/v_i.
b. Determine i_o/i_i.
c. Determine R_i.
d. Determine v_o/v_s.

CHAPTER NINE

Power Amplifiers

9-1 Instructional Objectives

This chapter considers the analysis and design of power amplifiers, whose main function is to provide high power to a load. After reading the chapter, the student should be able to:

1. Find the power dissipated in each component of a power amplifier.
2. Calculate the efficiency of a power amplifier.
3. Determine if a power transistor is operating within its safe operating region.
4. Determine the temperature at the junction and case of a transistor for a given power dissipation.
5. Determine the type of heat sink required for a particular application.
6. Analyze and design choke and transformer-coupled amplifiers.
7. Analyze and design Class B amplifiers.
8. Analyze complementary circuits.

9-2 Self-Evaluation Questions

Watch for the answers to the following questions. They should help you to understand the material presented.

1. Why is efficiency important in a power amplifier?
2. Why are heat sinks required? How are they constructed? Why do they have fins?

3. What is thermal capacitance? Why is it important?
4. What is the difference between Class A, Class B, and Class AB?
5. What is the advantage of using a transformer-coupled amplifier?
6. What are the advantages of Class B operation?
7. What are the causes of distortion? How is distortion minimized?
8. Why are Darlingtons and bootstrap capacitors used in complementary amplifiers?

9-3 INTRODUCTION

Many amplifiers require high power outputs rather than voltages or currents of a particular amplitude. This is generally true of amplifiers that must convert their electrical *energy* into mechanical *motion,* such as amplifiers that drive electric motors or loudspeakers that must vibrate the air to create sound. The last stage of these amplifiers must be a *power amplifier,* whose main function is to provide the electrical power needed to drive these loads.

The most common use of a power amplifier is to drive a loudspeaker in an audio system. Figure 9-1 shows such a system. It consists of

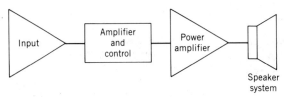

FIGURE 9-1
A typical audio system.

1. The input element. Typical inputs to an audio system are a microphone, a phonograph pickup, a tape head, or a radio antenna.
2. The amplifier and control section. The circuit amplifies the minute signal it receives from the input until it has sufficient volume to drive the power amplifier. The control consists of the volume control and bass or treble boosts that can be set by the user. Sophisticated systems also include a *preamplifier* to amplify and condition the output. In many cases the frequency response of the preamplifier is not uniform; it is designed instead to compensate for irregularities in the frequency response of the input element.[1]

---

[1] A discussion of preamplifiers is beyond the scope of this book. They are discussed in Chapter 12 of Pierce and Paulus (see References).

3. The power amplifier. The power amplifier must produce the high power needed by the loudspeakers.
4. The loudspeakers themselves.

The design of voltage amplifiers has been covered in previous chapters. This chapter will concentrate on the analysis and design of power amplifiers and the problems of interfacing them to the speaker system.

9-4 POWER DISSIPATION AND POWER TRANSISTORS

The transistors discussed in the previous chapters were used in voltage amplifiers. They were low-power circuits and the power considerations were ignored. We will show that to produce high-power outputs, the transistors themselves must be capable of dissipating significant amounts of power; or, to put it succinctly, you can't build a "blaster" with tiny transistors like the **2N3904**. Heavier power transistors are required.

9-4.1 Transistor Power Dissipation

The power, P_D, dissipated in a transistor during any time interval, T, is given by

$$P_D = \frac{i}{T}\int_0^T V_{ce}I_c\,dt \qquad (9\text{-}1)$$

If the transistor is quiescent (no ac signal applied), equation 9-1 reduces to

$$P_D = V_{CEQ}I_{CQ}$$

Figure 9-2 is a simple circuit that can be analyzed for power dissipation.

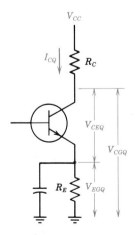

FIGURE 9-2
The basic transistor amplifier under quiescent conditions.

The quiescent (no signal) conditions of the circuit are:

I_{CQ} The quiescent current.
V_{CGQ} The quiescent collector-to-ground voltage.
V_{EGQ} The quiescent emitter-to-ground voltage.
V_{CEQ} The quiescent collector-to-emitter voltage across the transistor ($V_{CEQ} = V_{CGQ} - V_{EGQ}$).

When power is being considered, the amount of power delivered by the power supply, P_{CC}, must be considered. For this circuit

$$P_{CC} = V_{CC}I_{CQ}$$

The quiescent power absorbed by each component in Figure 9-2 is

$$P_{R_C} = I_{CQ}^2 R_C$$
$$P_{R_E} = I_{CQ}^2 R_E$$
$$P_D = V_{CEQ}I_{CQ}$$

The sum of these powers must equal the input power, P_{CC}.

EXAMPLE 9-1

In Figure 9-2, assume the transistor is biased so that $I_{CQ} = 1$ A. If $R_C = 20\ \Omega$, $R_E = 2\ \Omega$ and $V_{CC} = 40$ V, find the power in each element.

Solution
The power can be found as follows:

$$P_{CC} = V_{CC}I_{CQ} = 40\text{ V} \times 1\text{ A} = 40\text{ W}$$
$$P_{R_C} = I_{CQ}^2 R_C = (1)^2 \times 20\ \Omega = 20\text{ W}$$
$$P_{R_E} = I_{CQ}^2 R_E = (1)^2 \times 2\ \Omega = 2\text{ W}$$

To find the power dissipated by the transistor, V_{CEQ} must be found.

$$V_{CEQ} = V_{CGQ} - V_{EGQ} = 20\text{ V} - 2\text{ V} = 18\text{ V}$$
$$P_D = V_{CEQ}I_{CQ} = 18\text{ V} \times 1\text{ A} = 18\text{ W}$$

These calculations are checked by observing that $P_{CC} = P_{R_C} + P_{R_E} + P_D$.

If an ac source is applied to the base of the transistor, as shown in Figure 9-3, the output currents and voltages become

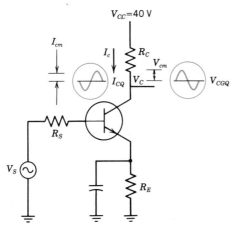

FIGURE 9-3
The basic power amplifier with an ac input. (The biasing resistors are not shown.)

$$I_C = I_{CQ} + I_{cm} \sin \omega t \qquad (9\text{-}2)$$

$$V_C = V_{CEQ} - V_{cm} \sin \omega t \qquad (9\text{-}3)$$

where I_{cm} is the maximum ac current swing and V_{cm} is the maximum ac voltage swing. The minus sign in equation 9-3 accounts for the phase difference between the collector voltage and collector current.

The power in each component can now be found:

$$P_{R_C} = \frac{1}{T}\int_0^T I_C^2 \, R_C \, dt = \frac{1}{T}\int_0^T R_C \, (I_{CQ} + I_{cm} \sin \omega t)^2 \, dt$$

$$P_{R_C} = \frac{1}{T}\int_0^T I_{CQ}^2 R_C \, dt + \frac{1}{T}\int_0^T 2R_C I_{CQ} \, I_{cm} \sin \omega t \, dt$$

$$\qquad + \frac{1}{T}\int_0^T R_C I_{cm}^2 \sin^2 \omega t \, dt$$

$$P_{R_C} = I_{CQ}^2 \, R_C + 0 + \frac{I_{cm}^2 \, R_C}{2} \qquad (9\text{-}4)^2$$

---

[2]Over any complete period $\int_0^T \sin \omega t \, dt = 0$.

The first term in this expression, $I_{CQ}^2 R_C$ is the dc power dissipated in the collector resistor, and the term $(i_{cm}^2 R_C)/2$ is the ac or useful power dissipated in the collector resistor.

The power in the transistor is

$$P_D = \frac{1}{T}\int_0^T (V_{CGQ} - V_{EGQ} - V_{cm} \sin \omega t)(I_{CQ} - I_{cm} \sin \omega t)\, dt$$

$$P_D = \frac{1}{T}\int_0^T (V_{CEQ} - V_{cm} \sin \omega t)(I_{CQ} + I_{cm} \sin \omega t)\, dt$$

After a bit of algebra, this becomes:

$$P_D = V_{CEQ} I_{CQ} - \frac{V_{cm} I_{cm}}{2} \qquad (9\text{-}5)$$

Because the voltage across the emitter-resistor is held constant by the capacitor

$$P_{R_E} = V_{EGQ} I_{CQ} = I_{CQ}^2 R_E$$

The efficiency, η, of a power amplifier is defined as the ratio of the useful power or ac output power to the input power drawn from the power supply.

$$\eta = \frac{P_{(ac)}}{P_{(dc)}} = \frac{P_D}{P_{CC}} = \frac{\frac{V_{cm} I_{cm}}{2}}{V_{CC} I_C} \qquad (9\text{-}6)$$

EXAMPLE 9-2

If the circuit of Figure 9-3 has $I_{CQ} = 1$ A and the applied ac signal creates an I_{cm} of 0.6 A, find the power in each component and the efficiency. Assume again that $R_C = 20\,\Omega$ and $R_E = 2\,\Omega$.

Solution

The solution uses the formulas developed before.

$$P_{R_C} = \left(I_{CQ}^2 R_C + \frac{I_{cm}^2 R_C}{2}\right) = 20\text{ W} + 3.6\text{ W} = 23.6\text{ W}$$

This output consists of 20 W of dc power and 3.6 W of ac power.

To find P_D, V_{cm} must first be found. Because the emitter is held at a constant voltage, V_{cm} at the collector is the same as V_{cm} across the transistor.

Here

$$V_{cm} = I_{cm} R_C = 0.6 \times 20 \, \Omega = 12 \text{ V}$$

$$P_D = V_{CEQ}I_C - \frac{V_{cm}I_{cm}}{2} = 18 \text{ V} \times 1 \text{ A} - \frac{12 \text{ V} \times 0.6 \text{ A}}{2} = 14.4 \text{ W}$$

$$P_{R_E} = I_{CQ}^2 R_E = 2 \text{ W}$$

$$P_{CC} = V_{CC} I_{CQ} = 40 \text{ W}$$

Again, $P_{CC} = P_{R_C} + P_D + P_{R_E}$
The efficiency is

$$\eta = \frac{P_{ac}}{P_{dc}} = \frac{3.6 \text{ W}}{40 \text{ W}} = 9\%$$

If the emitter-resistor of Figure 9-3 is unbypassed, V_{cm}, the maximum ac voltage across the transistor, is affected by the change of voltages at both the emitter and the collector. It can be found by considering the circuit voltages at the maximum and minimum points of the waveform.

EXAMPLE 9-3

For the circuit of Example 9-2 find V_{cm} if the emitter-resistor is not bypassed.

Solution
In this circuit there is a 1 A quiescent current and a 0.6 A ac swing. Therefore the current varies from 1.6 A to 0.4 A. At 1.6 A

$$V_c = V_{CC} - I_c R_c = 40 \text{ V} - 1.6 \text{ A} \times 20 = 8 \text{ V}$$

$$V_E = I_c R_E = 1.6 \text{ A} \times 2 = 3.2 \text{ V}$$

The minimum voltage across the transistor is the difference between these two voltages or 4.8 V. The ac signal causes the voltage across the transistor to swing from its quiescent value of 18 V to 4.8 V. The maximum ac voltage swing is

$$V_{cm} = V_{CEQ} - V_{min} = 18 \text{ V} - 4.8 \text{ V} = 13.2 \text{ V}$$

The new power dissipation in the transistor is slightly less than before, but this is compensated for by an ac power component in the emitter resistor (see problem 9-1).

Examples 9-2 and 9-3 show that as the ac output power increases, the power dissipated by the transistor decreases. The *worst case power dissipation* for a transistor occurs under *no signal* or quiescent conditions, and the prudent engineer designs power amplifiers so that they will not be destroyed under these conditions.

9-4.2 Maximum Efficiency

The efficiencies of resistive coupled circuits is inherently low. This is illustrated by Example 9-2 where the efficiency of the amplifier was only 9%. An examination of the load line for a resistive-coupled amplifier, as shown in Figure 9-4, shows why this is so. The maximum voltage swing V_{cm} can be no

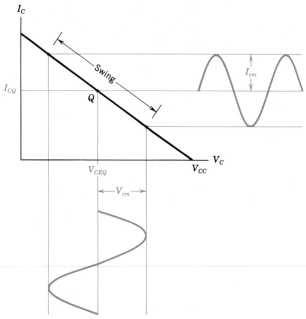

FIGURE 9-4
The swing of a resistive-coupled power amplifier.

greater than $V_{CC}/2$, and the maximum current swing can only equal I_{CQ}. If maximum swing is assumed

$$P_D = \frac{V_{cm}I_{cm}}{2} = \frac{V_{CC}I_{CQ}}{4} = \frac{P_{CC}}{4} \qquad (9\text{-}7)$$

Equation 9-7 shows that the maximum theoretical efficiency of a resistive-coupled amplifier is 25%. Actually an efficiency that high is never achieved. The load line in Figure 9-4 is idealized, by assuming that the ac and dc load lines are the same. Power losses in the emitter-resistor or other circuit components were ignored. Furthermore, speaker amplifiers can never be continuously operated at maximum swing; this means maximum volume, but music has quiet passages also.

If the emitter voltage is held constant (either because the emitter is bypassed or connected to ground), $V_{cm} = I_{cm} R_C$. Then the ac power in the collector resistor is

$$P_{ac} = \frac{I_{cm}^2 R_C}{2} = \frac{V_{cm}^2}{2 R_C} \qquad (9\text{-}8)$$

EXAMPLE 9-4

If 10 W of ac power are to be developed across a 10 Ω resistor, find V_{cm}, I_{cm}, and V_{CC} for the circuit.

Solution
Equation 9-8 can be transposed to

$$V_{cm} = \sqrt{2 P_{ac} R_C} = \sqrt{200} = 14.14 \text{ V}$$

$$I_{cm} = \frac{V_{cm}}{R_C} = 1.414 \text{ A}$$

$$V_{CC} = 2 V_{cm} = 28.28 \text{ V}$$

The efficiency of a power amplifier will also decrease if the power losses in the biasing components are considered.

EXAMPLE 9-5

Design a power amplifier with $V_{CC} = 20$ V, $I_{CQ} = 1$ A, $R_C = 10$ Ω, and $R_E = 1$ Ω. Design it for maximum power and find its efficiency. The biasing components must be taken into account. Use self-bias (see section 3-5.2) and assume the power transistor has an h_{fe} of 100.

Solution
The biasing components can be calculated as shown in section 3-5.2 and section 3-5.3.

$$R_B = \frac{h_{fe} R_E}{10} = \frac{100 \times 1 \text{ Ω}}{10} = 10 \text{ Ω}$$

$$I_B = \frac{I_{CQ}}{h_{fe}} = \frac{1 \text{ A}}{100} = 10 \text{ mA}$$

$$V_{BB} = V_{EGQ} + V_{BE} + i_b R_B = 1 + 0.7 + 0.1 = 1.8 \text{ V}$$

$$R_{B1} = R_B \times \frac{V_{CC}}{V_{BB}} = 10 \text{ }\Omega \times \frac{20 \text{ V}}{1.8 \text{ V}} = 111 \text{ }\Omega$$

$$R_{B2} = \frac{R_B}{1 - \frac{V_{BB}}{V_{CC}}} = \frac{10 \text{ }\Omega}{0.91} = 11.1 \text{ }\Omega$$

The circuit is shown in Figure 9-5. From it we see that $V_{CEQ} = 9$ V. At best, therefore, $V_{cm} = 9$ V and $I_{cm} = v_{cm}/R_C = 0.9$ A. The ac power in the load resistor is

$$P_{ac} = \frac{V_{cm} I_{cm}}{2} = \frac{8.1 \text{ W}}{2} = 4.05 \text{ W}$$

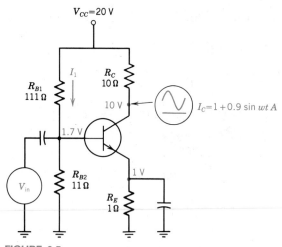

FIGURE 9-5
The circuit for Example 9-5.

The power dissipated in the transistor is:

$$P_D = V_{CEQ} I_{CQ} - \frac{V_{cm} I_{cm}}{2} = 9 \text{ W} - 4.05 \text{ W} = 4.95 \text{ W}$$

The power taken from the supply is $V_{CC} I_{CQ}$ (20 W), plus the power used by the biasing circuit. Figure 9-5 shows that

$$I_1 = \frac{20 \text{ V} - 1.7 \text{ V}}{111} = \frac{18.3 \text{ V}}{111} = 16.5 \text{ mA}$$

$$P_{bias} = V_{CC} I_1 = 20 \text{ V} \times 16.5 \text{ mA} = 0.33 \text{ W}$$

$$\eta = \frac{P_{ac}}{P_{dc}} = \frac{4.05 \text{ W}}{20.33 \text{ W}} = 20\%$$

This design is very close to optimal, but the efficiency is still only 20%.

Because of their low efficiencies, resistive-coupled power amplifiers are rarely used, but they do serve to introduce the topic, its concepts, and its problems. More practical power amplifiers are discussed in sections 9-6, 9-7, and 9-9.

9-4.3 Power Transistors

Transistors used in power amplifiers must dissipate large amounts of heat and power. They must also allow large voltage and current swings in order to provide a large power output. A class of transistors called *power transistors* is available with these characteristics. The primary function of a power transistor is to provide a large power output rather than a large current gain.

The **2N3055** is a very popular and commonly used *npn* power transistor. Figure 9-6 is the first page of the specifications for the **2N3055** and its *pnp* complement, the **MJ2955**. Table 9-1 compares the maximum ratings of the **2N3055** and the **2N3904** switching and amplifier transistor. From it we see that

1. The **2N3055** has a somewhat higher breakdown voltage (V_{CEO} = 60 V vs. 40 V).
2. The **2N3055** has a much higher allowable collector current (15 A vs. 200 mA).
3. The **2N3055** can dissipate much more power. The specification of 115 W is a huge amount of power, but must be taken with extreme caution. The **2N3055** can dissipate 115 W, but only if its *case temperature* is kept at 25°C. If a **2N3055** is dissipating significant power, there is no practical way to keep the case at 25°C. Included in the specifications of Figure 9-6 is a *derating curve*. It shows that the allowable power dissipation of the transistor drops off as the case temperature rises. For example, the derating curve shows that the **2N3055** can dissipate about 63 W at 100°C.
4. The junction temperature. This is the temperature that will damage the junction. It cannot be exceeded, but, because it is within the transistor, it cannot be directly measured and must be inferred from the external conditions.
5. The *thermal resistance*, θ_{JC}. The thermal resistance, power dissipation, and maximum junction temperature are all interrelated and are discussed in section 9-5.

NPN 2N3055   PNP MJ2955

1.3

COMPLEMENTARY SILICON POWER TRANSISTORS

...designed for general-purpose switching and amplifier applications.

- DC Current Gain – h_{FE} = 20-70 @ I_C = 4 Adc
- Collector-Emitter Saturation Voltage –
  $V_{CE(sat)}$ = 1.1 Vdc (Max) @ I_C = 4 Adc
- Excellent Safe Operating Area

15 AMPERE POWER TRANSISTORS COMPLEMENTARY SILICON

60 VOLTS
115 WATTS

MAXIMUM RATINGS

Rating	Symbol	Value	Unit
Collector-Emitter Voltage	V_{CEO}	60	Vdc
Collector-Emitter Voltage	V_{CER}	70	Vdc
Collector-Base Voltage	V_{CB}	100	Vdc
Emitter-Base Voltage	V_{EB}	7	Vdc
Collector Current — Continuous	I_C	15	Adc
Base Current	I_B	7	Adc
Total Power Dissipation @ T_C = 25°C Derate above 25°C	P_D	115 0.657	Watts W/°C
Operating and Storage Junction Temperature Range	T_J, T_{stg}	-65 to +200	°C

THERMAL CHARACTERISTICS

Characteristic	Symbol	Max	Unit
Thermal Resistance, Junction to Case	$R_{\theta JC}$	1.52	°C/W

NOTE:
1. DIM "Q" IS DIA.

STYLE 1:
PIN 1. BASE
2. EMITTER
CASE: COLLECTOR

DIM	MILLIMETERS		INCHES	
	MIN	MAX	MIN	MAX
A	–	39.37	–	1.550
B	–	21.08	–	0.830
C	6.35	7.62	0.250	0.300
D	0.99	1.09	0.039	0.043
E	–	3.43	–	0.135
F	29.90	30.40	1.177	1.197
G	10.67	11.18	0.420	0.440
H	5.33	5.59	0.210	0.220
J	16.64	17.15	0.655	0.675
K	11.18	12.19	0.440	0.480
Q	3.84	4.09	0.151	0.161
R	–	26.67	–	1.050

Collector connected to case.
CASE 11-01
(TO-3)

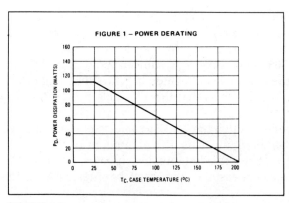

FIGURE 1 – POWER DERATING

FIGURE 9-6
The first page of the specifications for the **2N3055** and **MJ2955**. (Copyright by Motorola, Inc. Used by permission.)

Table 9-1
Maximum Ratings of the 2N3055 and 2N3904 Transistors

Characteristic		2N3055	2N3904	Unit
Breakdown voltage	V_{CEO}	60	40	V
Maximum collector current	I_c	15	0.2	A
Power dissipation	P_D	115	1	W
Junction temperature	T_J	200	150°	°C
Thermal resistance	θ_{JC}	1.52	125	°C/W

6. The h_{fe} of the power transistor is less than the small-signal transistor. For the **2N3055**, h_{fe} is specified to be between 20 and 70 at a current of 4 A. However, h_{fe} tends to increase as the current decreases.

In addition to the listed characteristics, power transistors are usually larger than switching transistors and are packaged in a metal case to facilitate heat sinking. The **2N3055** is packaged in the type TO-3 case shown in Figure 9-6, instead of the small ceramic package that houses most small signal transistors.

9-4.4 The Safe Operating Range for Power Transistors

The previous sections have shown that a transistor should be biased so that it does not exceed its maximum voltage, maximum current, or maximum power dissipation. The region of the transistor characteristics that meets these criteria is called the *safe operating region* for the transistor. For maximum power dissipation

$$P_D = V_{CEQ}I_{CQ} \qquad (9\text{-}9)$$

If P_D is held constant at the maximum power dissipation for the transistor, equation 9-9 is a hyperbola on the transistor characteristics. Figure 9-7

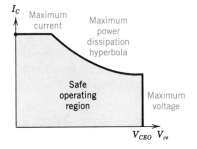

FIGURE 9-7
The safe operating region for a power transistor.

shows the safe operating region for a transistor. It is the region below I_{Cmax}, V_{CEO}, and the maximum power dissipation hyperbola.

Figure 9-8 shows the safe operating region for a **2N3055**. The dc line is the curve we have been discussing. It shows that the safe operating region is

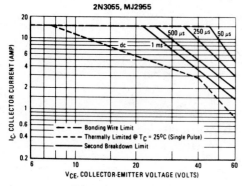

FIGURE 9-8
Active region safe operating area. (Copyright by Motorola, Inc. Used by permission.)

below I_{Cmax} (15 A), V_{CEO} (60 V), and P_{Dmax} (115 W). Above 40 V, regions of local heating called *hot spots* may develop that could destroy the transistor. Consequently, the transistor is more severely restricted in this region.

EXAMPLE 9-6

Find the maximum current for a **2N3055** if $V_{CEQ} = 20$ V.

Solution
Figure 9-8 shows that the $V_{CE} = 20$ V line intersects the dc line at $I_C = 5.8$ A. This is the maximum current at $V_{CE} = 20$ V. Notice also that the $V_{CE}I_C$ product is 116 W, which is just about P_{Dmax}, so this point is on the maximum power dissipation hyperbola.

A word of caution is again necessary. The curve of Figure 9-8 is for a transistor with a case temperature of 25°C. In most cases the case temperature will be considerably higher than the typical ambient temperature of 25°C, which causes the maximum power hyperbola to drop to a lower value. The transistor must be derated above 25°C (see problem 9-6).

9-4.5 Pulsed Operation
The **2N3055** may dissipate more than rated power if the power is applied in short, infrequent bursts. The curves in the upper right-hand corner of Figure

9-8 give the power limitations for pulses of 1 ms, 500 μs, 250 μs, and 50 μs duration. The duty cycle of these pulses must not exceed 10%.

EXAMPLE 9-7

If a pulse is to be applied to a **2N3055** for 500 μs and $V_{CE} = 40$ V, what is the maximum current and power this pulse may have?

Solution
Figure 9-8 shows that the 40 V line intersects the 500 μs line at $I_C = 5$ A. Therefore the maximum current is 5 A and the maximum power of the pulse is 200 W.

9-5 THERMAL RESISTANCE AND HEAT SINKS

The examples of the previous section show that the transistors in a power amplifier must dissipate a considerable amount of power. This power is generated at the collector-to-base junction within the transistor because that is where the voltage drop occurs. Naturally this power, and the energy associated with it, raises the temperature at the junction. Transistor manufacturers specify a maximum junction temperature that may not be exceeded; for silicon transistors this temperature is 175°C to 200°C. One of the major problems with a power transistor is *how to conduct the heat away from the junction*.

9-5.1 Thermal Resistance

Heat transfer in a power transistor is generally calculated by using the concept of *thermal resistance*. Thermal resistance is a measure of the ease of heat transfer between two points. If it is difficult to transfer heat, the thermal resistance is high. If the heat cannot be easily transferred away from the junction, the junction temperature will rise. This can severely limit the operating region of the transistor.

In an electrical circuit, power dissipation causes a rise in temperature between two points. This results in a power equation that is analogous to Ohm's law, where the temperature difference is similar to the voltage difference, the power dissipation plays the role of the current, and the thermal resistance is analogous to the electrical resistance.

$$\Delta T = P_D \theta \quad (9\text{-}10)$$

where ΔT is the temperature difference, in °C, between the points, P_D is the power dissipation, and θ is the thermal resistance between the points. The unit of thermal resistance is °C/W.

Within a transistor the thermal resistance is specified between two points: The p-n junction and the case. This thermal resistance, θ_{JC}, depends on the construction of the transistor and is generally specified by the manufacturer. It can also be calculated if the transistor's allowable power dissipation is known.

EXAMPLE 9-8

A **2N3055** power transistor can dissipate 115 W if its case temperature is maintained at 25°C. Its maximum junction temperature is 200°C. Find its thermal resistance.

Solution
The temperature difference from junction to case is

$$\Delta T = T_j - T_c = 200°C - 25°C = 175°C$$

From equation 9-10 we have

$$\theta_{JC} = \frac{\Delta T}{P_D} = \frac{175°C}{115\ W} = \frac{1.52°C}{W}$$

The thermal resistance can also be found from the manufacturer's specifications (Figure 9-6), and agrees with the results of Example 9-8.

9-5.2 Case-to-Air Thermal Resistance

The heat conduction between a transistor's case and the surrounding air is usually very poor, resulting in a large thermal resistance. Table 9-2 is a table of case-to-air thermal resistances for various transistor cases. Power transistors are usually packaged in larger cases, like the TO-3 type, that have lower thermal resistances, but the case-to-air thermal resistance is still much larger than θ_{JC}.

Thermal resistances are analogous to electrical resistances in that they act in series, and the temperature across each thermal resistance can also be found in the way voltage drops are found across series resistors.

EXAMPLE 9-9

If a **2N3055** transistor is in free air with the ambient temperature of a typical laboratory, 25°C, find
a. The maximum power the transistor can dissipate.
b. The temperature of the transistor case under these conditions.

Table 9-2
Thermal Resistance and Thermal Capacity of Common Transistors

Case-to-Free-Air Thermal Resistance for Popular JEDEC Cases	
CASE	θ_{CA} (°C/W)
TO-18	300
TO-46	300
TO-5	150
TO-39	150
TO-8	75
TO-66	60
TO-60	70
TO-3	30
TO-36	25

Package	Thermal Capacitance (Joules/°C)	Thermal Time Constant (Seconds)
TO-5	0.58	69
TO-66 (no button)	2.56	128
TO-8	1.84	110
TO-3 (Cu button)	6.8	204
TO-3 (Mod, 2N5575)	7.8	117

Solution

The specifications (Figure 9-6) for a **2N3055** show that its maximum junction temperature is 200°C, its θ_{JC} is 1.52 °C/W, and it is packaged in a TO-3 can, which has a thermal resistance of 30°C/W as given in Table 9-2.

a. The maximum power the transistor can dissipate is given by equation 9-10:

$$P_D = \frac{\Delta T}{\theta_{JC} + \theta_{CA}} = \frac{200°C - 25°C}{1.52°C/W + 30°C/W} = \frac{175°C}{31.52°C/W} = 5.56 \text{ W}$$

b. The temperature between the case and the air can now be found:

$$\Delta T = P_D \theta_{CA} = 5.56 \text{ W} \times 30°C/W = 167°C.$$

Example 9-9 illustrates some of the problems involved with power transistors. A 115 W transistor can dissipate no more than 5.56 W when operating in free air. In addition, the case temperature will then be 192°C (at an ambient of 25°C) so it may damage whatever it sits on, or whoever comes in contact with it.

9-5.3 Heat Sinks

The previous section showed that most power transistors cannot dissipate very much power when operating in free air. To reduce the case-to-air thermal resistance, most transistors are mounted on *heat sinks*. Figure 9-9a shows several types of heat sinks and lists the case types that these heat sinks

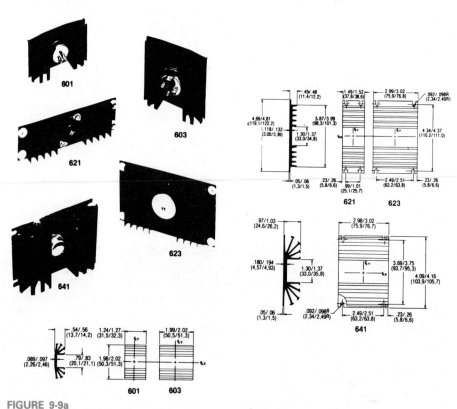

FIGURE 9-9a

(*a*) Various heat sinks. (Courtesy of EG&G Wakefield Engineering.) (*b*) Details for mounting a transistor (**TO-66** can) to a heat sink. (Copyright by Motorola, Inc. Used by permission.)

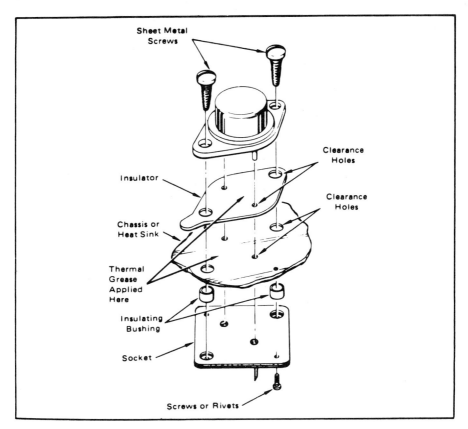

FIGURE 9-9b

will accommodate. Figure 9-9b shows how a transistor is mounted to a heat sink.

When heat sinks are used, transistors are tightly screwed to them. The heat sinks are metal to provide good thermal conductivity between the sink and the transistor, and have fins so that they can provide a large surface area between the sink and the surrounding air. Two new thermal resistances are introduced by the use of the heat sink:

1. The thermal resistance between the case and the sink, θ_{CS}. This is the small thermal resistance created by bonding the case and heat sink together. Table 9-3 gives typical values of θ_{CS} for various bonding conditions.
2. The thermal resistance between the sink and the air, θ_{SA}. This thermal resistance is normally specified by the manufacturer and typically is between 1 and 5°C/W. It is discussed further in the following section.

9-5 THERMAL RESISTANCE AND HEAT SINKS **415**

Table 9-3
Typical ranges of θ_{CS} for a TO-3 Case Under Various Mounting Conditions

Insulator	Thermal Compound Used?	θ_{CS} (°C/W)
None	No	.05–.20
None	Yes	.005–.10
Beryllium Oxide wafer	No	.10–.40
Hard Anodized Al	No	.35–.70
2 mil Plastic Film	No	.55–.80

Figure 9-10a shows a transistor mounted to a heat sink, and Figure 9-10b shows the electrical equivalent of the power circuit. It can be seen that the thermal resistances are in series and are additive.

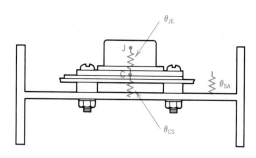

(a) Physical circuit

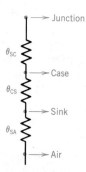

(b) Electrical equivalent

FIGURE 9-10
The thermal resistances of a transistor attached to a heat sink.

EXAMPLE 9-10

How much power can a **2N3055** dissipate if it is mounted on a heat sink whose θ_{SA} is 3.3°C/W and θ_{CS} is 0.18°C/W? Assume the transistor is in the laboratory where the ambient temperature is 25°C.

Solution
Here $\Delta T = 175°C$ and

$$\theta_{total} = \theta_{JC} + \theta_{CS} + \theta_{SA} = 1.52°C/W + 0.18°C/W + 3.3°C/W = 5°C/W$$

$$P_D = \frac{\Delta T}{\theta_{total}} = \frac{175°C}{5°C/W} = 35\ W$$

The addition of the heat sink has increased the allowable power dissipation of the **2N3055**.

Heat sink manufacturers supply catalogs giving the characteristics of their various heat sinks. Figure 9-9 is an example of a page from one of the catalogs. The larger the heat sink is, and the more fins it has, the lower θ_{SA} will be. The general engineering procedure for selecting a heat sink is first to determine the required θ_{SA}, and then to search the catalogs for a heat sink with an equal or lower θ_{SA}. Of course cost and mounting space may also have to be considered.

9-5.4 Natural Convection and Forced Convection

Heat sinks operate in one of two environments: *natural convection*, by which the heat is *passively* removed by the surrounding air, and *forced convection*, by which a fan or blower cools the sink. Most heat sink manufacturers do not give a θ_{SA} value for natural convection because θ_{SA} is not constant. At high temperatures heat sinks tend to become better radiators of heat and their θ_{SA} decreases. Instead manufacturers specify the temperature rise as a function of the power.

Figure 9-11 is a set of curves for a group of heat sinks in both natural and forced convection. The curves for natural convection tend to flatten at higher temperatures, reflecting a decreased θ_{SA}. Under forced convection θ_{SA} decreases as the air flow increases.

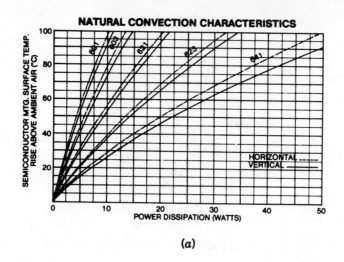

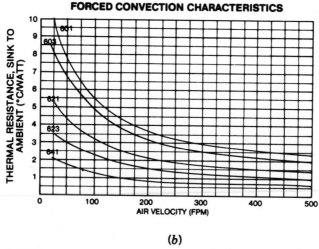

FIGURE 9-11
Characteristics of a series of heat sinks. (Courtesy of EG&G Wakefield Engineering.)

EXAMPLE 9-11

A **2N3055** transistor is mounted vertically on a type 623 heat sink. Find the junction temperature of the **2N3055** if $\theta_{CS} = 0.28°C/W$, the ambient temperature is 25°C, and the transistor power is 30 W.

a. Assume natural convection.
b. Assume forced convection with an air velocity of 300 feet per minute (fpm).

418   POWER AMPLIFIERS

Solution

a. Using natural convection, Figure 9-11a shows that the temperature rise at the sink with 30 W applied will be 90°C. In addition the temperature rise from the sink to the junction will be

$$T_{SJ} = (\theta_{CS} + \theta_{JC})P_D = 1.8°C/W \times 30 \text{ W} = 54°C.$$

The junction temperature will be

$$T_J = T_a + T_{sink} + T_{SJ}$$
$$= 25°C + 90°C + 54°C = 169°C$$

This is still below the allowable junction temperature of 200°C.

b. Figure 9-11b shows that θ_{SA} is 1.25°C/W at an air flow of 300 fpm. Therefore

$$\theta_{total} = \theta_{JC} + \theta_{CS} + \theta_{SA}$$
$$\theta_{total} = 1.52°C/W + 0.28°C/W + 1.25°C/W = 3.05°C/W$$
$$T_J = T_a + P_D \theta_{total} = 25°C + 30 \text{ W} \times 3.05°C/W = 116.5°C$$

As we would expect, the junction is much cooler when forced air is used.

9-5.5 Thermal Capacitance

When power is applied to a transistor it does not heat up instantly. It takes a finite time before the transistor will reach the temperature calculated in the previous sections.

Thermal capacitance is a measure of *how quickly* a transistor will heat up. If the transistor is coupled to a heat sink, the thermal capacitance of the circuit is primarily determined by the capacitance of the sink, which depends on its material and its weight. Larger heat sinks have the advantage of increasing the thermal capacity as well as reducing the thermal resistance.

Circuits where the power is not applied continuously, but is pulsed, can withstand slightly higher dissipations because the thermal capacity keeps the junction from getting too hot. A quantitative description of thermal capacity is beyond the scope of this book, but it is discussed further in some of the references.

9-6 CLASS A POWER AMPLIFIERS

A class A amplifier is defined as an amplifier in which there is always some current flowing in the transistor. This means the ac input signal can never drive the transistor into cutoff so that its current stops. It also implies that the transistor should not be driven into saturation, which would severely distort the output waveform. The voltage and power amplifiers discussed in the previous chapters and sections were class A to avoid distortion. In power amplifier design it is often advantageous *not* to use class A amplifiers. Other classes of amplifiers are discussed in section 9-7.

The resistive-coupled class A amplifier was discussed in section 9-4, and its efficiency proved to be very low. Two ways to significantly improve the efficiency of a class A amplifier are to use *choke-coupled* and *transformer-coupled* amplifiers. Transformer-coupled amplifiers are particularly important because they can also be used to match the impedance of the load (typically a speaker) with the impedance of the circuit. These amplifiers are discussed in this section.

9-6.1 The Choke-coupled Amplifier

Figure 9-12 shows a basic choke-coupled amplifier. It uses a *choke-coil*, otherwise known as a large inductor, between V_{CC} and the collector instead of the resistor. Ideally the choke-coil has infinite ac impedance and zero dc impedance so no dc or ac power is consumed in the coil, and the entire power output flows through the capacitor and into the load.

Figure 9-13 shows the ideal load line for a choke-coupled amplifier in black, and a practical load line in blue. The ideal dc load line for a choke-coupled amplifier is a vertical line at V_{CC} because there are no dc losses. The biasing then determines I_{CQ}, and the quiescent point is at the intersection of I_{CQ} and V_{CC}. The slope of the ac load line through the quiescent point is determined by the ac resistance (R_L in Figure 9-12).

The practical load line starts at V_{CC} and rises with a slope equal to $-1/(R_E + R_{CH})$ where R_{CH} is the dc resistance of the choke coil. If both R_E and R_{CH} are small, the dc load line is almost vertical. If R_E is properly bypassed so

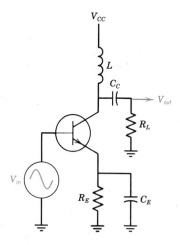

FIGURE 9-12
The basic choke-coupled amplifier.

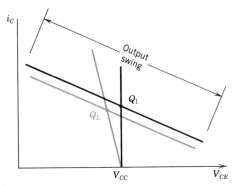

FIGURE 9-13
Ideal and actual swings for a choke-coupled amplifier.

that the bypass capacitor keeps the voltage across it constant, there will be no ac power dissipated in R_E. Similarly, if the inductance of the coil is large enough to keep the current in it constant, there will be no ac power loss in the coil. Then the slope of the ac load line will be determined solely by R_L, and will be the same as the slope of the ideal load line.

With ideal conditions, maximum power output will be attained when there is the largest current and voltage swing, but the maximum limits on voltage and current for the transistor must be observed. The largest voltage and current swings may be from their quiescent values to their maximum values and therefore

$$V_{CEQ} = \frac{V_{CEO}}{2} \qquad (9\text{-}11)$$

$$I_{CQ} = \frac{I_{Cmax}}{2} \qquad (9\text{-}12)$$

Ideally, maximum power output occurs when $V_{cm} = V_{CEQ} = V_{CC}$ and $I_{cm} = I_{CQ}$. Then

$$V_{cm} = I_{cm} R_L \qquad (9\text{-}13)$$

$$P_{ac} = \frac{V_{cm} I_{cm}}{2} = \frac{I_{cm}^2 R_L}{2} = \frac{V_{cm}^2}{2 R_L} \qquad (9\text{-}14)$$

At maximum output $V_{cm} = V_{CC}$, $I_{cm} = I_{CQ}$ and

$$P_{ac} = \frac{V_{CC} I_{CQ}}{2} = \frac{P_{CC}}{2} \qquad (9\text{-}15)$$

Equation 9-15 shows that the maximum efficiency of a choke-coupled amplifier is 50%. Of course, the ideal efficiency is never achieved in a real circuit because of the power losses in the biasing circuit, R_E, and so on, and the fact that the circuit is not always operated at maximum power output (maximum loudness), but the efficiencies of a choke-coupled amplifier are always much better than that of a resistive-coupled amplifier because the ac power is developed entirely across the load.

9-6.2 Analysis of a Choke-coupled Amplifier

A choke-coupled amplifier can be analyzed by using the methods of the previous sections. Again, the power taken from the source must equal the power dissipated by all the components in the circuit.

EXAMPLE 9-12

For the circuit of Figure 9-14 assume that the transistor is biased so that $I_{CQ} = 2$ A, the ac signal causes I_{cm} to be 2 A, and the dc resistance of the coil is 2 Ω. Find the power in each component and the efficiency of the circuit.

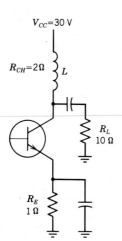

FIGURE 9-14
The circuit for Example 9-12.

Solution

First V_{CEQ} will be found. Then the power in each component can be found.

$$V_{CEQ} = V_{CC} - I_{CQ}(R_E + R_{CH}) = 30 \text{ V} - 2 \text{ A} \times 3 \text{ Ω} = 24 \text{ V}$$
$$V_{cm} = I_{cm}R_L = 2 \text{ A} \times 10 \text{ Ω} = 20 \text{ V}$$

The power in the load is

$$P_{ac} = \frac{I_{cm}^2 R_L}{2} = 20 \text{ W}$$

The other powers are

$$P_{choke} = I_{CQ}^2 R_{CH} = 8 \text{ W}$$
$$P_R = I_{CQ}^2 R_E = 4 \text{ W}$$
$$P_{transistor} = P_D = V_{CEQ} I_{CQ} - \frac{V_{cm} I_{cm}}{2}$$
$$= 24 \text{ V} \times 2 \text{ A} - 20 \text{ V} \times 2 \text{ A} = 28 \text{ W}$$
$$P_{CC} = V_{CC} I_{CQ} = 30 \text{ V} \times 2 \text{ A} = 60 \text{ W}$$

Check:

$$P_{CC} = P_{ac} + P_{CH} + P_R + P_D$$
$$= 20 \text{ W} + 8 \text{ W} + 4 \text{ W} + 28 \text{ W} = 60 \text{ W}$$

The efficiency is

$$\eta = \frac{P_{ac}}{P_{dc}} = \frac{20 \text{ W}}{60 \text{ W}} = 33\%$$

The efficiency of the circuit of Example 9-12, 33%, could be improved upon if the best value of V_{CC} were used instead of 30 V. The best value for V_{CC} is that voltage that is equal to V_{cm} plus the dc voltage drops across the coil and the emitter-resistor. For this circuit it is 26 V (see problem 9-19).

Of course the power dissipated in the transistor must not be larger than the transistor can handle.

EXAMPLE 9-13

In Example 9-10, a **2N3055** transistor mounted on a heat sink could dissipate 35 W. Can this transistor be used in the circuit of Example 9-12?

Solution
At first glance it would appear to be satisfactory, because the transistor dissipates only 28 W. This is incorrect, however, because the possibility of

quiescent or no signal operation must be considered. There will always be quiet intervals. Under quiescent conditions

$$P_D = V_{CEQ}I_{CQ} = 24 \text{ V} \times 2 \text{ A} = 48 \text{ W}$$

The **2N3055** would not work. Either a heavier transistor or a better heat sink is required.

Under ideal conditions the efficiency is 50%, and the entire power is dissipated in the transistor and the load. This leads to the conclusion that the power in the transistor and the load are each equal to $P_{CC}/2$. Under no signal conditions, however, the transistor must be able to dissipate the total power, P_{CC}. From these considerations we realize that *the transistor must be capable of dissipating at least twice the power that appears in the load.* This limitation must be taken into account when attempting to design a power amplifier.

9-6.3 Design of a Choke-coupled Amplifier

A typical design problem for a power amplifier is to put maximum power into a load based on two constraints:

1. The load resistance is fixed. This is often the case when the circuit is driving a loudspeaker. Most loudspeakers have resistances of 4 or 8 Ω.
2. The maximum power the transistor can dissipate. This was discussed in section 9-5.4, and depends on the rating of the power transistor, the heat sink selected, the ambient temperature, and whether a fan is used to cool the heat sink.

Once these constraints are established, the major design problem is to find the operating point. In section 9-6.2 it was shown that the transistor must dissipate P_D where

$$P_D = V_{CC}I_{CQ}$$

Under ideal circumstances, $V_{cm} = V_{CC} = V_{CEQ}$ and $I_{cm} = I_{CQ}$. Furthermore, $V_{cm} = I_{cm}R_L$ or $V_{CEQ} = I_{CQ}R_L$. Combining these equations we obtain

$$P_D = V_{CEQ}I_{CQ} = I_{CQ}^2 R_L = V_{CEQ}^2/R_L$$
$$V_{CEQ} = \sqrt{P_D R_L} \tag{9-16}$$

$$I_{CQ} = \sqrt{\frac{P_D}{R_L}} \tag{9-17}$$

Equations 9-16 and 9-17 can be used to determine best operating point for the transistor.

EXAMPLE 9-14

Design a power amplifier to put maximum power into a 10 Ω load. Assume the transistor and heat sink of Example 9-10 are used, that the **2N3055** has $h_{fe} = 100$, and that the coil has a dc impedance of 4 Ω. Find the biasing resistors, the power in the load, and the efficiency of the circuit.

Solution

The transistor in Example 9-10 had a maximum power dissipation of 35 W. Using equations 9-16 and 9-17 we have

$$V_{CEQ} = \sqrt{P_D R_L} = \sqrt{350} = 18.8 \text{ V}$$
$$I_{CQ} = \sqrt{P_D/R_L} = \sqrt{3.5} = 1.88 \text{ A}$$

Check: $I_{CQ} = V_{CEQ}/R_L = 1.88$ A

Now that the quiescent conditions have been found, V_{CC} and the biasing resistors can be determined. First a reasonable value of R_E must be selected. We will use 1 Ω ($R_L/10$).

$$V_{CC} = V_{CEQ} + I_{CQ}(R_E + R_{CH}) = 18.8 \text{ V} + 1.88 \text{ A} \times 5 \text{ Ω} = 28.2 \text{ V}$$

The biasing resistors can be found as in Chapter 3:

$$R_B = \frac{h_{fe} R_E}{10} = 10 \text{ Ω}$$

$$V_{BB} = I_{CQ} R_E + V_{be} + I_B R_B$$

where

$$I_B = \frac{I_{CQ}}{h_{fe}} = 18.8 \text{ mA}$$

$$V_{BB} = 1.88 \text{ V} + 0.7 \text{ V} + 0.18 \text{ V} = 2.76 \text{ V}$$

The biasing resistors can now be found by using equations 3-6 and 3-7. The results are

$$R_{B1} = 102 \text{ Ω}$$
$$R_{B2} = 11 \text{ Ω}$$

To find the bias current, the base voltage must first be found.

$$V_B = V_R + V_{be} = 1.88 \text{ V} + 0.7 \text{ V} = 2.58 \text{ V}$$

$$I_{bias} = \frac{V_{CC} - V_B}{R_{B1}} = \frac{28.2 \text{ V} - 2.58 \text{ V}}{102 \text{ }\Omega} = 0.25 \text{ A}$$

$$P_{CC} = V_{CC}I_{total} = 28.2 \text{ V} \times (1.88 + 0.25 \text{ A}) = 60 \text{ W}$$

The power in the load can be found as

$$P_L = \frac{V_{cm}^2}{2R_L} = \frac{350}{20} = 17.5 \text{ W}$$

Now the efficiency can be found:

$$\eta = \frac{P_L}{P_{CC}} = \frac{17.5 \text{ W}}{60 \text{ W}} = 29.1\%$$

9-6.4 Transformer Basics

Most class A amplifiers use transformer coupling instead of choke coils, because transformers allow the designer to match the actual impedance of the load with the impedance that is optimum for the transistor. The analysis of a transformer-coupled amplifier, however, is almost identical to the analysis of a choke-coupled amplifier, and the methods and equations developed in the previous sections apply here also. This section contains an introduction (or review) of the transformer. The transformer amplifier is analyzed in the next section.

The basic transformer is shown in Figure 9-15. It consists of two coils of wire, called *windings*, wrapped around the same iron core. If an ac voltage, e_1, is applied to the *primary winding*, which has N_1 turns of wire on it, the voltage produces a change of magnetic flux in accordance with Faraday's law

$$e_1 = N_1 d\Phi/dt$$

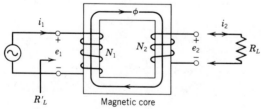

FIGURE 9-15
Transformer model used for developing the characteristics of the ideal transformer.

This change of flux is linked to the *secondary winding*, which has N_2 turns of wire on it, and causes the secondary voltage to be

$$e_2 = N_2 d\Phi/dt$$

It then follows that $e_1/e_2 = N_1/N_2 = N$, where N is defined as the *turns ratio* of the transformer. Because N_2 may be larger or smaller than N_1, the secondary voltage may be greater or less than the applied voltage.

A load resistance, R_L, is generally connected across the secondary winding, and a current $i_2 = e_2/R_L$ flows. This causes a corresponding current, i_1, to flow in the primary circuit, where $i_1/i_2 = 1/N$. Thus $e_1 i_1 = e_2 i_2$, which indicates that, in an *ideal* transformer, the power taken by the primary side is totally delivered to the secondary. An actual transformer has power losses due to the resistance of the windings and other causes so the primary power is greater than the power delivered to the load.

The impedance seen at the primary side of the transformer is called the *reflected impedance*, R'_L, where

$$R'_L = \frac{e_1}{i_1} = \frac{Ne_2}{i_{2_N}} = N^2 \frac{e_2}{i_2} = N^2 R_L$$

The advantage of using a transformer is that the load impedance can be transformed into any desired reflected impedance by choosing the proper turns ratio.

To recapitulate, the basic transformer equations are:

$$e_1 = Ne_2 \tag{9-18}$$

$$i_1 = i_2/N \tag{9-19}$$

$$R'_L = N^2 R_L \tag{9-20}$$

EXAMPLE 9-15

For the circuit of Figure 9-15, assume that the primary winding contains 100 turns, the secondary winding contains 50 turns, and $R_L = 10\ \Omega$. If 32 V is applied to the primary of the transformer, find e_2, i_2, and i_1.

Solution
The turns ratio is

$$N = \frac{100 \text{ turns}}{50 \text{ turns}} = 2$$

From equations 9-18, 9-19, and 9-20, we find

$$e_2 = \frac{e_1}{N} = \frac{32}{2} = 16 \text{ V}$$

$$i_2 = \frac{e_2}{R_L} = \frac{16 \text{ V}}{10} = 1.6 \text{ A}$$

$$i_1 = \frac{i_2}{N} = \frac{1.6 \text{ A}}{2} = 0.8 \text{ A}$$

The reflected impedance seen at the primary side of the transformer is

$$R'_L = \frac{e_1}{i_1} = \frac{32 \text{ V}}{0.8 \text{ A}} = 40 \text{ }\Omega$$

Check:

$$R'_L = N^2 R_L = 4 \times 10 \text{ }\Omega = 40 \text{ }\Omega$$

9-6.5 Analysis of the Transformer-coupled Amplifier

A class A transformer-coupled amplifier can be analyzed by following the steps listed here:

1. Using dc and bias considerations, find I_{CQ} and V_{CEQ}.
2. Reflect the load impedance into the primary to determine R'_L.
3. Analyze the primary circuit as a choke-coupled amplifier.
4. Using the transformer equations, the voltages and currents at the load can now be found. The maximum ac voltage and current in the secondary are called V_{lm} and I_{lm} respectively.

This procedure is best illustrated by an example.

EXAMPLE 9-16

In the circuit of Figure 9-16, the resistance of the transformer primary, R_P, is 2 Ω, the resistance of the transformer secondary, R_S, is 1 Ω, and the turns ratio is 2:1. Note that these resistances are proportional to the number of turns on the windings, as is often the case. If the transistor is biased so that $I_{CQ} = 1$ A, find the maximum voltage and current in the 8 Ω speaker load.

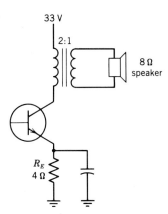

FIGURE 9-16
The transformer-coupled amplifier for Example 9-16.

Solution

First V_{CEQ} can be found using the dc conditions at the primary side.

$$V_{CEQ} = V_{CC} - I_{CQ}(R_E + R_P) = 33\text{ V} - 1\text{ A} \times 6\text{ }\Omega = 27\text{ V}$$

Both the load impedance and the resistance of the secondary winding are reflected into the primary:

$$R'_L = N^2(R_L + R_S) = 4 \times 9\text{ }\Omega = 36\text{ }\Omega$$

To find the maximum possible power delivered to the load, we can start by assuming a maximum collector current swing of $I_{cm} = I_{CQ} = 1$ A. Unfortunately, because $V_{cm} = I_{cm}R'_L$, this leads to a V_{cm} of 36 V. But V_{cm} can never become larger than V_{CEQ}, 27 V, so V_{cm} is the limitation on this circuit. If V_{cm} is at its theoretical maximum

$$I_{cm} = \frac{V_{cm}}{R'_L + R_P} = \frac{27\text{ V}}{38\text{ }\Omega} = 0.71\text{ A}$$

Now the secondary voltages and currents and the power in the load can be found using equations 9-18 and 9-19:

$$V_{lm} = V_{cm}/2 = \frac{27\text{ V}}{2} = 13.5\text{ V}$$

$$I_{lm} = NI_{cm} = 1.42\text{ A}$$

The power in the speaker is

$$P_L = \frac{I_{lm}^2 R_L}{2} = \frac{2.02 \times 8\text{ }\Omega}{2} = 8.08\text{ W}$$

The efficiency of this circuit is

$$\eta = \frac{P_L}{P_{CC}} = \frac{8.08 \text{ W}}{33 \text{ W}} = 24.5\%$$

The efficiency of the circuit of Figure 9-16 would have been higher if the value of the reflected load impedance were optimum. The optimum reflected load impedance is V_{CEQ}/I_{CQ}. This allows both V_{cm} and I_{cm} the maximum possible swing. In Example 9-16, the optimum R'_L was about 27 Ω, but the actual reflected R'_L was 36 Ω. Consequently, the current swing was restricted to 0.71 A, and the efficiency decreased (see problem 9-19).

9-6.6 Design of a Transformer-coupled Amplifier

The design of a transformer-coupled class A amplifier requires the engineer to make many decisions based on the material covered in the previous sections. Often the resistance of the load and the desired power in the load are specified. Using the fact that the power dissipated in the transistor must be at least twice the power in the load (see section 9-6.2), a transistor and heat sink must be selected that are capable of handling the required power. Then the power supply voltage and quiescent current can be specified, and the required turns ratio of the transformer can be calculated. Finally, the transformer catalogs must be searched to find a transformer that approximates the optimum. Once all the components have been selected, the circuit should be recalculated, to ascertain that it meets its specifications and that no limitations are exceeded.

EXAMPLE 9-17

Design a transformer-coupled class A amplifier to deliver 20 W to a 4 Ω speaker. Use a **2N3055** power transistor.

Solution

The **2N3055** must dissipate at least 40 W. This means that the heat sink discussed in Example 9-10 is inadequate. A larger heat sink, like the type 623 with forced air at 75 fpm should be selected.

Now V_{CC} can be chosen. As a rule of thumb, V_{CC} should not be larger than $V_{CEO}/2$, so we will use 30 V. To find a reasonable value of I_{CQ}, we should take an "educated guess" at the efficiency of the circuit (the education consists of reading the previous sections of this chapter). If we guess at an η of 33%, then $P_{CC} = 60$ W, and

$$I_{CQ} = P_{CC}/V_{CEQ} = 60 \text{ W}/30 \text{ V} = 2 \text{ A}$$

The emitter voltage should be between 10 and 20% of V_{CC}. A V_E of 4 V would be a good choice, which leads to an R_E of 2 Ω. If we assume that the dc resistance of the transformer is 1 Ω, then

$$V_{CEQ} = V_{CC} - I_{CQ}(R_E + R_P) = 30\text{ V} - 2\text{ A} \times 3\text{ Ω} = 24\text{ V}$$

The optimum reflected resistance at the primary is then

$$R'_L = V_{CEQ}/I_{CQ} = 12\text{ Ω}$$

This means that the transformer turns ratio should be $\sqrt{3}$, or 1.73. The circuit is shown in Figure 9-17.

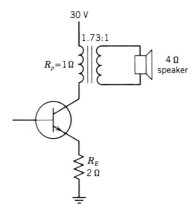

FIGURE 9-17
Circuit designed in Example 9-17.

Now a transformer must be found that approximates the calculated characteristics. If the circuit of Figure 9-17 were actually implemented, the worst-case power dissipation in the transistor would be $V_{CEQ}\, I_{CQ}$ = 24 V × 2 A = 48 W. If the transistor cannot absorb 48 W, V_{CC} could be changed or some other circuit adjustment made. V_{CC} might also have to be changed if the best available transformer does not have exactly the characteristics that were calculated or assumed. For the circuit under consideration, the primary resistance of the transformer could not be larger than 3 Ω if V_{CC} is to remain at 30 V. If it were, V_{CEQ} would be less than 20 V and the circuit could not deliver 20 W to the load.

9-7 CLASS B AMPLIFIERS

The class A amplifier, in which the transistor is constantly conducting current and operating in its linear region, has several disadvantages.

1. Its efficiency is low.
2. It dissipates considerable power in the quiescent (no signal) condition.

3. There is a significant dc current flowing in the transformer primary. This tends to saturate the transformer, making it less efficient.

Consequently, class A amplifiers are usually used only for relatively low-power circuits. Circuits that require higher power outputs use class B amplifiers.

Class B amplifiers reduce or eliminate these problems. In a class B amplifier, the transistor current flows for *half* the cycle, or 180°. The current waveform is shown in Figure 9-18. Of course, such a waveform is very distorted. In a class B amplifier, a second transistor is added to conduct during the other half of the waveform. The waveforms are then combined to produce a sinusoidal output.

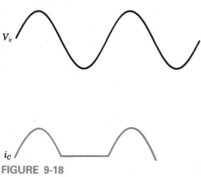

FIGURE 9-18
The current waveform in one transistor of a class B amplifier.

The classic class B amplifier is shown in Figure 9-19a. It consists of two transistors and two center-tapped transformers. The bias network, consisting of resistors R_{B1} and R_{B2}, is designed so that the voltage at their junction, which is also the dc voltage at the base of both transistors, is slightly less than 0.7 V. When the input is quiescent both transistors are on the verge of turning on, but neither transistor is drawing any current.

The current waveforms are shown in Figure 9-19b. When V_S goes positive the base voltage of Q_1 increases and Q_1 conducts. At the same time the base voltage of Q_2 decreases and is driven further into cutoff. The current i_1 flows through the top half of transformer T_2 and produces a positive voltage pulse across R_L. When V_S goes negative, i_2 flows through the lower half of transformer and produces a negative voltage pulse across R_L. Thus R_L receives alternate positive and negative voltage pulses, which constitutes an ac output. Because transistors Q_1 and Q_2 conduct alternately, this circuit is often called a *push-pull* circuit.

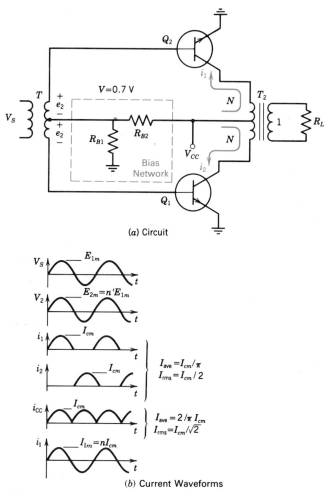

FIGURE 9-19
The basic class B push–pull circuit.

9-7.1 The Collector Characteristics of a Class B Amplifier

The collector or voltage–current characteristics of a transistor in a push–pull circuit are shown in Figure 9-20. When the transistor is quiescent, no current flows and the collector voltage is V_{CC}. The quiescent point is shown in Figure 9-20. When the transistor is conducting, the current flows through a load resistance R'_L, where $R'_L = N^2 R_L$. Here N is the ratio of the number of turns between the primary (from end to center-tap) and the secondary of T_2. During conduction, the collector current rises to a peak value of I_{cm} and the voltage falls to $V_{CC} - V_{cm}$, where V_{cm} equals $I_{cm} R'_L$. Figure 9-20 also shows that if the

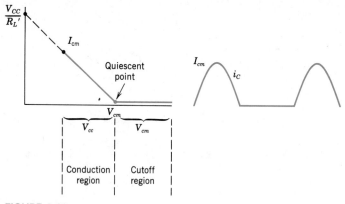

FIGURE 9-20
The collector characteristics of a transistor in a class B circuit.

voltage swing is maximum ($V_{cm} = V_{CC}$), then I_{cm} will be maximum and equal to V_{CC}/R'_L.

When Q_1 is not conducting, Q_2 is. The voltage across both halves of the transformer primary must be the same. During this time Q_1 is in the nonconducting region of Figure 9-20, but its collector voltage is higher than V_{CC} and rises to a maximum of $V_{CC} + V_{cm}$. This voltage must still be less than BV_{CEO}, and conservative designers will therefore limit V_{CC} to $BV_{CEO}/2$. The voltages and currents at the collector are shown in greater detail in Figure 9-21.

FIGURE 9-21
The primary side of a transformer used in a class B amplifier.

9-7.2 Class B Power Dissipation and Efficiency

One of the major disadvantages of class A power amplifiers is that they absorb maximum power when no ac signal is applied. A properly biased class B amplifier dissipates only the small amount of power necessary to bias the transistors when power is applied, because there is no collector current in the power transistors. When an input signal is applied, the current waveform in the power supply, i_{cc}, is a series of half sine waves as shown in Figure 9-19b. The average value of this waveform is $2I_{cm}/\pi$. The power provided by the supply is therefore

$$P_{CC} = \frac{2V_{CC}I_{cm}}{\pi} \tag{9-21}$$

The power dissipated by the load (neglecting transformer losses) is

$$P_L = \frac{I_{cm}^2 R_L'}{2} \tag{9-22}$$

where R_L' is the reflected load impedance.

The power delivered to the load will be maximum when

$$V_{cm} = I_{cm} R_L' = V_{CC}$$

Then $P_{L(max)} = \dfrac{V_{CC}^2}{2R_L'}$ and the efficiency at this operating point is

$$\eta = \frac{P_L}{P_{CC}} = \frac{\dfrac{V_{CC}^2}{2 R_L'}}{\dfrac{2}{\pi} V_{CC} I_{cm}} = \frac{\dfrac{V_{CC} I_{cm}}{2}}{\dfrac{2}{\pi} V_{CC} I_{cm}} = \frac{\pi}{4} = 78.5\%$$

The efficiency of a class B amplifier starts at 0% and increases to 78.5% at maximum load.

Not only do class B amplifiers have a higher theoretical maximum efficiency, but the power dissipated in each transistor is much less than that of a class A amplifier. If we assume the power dissipated in the transistors is the supply power minus the load power, then

$$2 P_D = \frac{2}{\pi} V_{CC} I_{Cm} - \frac{I_{cm}^2 R_L'}{2} \tag{9-23}$$

The maximum power dissipated in the transistors does not occur at maximum load. It can be found by differentiating equation 9-23 and occurs when

$$I_{cm} = \frac{2}{\pi} \frac{V_{CC}}{R_L'} \tag{9-24}$$

At this worst-case point, the power dissipated in each transistor is approximately 10% of the maximum load power (see problem 9-23). In a class A amplifier the single transistor must dissipate twice the maximum load power, so for the same output power, the heat sink problems are much simpler in a class B amplifier.

EXAMPLE 9-18

For the circuit of Figure 9-19, the load is a 4 Ω speaker, the transformer turns ratio is 2:1, and $V_{CC} = 25$ V. Assume the transistors have a saturation voltage $V_{CE(sat)}$ of 1 V, and neglect any transformer losses.

If the circuit is to deliver maximum power to the load, find

a. V_{cm}
b. I_{cm}
c. V_{lm}
d. I_{lm}
e. The power in the load
f. P_{CC}
g. The power dissipated in each transistor
h. The efficiency

Solution

a. The maximum swing is from V_{CC} to $V_{CE(sat)}$

$$V_{cm} = V_{CC} - V_{CE\ (sat)} = 25\ V - 1\ V = 24\ V$$

b. $R'_L = N^2 R_L = (2)^2 \times 4 = 16\ \Omega$

$$I_{cm} = \frac{V_{cm}}{R'_L} = \frac{24\ V}{16} = 1.5\ A$$

c. $V_{lm} = \dfrac{V_{cm}}{N} = \dfrac{24}{2} = 12\ V$

d. $I_{lm} = N I_{cm} = 3\ A$

(Check: $R_L = \dfrac{V_{lm}}{I_{lm}} = \dfrac{12\ V}{3\ A} = 4\ \Omega$)

e. $P_L = \dfrac{V_{lm} I_{lm}}{2} = \dfrac{12 \times 3}{2} = 18\ W$

Note that P_L also equals $V_{cm} I_{cm}/2$ or $(V_{lm})^2/2R_L$.

f. $P_{CC} = \dfrac{2}{\pi} V_{CC} I_{cm} = 2 \times 25\ V \times 1.5\ A = 23.9\ W$

g. The power dissipated in each transistor is

$$P_D = \frac{P_{CC} - P_L}{2} = \frac{23.9\ W - 18\ W}{2} = 2.95\ W$$

h. $\eta = \dfrac{P_L}{P_{CC}} = \dfrac{18\ W}{23.9\ W} = 75.3\%$

Because the power amplifier is operating at maximum output, the efficiency is close to the theoretical maximum.

9-8 DISTORTION AND CLASS B BIASING

When a signal is applied to an amplifier, the output waveform will not be an exact replica of the input: it will be somewhat distorted. In an audio amplifier this distortion affects the quality of the sound. Thus power amplifiers must be designed to provide the required power output and to minimize the distortion of the output signal.

When a sine wave, $V_S \sin \omega t$, is applied to an amplifier, the distortion can be represented by signals introduced at *multiples* of the input frequencies, call *harmonics*. The output waveform can be expressed as

$$V_{out} = A_0 + A_1 \sin \omega t + A_2 \sin 2\omega t + A_3 \sin 3\omega t + \cdots \quad (9\text{-}25)$$

where A_0 is any dc component introduced by the amplifier and A_2, A_3, ... are the magnitudes of the second, third ... harmonics. In equation 9-25, the ideal, undistorted output would have all A's except A_1 equal to zero. In general, as the harmonic increases, the amount of distortion introduced decreases, so the lower harmonics provide most of the distortion and are most important.

If the output waveform can be accurately determined, the various A's in equation 9-25 can be found. Pierce and Paulus (see references), for example, give equations for determining the values of the A's from an accurate representation of the waveform. This is the graphical method. Once the values for the A's are found, the total harmonic distortion can be found from the equation

$$\%D = \sqrt{\frac{A_2^2 + A_3^2 + A_4^2 + }{A_1}} \quad (9\text{-}26)$$

Unfortunately, the graphical method is usually impractical. Distortion is most often determined in the laboratory by using a *spectrum analyzer* or *distortion analyzer*. The spectrum analyzer can measure the amount of output power at the fundamental and each of its harmonics, while the distortion analyzer filters out the fundamental and finds the distortion as a function of the output at the higher frequencies.

9-8.1 Distortion in Power Amplifiers

There are three major causes of distortion in power amplifiers: harmonic distortion, class B distortion, and intermodulation distortion. *Harmonic distortion* is caused by the nonlinearities in the V-I characteristics of the power transistors. It is generally reduced by adding feedback to the amplifier.

Class B distortion is shown in Figure 9-22; it occurs when the bias point is not exactly at the cutoff point of the transistor. If the transistors are biased

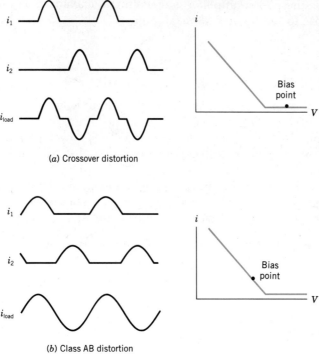

FIGURE 9-22
Class B distortion.

so that they conduct for slightly less than a half cycle, *crossover distortion* results, as shown in Figure 9-22a. Crossover distortion is characterized by intervals during each cycle when no current flows. If the transistors are biased so that they each conduct for more than a half cycle, as shown in Figure 9-22b, then *class AB distortion* occurs. The transistors are operating in class AB because they conduct for more than 180° of the cycle (class B), but less than 360° of the cycle (class A). Class AB distortion results in slightly steeper sides to the sine wave. It also has the disadvantage of using power when no signal is applied.

Intermodulation distortion occurs when the input consists of two or more frequencies. If the amplifier distorts, the output can have components at frequencies that are the *sum* and *difference* of the applied frequencies. Because these frequencies do not exist at the input, they are another form of distortion.

9-8.2 Biasing the Class B Amplifier

The previous section has shown that the ideal point at which to bias a class B amplifier is where the transistors conduct no current, but are just on the verge of turning on so that any ac input signal will result in current flow in the push–pull circuit. If the emitters are grounded, the base of the transistors should therefore have a dc voltage of slightly less than 0.7 V applied. This quiescent voltage can be obtained from a resistive divider, but perhaps the simplest and best way is to use a combination of a resistor and diode as shown in Figure 9-23. The current flow through the diode will keep the dc voltage at the transistor bases at 0.7 V. This circuit has an additional advantage; the voltage variations of the base-to-emitter junctions due to temperature changes can be approximately matched by the changes across the diode. Thus the diode can be made to "track" the transistor as the temperature changes and keep the bias point at its proper place. The change in performance due to temperature changes is an important consideration in the design of power amplifiers.

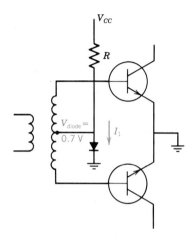

FIGURE 9-23
A bias circuit for a class B amplifier.

In the circuit of Figure 9-23, there is a maximum value for the resistor that depends on the output power. This is shown in Example 9-19.

EXAMPLE 9-19

In Example 9-18 we used a circuit where $V_{CC} = 25$ V, $R'_L = 16\ \Omega$, and $I_{cm} = 1.5$ A. If the circuit of Figure 9-23 is used to bias this circuit, and the transistors have an h_{fe} of 75, what is the maximum value of the resistor, R?

Solution
The current through the resistor must provide sufficient base current for the *on* transistor, and enough additional current to keep the diode turned on. In this circuit the maximum collector current is 1.5 A, and therefore the required base current is 1.5 A/75 or 20 mA. If we also assume that there must be at least 2 mA in the diode, then the resistor must allow 22 mA to flow through it.

$$R_{(max)} = \frac{V_{CC} - V_{BE}}{I} = \frac{24.3\ \text{V}}{22\ \text{mA}} = 1100\ \Omega$$

In a practical circuit, a 1 kΩ resistor (slightly less than $R_{(max)}$) would be chosen for R.

9-9 COMPLEMENTARY SYMMETRY AMPLIFIERS

The class B transformer-coupled amplifier originated in the vacuum tube era, when transformers were used to match the relatively high impedance of the amplifier to the low impedance of the loudspeaker or other load. Unfortunately, transformers are often bulky, heavy, and costly and have limited frequency response. The availability of both *npn* and *pnp* power transistors has allowed engineers to design *transformerless* or *complementary symmetry* amplifiers that operate without transformers. Most modern power amplifiers use complementary symmetry and operate in class B mode so the equations of section 9-7 can also be used.

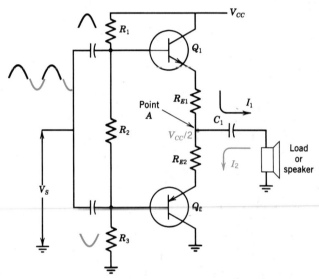

FIGURE 9-24
The basic complementary symmetry amplifier.

The basic complementary amplifier is shown in Figure 9-24. It uses an *npn* and a *pnp* transistor to drive the load. Both transistors are connected in the common-collector or emitter-follower configuration, so the circuit provides no voltage gain. Resistors R_{E1} and R_{E2} are generally very small and are used to stabilize the amplifier. Resistors R_1, R_2, and R_3 are used to bias the transistors. With no signal applied they are both turned off, but on the verge of turning on. The dc voltage at point A, between the emitter resistors, is $V_{CC}/2$. Capacitor C_1 blocks this dc level and leaves the quiescent voltage at the load at ground.

When a sine wave is applied to the input, the positive half cycle causes Q_1 to conduct, but drives Q_2, the *pnp* transistor, further into cutoff. Thus the positive half cycle current pulse reaches the load. This is shown in black on Figure 9-24. On the negative half cycle, only the *pnp* transistor conducts and the negative current pulse is applied to the load. This is shown in blue. The load, therefore, receives alternating current pulses that provide the ac output.

EXAMPLE 9-20

For the circuit of Figure 9-25, assume $V_{BE} = 0.7$ V, $V_{CE(sat)} = 0$ V. Find

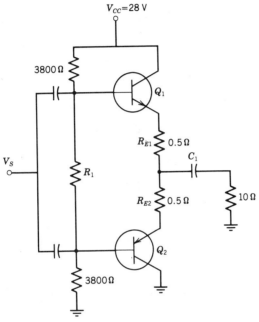

FIGURE 9-25
Circuit for Example 9-20.

a. R_1
b. The maximum power in the 10 Ω resistor
c. P_{CC}
d. η

Solution

a. This circuit should be biased so that each transistor is on the verge of turning on. The voltage at point A should be $V_{CC}/2$ or 14 V. Therefore the voltage at the base of Q_1 should be 14.7 V and the voltage at the base of Q_2 should be 13.3 V. The voltage across each 3.8 kΩ resistor is 13.3 V, and the bias current is therefore

$$I_b = \frac{13.3 \text{ V}}{3800} = 3.5 \text{ mA}$$

The voltage across R_1 must be 1.4 V. Therefore

$$R_1 = \frac{1.4 \text{ V}}{3.5 \text{ mA}} = 400 \text{ }\Omega$$

Note that with these bias voltages, the quiescent voltage at point A must remain at 14 V. If it tried to go lower, Q_1 would turn on and charge C_1, and if it tried to go higher, Q_2 would turn on and discharge C_1.

b. Because point A is at 14 V, V_{cm} is limited to 14 V.

$$I_{cm} = \frac{14 \text{ V}}{10.5 \text{ }\Omega} = 1.33 \text{ A}$$

$$P_L = \frac{I_{cm}^2 R_L}{2} = \frac{(1.33)^2 \times 10}{2} = 8.845 \text{ W}$$

c. In this circuit the power supply delivers current only when transistor Q_1 turns on. This supplies the load resistor and also charges the capacitor. The capacitor discharges to supply the power when Q_2 turns on. The current waveform at the power supply is therefore a half sine wave and the average current is I_{cm} and

$$P_{CC} = \frac{V_{CC} I_{cm}}{\pi} = \frac{28 \text{ V} \times 1.33 \text{ A}}{\pi} = 11.86 \text{ W}$$

d. $\eta = \dfrac{P_L}{P_{CC}} = \dfrac{8.845 \text{ W}}{11.86 \text{ W}} = 74.6\%$

The circuits of Figure 9-24 and 9-25 show the basics of a complementary power amplifier, but they are not practical circuits. The following sections show how this basic circuit is modified and expanded upon to produce a useful power amplifier.

9-9.1 Diodes and Double-ended Power Supplies

The biasing resistor (R_1 in Figure 9-25) is often replaced by two forward-biased diodes. They also produce the required drop of 1.4 V. Diodes have the advantage of being able to track the base-to-emitter voltage and stabilize the amplifier against temperature variations. If resistors are used, however, the biasing resistor can be variable and can be adjusted to make the circuit operate at its optimum point. In some power amplifiers, a combination of diodes and resistors is used.

The circuit of Figure 9-24 uses a single power supply. This places the voltage at point A at $V_{CC}/2$ and requires a capacitor, C_1, to remove the dc component of the load. In this circuit the voltage drop across the capacitor must be low, compared to the inherently low impedance of the load, and this must occur at the lowest frequency the power amplifier must pass, which can be as low as 20 Hz for an audio system. These factors cause C_1 to be a very large capacitor; in some circuits it is as high as 2000 µF.

C_1 can be eliminated by using a positive and a negative power supply, and this is often done in practical power amplifiers. The positive supply, $+V_{CC}/2$, is connected to the collector of the *npn* power transistor, and the negative supply, $-V_{CC}/2$, is connected to the collector of the *pnp* transistor. The bias circuits can now cause the quiescent voltage at point A to be at ground potential, which eliminates C_1. One of the decisions a designer must make is whether to use C_1 or two power supplies.

9-9.2 The Input Transistor

Most power amplifiers are driven by an input transistor, Q_3, as shown in Figure 9-26. The collector current of Q_3 provides the bias voltages and Q_3 also gives the circuit some voltage gain.

Q_3 could be biased by a resistor connected to V_{CC}, but is usually biased

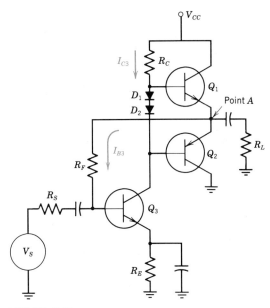

FIGURE 9-26
A complimentary symmetry amplifier driven by an input stage.

by a resistor, R_F, connected to point A, as shown in Figure 9-26. R_F provides both dc and ac feedback and therefore stabilizes the circuit.

EXAMPLE 9-21

In the circuit of Figure 9-26, the collector current of Q_3 is to be 10 mA. If $V_{CC} = 30$ V, $R_E = 200$ Ω, and Q_3 has an $h_{fe} = 100$, find R_C and R_F.

Solution

The voltage at point A must be $V_{CC}/2 = 15$ V. Therefore the voltage at the base of Q_1 is 15.7 V.

$$R_C = \frac{V_{CC} - V_{BQ1}}{I_C} = \frac{30 \text{ V} - 15.7 \text{ V}}{10 \text{ mA}} = 1.43 \text{ k}\Omega$$

To find R_F, the voltage at the emitter of Q_3 is

$$V_{RE} \approx 10 \text{ mA} \times 200 \text{ }\Omega = 2 \text{ V}$$

Therefore the voltage at the base of Q_3 is 2.7 V. The quiescent base current of Q_3 is

$$I_{C3/h_{fe}} = 0.1 \text{ mA}$$

$$R_F = \frac{V_{(\text{point } A)} - 2.7 \text{ V}}{I_{b3}} = \frac{12.3 \text{ V}}{0.1 \text{ mA}} = 123 \text{ k}\Omega$$

The use of R_F in Figure 9-26 provides dc feedback for biasing. If the quiescent voltage at point A tends to increase, the base current of Q_3 will also increase, which will increase I_{b3} and lower the voltage at the base of Q_1. This will decrease the voltage at point A, so R_F serves to stabilize the circuit.

The use of R_F also introduces ac feedback into the circuit. This feedback is desirable because it helps reduce the harmonic distortion of the amplifier. The voltage gain of the circuit of Figure 9-26, with feedback, is shown in Appendix F4 to be

$$A_{vf} = \frac{h_{fe} R_C}{h_{ie} + R_S\left(1 + \frac{h_{fe} R_C}{R_F}\right)} \tag{9-27}$$

EXAMPLE 9-22

a. For the circuit of Example 9-21, assume $R_S = 1$ kΩ. What is the ac voltage gain of the circuit?
b. If the output ac voltage is to be 12 V, what must V_S be?

Solution

a. First, h_{ie} must be found.

$$h_{ie} = \frac{30\, h_{fe}}{I_C} = \frac{30 \times 100}{10} = 300\ \Omega$$

$$A_{vf} = \frac{100 \times 1.43\text{ k}\Omega}{300\ \Omega + 1\text{ k}\Omega\left(1 + \frac{100 \times 1.43\text{ k}\Omega}{123\text{ k}\Omega}\right)}$$

$$= \frac{100 \times 1.43\text{ k}\Omega}{300\ \Omega + 1\text{ k}\Omega(2.16)} = \frac{143{,}000\ \Omega}{2460\ \Omega} = 58$$

b. If the load voltage is to be 12 V, then

$$V_S = \frac{V_{out}}{A_{vf}} = \frac{12}{58} = 207\text{ mV}$$

Therefore the source must supply 207 mV to drive the amplifier.

The use of two power supplies does not affect the calculations. If two power supplies are used, the voltage at point A becomes 0 V, but the emitter-resistor of Q_3 is connected to the negative supply rather than ground, and the calculations are about the same.

9-9.3 Darlingtons

High-power amplifiers require a peak output current of several amperes. If a single power transistor is used, the amount of base current required to support the large collector current can be excessive. For example, if $I_{cm} = 2$ A and the transistor has an $h_{fe} = 50$, then I_b must be 40 mA. The biasing circuits of Figure 9-26 do not allow a base current this large.

Darlingtons, which have a high current gain, were introduced in section 6-6. They are generally used in power amplifiers to replace the single power transistor. Figure 9-27 shows a Darlington output stage, with an amplifier with two power supplies. Matched *npn* and *pnp* power Darlingtons are available for use in this type of circuit.

Figure 9-28 is a variation on the Darlington circuit called a *quasi-complementary* circuit. In this configuration, only Q_4 is a *pnp* transistor, and both Q_1 and Q_3 can be the same type of power transistor.

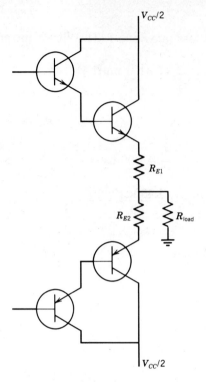

FIGURE 9-27
A complimentary amplifier using Darlingtons.

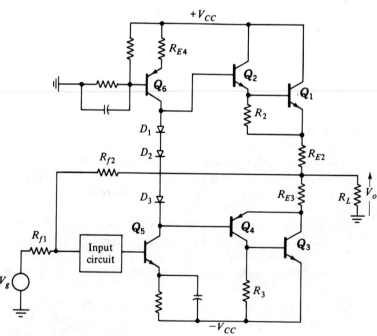

FIGURE 9-28
A quasi-complementary power amplifier.

9-9.4 Bootstrapping

Even when Darlingtons are used, the circuit of Figure 9-26 still has drive problems. If V_{CC} is 30 V, capacitor C_1 will charge up to 15 V. If V_{cm} is required to be 12 V, for example, this means the voltage at the emitter of Q_1 is 27 V and the voltage at the base of Q_1 will be 27.7 V or 28.4 V if Q_1 is replaced by a Darlington. In either case, there is not enough voltage left to drive the base current.

A *bootstrap* circuit is a circuit used to provide additional base current when it is needed at the peak of the output voltage. The circuit is shown in Figure 9-29. In most cases, the biasing resistor R_C is split in two and the

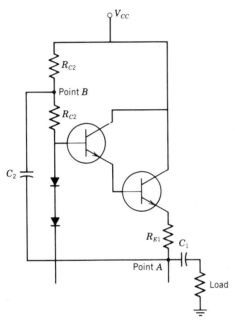

FIGURE 9-29
A bootstrap circuit.

bootstrap capacitor, C_2, is placed between point A and point B. This capacitor should be large enough to retain approximately a constant voltage across it throughout each cycle. When the voltage at point A peaks, the voltage at point B also peaks and C_2 discharges through $R_C/2$ and the base of the Darlington, thus providing the additional current needed at this time.

EXAMPLE 9-23

In the circuit of Figure 9-29, if V_{CC} is 30 V, what is the quiescent voltage across C_2? If a V_{cm} of 12 V is developed across the load, what is the voltage at point B at its peak?

Solution
The voltage at point A will be 15 V. ($V_{CC}/2$) If the circuit is biased properly, the dc voltage at the input to the Darlington will be two base-to-emitter drops higher or 16.4 V. The voltage at point B will be halfway between 16.4 V and V_{CC} or at 23.2 V. Therefore the voltage across C_2 is $V_B - V_A = 8.2$ V.

At the instant that $v_{in} = 12$ V, the voltage at point A is 27 V and the voltage at point B will be 35.2 V (assuming neither C_1 nor C_2 loses any significant charge). This voltage is higher than V_{CC} and is responsible for the base current into the Darlington.

Some manufacturers use constant-current sources (see section 6-7) instead of Darlingtons to provide the required drive at the peak of the output swing.

9-9.5 A Practical Amplifier
Figure 9-30 shows an amplifier manufactured by Southwest Technical Products Corp. that includes the features discussed in this chapter. It uses a single power supply with *pnp* and *npn* Darlingtons. The load is an 8 Ω speaker.

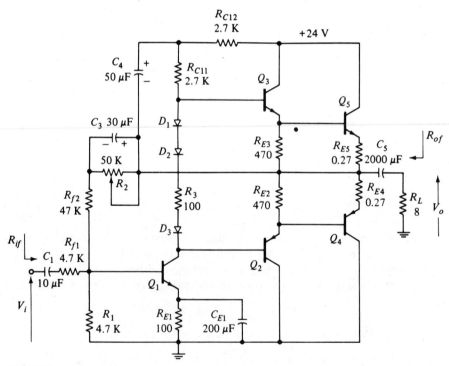

FIGURE 9-30
One example of a complementary power amplifier. (From *Applied Electronics* by Pierce and Paulus, 1972.)

Notice the 2000 μF capacitor in series with the load. The biasing resistors are R_{C12} and R_{C11} and C_4 is the bootstrap capacitor. The 2.8 V drop that should be between the Darlington inputs are developed by diodes D_1, D_2, and D_3, and R_3.

The dc bias for Q_1 is developed by R_2 and R_{f2}. R_2 is adjustable for proper biasing. For ac it is shorted out by C_3 to provide more ac feedback.

The ac gain can be found from equation 9-27. In some circuits such as this one,

$$\frac{h_{fe} R_C}{R_{f2}} \gg 1$$

and

$$\frac{R_S h_{fe} R_C}{R_{f2}} \gg h_{ie}$$

If these assumptions are valid, equation 9-27 reduces to R_{f2}/R_S. For this circuit, R_{f2} = 47 kΩ and R_S is R_{f1} here or 4.7 kΩ. Thus the ac gain of this circuit is 10.

Practical power amplifiers contain many variations of the circuits presented here, but the principles involved should help the reader understand them.

One last feature of a complementary amplifier should be mentioned; the designer can usually get more power into a 4 Ω speaker than an 8 Ω speaker for the same V_{CC}. This is because V_{cm} is usually limited by V_{CC} and the output power is $V_{cm}^2/2 R_L$.

9-10 INTEGRATED-CIRCUIT POWER AMPLIFIERS

A large variety of integrated-circuit (IC) power amplifiers exist, and many engineers simply select a packaged amplifier rather than designing one. The basic problem of dissipating the heat generated by the power amplifier remains, however, and integrated-circuit packaging does not simplify this problem. Therefore all IC amplifiers can be mounted on heat sinks to help cool them. IC amplifiers are used when lower power outputs are required, but for higher power outputs the discrete class B amplifiers discussed in sections 9-7 and 9-9 are often used.

The **2002** is a widely used type of IC power amplifier. Figure 9-31 contains some of the specifications on the **TDA2002**. Figure 9-31a is the first page of the specifications. It shows that the **TDA2002** is a class B amplifier packaged in a TO-220 style case with overload protection. The maximum ratings for the device are also given. The drawing of the package shows that there are five terminals:

ULN-3701Z / TDA2002
5 TO 10-WATT AUDIO POWER AMPLIFIER

FEATURES
- Low External Parts Count
- Low Distortion
- Class B Operation
- Short-Circuit Protected
- Thermal Overload Protected
- Low Noise
- High Output-Voltage Swing
- TO-220 Style Package
- Direct Replacement for LM383 and CA2002

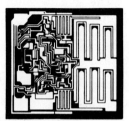

DESIGNED specifically to drive low-impedance loads down to 1.6 Ω, the Type ULN-3701Z / TDA2002 audio power amplifier is ideal for automotive radio, tape player, and CB applications.

It can deliver 5 W to 10 W of audio in the single-ended mode. Operating in the extremely harsh automotive environment, this device is capable of withstanding high ambient temperatures, output overloads, and repeated power supply transient voltages without damage.

The Type ULN-3701Z/TDA2002 amplifier is rated for continuous operation with supply voltages of up to 18 V. With the application of increased voltages (to 28 V, maximum), a high-voltage protective circuit becomes operative, disabling the device. Devices without this internal high-voltage shutdown are available as Type ULN-3702Z/TDA2008 and are recommended for use where more than 10 W of audio power is required with higher impedance loads and supply voltages to 28 V. In all other respects, Types ULN-3701Z and ULN-3702Z are identical.

Type ULN-3701Z/TDA2002 is supplied in a modified 5-lead JEDEC Style TO-220 plastic package. The heat-sink tab is at ground potential; no insulation is required.

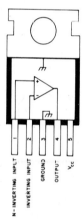

FIGURE 9-31a
(a) Specifications of the TDA 2002 A. (b) The ULN-3702Z/TDA2002A 12 watt power amplifier. (c) The ULN-3701Z/TDA2002 5- to 10-watt audio power amplifier.

Power
Ground
Noninverting input
Inverting input
Output

The drawing also shows the tab on the package, which is designed to be connected to a heat sink.

ULN-3701Z/TDA2002 AUDIO POWER AMPLIFIER

ELECTRICAL CHARACTERISTICS at $T_A = +25°C$, $V_{CC} = 14.4V$, $R_L = 4\Omega$, $f = 1$ kHz, $R_{fb} = \infty$ (unless otherwise noted)

Characteristic	Symbol	Test Conditions	Min.	Typ.	Max.	Units
Supply Voltage Range	V_{CC}		8.0	14.4	18	V
Quiescent Supply Current	I_{CC}	No signal applied	—	45	80	mA
Quiescent Output Voltage	V_4	No signal applied	6.4	7.2	8.0	V
Open Loop Gain	A_v		—	80	—	dB
Closed Loop Gain	A_v		39.5	40	40.5	dB
Total Harmonic Distortion	THD	$P_{OUT} = 0.05$ to 3.5W	—	0.2	—	%
		$P_{OUT} = 0.05$ to 5.0W, $R_L = 2\Omega$	—	0.2	—	%
Audio Power Output	P_{OUT}	THD = 10%	4.8	5.2	—	W
		THD = 10%, $R_L = 2\Omega$	7.0	8.0	—	W
		THD = 10%, $V_{CC} = 16V$	—	6.5	—	W
		THD = 10%, $R_L = 2\Omega$, $V_{CC} = 16V$	—	10	—	W
Efficiency	η	$P_{OUT} = 5.2W$	—	68	—	%
		$P_{OUT} = 8.0W$, $R_L = 2\Omega$	—	58	—	%
Input Impedance	Z_i		70	150	—	kΩ
Power Supply Rejection	PSR	$f_{ripple} = 120$Hz, $V_{ripple} = 0.5V$	30	35	—	dB
Equiv. Input Noise Voltage	v_N	$f = 40$Hz to 15kHz	—	4.0	—	μV
Equiv. Input Noise Current	i_N	$f = 40$Hz to 15kHz	—	60	—	pA
Input Sensitivity	v_{in}	$P_{OUT} = 0.5W$	—	15	—	mV
		$P_{OUT} = 0.5W$, $R_L = 2\Omega$	—	11	—	mV
		$P_{OUT} = 5.2W$	—	55	—	mV
		$P_{OUT} = 8.0W$, $R_L = 2\Omega$	—	50	—	mV
Input Saturation Voltage	v_{in}		—	600	—	mV
Frequency Response (−3dB)		$C_{fb} = 0.039\mu F$, $R_{fb} = 39\Omega$	40	—	15k	Hz
Thermal Resistance	$R_{\Theta JT}$		—	—	4.0	°C/W

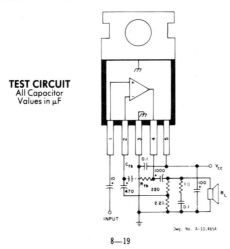

TEST CIRCUIT
All Capacitor Values in µF

FIGURE 9-31b

Figure 9-31*b* gives the electrical characteristics, test circuit, and a typical low-cost application. This shows the **TDA2002** connected to a speaker. The inverting input is connected to a small resistor in series with the speaker to provide a small amount of feedback for stability.

9-10 INTEGRATED-CIRCUIT POWER AMPLIFIERS

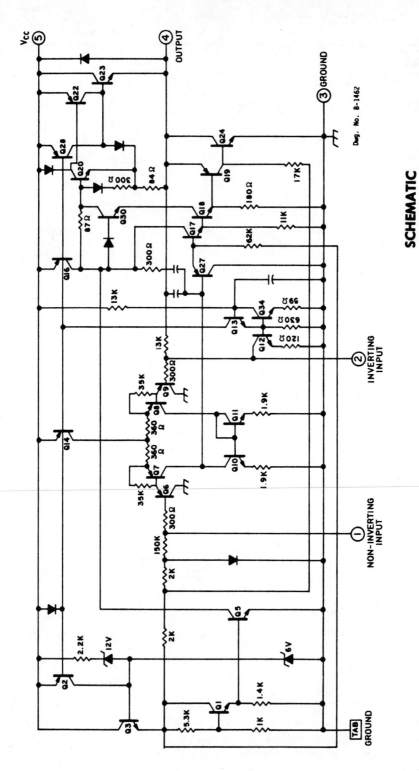

FIGURE 9-31c

Figure 9-31c is the schematic diagram of the **TDA2002**. Transistors Q1–Q5 are used for the short-circuit and thermal overload protection. Transistors Q6–Q11 form a difference amplifier for the inverting and noninverting inputs. The positive output is provided by transistors Q22 and Q23, and the negative output is provided by transistors Q19 and Q24. Both of these pairs form a quasi-complementary Darlington.

The **TDA2002** is only one of many available IC power amplifiers. Figure 9-31 shows only part of the device specifications. A set of complete specifications also includes a series of curves on the operation of the device. The most important curves are device dissipation versus power output, and total harmonic distortion (THD) versus frequency at various levels of output power. Complete specifications also include many circuits that are typically used with the device.

Some manufacturers provide an *application note* on their IC. These application notes go into much greater detail on the design and the uses of the device.

9-10.1 A Practical IC Power Amplifier

Figure 9-32 is the schematic for a practical power amplifier designed by Scientific Radio Systems Corporation of Rochester, N.Y. This amplifier uses the **TDA2002** IC. The inverting input is connected between R_1 and R_2, which form a feedback network where $\beta = R_2/(R_1 + R_2) = 0.01$. The gain of this amplifier is $1/\beta$ or 100 (see section 8-3.4). The function of R_3 and C_1 is to roll off the response at frequencies above 3 kHz, because this amplifier is for voice

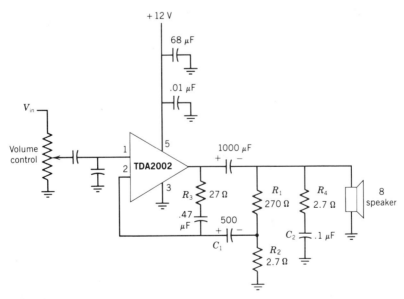

FIGURE 9-32
Schematic for a practical power amplifier.

communications rather than high fidelity, and the higher, unneeded frequencies produce a hiss in the speaker. Resistor R_4 and capacitor C_2 are required to suppress a type of low-frequency oscillation known as "motor-boating."

9-11 SUMMARY

This chapter discussed the principles of power amplifiers and presented several examples of their use. Class A amplifiers were described first, in order to establish the principles of operation. Then the basic question of how to dissipate the heat generated was discussed and heat sinks were introduced. After that the more practical power amplifiers, such as the class A transformer-coupled, the class B, and the complementary amplifier were presented. Both the analysis and the design of these amplifiers were discussed. Finally, commercially available integrated-circuit power amplifiers were discussed and an example was presented.

9-12 GLOSSARY

Bootstrap circuit. A circuit to provide current into the *npn* transistor or Darlington at the peak of the output waveform.

Choke coil. A large inductor. Ideally its ac impedance is infinite and its dc impedance is zero.

Class A amplifier. An amplifier in which current flows in the transistor throughout the entire cycle (360°).

Class B amplifier. An amplifier in which current flows in each transistor for half of each cycle (180°).

Class AB amplifier. An amplifier in which current flows in the transistor for more than half but less than a full cycle (between 180° and 360°).

Complementary amplifier. The use of both *npn* and *pnp* transistors to provide an ac output.

Efficiency. The ratio of useful output power to input power.

P_{CC}. The dc power supplied to an amplifier.

P_D. The power dissipated in the transistor of an amplifier.

Reflected impedance. The impedance seen on the primary side of a transformer due to the load on the secondary side.

Thermal resistance. A measure of the ease of heat transfer between two points.

9-13 REFERENCES

ROBERT BOYLESTAD and LOUIS NASHELSKY, *Electronic Devices and Circuit Theory*, 4th Edition, Prentice-Hall, Inc., Englewood Cliffs, N.J., 1987.

RODNEY B. FABER, *Essentials of Solid State Electronics*, John Wiley, New York, 1985.

JACOB MILLMAN, *Microelectronics*, 2nd Edition, McGraw-Hill, New York, 1987.

J. F. PIERCE and T. J. PAULUS, *Applied Electronics*, Charles E. Merrill Co., Columbus, Ohio, 1972.

RCA *Designer's Handbook, Solid State Power Circuits*, RCA Corp. Somerville, N. J., 1971.

DONALD L. SCHILLING and CHARLES BELOVE, *Electronic Circuits, Discrete and Integrated*, 2nd Edition, McGraw-Hill, New York, 1979.

J. D. SHERRICK, *Notes for ITEE 532 Power Amplifier Design*. This is an unpublished note. For further information, contact Professor John D. Sherrick, Rochester Institute of Technology, Rochester, N. Y. 14623.

Note: The last four references contain advanced topics not thoroughly covered in this chapter.

9-14 PROBLEMS

9-1 Repeat Example 9-2 if the emitter-resistor is not bypassed.

9-2 Design a self-bias circuit for the circuit of Example 9-2. Find the power absorbed by this circuit and the new efficiency if this power is taken into account. Assume the emitter resistor is bypassed.

9-3 Draw the ac and dc load lines for the circuit of Figure P9-3. Assume the circuit is biased in the middle of the dc load line and that the input signal causes the current to swing by 0.8 A.

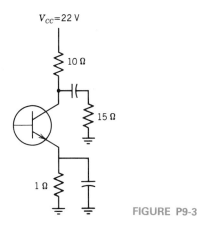

FIGURE P9-3

  a. Find the ac and dc power in each resistor.
  b. Find the quiescent power dissipated in the transistor and the power dissipated when the ac signal is applied.
  c. If the output power is defined as the power in the 15 Ω resistor, find the efficiency.

9-4 Repeat problem 9-3 if the emitter-resistor is unbypassed.

9-5 Design a resistive-coupled power amplifier to provide 7 W of ac power to a 14 Ω load resistor. Specify R_E and V_{CC}. Find the efficiency of the circuit.

9-6 For the amplifier shown in Figure P9-6
  a. Determine R_{dc} and R_{ac}.
  b. Sketch the load lines showing all intercepts assuming the Q point is located in the middle of the dc load line.

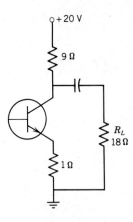

FIGURE P9-6

c. Determine P_{CC}, P_{DC}, P_{ac}, P_o, and calculate η_t under maximum unclipped output conditions.

d. Determine the minimum power ratings required for all components assuming maximum unclipped sinusoidal operation.

9-7 A circuit employing a **2N3055** with a case temperature of 25°C is shown in Figure P9-7. Determine from its electrical characteristics

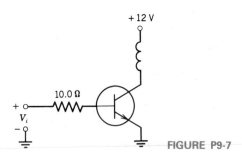

FIGURE P9-7

a. The minimum acceptable value of V_i to ensure a collector current of at least 4 A.

b. The value of V_i that would be acceptable if the transistor were "typical."

9-8 A **2N3055** is dissipating 10 W at a case temperature of 45°C. What is its junction temperature?

9-9 A transistor is rated at 30 W at $T_c = 25°C$. It is to be derated 0.35 W/°C above 25°C. Sketch the transistor's derating curve and determine θ_{jc} and T_{jm}.

9-10 A transistor is rated at 100 W to case temperatures of 70°C. Its maximum rated junction temperature is 200°C. What power may it dissipate at a case temperature of 120°C?

9-11 If the case temperature of a **2N3055** is 60°C, draw the maximum power hyperbola on the curve of Figure 9-8. Find the maximum allowable current if V_{CE} is 20 V.

POWER AMPLIFIERS

9-12 From the specification sheet for the **2N3904** find T_J, the maximum power dissipation, θ_{JC}, and θ_{CA}.

9-13 Figure P9-13 is a set of curves for the **TIP31** series of power transistors

TYPICAL CHARACTERISTICS

STATIC FORWARD CURRENT TRANSFER RATIO
VS
COLLECTOR CURRENT

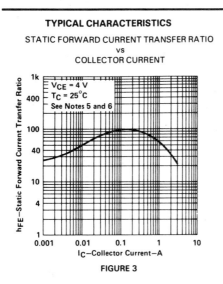

FIGURE 3

NOTES: 6. These parameters must be measured using pulse techniques, t_w = 300 μs, duty cycle ≤ 2%.

7. These parameters are measured with voltage-sensing contacts separate from the current-carrying contacts.

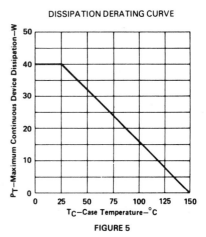

THERMAL INFORMATION

DISSIPATION DERATING CURVE

FIGURE 5

MAXIMUM SAFE OPERATING AREA

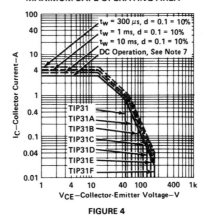

FIGURE 4

NOTE 8: This combination of maximum voltage and current may be achieved only when switching from saturation to cutoff with a clamped inductive load.

FIGURE P9-13
(Courtesy of Texas Instruments, Inc.)

manufactured by Texas Instruments, Inc. From the curves for the TIP31B find:
a. $T_{J(max)}$
b. θ_{JC}
c. $P_{D(max)}$
d. $I_{C(max)}$
e. V_{CEO}

9-14 A transistor with a maximum junction temperature of 175°C and a θ_{JC} of 2°C/W is housed in a TO-66 can. Find its maximum power dissipation
a. if it is in the laboratory ($T_{ambient} = 25°C$).
b. If it is in an electrical cabinet with other components ($T_{ambient} = 60°C$).

9-15 Repeat problem 9-14 if the transistor is mounted on a heat sink ($\theta_{CS} = 0.2°C/W$ and $\theta_{SA} = 2°C/W$).

9-16 A transistor is rated at 100 W at 25°C case temperature. Its maximum junction temperature is 175°C. If the transistor is mounted to a heat sink with $\theta_{CS} = 0.2°C/W$, determine
a. The maximum allowed θ_{SA} if the transistor is to dissipate 50 W at $T_a = 25°C$.
b. The temperature of the sink in part a.

9-17 A transistor has a maximum junction temperature of 180°C, and a θ_{JC} of 1.8°C/W. It is mounted on a vertical type **621** heat sink, with $\theta_{CS} = 0.2°C/W$. Find the maximum ambient temperature
a. In still air if it must dissipate 20 W.
b. With a fan blowing air at 225 fpm.

9-18 Repeat Example 9-12 if the emitter resistor is unbypassed. Find the power in each component of Figure 9-14 and the efficiency of the circuit.

9-19 Repeat Example 9-12 if $V_{CC} = 26$ V.

9-20 A choke-coupled amplifier is to have an output voltage of $V_{cm} = 15$ V across a 12 Ω load resistor. Design the amplifier. Assume $R_E = 2$ Ω and the dc coil resistance is 1.5 Ω. Find the best value for V_{CC}, the biasing resistors, and the efficiency. Assume $h_{fe} = 80$.

9-21 A **2N3055** is to drive an 8 Ω speaker using a choke-coupled circuit with $R_E = 1$ Ω and the dc coil resistance is 1.5 Ω. If the ambient temperature is 35°C and the transistor is mounted on a type **623** heat sink with 200 fpm of forced air, design the circuit for maximum power output. Find V_{CEQ}, I_{CQ}, the output power, the biasing resistors, and the efficiency.

9-22 The amplifier of Figure P9-22 is biased at I_{CQopt}. Assume the transformer and transistor are ideal. Determine P_{CC}, max P_o, and the maximum collector circuit efficiency. What average power will the transistor have to dissipate?

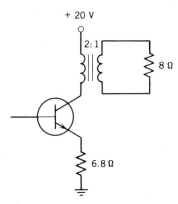

FIGURE P9-22

9-23 Repeat the calculations of problem 7.4 if the emitter resistance is by-passed. Assume $h_{fe} = 40$.

9-24 A transformer is required to make a 4 Ω speaker look like a 64 Ω impedance at the primary. Find its turns ratio.

9-25 In the circuit of Figure P9-25, there must be 40 W in the 10 Ω resistor. If the resistance of the transformer primary and secondary are both 1 Ω, find v_2, i_2, v_1, i_1, and the efficiency of the transformer.

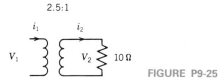

FIGURE P9-25

9-26 Redesign the circuit of Example 9-16 so that $V_{CEQ} = 27$ V, and $I_{CQ} = 0.75$ A. Find the new value of V_{CC} and the new efficiency.

9-27 Find the power dissipated in each component of Example 9-16.

9-28 For the circuit of Figure P9-28, the dc impedance of the primary is 1.5 Ω and the secondary impedance is negligible. Find

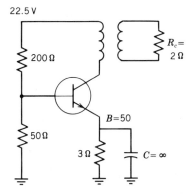

FIGURE P9-28

9-14 PROBLEMS 459

a. I_{CQ}
b. V_{CEQ}
c. The turns ratio for maximum power to the load
d. V_{cm}
e. I_{cm}
f. V_{lm}
g. I_{lm}
h. The efficiency

9-29 For the circuit and load line of Figure P9-29 find

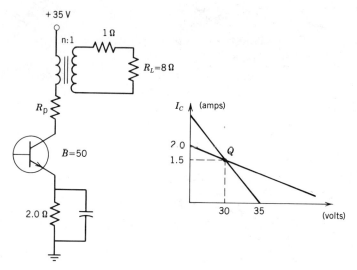

FIGURE P9-29

a. R_P
b. The transformer turns ratio.
c. The maximum power that can be delivered to the load.
d. The power dissipated by the transistor, both at quiescent conditions and at maximum output.

9-30 Show that equation 9-24 is correct by differentiating equation 9-23. Also show that the power dissipated in each transistor is approximately 10% of the maximum power delivered to the load.

9-31 Repeat Example 9-18 if I_{cm} is held to 1 A. Let V_{cm} remain at 24 V.

9-32 A loudspeaker is rated at 2 W, 4 Ω. The power supply is 20 V, but V_{cm} is to be limited to a 16 V swing. The loudspeaker is to be driven by a class B amplifier. Find
a. R_L
b. I_{cm}

c. N
d. V_{lm}
e. I_{lm}
f. P_{CC}
g. Efficiency.
h. The power dissipated in each transistor.

9-33 The specifications on a pair of power transistors are $BV_{CEO} = 50$ V, $I_{max} = 2.5$ A and $V_{CE(sat)} = 1$ V. Design a class B amplifier to deliver as much power as possible to a 4 Ω speaker. Find the transformer turns ratio, and the power, voltage, and current in the load.

9-34 A class B push–pull amplifier has $V_{CC} = 20$ V, and $R_L = 5$ Ω, and uses transistors with $h_{fe} = 50$. If it is biased by the circuit of Figure 9-23 and the base is to be at 0.7 V, what is the maximum value the resistor can have if the circuit is to deliver maximum power to the load? Allow 5 mA to flow in the diode.

9-35 For the circuit of Figure P9-35, if $V_{in} = 0$, find V_{b1}, V_{b2}, and I_{cq}. For maximum power to the load, find V_{cm}, I_{cm}, P_L and the efficiency. Assume V_{be} (SAT) = 0. Ignore the bias current for your efficiency calculation. Sketch the waveform of the current through Q_1.

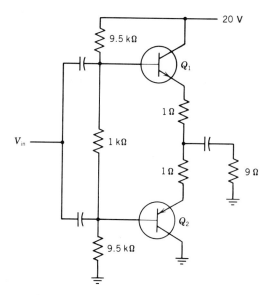

FIGURE P9-35

9-36 Repeat Example 9-21 if V_{CC} is changed to 40 V.

9-37 For the circuit of Figure P9-37, find R_D, R_F, and the voltage gain if $R_S = 500\ \Omega$.

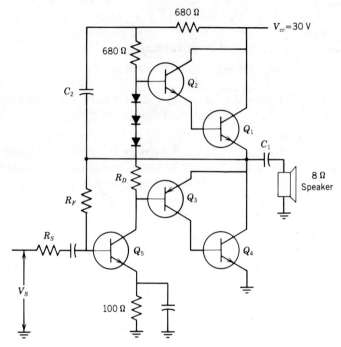

FIGURE P9-37

CHAPTER TEN

Power Supplies

10-1 Instructional Objectives

Power supplies are required to transform ac voltages into dc to provide power to electronic circuits. These power supplies and their component parts are considered in this chapter. After reading it, the student should be able to:

1. List the components of a power supply and explain their function.
2. Calculate the rms value and ripple factor for a given waveform.
3. Use curves to find V_{dc}/V_m and the ripple factor for a capacitive filter.
4. Calculate the conduction angle and percent regulation of a capacitor filtered load.
5. Design and analyze Zener diode regulators.
6. Analyze a linear voltage regulator.
7. Use IC voltage regulators.
8. Analyze a switching regulator and determine the proper values for the inductor and capacitors.

10-2 Self-Evaluation Questions

Watch for the answers to the following questions as you read the chapter. They should help you to understand the material presented.

1. What is peak inverse voltage? Why is it important?
2. Does the conduction angle increase or decrease as the load increases? Why?

3. What is nonrepetitive surge current? When does it occur?
4. What is a bleeder resistor? What is its function?
5. What are the advantages and disadvantages of an R-C filter?
6. What is a common method for temperature compensating a Zener diode? Explain how it works.
7. What is the function of a preregulator?
8. How are regulators protected against overvoltage? Against overcurrent?
9. What is the advantage of switching regulators?

10-3 INTRODUCTION TO POWER SUPPLIES

The function of a power supply is to transform ac voltages into the dc voltage required by a particular circuit. By far the most common source of ac voltage is the standard 117 V, 60 Hz available at the nearest wall outlet. Power supplies transform this voltage into a dc voltage that supplies the direct current required by the circuit. This direct current is often called the *load* on the supply. Those power supplies that provide an adjustable output voltage, such as those found in typical laboratories, are known as *variable-voltage* power supplies. Those power supplies that are designed to power integrated circuits are usually *fixed-voltage* supplies whose output is 5 V dc.

A power supply consists of four basic parts, as shown in Figure 10-1. The four parts are

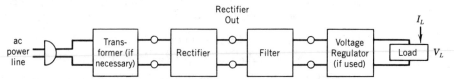

FIGURE 10-1
Power-supply block diagram. (From *Semiconductor Devices and Circuits* by Henry Zanger. Copyright © 1984 by John Wiley & Sons, Inc. Reprinted by permission of John Wiley & Sons, Inc.)

1. *The transformer.* The optimum ac input voltage is usually not the 117 V supplied by the power company. If the input voltage is not optimum, the power supply itself will consume more power and generate more heat than necessary. Fortunately, transformers can change the line voltage to the desired ac voltage with little loss of power.
2. *The rectifier.* The function of the rectifier is to transform the ac voltage into a dc voltage.
3. *The filter.* The filter smooths the output of the rectifier so that the dc voltage is relatively free of ripple.
4. *The regulator.* The function of the regulator is to hold the dc output voltage constant, regardless of changes in the load or the input voltage.

Each of these components will be discussed in the succeeding sections. If only a low-current supply is needed, and expense is a major consideration, the designer may be able to use a half-wave rectifier, a capacitive filter, and no regulator. Most modern electronic systems have more severe requirements and need a better power supply.

10-4 RECTIFIERS

Rectifiers use diodes to convert ac to dc. Diode characteristics were introduced in section 1-7.1. The engineer must consider such diode characteristics as the average dc current, the peak inverse voltage, and the maximum surge current, when analyzing or designing a power supply.

10-4.1 The Half-wave Rectifier

The half-wave rectifier is the simplest and cheapest rectifier to build, and can be used in inexpensive circuits where the purity of the output voltage is not a major consideration. The half-wave rectifier is shown in Figure 10-2a and consists merely of a transformer (not needed if the ac source voltage is the voltage that can be used directly in circuit), a diode, and the load resistor. The waveforms are shown in Figure 10-2b. The load voltage is $V_m \sin \omega t$ during the half-cycle when the input ac is positive and the diode is conducting and zero when the input voltage is negative and the diode is cut off. Here V_m is the maximum value of the ac voltage at the secondary of the transformer.

The *peak inverse voltage* (PIV) across the diode must also be considered in any rectifier circuit. This is the maximum voltage placed across the diode in

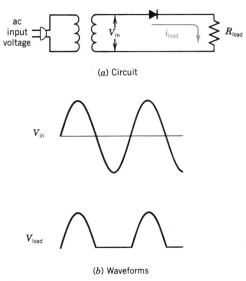

FIGURE 10-2
The half-wave rectifier.

the reverse direction. It must be less than the maximum PIV specified by the manufacturer for that diode. In the circuit of Figure 10-2, the inverse voltage peaks when the ac input voltage is $-V_m$. At this time there is no voltage across the load and the inverse voltage across the diode is $-V_m$. The peak inverse voltage will change if a different type of rectifier is used, or if the output is filtered (see section 10-4.3).

10-4.2 The Average and RMS Values of a Waveform

The waveform of the half-wave rectifier is shown in Figure 10-2b, and its equation can be expressed as

$$v = V_m \sin \omega t \quad 0 < \omega t < \pi$$
$$v = 0 \quad \pi < \omega t < 2\pi \tag{10-1}$$

where ω is 2π times the frequency of the ac input voltage. The frequency of the ac input voltage is generally 60 Hz, so $\omega = 377$ radians per second (rad/s).

The waveform can also be expressed by its Fourier series expansion:

$$V_{(t)} = A_0 + A_1 \sin \omega t + A_2 \sin 2\omega t + A_3 \sin 3\omega t \ldots \tag{10-2}$$

where A_0 is the dc or average value of the wave, and $A_1, A_2 \ldots$ are the values of the first, second . . . harmonics.

For any waveform

$$A_0 = \frac{1}{2\pi} \int_0^{2\pi} v(t)\, dt \tag{10-3}$$

For the half-wave circuit

$$A_0 = \frac{1}{2\pi} \int_0^{\pi} V_m \sin \omega t\, dt = \frac{V_m}{\pi} = 0.318\, V_m \tag{10-3}$$

The root-mean-square (rms) value of a waveform is a frequently used measure of the power of a wave.

$$P = \frac{V_{rms}^2}{R_L} = I_{rms}^2 R_L$$

For any wave

$$V_{rms} = \sqrt{\frac{1}{2\pi} \int_0^{2\pi} v^2\, dt} \tag{10-4}$$

For a sine wave equation 10-4, reduces to the familiar equation $V_{rms} = V_m/\sqrt{2}$. For the half-wave rectifier it becomes: $V_{rms} = V_m/2$.

EXAMPLE 10-1

If the house lines are applied directly to a rectifier, find
a. The average value of the output voltage.
b. The peak inverse voltage of the diode.
Assume the house voltage is 117 V.

Solution

The 117 V supplied by the power company is an rms value. The peak value is therefore

$$V_m = V_{rms} \sqrt{2} = 165 \text{ V}$$

The average or dc voltage is

$$V_{dc} = 0.318 \, V_m = 52.6 \text{ V}$$

The peak inverse voltage is the same as V_m or 165 V.

10-4.3 The Ripple Factor

The ideal output of a rectifier is a perfectly smooth dc voltage. If the ac component of an output waveform is large, the output exhibits a large amount of *ripple*. The waveform of Figure 10-2b shows the ripple very clearly.

The *ripple factor* is a measure of the ac content of a waveform. As equation 10-5 shows, it is the ratio of the rms ac content of the waveform to its dc or average value.

$$\text{R.F.} = \frac{\text{rms value of the ac harmonic content}}{\text{average or dc value}} \quad (10\text{-}5)$$

If the rms value of the total waveform can be found, then the ripple factor can be calculated from equation 10-6:

$$\text{R.F.} = \sqrt{\frac{I_{rms}^2 - I_{dc}^2}{I_{dc}}} = \sqrt{\frac{V_{rms}^2 - V_{dc}^2}{V_{dc}}} \quad (10\text{-}6)$$

EXAMPLE 10-2

Find the ripple factor of the 117 V ac wave after it is rectified by a half-wave rectifier.

Solution
For this wave

$$V_{rms} = \frac{V_m}{2} = 82.5 \text{ V}$$

$$V_{dc} = 52.6 \text{ V}$$

$$\text{R.F.} = \sqrt{\frac{V_{rms}^2 - V_{dc}^2}{V_{dc}}} = \sqrt{\frac{(82.5)^2 - (52.6)^2}{52.6}} = \frac{63.5}{52.6} = 1.21$$

The ripple factor for a half-wave rectifier is 1.21, regardless of the magnitude of the input.

Example 10-2 shows that the ripple is 20% greater than the dc output voltage. This is too high in most circuits, and the use of other circuits or filters reduces the ripple.

10-4.4 The Full-wave Rectifier

The full-wave rectifier is a major improvement because it produces a pulse of current on both the positive and negative half cycles of the ac input. The circuit is shown in Figure 10-3a and the waveforms are shown in Figure 10-3b.

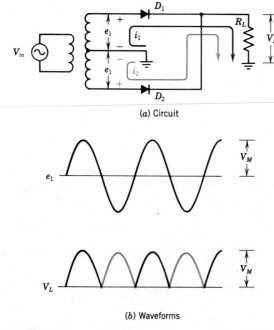

FIGURE 10-3
The full-wave rectifier.

Figure 10-3a shows that the full-wave rectifier uses a center-tapped transformer. The transformer produces two identical voltages, e_1, where e_1 is the voltage between either end of the transformer and the center-tap. Here $e_1 = V_m \sin \omega t$, where V_m is the maximum value of the ac voltage between ground and the center-tap. When e_1 is positive, diode D1 conducts and the current path is shown in black. On the alternate half-cycle, e_1 is negative and diode D2 conducts; this current path is shown in blue. In either case there is a current pulse through the load. The waveform of the load voltage and current is much closer to dc and much easier to filter than the half-wave waveform.

The average or dc value of the waveform can be calculated by equation 10-3. It is simpler, however, to observe that the full-wave circuit has twice as much current and voltage as the half-wave circuit. Therefore

$$V_{dc} = 0.636\, V_m \qquad (10\text{-}7)$$

The rms value can be calculated using equation 10-4. Again, it is simpler to observe that the rms value is the same as for a sine wave (polarity does not matter in an rms calculation because the waveform is squared). Therefore

$$V_{rms} = 0.707\, V_m \qquad (10\text{-}8)$$

The ripple factor for a full-wave rectifier is 0.48 (see problem 10-1). Thus the ripple has been considerably reduced.

The peak inverse voltage (PIV) for the diodes in the full-wave circuit can be found by considering the circuit when D1 is fully conducting. At this instant $e_1 = V_m$, and the voltage at the load is V_m if the diode drop is neglected. But this is also the voltage at the cathode of D2. The voltage at the anode of D2 is $-V_m$, so the PIV for the diode is $2V_m$. Table 10-1 summarizes the characteristics of the diode circuits.

Table 10-1
Characteristics of Rectifier Circuits

	V_{dc}	V_{rms}	Ripple factor	PIV
Half-wave	$0.318V_m$	$0.5V_m$	1.21	V_m
Full-wave	$0.636V_m$	$0.707V_m$	0.48	$2V_m$
Bridge	$0.636V_m$	$0.707V_m$	0.48	V_m

10-4.5 The Bridge Rectifier

The bridge rectifier is a configuration of four diodes as shown in Figure 10-4. The output is a full-wave waveform similar to Figure 10-3b. The bridge rectifier is very popular because it provides a full-wave output but eliminates the need for a center-tapped transformer.

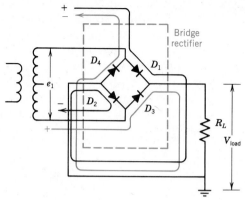

FIGURE 10-4
The bridge rectifier.

In Figure 10-4, if e_1 is positive the current flows through diode $D1$. It is then blocked by $D3$, so it flows through the load and returns to the transformer through $D2$. This current path is shown in black. When e_1 is negative, $D3$ and $D4$ conduct and there is a current pulse through the load. This is shown in blue.

The peak load voltage is V_m (neglecting diode drops), where V_m is the peak voltage across the secondary of the input transformer (or the peak of the input ac if no transformer is used). The PIV for each diode is also V_m, as can be seen by considering diode $D4$. At the instant that the load voltage is V_m, the cathode voltage at $D4$ is also V_m, but the anode voltage is 0, so its PIV is V_m.

EXAMPLE 10-3

A bridge is to be used to transform the 117 V ac into a dc voltage of 25 V. Find the transformer ratio. Do not neglect diode drops.

Solution
Because the output is full-wave, $V_{dc} = 0.636\ V_m$. If $V_{dc} = 25$ V then $V_m = 39.3$ V. To find the voltage at the secondary of the transformer, the two diode drops (1.4 V) must be added, so the required secondary voltage is 40.7 V. The peak primary voltage is 165.4 V (1.414 × 117 V), so the transformer turns ratio must be

$$N = 165.4/40.7 = 4.06$$

POWER SUPPLIES

Bridge rectifiers are very popular and are available as packaged units from many manufacturers. Motorola, among others, makes a series of bridge rectifiers at various current ratings. Figure 10-5 is the specification sheet for the **MDA970A1** through **MDA970A6** bridge rectifiers. These are all rated at

MDA970A1 thru MDA970A6

Designers Data Sheet

INTEGRAL DIODE ASSEMBLIES

...diffused silicon dice interconnected and transfer molded into rectifier circuit assemblies for use in application where high output current/size ratio is of prime importance. These devices feature:

- Void-free, Transfer-molded Encapsulation to Assure High Resistance to Schock, Vibration, and Temperature Extremes
- High Dielectric Strength
- Simple, Compact Structure for Trouble-free Performance
- High Surge Capability — 100 Amps

Designers Data for "Worst Case" Conditions
The Designers Data Sheet permits the design of most circuits entirely from the information presented. Limit curves — representing boundaries on device characteristics — are given to facilitate "worst case" design.

SINGLE-PHASE FULL-WAVE BRIDGE

4 AMPERES 50-600 VOLTS

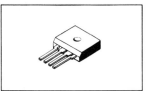

MAXIMUM RATINGS (T_A = 25°C unless otherwise noted)

Rating	Symbol	MDA970A1	MDA970A2	MDA970A3	MDA970A5	MDA970A6	Unit
Peak Repetitive Reverse Voltage Working Peak Reverse Voltage DC Blocking Voltage	V_{RRM} V_{RWM} V_R	50	100	200	400	600	Volts
RMS Reverse Voltage	$V_{R(RMS)}$	35	70	140	280	420	Volts
DC Output Voltage Resistive Load Capacitive Load	 Vdc Vdc	 31 50	 62 100	 124 200	 248 400	 372 600	Volts
Average Rectified Forward Current T_A = 25°C T_C = 55°C	I_O	← 4.0 →		← 8.0 →			Amp
Nonrepetitive Peak Surge Current (surge applied at rated load conditions, T_J = 150°C)	I_{FSM}	← 100 →					Amp
Operating and Storage Junction Temperature Range	T_J, T_{stg}	← -65 to +150 →					°C

THERMAL CHARACTERISTICS

Characteristics	Symbol	Max (Per Die)	Unit
Thermal Resistance, Junction to Case Each Die	$R_{\theta JC}$	10	°C/W
Effective Bridge	$R_{\theta(EFF)}$	7.75	°C/W

ELECTRICAL CHARACTERISTICS

Characteristic	Symbol	Min	Max	Unit
Instaneous Forward Voltage (Per Diode) (i_F = 6.28 Amp, T_J = 25°C) (i_F = 6.28 Amp, T_J = 150°C)	v_F	— —	 1.1 1.0	Vdc
Reverse Current (Rated V_{RM} applied to ac terminals, + and − terminals open, T_A = 25°C)	I_R	—	1.0	mA

CASE: Transfer-molded plastic encapsulation.
FINISH: All external surfaces are corrosion-resistant. Leads are readily solderable.
POLARITY: Embossed symbols
  AC input = ~   DC output = +   DC output = −
MOUNTING POSITION: Any
WEIGHT (Approximately): 7.5 Grams
MOUNTING TORQUE: 5 in.-lb. Max

CASE 117A-02

FIGURE 10-5
The specification sheet for a bridge rectifier. (Copyright by Motorola, Inc. Used by permission.)

an average rectified current of 4 A at 25°C. The differences between the **MDA970A1, MDA970A2** . . . are the reverse voltage ratings.

The figure shows that the rectifier is simply four diodes encapsulated in a bridge configuration. There are four leads on each package—two for the ac input, and two for the dc output.

10-5 THE CAPACITIVE FILTER

A filter smooths the wave and reduces the ripple so the load voltage is a more uniform dc. The simplest type of filter is a capacitor placed directly across the load, as shown in Figure 10-6.

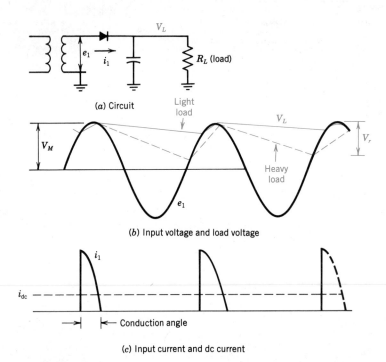

FIGURE 10-6
The half-wave rectifier with a capacitive filter.

10-5.1 Dc and Ripple Voltages in a Capacitive Filter

Figure 10-6 shows the circuit and waveforms for a half-wave rectifier with a capacitive filter. When the applied voltage, e_1, is near its peak, the diode conducts and charges the capacitor to V_{max}. When e_1 starts to decrease, the diode is reverse-biased and the capacitor discharges through the load resistor, R_L.

The waveforms for e_1 and the load voltage are shown in Figure 10-6b. A light load means a high value of R_L and a small current drain from the capacitor. This condition is shown by the solid blue line in Figure 10-6b. A smaller resistor, or heavier load, causes a greater decrease in the voltage, as shown by the dashed blue line. In either case, the capacitor discharges until the time in the next cycle when e_1 becomes greater than the capacitor voltage. Then the diode becomes forward-biased and the capacitor recharges.

While the diode is cut off, the capacitor discharges. The reduction in voltage that occurs until the capacitor charges again is called the ripple voltage, V_r, and is also shown in Figure 10-6b for the heavy load. Actually, this dropping voltage is exponential and given by the equation

$$v_L = V_m e^{-t/RC}$$

where v_L is the load voltage.

In most cases the voltage, v_L, can be approximated by a straight line when the diode is cut off, going from V_m to $V_m - V_r$. The dc or average voltage and current are approximately

$$V_{dc} = V_m - \frac{V_r}{2} \qquad (10\text{-}9)$$

$$I_{dc} = \frac{V_{dc}}{R_L} \qquad (10\text{-}10)$$

Figure 10-7 is a curve of the ratio of V_{dc}/V_m versus ωRC for both the half- and full-wave circuits. Certainly there will be less decay in V_L (V_{dc}/V_m will rise) if

1. ω increases. As the frequency of the applied voltage increases, there will be less time for the capacitor to discharge between voltage peaks.
2. R_L increases. This means the load takes less current.
3. C increases. This increases the ability of the capacitor to store charges.

As V_{dc}/V_m increases, V_r and the ripple factor decrease. A curve of the ripple factor versus ωRC is given in Figure 10-8. The ripple factor can also be approximated by the equation

$$\text{R.F.} = \frac{V_r/2\sqrt{3}}{V_{dc}} \qquad (10\text{-}11)$$

The curves of Figure 10-6 and 10-7 clearly show the advantage of the full-wave circuit over the half-wave circuit.

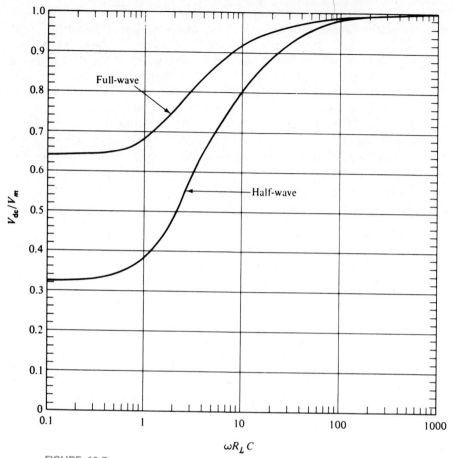

FIGURE 10-7
The ratio V_{dc}/V_m as a function of $\omega R_L C$ for a half-wave and a full-wave rectifier with a shunt capacitor filter. The resistance is in ohms, and the capacitance is in farads. (From *Applied Electronics* by Pierce and Paulus, 1972.)

EXAMPLE 10-4

For a capacitive filter circuit, such as shown in Figure 10-5, assume $e_1 = 10$ V, the applied frequency is 60 Hz, $C = 100$ μF and $R = 100$ Ω. Find V_{dc}/V_m, the ripple factor and V_r for both half-wave and full-wave inputs.

Solution
With the given values

$$\omega RC = 377 \times 100 \times 10^{-4} = 3.77$$

From Figure 10-7 we find $V_{dc}/V_m = 0.63$ for the half-wave circuit and 0.83 for the full-wave circuit.

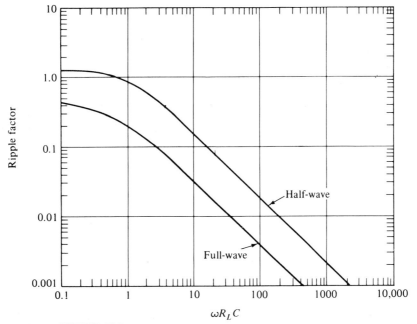

FIGURE 10-8
Ripple factor as a function of $\omega R_L C$ for a half-wave and a full-wave rectifier with a shunt capacitor filter. (From *Applied Electronics* by Pierce and Paulus, 1972.)

The ripple factors can be found from Figure 10-8 and are 0.4 for the half-wave circuit and 0.085 for the full-wave circuit.

The ripple factors can also be found approximately. From equation 10-9

$$V_{dc} = V_m - \frac{V_r}{2}$$

$$6.3 = 10 - \frac{V_r}{2}$$

$$V_r = 7.4 \text{ V}$$

From equation 10-11 we have

$$\text{R.F.} = \frac{V_R/2\sqrt{3}}{V_{dc}} = \frac{\frac{7.4}{2\sqrt{3}}}{6.3} = 0.34$$

This approximation differs from the values on the curves by about 15%.

For the full-wave circuit $V_r = 3.4$ V and

$$\text{R.F.} = \frac{\frac{3.4}{2\sqrt{3}}}{8.3} = 0.118$$

There is about a 30% discrepancy between the approximate equation and the curves.

10-5.2 The Current in the Diode

The diode in a capacitive-filter circuit conducts only when the applied voltage, V_m, is larger than the capacitor voltage. This occurs near the peak of the applied sinusoidal voltage. When the diode is conducting, the source must supply the current in the load and the current needed to recharge the capacitor. When the applied voltage is low or negative, the diode cuts off and the capacitor supplies the load current.

Figure 10-6c shows the current pulse through the diode. As shown in the figure, the time of conduction during each cycle is called the *conduction angle*. The peak value of the diode current is usually several times greater than the dc current, because the current to charge the capacitor while the diode is conducting must equal the capacitor discharge current, which is the current that flows in the load, during the rest of the cycle. Since the conduction angle is smaller than the nonconducting angle, more current must flow in the diode when it is conducting.

In most reasonably regulated circuits, the diode passes maximum current as soon as it turns on, and this current then decays to 0. When the diode is on, and neglecting diode drops, the equation for the diode current is

$$i = C\frac{dv}{dt} + \frac{V}{R} \tag{10-12}$$

Because $v = V_m \sin \omega t$, equation 10-12 becomes

$$i = \omega C V_m \cos \omega t + \frac{V_m \sin \omega t}{R} \tag{10-13}$$

This current is of the form $i = I_m \sin(\omega t + \psi)$, where

$$I_m = V_m \sqrt{\frac{1}{R_L^2} + w^2 C^2} \qquad \psi = \arctan \omega C R_L$$

Figure 10-9 shows the diode current along with the cut-in point, where the diode starts to conduct, and the cut-out point where the diode stops conducting. The cut-in point can be approximately determined by assuming

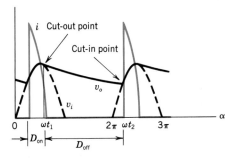

FIGURE 10-9
Theoretical sketch of diode current i and output voltage v_o in a half-wave capacitor-filtered rectifier. (Millman and Halkias, *Electronic Devices and Circuits*, McGraw-Hill, 1967.)

the diode starts to conduct when the input voltage is $V_m - V_r$, as shown in Figure 10-6b.

$$\theta_1 \text{ (cut-in angle)} = \sin^{-1} \frac{V_m - V_r}{V_m}$$

The cut-out angle, θ_2, can be found by setting $i = 0$ in equation 10-13:

$$\theta_2 = -\arctan \omega RC$$

The cut-out angle is $\pi - \arctan \omega RC$.

EXAMPLE 10-5

For the half-wave circuit of Example 10-4, find
a. The cut-in angle.
b. The cut-out angle.
c. The capacitive charging current of cut-in.
d. The diode current when $\omega t = \pi/2$.

Solution
a. In Example 10-4, we found that $V_r = 7.4$ V

$$\theta_{\text{cut-in}} = \sin^{-1} \frac{2.6 \text{ V}}{10 \text{ V}} = 15°$$

b. For cut-out

$$\arctan \omega RC = \arctan 3.77 = 75°$$
$$\theta_2 = 180° - 75° = 105°$$

c. At cut-in the diode current can be determined from equation 10-13 as

$$i = \omega C V_m \cos \omega t + \frac{V_m \sin \omega t}{R}$$

In point a we found that cut-in was at 15°. Therefore

$$i = 377 \times 10^{-4} \times 10 \times 0.965 + \frac{10}{100} \times 0.26$$

$$i = 0.365 \text{ A} + 0.026 \text{ A} = 0.391 \text{ A}$$

Most of the current (0.365 A) is charging the capacitor at this time.

d. At $\omega t = \pi/2$, $dv/dt = 0$, so the capacitor is neither charging nor discharging. The load current is 10 V/100 = 0.1 A. This current is supplied entirely by the source.

10-5.3 Regulation

As the load current in a circuit increases, the load voltage tends to fall off. The basic circuit is shown in Figure 10-10, where V_B, the battery voltage, is assumed to be perfectly constant, but there is also a small internal resistance, R_i, that causes a voltage drop at the output. The load voltage, V_L, is

$$V_L = V_B - IR_i \tag{10-14}$$

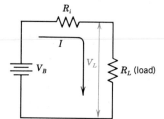

FIGURE 10-10
The basic power supply circuit.

The *percent regulation* is a measure of the load voltage drop as the current increases. The *no-load voltage*, V_{NL}, is the load voltage when the load current is 0, or when R_L is an open circuit. From equation 10-14, we can see that $V_{NL} = V_B$. For any circuit there is a maximum current that can flow. This current is called the *full-load current*, I_{FL}, and the corresponding *full-load voltage* is V_{FL}. The percent regulation, $\%R$, is given by equation 10-15.

$$\%R = \frac{V_{NL} - V_{FL}}{V_{FL}} \times 100 \tag{10-15}$$

Equation 10-14 shows that if the internal resistance, R_i, were 0, the regulation would be perfect (%$R = 0$). It is advantageous to have R_i as small as possible, but it can never be entirely eliminated.

EXAMPLE 10-6

For the half-wave circuit discussed in Example 10-4, $V_m = 10$ V and $V_{dc} = 6.3$ V. Assuming that this circuit is operating at full load, find its %R.

Solution
If the resistor were not present, the capacitor would charge up to V_m, and stay at that voltage. Therefore $V_{NL} = V_m$ and:

$$\%R = \frac{V_{NL} - V_{FL}}{V_{FL}} \times 100 = \frac{10 - 6.3}{6.3} \times 100 = 58.7\%$$

This large %R shows that the regulation of this circuit is poor.

If the resistance or load of a circuit is held constant while the capacitor is increased, the ripple factor and %R both decrease. The only disadvantage of increased capacitance is that the peak current in the diode also increases.

10-5.4 Surge Currents

When power is first applied to a rectifier circuit, the capacitor voltage is 0, and it acts as a short circuit for an instant. The worst-case current at this instant can be V_m/R_T, where R_T is the sum of the small resistance in the transformer and the wires. In some circuits this current surge may damage the diodes, and a small resistor might be needed to reduce the current pulse. In most cases, the circuit will survive without any additional resistance.

Manufacturers specify this current as the *nonrepetitive peak surge current*. The bridge rectifier whose specifications are given in Figure 10-5 is rated at 4 A, but the specifications also show that the nonrepetitive peak surge current is 100 A.

10-6 OTHER FILTERS AND THE VOLTAGE DOUBLER

By adding inductors, resistors, and additional capacitors, the simple capacitive filter of the previous section can be improved, and the ripple factor reduced further. Circuits with more elaborate filters are almost always driven by a full-wave input, usually a bridge rectifier, because it is poor engineering to employ additional components without taking advantage of the smoothing provided by full-wave rectifiers. The disadvantage of these circuits is that they require additional components, and their output dc voltage is often less than V_m.

10-6.1 The L-section Filter

The L-section filter is constructed simply by placing an inductor before the capacitor as shown in Figure 10-11. The inductor smooths out the current

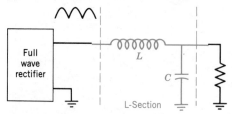

FIGURE 10-11
The L-section filter.

waveform and keeps current flowing at all times. The dc value of the load voltage is the same as the average for a full-wave rectifier, $0.636V_m$, and the ripple factor for this filter with a 60 Hz input is

$$\text{R.F.} = \frac{0.83}{LC} \times 10^{-6}$$

There is a minimum value the inductor must have in order to keep current flowing. It depends on the load resistor and the applied frequency. At 60 Hz the minimum inductance is

$$L \geq \frac{R_L}{1130}$$

EXAMPLE 10-7

For the circuit of Figure 10-11, assume $L = 10$ Hz, $C = 100$ μF, $R = 10$ Ω, and the applied voltage is 10 V (max) at 60 Hz. Find
a. The dc load voltage.
b. The dc load current.
c. The ripple factor.
d. The maximum value that R may change to and still keep the filter operating properly.

Solution

a. The dc load voltage is $0.636\ V_m$, or 6.36 V.
b. The dc current is simply V_{dc}/R_L or 636 mA.
c. The ripple factor is

$$\text{R.F.} = \frac{0.83}{LC} \times 10^{-6} = \frac{0.83}{10 \times 10^{-4}} \times 10^{-6} = 0.00083$$

d. The maximum value that R may change to before the inductor fails to operate properly is $1130R_L$ or $11{,}300\ \Omega$.

Part d of Example 10-7 shows that the circuit will function properly if the load resistor is increased to $11{,}300\ \Omega$. This corresponds to a load current of about 0.5 mA. Some circuits use a *bleeder resistor* across the load to be sure there is always enough load current for the inductor. In this circuit, the bleeder resistor would be $11{,}300\ \Omega$.

10-6.2 The Pi Filter
The pi filter is shown in Figure 10-12, and is essentially a capacitive filter followed by an L-section filter. Because it provides an additional level of filtering at little additional cost, it is often used in place of an L-section filter.

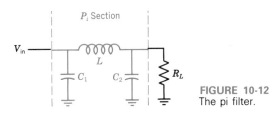

FIGURE 10-12
The pi filter.

10-6.3 The RC Filter
An RC filter is a filtler with the inductor replaced by a resistor, as shown in Figure 10-13. It is less expensive and bulky than the pi filter, but its filtering is not as good, it reduces the output voltage, it absorbs power, and its regulation is poor. This filter is generally limited to use in low-current power supplies.

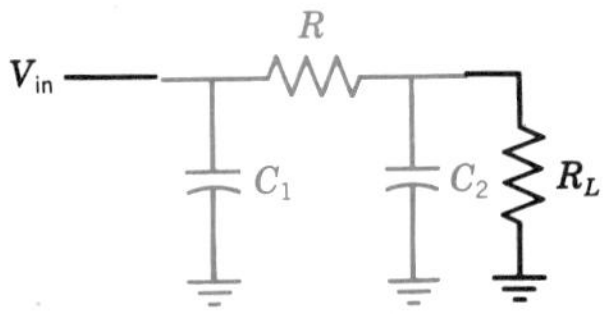

FIGURE 10-13
The RC filter.

10-6.4 Filter Summary
Figure 10-14 is a summary of the characteristics and equations for the various filters. The equations for V_{dc} as a function of V_m and the ripple factor are given. The simplified equations for the most common input frequency (60 Hz) are also given.

Comparison of filters.*

Filter	V_{dc}	RF
Shunt-Capacitance	$\dfrac{V_{s(max)}}{1 + 1/(4fR_LC)}$	$\dfrac{1}{4\sqrt{3}fR_LC}$
(Rectifier FW, C, R_L)	For 60 Hz: $\dfrac{V_{s(max)}}{0.00417/R_LC}$	For 60 Hz: $\dfrac{2.41 \times 10^{-3}}{R_LC}$
Pi	$\dfrac{V_{s(max)}R_L/(R_L + r_L)}{1 + 1/(4fR_LC_1)}$	$\dfrac{\sqrt{2}}{\omega^3 C_1 C_2 L R_L}$
(Rectifier FW, C_1, L, C_2, R_L)	For 60 Hz: $\dfrac{V_{s(max)}R_L/(R_L + r_L)}{1 + 0.00417/R_LC}$	For 60 Hz: $\dfrac{0.026}{C_1 C_2 R_L L} \times 10^{-6}$
RC	$\dfrac{V_{s(max)}R_L/(R_L + R)}{1 + 1/(4fR_LC_1)}$	$\dfrac{\sqrt{2}}{\omega^2 C_1 C_2 R_L R}$
(Rectifier FW, C_1, R, C_2, R_L)	For 60 Hz: $\dfrac{V_{s(max)}R_L/(R_L + R)}{1 + 0.00417/R_LC}$	For 60 Hz: $\dfrac{9.95}{C_1 C_2 R_L R} \times 10^{-6}$
L-Section	$0.636 V_{s(max)}$	$\dfrac{0.118}{(\omega^2 LC)}$
(Rectifier FW, L, C, R_L)	For 60 Hz: $0.636 V_{s(max)}$	For 60 Hz: $\dfrac{0.83}{LC} \times 10^{-6}$

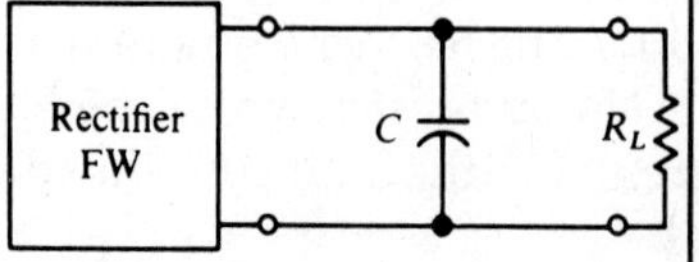

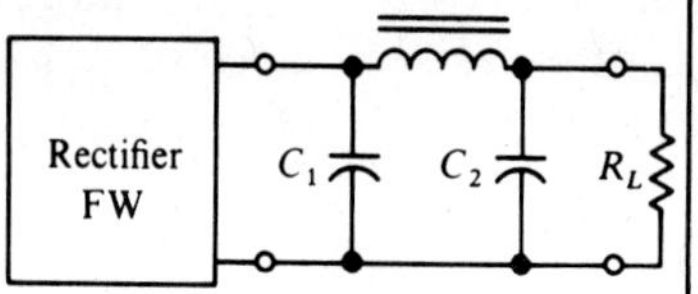

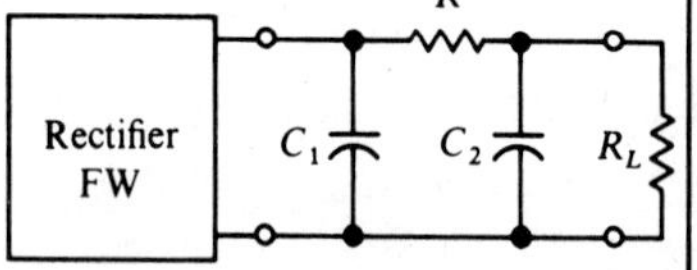

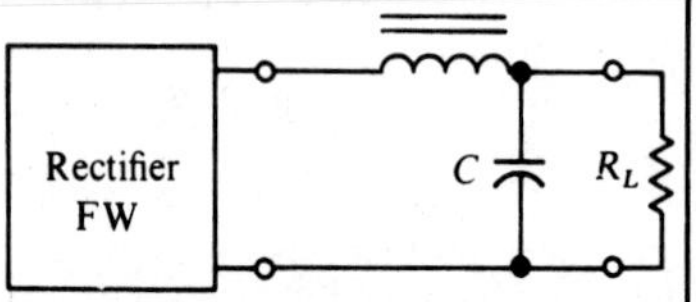

*Equations are given for any line frequency and for a 60-Hz frequency.

FIGURE 10-14
Comparison of filters. (From *Electronics* by Arthur Seidman and Jack Waintraub.)

10-6.5 The Voltage Doubler

By properly using capacitors, the output voltage can be doubled (be made twice V_m) or even multiplied further. Figure 10-15 shows the construction of a *voltage doubler*. When e_1 is positive, as shown in black, capacitor C_1 charges to

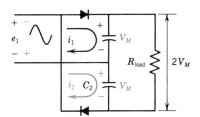

FIGURE 10-15
The voltage doubler.

V_m. When e_1 is negative, as shown in blue, C_2 charges to V_m. The voltage across the load is the sum of these voltages, or $2V_m$.

Voltage doublers have two disadvantages; their ripple and regulation are generally poor. Besides, the desired output voltage can often be obtained by using transformers. As a result, voltage doublers are used only for specialized, low-current circuits.

10-7 VOLTAGE REGULATORS

Many electronic circuits require a constant power supply output voltage regardless of variations in the current they draw. TTL integrated circuits, for example, must have a supply voltage between 4.75 and 5.25 V. Power supplies for these circuits must be very stable; they are nominally 5 V and can only vary by 0.25 V.

A *regulator* is an electronic circuit whose function is to maintain a constant output voltage at the power supply. Unfortunately, these regulators also dissipate power and the efficiency of a regulator can be expressed as

$$\eta = \frac{P_{\text{load}}}{P_{\text{in}}} \tag{10-16}$$

where P_{load} is the power in the load and P_{in} is the input power to both the regulator and the load. Despite their power consumption, regulators are required for most electronic circuits.

10-7.1 The Zener Regulator

The simplest type of regulator is the Zener-diode type. Zener diodes were introduced in section 1-10.1 and are reviewed briefly here. The characteristics of a Zener diode are shown in Figure 10-16a. Zeners are almost always reverse-biased. When the reverse voltage is less than the rated Zener voltage, V_Z, the diode does not conduct. When the reverse voltage reaches V_Z, the diode conducts and the voltage across the Zener is V_Z and is basically independent of the current in it.

Because the Zener diode maintains a constant voltage across itself, it can be placed in parallel with a load, and it will hold the load voltage constant. Figure 10-16b shows a Zener diode-regulating circuit. The ac input is rectified by the bridge and filtered by the capacitor to provide an unregulated dc voltage that is subject to variations due to changes in the input ac voltage and

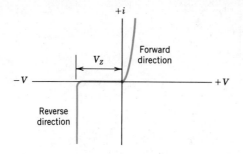

(a) Zoner characteristics

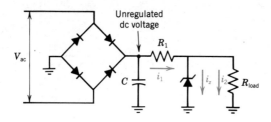

(b) Zoner circuit

FIGURE 10-16
The Zener-diode regulator circuit.

to ripple. The current I_1 flows through R_1 and divides into the Zener current, I_Z, and the load current, I_L. The voltage is V_Z and is approximately constant.

Resistor R_1 in Figure 10-16b is a necessary evil. It dissipates power, but it is needed to reduce the dc input voltage to the load voltage. R_1 should be as large as possible to limit the current and power out of the source, but the following inequality must be satisfied:

$$\frac{V_{in} R_{load}}{R_{load} + R_1} \geq V_Z \tag{10-17}$$

If it is not, the voltage across the load will be less than V_Z, the Zener will not conduct, and there will be no voltage regulation. The inequality leads to a maximum value for R_1

$$\frac{R_{load}(V_{in} - V_Z)}{V_Z} > R_1 \tag{10-18}$$

EXAMPLE 10-8

In Figure 10-16b, the Zener diode is rated at 10 V, and the load is 10 Ω. If the dc input voltage varies between 15 and 20 V, find

a. The best value for R_1.
b. The currents when $V_{in} = 15$ V.
c. The currents when $V_{in} = 20$ V.
d. The efficiency in each case.

Solution

a. To find R_1, we can use equation 10-18:

$$R_1 < \frac{R_{load}(V_{in} - V_Z)}{V_Z} = \frac{10 \times (15 - 10)}{10} = 5 \, \Omega$$

b. When V_{in} is 15 V, the current in R_1 is

$$I_1 = \frac{(V_{in} - V_Z)}{R_1} = \frac{15 \text{ V} - 10 \text{ V}}{5 \, \Omega} = 1 \text{ A}$$

The load current is also 1 A, and the Zener is just on the verge of turning on. In practical circuits, R_1 should be reduced slightly to provide some current in the Zener diode.

c. When V_{in} is 20 V, $I_1 = 2$ A. Now 1 A flows in the load and 1 A flows in the Zener. The Zener must be capable of handling 1 A, and also capable of dissipating 10 W. A heat sink may be needed.

To find the efficiency, we realize that the load voltage is constant, so the load current is also constant and

$$P_L = V_L I_L = 10 \text{ V} \times 1 \text{ A} = 10 \text{ W}$$

When $V_{in} = 15$ V, $P_{in} = 15$ V $\times$ 1 A = 15 W, and

$$\eta = \frac{10 \text{ W}}{15 \text{ W}} = 66.7\%$$

When $V_{in} = 20$ V, $P_{in} = 20$ V $\times$ 2 A = 40 W, and

$$\eta = \frac{10 \text{ W}}{40 \text{ W}} = 25\%$$

The efficiency varies greatly depending on the input voltage. In this case the load dissipated 10 W. The resistor, R_1, must dissipate the other 20 W. Therefore it must be physically large, and it should be well ventilated.

10-7.2 Secondary Effects

The circuit of Example 10-8 assumed an ideal Zener diode. For precise analysis, secondary effects on the Zener should be considered. These cause small variations in the Zener voltage and the output voltage.

One secondary effect is the *dynamic resistance* of the diode. A Zener may be represented as a constant dc voltage source in series with a small internal resistance. As the current in the Zener increases, the voltage across it will also increase. Unfortunately, high load circuits require large currents and small load resistances. In these cases the Zener resistance may be appreciable and may prevent the Zener from regulating as well as it should.

Another secondary effect is the variation of the Zener voltage with temperature. Most Zener diodes tend to have a positive temperature coefficient, which means the output voltage will rise as their temperature rises. If the change in Zener output voltage with temperature is unacceptable, there are various compensation circuits that will reduce the voltage change. Perhaps the simplest of them is shown in Figure 10-17. It consists merely of placing a forward-biased diode in series with the Zener. The effective Zener voltage is now $V_z + V_d$ ($\approx V_z + 0.7$ V), but the forward-biased diode has a negative temperature coefficient so that, as the temperature rises, the voltage across the diode decreases while the voltage across the Zener increases, and the changes tend to balance each other.

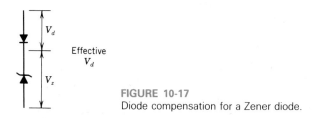

FIGURE 10-17
Diode compensation for a Zener diode.

Many manufacturers specify the Zener temperature coefficient as %/°C, or the percent of the Zener voltage for each degree C rise in temperature.

EXAMPLE 10-9

The **1N2808** is a 10 V, 50 W Zener diode manufactured by Motorola, Inc. Its temperature coefficient is 0.055 %/°C. If its case temperature changes by 40°C, by how much will its output voltage change?

Solution
The voltage change of the diode is .055%/°C of the 10 V. To find the voltage change per °C, this must be multiplied by the 10 V rating of the Zener.

$$\frac{dV}{dT} = 10 \text{ V} \times 0.055\%/°C = 10 \text{ V} \times 0.00055/°C = 5.5 \text{ mV}/°C$$

Thus a change of 40°C results in a change of 220 mV in the output voltage. For many applications, this large a change is prohibitive.

To help overcome this problem, some manufacturers provide *temperature-compensated* Zener diodes that have very low temperature coefficients.

10-7.3 Disadvantages of the Zener Regulator

The Zener regulator is simple and inexpensive, and it is used in some circuits, but it has several disadvantages as listed here:

1. It is often inefficient. The power dissipated in R_1 is wasted and may be too high.
2. The Zener is subject to wide changes in current, which may change its output voltage. In Example 10-8, the Zener current varied from 0 to 1 A.
3. The output voltage is fixed at the Zener voltage and not adjustable.
4. The internal resistance of the Zener may lead to a larger voltage change than can be tolerated.
5. The change of the Zener voltage with temperature may be unacceptable.

Transistor regulators, described in the next section, are used when more effective regulators are required.

10-8 TRANSISTOR VOLTAGE REGULATORS

In order to achieve a precise and highly regulated output voltage, most modern power supplies use a transistor voltage regulator. In this type of regulator, a portion of the output voltage is fed back through a transistor amplifier to control the output voltage and current.

Figure 10-18 is a block diagram of a regulated power supply. The power elements are shown in black and the control elements are shown in blue. The power elements are an unregulated dc power source, a pass element, and the load. The unregulated power source typically consists of a transformer (to reduce the ac line voltage), a bridge rectifier, and a large capacitor for filtering. The pass element is a power transistor or power Darlington; the load current

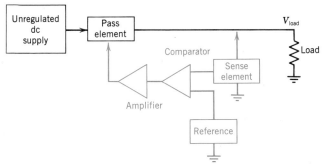

FIGURE 10-18
A block diagram of an electronic voltage regulator.

flows through the pass element. The load is an amplifier, a motor, or another electronic circuit. It is represented by a resistor in Figure 10-18.

The control elements, shown in blue, are a sense element, a reference, a comparator, and an amplifier. The sense element is typically a resistive divider circuit, whose output is a fixed portion of the load voltage. The reference is a constant dc voltage. Typically, the constant dc voltage is obtained by using a Zener diode. Any difference between the output of the sense element and the reference voltage is detected by the comparator, amplified by the amplifier, and used to control the pass element so that the load voltage remains constant. This is basically a feedback circuit (see Chapter 8). Any change in the nominal output voltage is fed back and the input is adjusted so that the change is minimized.

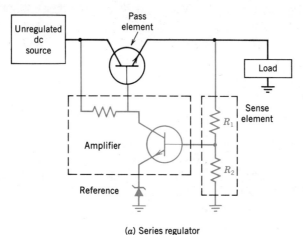

(a) Series regulator

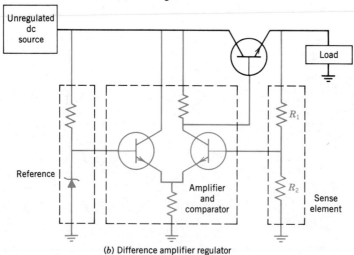

(b) Difference amplifier regulator

FIGURE 10-19
Two types of voltage regulators.

Figure 10-19 shows two types of commonly used voltage regulators. The series regulator of Figure 10-19a uses a single transistor with a Zener diode in its emitter circuit for reference. The circuit of Figure 10-19b uses a difference amplifier (see section 6-9). The reference voltage established by the Zener diode is connected to the base of one transistor and the sense output is connected to the base of the other transistor. Any variation in the sense output is amplified by the difference amplifier and affects the pass transistor. Both circuits use a pass transistor connected to the load in an emitter-follower configuration, a resistive divider for a sense element, and a Zener diode for a reference. In either circuit the current through the Zener diode is much less than it is in the Zener regulators. In the difference amplifier, this current is practically constant, but even in the series regulator its variation is small.

EXAMPLE 10-10

In Figure 10-19 (a or b), if R_2 is 10 kΩ, and the reference voltage is 10 V, find the value of R_1 if the load voltage (V_L) is to be
a. 15 V
b. 20 V
c. 25 V

Solution
In either circuit the output of the resistive divider must equal the reference voltage. This output must therefore be 10 V, which is across R_2. If the base currents of the transistor are neglected (a reasonable first assumption) R_1 and R_2 form a voltage divider and

$$V_{ref} = \frac{V_{load} \, R_2}{R_1 + R_2}$$

Using this equation, if $V_L = 15$ V, $R_1 = 5$ kΩ; if $V_L = 20$ V, $R_1 = 10$ kΩ; and if $V_L = 25$ V, $R_L = 15$ kΩ. This example shows that the output voltage can be adjusted by changing R_1. Note also that, for a 10 V reference, the series regulator would require a 9.3 V Zener, because the reference is the voltage across the Zener plus the base-to-emitter drop of the transistor, but the difference amplifier would require a 10 V Zener.

10-8.1 Analysis of the Series Regulator
The operation of a series regulator can be precisely analyzed using the current and voltage equations developed in the previous chapters. The circuit is shown in Figure 10-20. The input is represented by an ideal, though unregulated, source, V_S, in series with an internal impedance represented by R_4. The transistor of Figure 10-19 has been replaced by a power Darlington (see section 9-9.3). The current gain of Q_1 is h_{fe1}, the current gain of the Darlington

is h_{fe2}, and its base-to-emitter voltage is V_{BE}. The currents in R_1 and R_2 are $I_1 + I_B$, and I_1, respectively, and the current in R_3 is called I_R. Because this is a power supply, I_L is assumed to be in the order of amperes, while I_1 and I_R are about 1 mA or less. The equations will be simplified by neglecting these currents with respect to I_L.

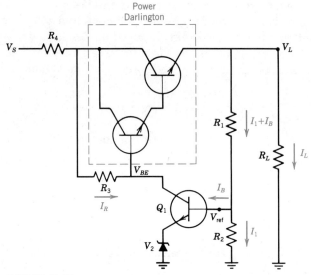

FIGURE 10-20
Detail of the series regulator.

Several equations can be derived from Figure 10-20. Considering the voltage drops across R_1 and R_2, we have

$$V_L = (I_1 + I_B) R_1 + I_1 R_2$$

But

$$V_{ref} = I_1 R_2$$

$$V_L = \left(\frac{V_{ref}}{R_2} + I_B\right) R_1 + V_{ref}$$

or

$$I_B = \frac{V_L - V_{ref}\left(1 + \frac{R_1}{R_2}\right)}{R_1} \qquad (10\text{-}19)$$

It can also be seen that

$$I_R = h_{fe1}I_B + \frac{I_L + I_1}{h_{fe2}}$$

If $I_1 \ll I_L$ this becomes

$$I_R = h_{fe1}I_B + \frac{I_L}{h_{fe2}} \quad (10\text{-}20)$$

Taking a voltage loop through the source, the load, and the Darlington, we have

$$V_S - (I_R + I_L)R_4 - I_R R_3 = V_{BE} + V_L$$

Assuming $I_R \ll I_L$

$$V_S - I_L R_4 - I_R R_3 = V_{BE} + V_L \quad (10\text{-}21)$$

and, of course

$$I_L = \frac{V_L}{R_L} \quad (10\text{-}22)$$

Equations 10-19, 10-20, 10-21, and 10-22 can be combined to eliminate I_B, I_R, and R_L. After some algebra we obtain

$$V_S + h_{fe1}\frac{R_3}{R_1}\left[V_{ref}\left(1 + \frac{R_1}{R_2}\right)\right] - V_{BE}$$
$$= V_L\left[1 + \frac{R_3}{R_1}h_{fe1}\right] + I_L\left[\frac{R_3}{h_{fe2}} + R_4\right] \quad (10\text{-}23)$$

We define equation 10-23 as the equation for the circuit. It is of the form $AV_L + BI_L = C$ where A, B, and C are constants that depend upon the parameters of the circuit. Once the equation is in this form it tells the user everything about the operation of the circuit. It applies whether a single power transistor or a Darlington is used. For a single transistor, $V_{BE} \approx 0.7$ V and h_{fe1} varies from about 20 to 100. For a Darlington, $V_{BE} \approx 1.4$ V and h_{fe1} can be several thousand.

EXAMPLE 10-11

For the circuit of Figure 10-21, the specifications for the Darlington are a current gain of 4000 and $V_{BE} = 1.4$ V. The transistor has a current gain of 100.

Find
a. The equation for the circuit.
b. The % regulation if $I_{FL} = 2$ A.
c. The internal impedance of the regulator.
d. The output voltage at $I = 2$ A if V_S becomes 35 V.
e. The power dissipated in the Darlington under these conditions.

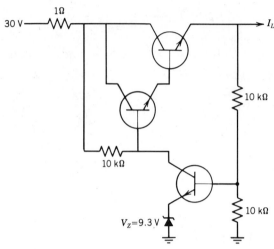

FIGURE 10-21
Circuit for Example 10-11.

Solution
a. Using equation 10-23, with $V_{ref} = 10$ V, we have

$$h_{fe1} \frac{R_3}{R_1}\left[V_{ref}\left(1 + \frac{R_1}{R_2}\right)\right] = 100 \times \frac{10\text{ k}\Omega}{10\text{ k}\Omega} \times 10\text{ V}\left(1 + \frac{10\text{ k}\Omega}{10\text{ k}\Omega}\right)$$

$$= 2000$$

$$30 + 2000 - 1.4 = V_L(101) + I_L\left(\frac{10{,}000}{4000} + 1\right)$$

This reduces to

$$101\, V_L + 3.5\, I_L = 2028.6 \qquad (10\text{-}24)$$

This is the equation for the circuit.

b. Percent regulation was defined by equation 10-15 (section 10-5.3) as

$$\%\,R = \frac{V_{NL} - V_{FL}}{V_{FL}} \times 100$$

492 POWER SUPPLIES

The no load voltage occurs when $I_L = 0$. Then equation 10-24 becomes

$$101 \, V_L = 2028.6 \qquad V_{NL} = 20.085 \text{ V}$$

At full load $I = 2$ A and the equation for the circuit becomes

$$101 \, V_L = 2021.6 \qquad V_{FL} = 20.016$$

$$\%R = \frac{20.085 - 20.016}{20.016} \times 100 = \frac{0.069 \times 100}{20.016} = 0.345\%$$

c. The internal impedance of the regulator is dV/dI. Differentiating equation 10-24, we have

$$\frac{dV}{dI} = \frac{3.5}{101} = 0.0346 \, \Omega$$

d. If V_S goes to 35 V (perhaps due to a variation in the input ac voltage), the additional 5 V is simply added to the right side of the equation and it becomes

$$101 \, V_L + 3.5 \, I_L = 2033.6$$

At

$$I_L = 2 \text{ A} \qquad V_{FL} = \frac{2026.6}{101} = 20.065$$

Thus a 5 V change in the input voltage resulted in a 0.049 V (20.065 − 20.016) change in the output voltage. This is a measure of *line regulation*, the change of load voltage due to a change of input voltage.

e. To find the power dissipated by the Darlington at $V_S = 35$ V and $I_L = 2$ A, the drop across R_4 is 2 A × 1 Ω = 2 V, so the collector voltage of the Darlington is 33 V. The load voltage is approximately 20 V, so the Darlington dissipates

$$P = 13 \text{ V} \times 2 \text{ A} = 26 \text{ W}$$

This is not an excessive amount of power, but the Darlington will require a heat sink.

This example shows that the regulator is highly effective, its impedance is low, and it maintains an output of about 20 V despite variations in the load current and the input voltage.

10-8.2 Preregulation

Many power supplies use a *preregulator* to further improve the regulation. A preregulator replaces R_3 with a constant current source, as shown in Figure 10-22. The constant current source (see section 6.7) consists of the Zener diode, V_{Z2}, the *pnp* transistor Q_2, and R_E.

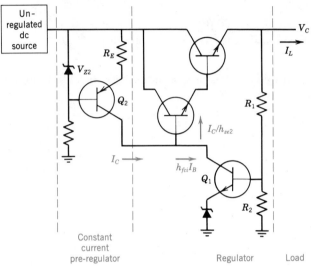

FIGURE 10-22
Regulator with a constant current preregulator.

The analysis of a power supply with a preregulator is very simple. It relies on the fact that the constant current, I_C, equals the current in the Darlington, plus the collector current in Q_1.

$$I_C = \frac{I_L}{h_{fe2}} + h_{fe1} I_B \tag{10-25}$$

If equation 10-19 is used to eliminate I_B, the equation for this circuit becomes

$$V_{ref}\left(1 + \frac{R_1}{R_2}\right) + \frac{R_1 I_C}{h_{fe1}} = V_L + \frac{R_1 I_L}{h_{fe1} h_{fe2}} \tag{10-26}$$

The value of I_C is typically chosen to be slightly more than the base current required by the Darlington at full load, plus a reasonable collector current in Q_1, perhaps 1 mA.

EXAMPLE 10-12

Assume the circuit of Figure 10-22 has the same values as Figure 10-21, and that the constant current is selected to be 2 mA.
a. If $V_{Z2} = 4.7$ V, find R_E.
b. Find the regulation of this circuit.

Solution
a. With $V_{Z2} = 4.7$ V, the voltage across R_E is 4 V. To provide 2 mA across 4 V a 2 kΩ resistor is needed.
b. In this circuit equation 10-26 becomes

$$20 + 100\, I_C = V_L + \frac{I_L}{40}$$

At no load $I_L = 0$ and we have

$$20.2 = V_{NL}$$

At full load (2 A) this equation becomes

$$20.2 = V_{FL} + 0.05$$
$$V_{FL} = 20.15$$

$$\%\text{Regulation} = \frac{20.2 - 20.15}{20.15} \times 100 = \frac{0.05}{10.15} \times 100 = 0.248\%$$

This is about 30% less than the regulation found in Example 10-11, where the circuit did not have a preregulator.

10-8.3 Overvoltage Protection

The circuits that are driven by regulated power supplies should be protected against power supply malfunctions. One possible malfunction is *overvoltage,* whereby the voltage of the regulator suddenly increases. In Figure 10-21 or Figure 10-22, for example, if the pass transistor shorts out, the 30 V input will be applied to the load instead of the 20 V regulated output, and the results may be devastating. Overvoltage is particularly dangerous when a power supply is driving a set of IC boards. ICs are not generally fragile, but they are sensitive to overvoltage, and a malfunctioning power supply can destroy hundreds of ICs.

Perhaps the most common method of protecting circuits against overvoltage is the combination of the fuse and crowbar shown in Figure 10-23a. A fuse is not effective protection for an electronic circuit; it takes too long to blow. Many electronic components will be destroyed before the fuse goes.

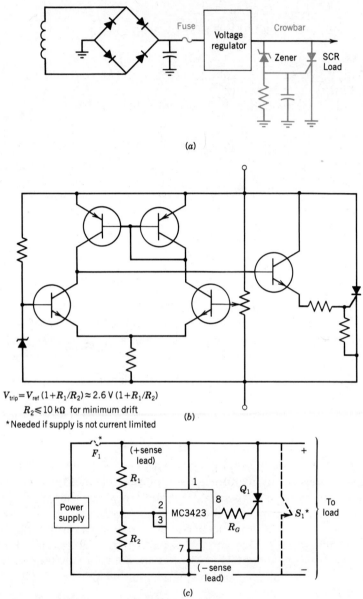

$V_{trip} = V_{ref}(1+R_1/R_2) \approx 2.6\,\text{V}\,(1+R_1/R_2)$
$R_2 \leq 10\,\text{k}\Omega$ for minimum drift
*Needed if supply is not current limited

FIGURE 10-23
A "crowbar" overvoltage protection circuit.

The *crowbar* circuit, shown in blue in Figure 10-23a, is designed to react quickly and protect the load against an overvoltage condition. It consists of an SCR (silicon controlled rectifier), a Zener diode, and a capacitor. If the voltage on the gate of the SCR becomes large enough, the SCR *fires,* which means that it becomes essentially a short circuit, and remains that way until the dc voltage across it is removed. The Zener diode is adjusted so that the gate voltage of the SCR is slightly less than the firing voltage. Under normal operation the SCR is an open circuit and does not affect the operation. The capacitor is used to prevent short spikes or glitches from triggering the capacitor. Should the output voltage of the regulator become too high, the gate voltage on the SCR will increase, and the SCR will fire, placing a short circuit across the load to protect it. Under these conditions, however, a large current flows through the fuse, the pass transistor, and the SCR. The fuse should blow to protect these components.

The circuit of Figure 10-23a is perhaps the simplest crowbar circuit, and illustrates the principles behind crowbar protection, but Motorola[1] warns against using it because it provides poor gate drive and the trip point cannot be adjusted except by changing the Zener diode. Figure 10-23 shows a more sophisticated discrete sense circuit that uses a difference amplifier to turn on the SCR. Motorola also manufactures an IC, the **MC3423** that is designed to drive an SCR gate. A circuit using this IC is shown in Figure 10-23c. The resistors R_1 and R_2 sense the output and become the input to the **MC3423**. Its output controls the gate of the SCR.

10-8.4 Current Limiting

Trouble can also occur if the current drain becomes excessive. This could happen if the load demands too much current, or, in worst case, shorts out. Excessive current can raise the power in the pass transistor and eventually destroy it. One of the most common causes of power supply failures is blown pass transistors.

Current-limiting circuits are often used to prevent the output current of a regulator from becoming excessive. One variation of the current-limiting circuit is shown in Figure 10-24a. Its major components are transistor, Q_4, and the variable resistor, R_4. During normal operation Q_4 is saturated and is approximately a short circuit. Should the load current become excessive, the voltage drop across R_4 will rise and drive Q_4 out of saturation. Its impedance then becomes significant, and it limits the current in the circuit. Figure 10-24b shows the current in this circuit. The current limiting prevents the current in this circuit from exceeding 0.75 A.

---

[1]From the Motorola *Linear/Switchmode Voltage Regulator Handbook* (see references).

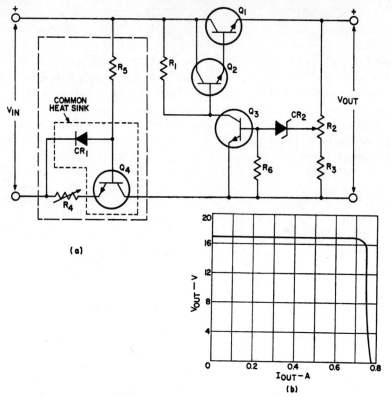

FIGURE 10-24
Series voltage regulator with transistor current-limiting circuit (inside dashed lines) added: (a) schematic diagram; (b) response characteristics. (GE/RCA Solid State Division.)

10-9 INTEGRATED CIRCUIT VOLTAGE REGULATORS

Voltage regulators are available in integrated-circuit (IC) packages from many manufacturers. These are called *three-terminal regulators* because they have only an input terminal, an output terminal, and a ground terminal. A typical circuit is shown in Figure 10-25. It takes a poorly regulated input voltage and transforms it into a highly regulated output voltage. The manufacturers also recommend the use of an input capacitor, C_{in}, and an output capacitor, C_{out}. C_{in} is used to damp out any transients on the line between the power supply

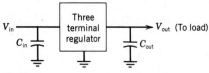

FIGURE 10-25
The basic three-terminal regulator.

and the regulator. A typical value for C_{in} is 1 µF. C_{out} is used to reduce load transient spikes, such as are often created by fast switching digital logic, and is typically 0.1 µF or greater.

10-9.1 Characteristics of IC Voltage Regulators

A user selects a voltage regulator primarily to satisfy the voltage and current requirements of the load. One typical voltage regulator is the **LM123**, manufactured by National Semiconductor, Inc. It provides a regulated 5 V output at load currents up to 3 A. Figure 10-26 is two pages of the specifications for the **LM123**. The specifications give the circuit diagram, which is far too complex for analysis in this book, the connection diagram, maximum ratings, and typical characteristics.

Besides the output voltage and current, many other voltage regulator characteristics are also important. They are

1. *The input voltage* V_{in}. The input voltage to a regulator can be a varying dc voltage. Typically its minimum value is 2 to 3 V above V_{out}. The maximum input voltage is also specified. For the **LM123**, V_{in} must be between 7.5 V and 20 V.
2. *The line regulation.* This is the variation of the output voltage with respect to changes in the input voltage. For an **LM123**, if the input voltage changes from 7.5 V to 15 V, the output will typically change by 5 mV.
3. *The load regulation.* This is the variation of the output voltage as the current changes. For a typical **LM123** this is 25 mV, when the load current goes from 0 to 3 A.
4. *The output impedance.* This should be as small as possible. For an **LM123**, it is 0.01 Ω.
5. *Power dissipation.* For most IC regulators, the input current is approximately the same as the output current. Therefore the power dissipation in the regulator is

$$P_D = (V_{in} - V_{out})I$$

This can be significant. If an **LM123** is operating at full current with an input of 10 V, for example, it must dissipate 15 W. The **LM123** comes in a metal case that can be used with a heat sink to help dissipate this power.
6. *Protection.* The **LM123** is internally protected against current and thermal overloads. It will shut itself down if the current or thermal limitations are exceeded; therefore it is virtually blowout-proof.

10-9.2 Fixed and Adjustable Voltage Regulators

The **LM123** is an example of a positive, fixed-voltage regulator. There are many fixed-voltage regulators available for a variety of voltages and currents. Negative voltage regulators are also available.

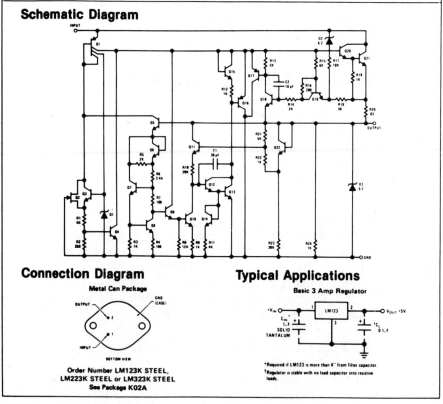

FIGURE 10-26
The **LM123** voltage regulator. (From the *Linear Databook* published by National Semiconductor, Inc., 1982.)

Absolute Maximum Ratings

Input Voltage	20V
Power Dissipation	Internally Limited
Operating Junction Temperature Range	
LM123	−55°C to +150°C
LM223	−25°C to +150°C
LM323	0°C to +125°C
Storage Temperature Range	−65°C to +150°C
Lead Temperature (Soldering, 10 sec)	300°C

Preconditioning

Burn-In in Thermal Limit — 100% All Devices

Electrical Characteristics (Note 1)

PARAMETER	CONDITIONS	LM123/LM223 MIN	TYP	MAX	LM323 MIN	TYP	MAX	UNITS
Output Voltage	$T_j = 25°C$, $V_{IN} = 7.5V$, $I_{OUT} = 0$	4.7	5	5.3	4.8	5	5.2	V
Output Voltage	$7.5V \leq V_{IN} \leq 15V$, $0 \leq I_{OUT} \leq 3A$, $P \leq 30W$	4.6		5.4	4.75		5.25	V
Line Regulation (Note 3)	$T_j = 25°C$, $7.5V \leq V_{IN} \leq 15V$		5	25		5	25	mV
Load Regulation (Note 3)	$T_j = 25°C$, $V_{IN} = 7.5V$, $0 \leq I_{OUT} \leq 3A$		25	100		25	100	mV
Quiescent Current	$7.5V \leq V_{IN} \leq 15V$, $0 \leq I_{OUT} \leq 3A$		12	20		12	20	mA
Output Noise Voltage	$T_j = 25°C$, $10\,Hz \leq f \leq 100\,kHz$		40			40		µVrms
Short Circuit Current Limit	$T_j = 25°C$, $V_{IN} = 15V$, $V_{IN} = 7.5V$		3 4	4.5 5		3 4	4.5 5	A A
Long Term Stability				35			35	mV
Thermal Resistance Junction to Case (Note 2)			2			2		°C/W

Note 1: Unless otherwise noted, specifications apply for $-55°C \leq T_j \leq +150°C$ for the LM123, $-25°C \leq T_j \leq +150°C$ for the LM223, and $0°C \leq T_j \leq +125°C$ for the LM323. Although power dissipation is internally limited, specifications apply only for $P \leq 30W$.

Note 2: Without a heat sink, the thermal resistance of the TO-3 package is about 35°C/W. With a heat sink, the effective thermal resistance can only approach the specified values of 2°C/W, depending on the efficiency of the heat sink.

Note 3: Load and line regulation are specified at constant junction temperature. Pulse testing is required with a pulse width $\leq 1\,ms$ and a duty cycle $\leq 5\%$.

Typical Applications (cont'd.)

Adjustable Output 5V − 10V 0.1% Regulation

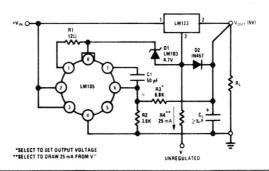

*SELECT TO SET OUTPUT VOLTAGE
**SELECT TO DRAW 25 mA FROM V⁻

FIGURE 10-26 continued

Manufacturers can also supply *adjustable voltage regulators,* where the output voltage is adjustable by resistors connected to various pins on the regulator. Figure 10-27 shows the **MC1723** adjustable voltage regulator designed to deliver 30 mA at 15 V. Equations for calculating R_1, R_2, R_3, and R_{SC} are also given. R_{SC} is the resistor used to set the current limiting; its value depends on the short-circuit current (maximum rated current) of the regulator.

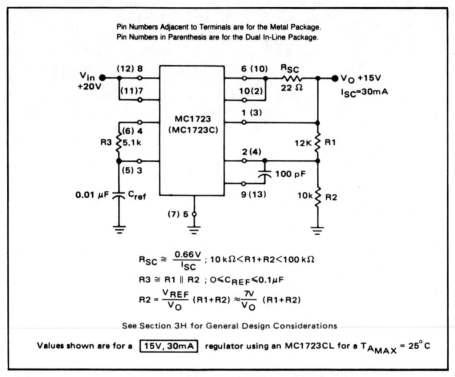

FIGURE 10-27
MC1723 basic circuit configuration for $V_{ref} \leq V_o \leq 37V$. (Copyright by Motorola, Inc. Used by permission.)

EXAMPLE 10-13

A 20 V, 25 mA regulator is to be designed using an **MC1723**. If R_2 is selected as 10 kΩ, find R_1, R_3, and R_{SC}.

Solution
First R_1 can be found. From Figure 10-27 we see

$$R_2 = \frac{7\,V}{V_o}(R_1 + R_2)$$

$$R_1 + R_2 = 10 \text{ k}\Omega \times \frac{20 \text{ V}}{7 \text{ V}} = 28.6 \text{ k}\Omega$$

$$R_1 = 18.6 \text{ k}\Omega$$

$$R_3 \approx R_1 \parallel R_2 = 10 \text{ k}\Omega \parallel 18.6 \text{ k}\Omega = 6500 \text{ }\Omega$$

$$R_{SC} = \frac{0.66 \text{ V}}{I_{SC}} = \frac{0.66 \text{ V}}{0.025 \text{ A}} = 26.4 \text{ }\Omega$$

The *tracking regulator* is another type of adjustable voltage regulator. It provides both a positive and a negative output voltage of equal magnitude. The tracking feature means that as the output voltage is varied, the positive and negative voltages will track; they will always remain equal in magnitude.

10-9.3 Boost Regulators

Because many adjustable voltage regulators cannot supply the current needed in higher power circuits, they are often used as *boost regulators*, where they provide the proper voltage to the base of a *pass transistor* that supplies the current to the load and absorbs the power. Figure 10-28 shows a boost circuit,

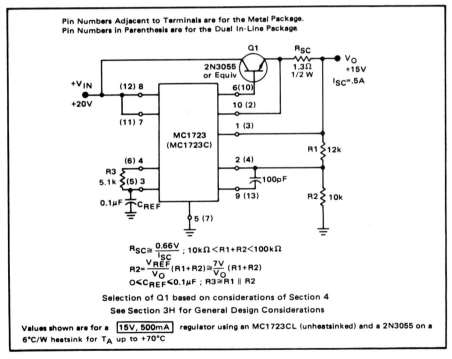

FIGURE 10-28
MC1723C *npn* boost configuration. (Copyright by Motorola, Inc. Used by permission.)

where an **MC1723** drives a **2N3055** power pass transistor. The **2N3055** must be mounted on a heat sink to dissipate the power it absorbs.

EXAMPLE 10-14

For Figure 10-28 find the maximum power that must be absorbed by the pass transistor.

Solution
The output is given as 15 V at 0.5 A. Therefore

$$P_D = (V_{in} - V_{out})I = 5 \text{ V} \times 0.5 \text{ A} = 2.5 \text{ W}$$

10-9.4 Practical Voltage-Regulator Circuits

One example of the use of three-terminal regulators is the S-100 bus (also called the IEEE-696 bus), which has 100 pins on it. This is a bus connected to a backplane or motherboard, and logic cards are plugged into the backplane and communicate with the bus. The bus is powered by $+8$, $+16$, and -16 V power supplies. The $+8$ V comes in to the cards on pins 1 and 51, $+16$ V comes in on pin 2, and -16 V comes in on pin 52.

In typical operation, most cards contain several integrated circuits (ICs) that require 5 V. Each of these cards must also have a three-terminal regulator mounted on it that takes the $+8$ V from the bus and regulates it down to the 5 V for the ICs.

A second example is the use of a three-terminal regulator in a basic power supply, as shown in Figure 10-29. The line voltage is stepped-down by the transformer, rectified by the bridge, and regulated. The problem with this circuit is that the output of the bridge is a half sine wave that goes to 0 at 0° and 180°. C_{in} must be large enough to filter this voltage so that it never goes below the minimum required for regulation. Now C_{in} must be large. It can be calculated by equation 10-27.

$$C_{min} = \frac{I_{max}\left[\frac{1}{4f} + \frac{1}{2\pi f}\arcsin\left(\frac{V_{min}}{V_{max}}\right)\right]}{V_{max} - V_{min}} \qquad (10\text{-}27)$$

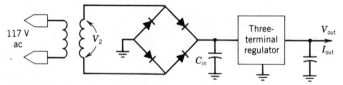

FIGURE 10-29
A basic power supply using a three-terminal regulator.

where i_{max} is the maximum current in the regulator, f is the line frequency, V_{max} is the maximum voltage at the output of the bridge, and V_{min} is the minimum voltage that can be applied to the regulator.

EXAMPLE 10-15

For the circuit of Figure 10-29 assume the regulator is an **LM123** operating with V_{max} = 15 V, and V_{min} = 8 V. Find
a. C_{in}.
b. The transformer turns ratio if the transformer and wiring resistance is 2 Ω.
c. The power dissipated by the regulator.

Solution
a. The specifications for the **LM123** show that it delivers 3 A at 5 V. From equation 10-27 we have

$$C_{in} = \frac{3\left[\frac{1}{240} + \frac{1}{2\pi \times 60} \arcsin\left(\frac{8}{15}\right)\right]}{7}$$

$$C_{in} = 2392 \ \mu F$$

b. The voltage at the output of the transformer must be V_{max} plus the two diode drops, plus the IR drop in the resistance.

$$V_2 = 15 \text{ V} + 2(0.7 \text{ V}) + 3 \text{ A} \times 2 = 22.4 \text{ V}$$

The voltage at the primary of the transformer is 117 V × $\sqrt{2}$ = 165.5 V. Therefore N = 165.5:22.4 or 7.4:1.

c. The input voltage to the regulator varies from 15 V to 8 V. It is reasonable to assume that the average voltage is 11.5 V. Then

$$P_D = (11.5 \text{ V} - 5\text{V}) \times 3 \text{ A} = 19.5 \text{ W}$$

The regulator will require a heat sink to dissipate this power.

10-10 SWITCHING REGULATORS

The linear regulators described in sections 10-8 and 10-9 are *dissipative*. Because the pass transistors operate in the *active region* (see section 3-4.1), they dissipate power and decrease the efficiency of the power supply. In many applications, linear regulators are being replaced by more efficient *switching regulators*.

In a switching regulator, the pass transistor oscillates between cutoff and saturation. Thus it acts as a switch that is rapidly being turned on and off, but it consumes very little power in either case. This is why switching regulators are highly efficient. Of course, energy must be supplied to the load when the transistor is off. This energy must be produced during the on period of the transistor and must be stored for use during the off period. This indicates the need for energy-storing elements, such as inductors and capacitors.

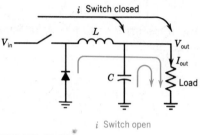

FIGURE 10-30
The basic switching regulator.

The basic switching regulator is shown in Figure 10-30, where the pass transistor has been represented as a switch. This is a step-down circuit, where V_{out} must be less than V_{in}. When the "switch" is closed (the pass transistor is on), the current flows through the inductor, supplies the load, and also charges the capacitor. This is shown in black on the figure.

When the switch is open, the current through the inductor does not change instantly. It continues to supply current to the load. The return path for this current is through the diode. The capacitor also supplies current to the load at this time.

10-10.1 Control of the Switching Regulator

In most circuits the switching regulator is controlled by a *pulse-width modulated square-wave oscillator*. The waveforms for this oscillator are shown in Figure 10-31. The switch or pass transistor is on when the oscillator output is high.

Figure 10-31 also shows how control is achieved in a switching regulator. The oscillator produces a fixed frequency output, whose period is T. The oscillator is modulated, however, to produce the desired value of V_{out}. If V_{out} is to be relatively high, as shown in Figure 10-31a, the switch is kept on for a relatively long time, τ_1. If V_{out} is to be less, as shown in Figure 10-31b, the switch is kept on for a shorter time, τ_2.

For a repetitive square wave, as shown in Figure 10-31, the *duty cycle* is defined as

$$\%D = \frac{\tau}{T} \times 100 \qquad (10\text{-}28)$$

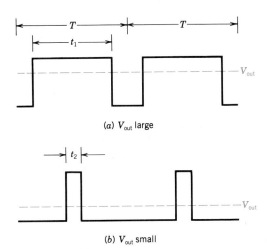

FIGURE 10-31
Waveform for a switching regulator oscillator.

The duty cycle also relates V_{out} to V_{in}:

$$V_{out} = V_{in} \times \frac{\tau}{T} \qquad (10\text{-}29)$$

EXAMPLE 10-16

a. If the input voltage to a switching regulator is 50 V, and the output is to be 15 V, find the duty cycle.
b. If the frequency of the oscillator is 20 kHz, how long is the oscillator high and how long is it low?

Solution
a. From equation 10-29 we have

$$\frac{\tau}{T} = \frac{V_{out}}{V_{in}} = 0.3$$

and the duty cycle is 30%.

b. The period of the oscillator is 50 μs. Because the duty cycle is 30%, the output will be high for 15 μs and low for 35 μs.

Figure 10-32 shows the block diagram of a basic switching regulator. The power components are shown in black and include a *pnp* pass transistor. The *control* components are shown in blue. The output is monitored by the sense

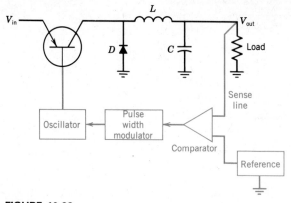

FIGURE 10-32
Diagram of a basic switching regulator.

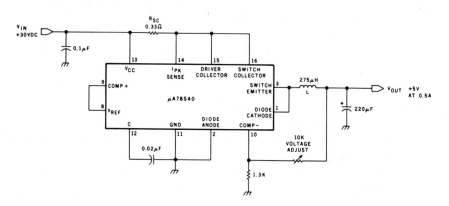

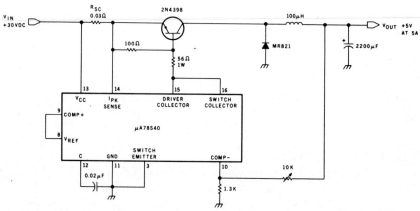

FIGURE 10-33
(a) Schematic diagram of a 5 V, 0.5 A step-down switching voltage regulator. For low currents, the **78S40**'s internal switching transistor and diode may be used. (Copyright Steven A. Ciarcia, used by permission.) (b) Schematic diagram of a 5 V, 5 A step-down switching regulator. An external heat-sinked transistor and diode are needed for the higher current. (Copyright Steven A. Ciarcia, used by permission.)

line and compared to the reference by the comparator. The output of the comparator feeds the pulse-width modulator, which determines the pulse width of the oscillator. Any change in the output can be compensated by a change in the pulse width.

Oscillators, comparators, and pulse-width modulators can be built using transistors and other components, but most modern switching regulators use a single integrated circuit to perform these functions. The **78S40**, manufactured by Fairchild, Inc., is a readily available integrated regulator. Figure 10-33 shows two switching regulators that can transform 30 V to 5 V using it. In Figure 10-33a the output current is only 0.5 A, and the IC can handle it. In Figure 10-33b the output is 5 A. This is too much for the IC, and a *pnp* power transistor is added to supply the current.

10-10.2 The Oscillator Frequency

The operating frequency of most modern switching regulators is about 20 kHz. Higher frequencies tend to reduce the size of the capacitors and inductors, and also reduce ripple, but because of the frequent switching, they radiate electrical noise, and often have to be shielded by an RFI (Radio Frequency Interference) shield. Frequencies less than 20 kHz are more efficient, but they require larger components, have greater ripple, and also produce noise that can be heard by the human ear and can be very annoying.

10-10.3 Mathematical Analysis of a Switching Regulator

The switching regulator can be analyzed to determine the proper values of L and C for a given load. The basic step-down circuit is shown in Figure 10-34a, and the waveforms are shown in Figure 10-34b. The circuit can be analyzed by making the following assumptions: V_{in} is constant, V_{out} is constant (it would be a poor regulator if this were not true), and the transistor is driven into saturation when the control input is high and cut off when the control voltage is low.

When the control voltage is high (during time τ in Figure 10-34), the voltage at point A is $V_{in} - V_S$, the saturation voltage of the transistor. This voltage is small (see section 2-10) and can be ignored for this approximate analysis. The voltage across the inductor is

$$V_L = V_{in} - V_{out}$$

Thus V_L is constant during this interval. But, for any coil, $V_L = L\, di/dt$. It therefore follows that di/dt is constant, and the inductor current i_L has a *ramp* waveform that goes from a minimum value, I_{min}, to a maximum value, I_{max}.

During the time interval $T - \tau$, the transistor is cut off, but current flows through the diode and the voltage at point A is $-V_D$.

$$V_L = V_{out} + V_D$$

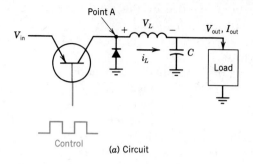

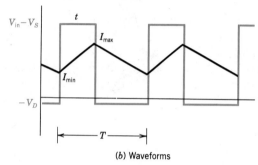

FIGURE 10-34
Analysis of the step-down switching regulator.

For simplicity V_D will be ignored compared to V_{out} in the following analysis, so that $V_L = V_{out}$. V_L is also constant during this interval, so the current ramps downward from I_{max} to I_{min}. The current waveforms are shown in Figure 10-34b.

When the current is increasing we have

$$V_L = L \frac{di}{dt}$$

$$V_{in} - V_{out} = \frac{L}{\tau}(I_{max} - I_{min}) = \frac{L}{\tau}\Delta I \qquad (10\text{-}30)$$

where $\Delta I = I_{max} - I_{min}$.

When the current is decreasing, we have

$$V_{out} = \frac{L\Delta I}{T - \tau} \qquad (10\text{-}31)$$

Combining equations 10-30 and 10-31 to eliminate τ gives

$$\Delta I = \frac{V_{out} T(V_{in} - V_{out})}{L V_{in}} \qquad (10\text{-}32)$$

The output or load current, I_o, must be equal to the average current through the inductor. From this we have

$$I_o = \frac{V_{out}}{R_L} = \frac{I_{max} + I_{min}}{2} \tag{10-33}$$

or

$$I_{max} = \frac{2 V_{out}}{R_L} - I_{min} \tag{10-34}$$

where R_L is the resistance of the load. But equation 10-30 can also be rewritten as

$$I_{max} = I_{min} + (V_{in} - V_{out})\frac{\tau}{L}$$

Combining these equations to eliminate I_{max}, and using the fact that $V_{out} T = V_{in} \tau$ from equation 10-29, gives

$$I_{min} = \frac{V_{out}}{R_L} - \frac{V_{out}(T - \tau)}{2L} \tag{10-35}$$

or

$$I_{min} = \frac{V_{out}}{R_L} - \frac{V_{out}(V_{in} - V_{out}) T}{2 V_{in} L} \tag{10-36}$$

In order for these equations to operate, I_{min} must be positive. Setting $I_{min} = 0$ in equation 10-36 gives

$$\frac{L}{R_L} > \frac{(V_{in} - V_{out}) T}{2 V_{in}} \tag{10-37}$$

Equation 10-36 shows that as the load current increases, V_{out} remains constant, but R_L decreases, and therefore I_{min} and I_{max} increase to provide the additional current. Equation 10-37 shows that R_L must be less than a maximum (depending on L) to allow regulation.

It can also be shown (see Professor Sherrick's notes in the References) that the peak-to-peak ripple voltage is given by

$$V_r = \frac{V_{out} T^2 (V_{in} - V_{out})}{8 LC V_{in}} \tag{10-38}$$

EXAMPLE 10-17

A step-down power supply has the following specifications:

$$V_{in} = 160 \text{ V}$$
$$V_{out} = 10 \text{ V}$$
$$I_{out} = 10 \text{ A}$$
$$V_{ripple} \text{ (peak-to-peak)} = 20 \text{ mV}$$
$$I_{out(min)} = 0.1 \text{ A}$$

Note that V_{in} is approximately the voltage obtained if the 117 V, 60 Hz is passed through a full-wave rectifier and filter.

If the switching frequency is 20 kHz, find
a. The L and C required.
b. The voltages on the control input of the *pnp* transistor.

Solution
a. The maximum value of R is

$$R = \frac{10 \text{ V}}{I_{out(min)}} = \frac{10 \text{ V}}{0.1 \text{ A}} = 100 \text{ }\Omega$$

From equation 10-37 we have

$$L > \frac{(V_{in} - V_{out}) T}{2 V_{in}} \times R$$

At 20 kHz, $T = 50 \times 10^{-6}$ and

$$L > \frac{150}{320} \times 50 \times 10^{-6} \times 100 = 2.34 \text{ mH (milliHenries)}$$

A 3 mH inductor would be a reasonable choice.
The value of the capacitor is given by equation 10-38

$$20 \times 10^{-3} = \frac{V_{out} T^2 (V_{in} - V_{out})}{8 \, LC \, V_{in}}$$

$$C_{in} = \frac{10 \times (2500 \times 10^{-12}) \times 150}{8 \times 160 \times 3 \times 10^{-3} \times 20 \times 10^{-3}} = \frac{3.75 \times 10^{-6}}{0.077} = 48.7 \text{ }\mu\text{F}$$

b. The driving voltage for the transistor must be large enough to saturate it when it is to be on and to cut it off. To cut off the transistor the drive voltage should be at least the same as V_{in} or 160 V. Anything less than that will turn the transistor on.

10-10.4 Other Switching Regulators

The step-down switching regulator, discussed in the previous sections, is the most commonly used switching regulator. There are several other types of switching regulators in use, and many variations exist for each circuit. Several regulators will be discussed in this paragraph, but space limitations preclude a detailed analysis of them.

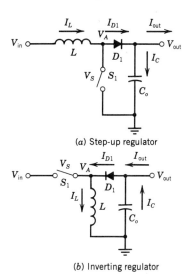

FIGURE 10-35
Other types of regulators.

The circuit of Figure 10-35a is a *step-up* or *boost* switching regulator. It steps up the output voltage so that V_{out} is greater than V_{in}. As before, the switch in the circuit represents a controlled transistor operating in cutoff or saturation. The reader must remember that this switch is constantly and rapidly opening and closing.

When the switch is closed, the diode is reverse-biased because $V_{out} > V_{in}$. At this time V_A is ground and the current through the coil is increasing. When the switch opened, this current flows through the diode, charges the capacitor, and supplies the load. As before, the output voltage is controlled by regulating the on time of the switch.

The circuit of Figure 10-35b is an *inverting* switching regulator that produces a negative output voltage. When the switch is closed, the diode is reverse-biased because V_{out} is negative and the current in the inductor increases. When the switch is open the current in the inductor flows through the capacitor and the diode, as shown in the figure, and places a negative charge on the capacitor.

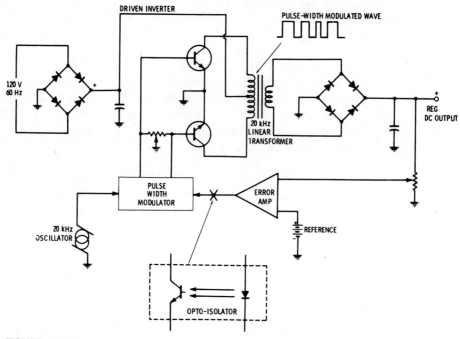

FIGURE 10-36
An off-line regulator. (Reproduced by permission of the Publisher, Howard W. Sams & Co. *Regulated Power Supplies*, 3rd ed., by Irving Gottlieb, 1984.)

The *off-line* switching regulator is one of the most efficient regulators. Figure 10-36 shows a complete power supply using this regulator. It operates as follows:

1. The 120 V, 60 Hz line voltage is rectified by a bridge and fed to a transformer.
2. The current in this transformer is controlled by the two transistors, operating as a push–pull pair. These transistors operate in switching mode; they are either saturated or off.
3. The ac output of the transformer is rectified by a second bridge, filtered by the capacitor, and drives the load.
4. The output voltage is sampled, compared to a reference, and drives the pulse-width modulator that controls the transistors. This feedback loop keeps the output voltage constant.

This circuit uses a transformer and two switching transistors in place of the inductor, switching transistor, and diode of the step-down regulator. It does require two bridge rectifiers, but bridges have high efficiency and are inexpensive. The major saving occurs because the transformer is operated at the switching frequency, perhaps 20 kHz, instead of 60 Hz. This allows the use of a smaller, lighter, and less expensive transformer.

Two pulse-width modulators that can be used with this circuit are the **MC3240,** whose diagram is shown in Figure 10-37a, and the **SG1524** manufactured by Silicon General, Inc. The output waveforms are shown in Figure 10-37b. The circuit is controlled by controlling the *dead time* in the output waveform.

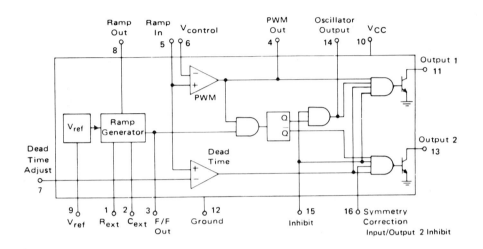

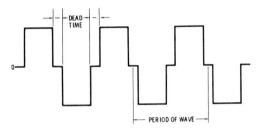

FIGURE 10-37
The **MC3420** regulator controller. (Reproduced by permission of the Publisher, Howard W. Sams & Co. *Regulated Power Supplies*, 3rd ed., by Irving Gottlieb, 1984.)

10-11 SUMMARY

This chapter considered the power supplies that are needed to provide power for amplifiers and other electronic circuits. Rectifiers and filters were introduced in the first part of the chapter, and their limitations were discussed. Then regulated power supplies were introduced and their operation was explained. A set of equations was presented that analyzed the linear regulator.

In the latter part of the chapter more modern power supplies were presented. These depend on integrated circuits for their control. Integrated circuit regulators and switching power supplies with IC modulators were discussed.

Power supply technology will continue to evolve as more ICs are developed. The operating principles and some of the ICs were presented here, but the user who is actually designing power supplies should consult the references and the manufacturers' specifications to find the newest devices and most modern approach to the problem.

10-12 GLOSSARY

Bridge rectifier. A full-wave rectifier consisting of four diodes.

Crowbar. An SCR circuit that provides overvoltage protection for a regulator.

Filter. A circuit, generally consisting of capacitors and inductors, that reduces the ac component of a rectified wave.

L-section. A filter section consisting of an inductor followed by a capacitor.

Pass transistor. The transistor that conducts the load current.

Peak inverse voltage. The peak or maximum reverse voltage across a diode.

Pi filter. A filter consisting of two capacitors and an inductor.

Preregulator. A constant current source that replaces R_3 in a linear voltage regulator.

Rectifier. A circuit that converts ac voltage to dc voltage.

Regulation. A measure of the ability of a power supply to maintain a constant dc output voltage.

Ripple factor. A measure of the amount of ac in a dc wave.

Switching regulator. A voltage regulator where the pass transistor is either saturated or cut off.

Tracking regulator. An adjustable regulator that provides equal positive and negative output voltages.

Voltage regulator. A circuit that holds the output voltage of a power supply constant.

10-13 REFERENCES

Russell J. Apfel and David B. Jones, Universal Switching Regulator Diversifies Power Subsystem Applications, *Computer Design Magazine*, 1978.

Robert Boylestad and Louis Nashelsky, *Electronic Devices and Circuit Theory*, 4th Edition, Prentice-Hall, Englewood Cliffs, N.J., 1987.

Steve Ciarcia, Switching Power Supplies, An Introduction, *Byte Magazine*, November, 1981.

Rodney B. Faber, *Essentials of Solid State Electronics*, John Wiley, New York, 1985.

Irving M. Gottlieb, *Regulated Power Supplies*, 3rd Edition, Howard W. Sams & Co, Indianapolis, 1981.

Linear and Interface Circuits Applications, Volume 1, Texas Instruments, Inc., 1985.

Linear/Switchmode Voltage Regulator Handbook, Motorola, Inc., 1982.

Jacob Millman, *Microelectronics*, 2nd Edition, McGraw-Hill, New York, 1987.

J. F. Pierce and T. J. Paulus, *Applied Electronics*, Charles E. Merrill, Columbus, Ohio, 1972.

RCA Designer's Handbook, Solid State Power Circuits, RCA Corp., Somerville, N.J., 1971.

ARTHUR H. SEIDMAN and JACK L. WAINTRAUB, Electronics, Devices, Discrete and Integrated Circuits, Charles E. Merrill, Columbus, Ohio, 1977.

J. D. SHERRICK, Unpublished Notes for Power Amplifier Design, Rochester Institute of Technology, Rochester, N.Y., 1981.

JOHN THOMAS, Calculating Filter Capacitor Values for Computer Power Supplies, *Byte Magazine*, April, 1980.

HENRY ZANGER, *Semiconductor Devices and Circuits*, John Wiley, New York, 1984.

10-14 PROBLEMS

10-1 Using equations 10-3 and 10-4, find the dc and rms values of a full-wave rectifier as a function of V_m. Using equation 10-6 find the ripple factor.

10-2 Show that the rms value of the wave given by equation 10-2 is

$$V = \sqrt{A_0^2 + \frac{(A_1^2 + A_2^2 + \cdots)}{2}}$$

10-3 The waveform of Figure P10-3 can be expressed as $4 + 1\sin\omega t$

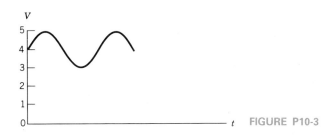

FIGURE P10-3

  a. Find its rms value.
  b. Find its ripple factor using equation 10-5.
  c. Find its ripple factor using equation 10-6.

10-4 Show that the rms value of the waveform of Figure P10-4 is

$$v_{rms} = \frac{V}{2\sqrt{3}}$$

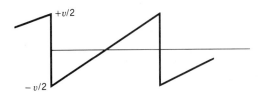

FIGURE P10-4

10-5 Find the ripple factor for the waveform of Figure P10-5. (*Hint:* The results of problem 10-4 can help.)

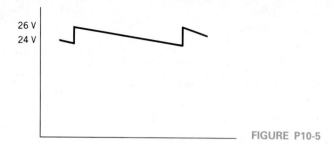

FIGURE P10-5

10-6 **a.** The transformer of Figure P10-6 is connected to a half-wave rectifier. Find the dc value of V_{out}
   b. Repeat for a full-wave rectifier.
   c. Repeat for a bridge.
   d. Repeat parts a, b, and c if a capacitor filters the load.
   Assume the load takes so little current that the capacitor voltage is constant.
   For parts a and c assume the transformer is connected end-to-end.

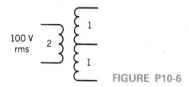

FIGURE P10-6

10-7 Using the house lines as input, an output with an average value of 40 V must be produced. Find the transformer turns ratio if
   a. A half-wave rectifier is used.
   b. A full-wave rectifier is used.
   c. A bridge is used.
   In each case assume that all diode drops are 0.7 V.

10-8 Repeat problem 10-7 if a capacitor is used to filter the output.

10-9 For the circuit of Figure P10-9 find the dc output voltage and the ripple factor.

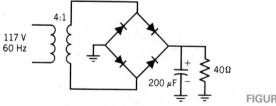

FIGURE P10-9

10-10 Design a circuit to produce a 10 V, 100 mA dc output from the house lines. Assume a ripple factor of 0.1 is acceptable. Find the turns ratio of the transformer and the value of the capacitor you need.

a. Use a half-wave rectifier.
b. Use a full wave rectifier.
c. Use a bridge.

10-11 Find the conduction angle and percent regulation (%R) for each part of problem 10-10.

10-12 An L-section filter has a 10 Hz coil and must operate with a ripple factor of .001. Find the required value of C and the maximum load resistance.

10-13 For each circuit in Figure 10-13, assume the input is a full-wave rectified waveform with a peak of 30 V, and the R_L is 10 Ω. If the ripple factor is to be .001, select values for the other components.

10-14 For the Zener regulator of Figure P10-14, the input voltage can vary from 25 V to 40 V. Find the value of R_1 and the power ratings of both resistors and the Zener diode.

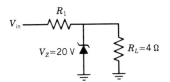

FIGURE P10-14

10-15 Design a Zener diode regulator to provide 5 V at 1 A if the input voltage can vary between 10 V and 20 V. Find the power ratings of your components.

10-16 How would you temperature compensate the Zener diode of Example 10-9. What would be the change in output voltage for your circuit if the ambient temperature changed by 40°C?

10-17 **a.** If $V_{in} = 30$ V for the regulator of Figure P10-17, find V_{out} if $I_L = 0$ and if $I_L = 5$ A.
 b. If $I_L = 2$ A, find V_{out} if V_{in} varies from 20 to 30 V.

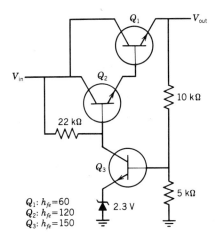

FIGURE P10-17

10-18 For the circuit of Figure P10-18
  a. Develop the equation for V_{out} and I_L.
  b. Find the percent load regulation if the full load current is 2 A.

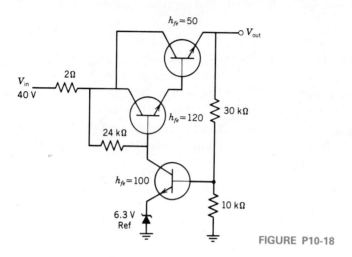

FIGURE P10-18

10-19 In the circuit of Figure P10-18 the 24 kΩ resistor is to be replaced by a constant current source of 2.5 mA.
  a. Add the constant current source to the figure.
  b. Find the new equation for the circuit.
  c. Find the percent load regulation if the full load current is 2 A.

10-20 Figure P10-20 shows a series regulator with a difference amplifier. Show that the following equation holds:

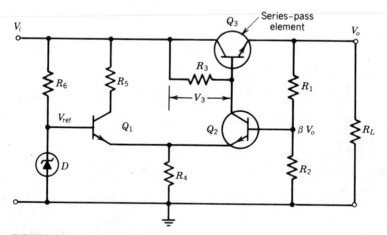

FIGURE P10-20
(From *Essentials of Solid State Electronics* by Rodney B. Faber. Copyright © 1985 by John Wiley & Sons, Inc. Reprinted by permission of John Wiley & Sons, Inc.)

$$I_R R_3 = V_{in} - (V_{out} + V_{be})$$

Using this equation and equations 10-19 and 10-20, eliminate I_R and I_B, and derive an equation for the circuit.

10-21 The input to the regulator of Figure 10-20 is 32 V. If it is to supply a regulated output of 25 V at 2 A, determine R_1, R_2, and R_3. Assume the Darlington has an h_{fe} of 1500, and a V_{be} of 1.5 V, and the transistor has an h_{fe} of 100. The diode is a 6.3 V Zener. Calculate the resulting regulator characteristics.

10-22 An **LM123** is operating from a 20 V source, and supplying 2 A to the load. How much power must it dissipate?

10-23 Design an **MC1723** regulator for an output of 10 V at 40 mA.

10-24 A three-terminal regulator is connected as shown in Figure 10-28. Find the value of C_{in} if $V_{max} = 30$ V, $V_{min} = 20$ V, and $I_1 = 2$ A. If the output voltage is 12 V, find the power dissipated in the regulator.

10-25 Repeat Example 10-16 if the output voltage is 10 V.

10-26 A switching power supply has the following specifications:

$$V_{in} = 100 \text{ V}$$
$$V_{out} = 20 \text{ V}$$
$$I_{load} = 5 \text{ A}$$
$$V_{ripple} = 10 \text{ mV}$$
$$I_{out(min)} = 50 \text{ mA}$$

If the switching frequency is 25 kHz, find
a. The required values of L and C.
b. I_{min} and I_{max}.

CHAPTER ELEVEN

Tuned Amplifiers and Oscillators

11-1 Instructional Objectives

This chapter explains the design and construction of frequency-selective tuned amplifiers. It also discusses several types of oscillators.

After reading the chapter the student should be able to:

1. Design a tuned circuit amplifier for a specific Q and resonant frequency.
2. Convert a coil with a given internal resistance into its equivalent parallel circuit.
3. Place the poles in a Butterworth amplifier and design the amplifier.
4. Design Hartley and Colpitts oscillators.
5. Construct crystal-controlled oscillators.
6. Design phase-shift and Wien bridge oscillators.
7. Construct a digital oscillator using a 555 timer or a crystal.

11-2 Self-Evaluation Questions

Watch for the answers to the following questions as you read the chapter. They should help you to understand the material presented.

1. What is an intermediate frequency amplifier? What is the advantage of using it in radio and TV receivers?
2. How does a coil's Q affect the bandwidth of a tank circuit? How can this be used to control the bandwidth?

3. What is the advantage of using double-tuned amplifiers?
4. What is the Barkhausen criterion for oscillation?
5. What are the conditions of oscillation for a Colpitts oscillator? For a Hartley oscillator?
6. What are the operating principles of a phase-shift oscillator?
7. What are the differences between digital oscillators and sine wave oscillators?

11-3 THE TUNED-CIRCUIT AMPLIFIER

The amplifiers considered in the previous chapters are generally classified as *wide-band* or *video* amplifiers, because they amplify all frequencies up to f_h. Sometimes it is necessary to have an amplifier that amplifies only a small band of frequencies and rejects the rest. An example of this is the intermediate frequency (IF) amplifier in an AM or FM radio or a TV. An AM radio receives many stations, each at a different frequency, but these are *beat* or *heterodyned* against a local oscillator to produce an *intermediate frequency* that is always 455 kHz. This signal is then sent to the IF amplifier, which amplifies only a small band of frequencies (approximately 455 kHz ± 7.5 kHz). If a wide-band amplifier were used, all the incoming stations would be amplified and the output would be chaotic. The IF amplifier provides *selectivity*, because it selects only the desired station. FM radio and TV operate similarly. For FM radio the IF frequency is 10.7 MHz and for TV it is 4.5 MHz.

The *tuned-circuit amplifier*, shown in Figure 11-1a, is perhaps the simplest frequency-selective amplifier. The load on the transistor is a *tank circuit*, which consists of the inductor, capacitor, and resistor. The gain of this circuit depends on frequency and can be expressed as $-g_m Z(s)$, where g_m is the transconductance of the amplifier and $Z(s)$ is the impedance of the tank circuit as a function of frequency. The *admittance* of the circuit is

$$Y(s) = \frac{1}{R} + sC + \frac{1}{sL} \qquad (11\text{-}1)$$

for $s = j\omega$ this becomes

$$Y(j\omega) = \frac{1}{R} + j\left(\omega C - \frac{1}{\omega L}\right)$$

The admittance is minimum (and the impedance maximum) when

$$\omega C = \frac{1}{\omega L}$$

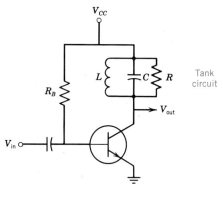

(a) Circuit

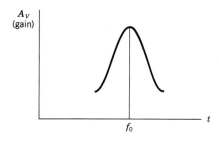

(b) Frequency response

FIGURE 11-1
The tuned-circuit amplifier.

or

$$f_0 = \frac{1}{2\pi\sqrt{LC}} \qquad (11\text{-}2)$$

Here f_0 is defined at the *resonant frequency*. At f_0 $A_v = -g_m R_L$. If the applied frequency goes above or below f_0, the gain drops off, as shown in Figure 11-1b.

EXAMPLE 11-1

For the circuit of Figure 11-1

$$L = 1 \text{ MH}$$
$$R = 20{,}000$$
$$h_{fe} = 100$$
$$h_{ie} = 1000$$

Find
a. The capacitor needed if the resonant frequency is to be 455 kHz.
b. The gain at f_0
c. The gain at 400 kHz

Solution

a.
$$f_0 = \frac{1}{2\pi\sqrt{LC}}$$

$$C = \frac{1}{4\pi^2 f_0^2 L} = \frac{1}{4\pi^2(455 \times 10^3)^2 \times 10^{-3}} = \frac{1}{8.16 \times 10^{12} \times 10^{-3}}$$

$$= \frac{1}{8.16 \times 10^9} = 122 \text{ pF}$$

b. The gain at f_0 is simply $-g_m R_L$

$$g_m = \frac{h_{fe}}{h_{ie}} = \frac{100}{1000 \, \Omega} = 0.1 \text{ S}$$

$$A_v = -g_m R_L = -0.1 \text{ S} \times 20{,}000 \, \Omega = -2{,}000$$

At 400 kHz

$$Y(j\omega) = \frac{1}{20{,}000}$$
$$+ j\left(2\pi \times 400 \times 10^3 \times 122 \times 10^{-9} - \frac{1}{2 \times 400 \times 10^{-3} \times 10^{-3}}\right)$$

$Y(j\omega) = 0.5 \times 10^{-4} + j(3.064 \times 10^{-4} - 3.98 \times 10^{-4})$

$Y(j\omega) = 0.5 \times 10^{-4} + j(-0.916 \times 10^{-4})$

$Y(j\omega) = 1.04 \times 10^{-4} \angle -61° \text{ S}$

$Z(j\omega) = \frac{1}{Y(j\omega)} = 9582 \angle 61° \, \Omega$

The gain at a 400 kHz is

$$A_v = -g_m Z(j\omega) = 0.1 \times 9582 = 958.2$$

This is the magnitude of the gain. There is also a 61° phase shift. At 400 kHz the impedance and gain are less than half of their mid-frequency values.

11-3.1 Coils and Quality Factor

A typical coil consists of a winding of wire over a powdered metal core, sometimes called a *slug*. In many cases, the core can be moved in and out of the winding. This affects the inductance of the coil and provides the user with adjustments and control.

A coil can be represented by an inductor in series with a resistor, as shown in Figure 11-2a. The impedance of the coil is

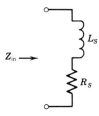

(a) Series circuit

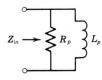

(b) Parallel circuit

FIGURE 11-2 Two representations of a coil and its impedance.

$$Z_{in} = R_S + j\omega L$$

where R_S is the series impedance of the coil.

Coils are generally more effective if their inductive reactance is high, and their internal resistance is low. The quality factor, Q, of a coil is the ratio of the inductive impedance to the resistance.

$$Q = \frac{\omega L}{R_S} \tag{11-3}$$

Coils can also be represented by a pure inductor, L_P, in parallel with a resistor, R_P, as shown in Figure 11-2b. The impedance of this circuit is

$$Z'_{in} = \frac{R_P \, j\omega L_P}{R_P + j\omega L_P}$$

To find the relationship between Z_{in} of the series circuit and Z'_{in} of the parallel circuit, we note that

$$Z'_{in} = \frac{R_P \, j\omega L_P}{R_P + j\omega L_P} \times \frac{R_P - j\omega L_P}{R_P - j\omega L_P} = \frac{j\omega L \, R_P^2 + \omega^2 L^2 R_P}{R_P^2 + \omega^2 L^2}$$

If $R_P \gg \omega L$, this simplifies to

$$Z'_{in} = \frac{\omega^2 L^2}{R_P} + j\omega L_P$$

If the coil has a high Q (generally 5 or greater), the series and parallel circuits are equivalent and the following approximate equations are valid:

$$L_P = L_S$$

$$Q_P = \frac{R_P}{\omega L} \qquad (11\text{-}4)$$

$$Q_P = Q_S$$

$$R_P = Q^2 R_S \qquad (11\text{-}5)$$

The mathematics behind these statements is given in Appendix G. Placing a resistor in parallel with a coil, as shown in Figure 11-3, effectively lowers its Q. The effective Q, or the Q for the circuit, can be found using equations 11-3, 11-4, and 11-5.

EXAMPLE 11-2

A coil has a Q of 20 and a resistance of 50 Ω. Find ωL. What is the effective Q of the circuits if a 30,000 Ω resistor is placed in parallel with the circuit, as shown in Figure 11-3?

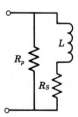

FIGURE 11-3
A coil with a parallel resistor.

Solution
From equation 11-3, $\omega L = QR_S = 1000$ Ω. To find the effective Q of the circuit, the equivalent parallel resistance of the coil must first be calculated.

$$R_P = Q^2 R_S = 400 \times 50 = 20{,}000 \text{ Ω}$$

The 30,000 Ω external resistance is in parallel with the 20,000 Ω effective resistance of the coil.

$$R'_P = R_P \| R_X = 20{,}000 \ \Omega \| 30{,}000 \ \Omega = 12{,}000 \ \Omega$$

The effective Q of the circuit depends on this resistance.

$$Q_{\text{eff}} = \frac{R'_P}{\omega L} = \frac{12{,}000 \ \Omega}{1000 \ \Omega} = 12$$

The resistor has lowered the Q of the circuit from 20 to 12.

Tank circuits, such as shown in Figure 11-1, have no collector resistor. If the Q of the coil is large, R_P can be very high. In this case the output admittance of the transistor, h_{oe}, can become significant and the effective resistance of the circuit becomes $R_P \| 1/h_{oe}$.

11-3.2 The Effect of Q on Bandwidth

The *bandwidth* of the single-tuned amplifier of Figure 11-1 is defined in the normal manner as $f_2 - f_1$, where f_2 and f_1 are the upper- and lower-half-power frequencies. It can be shown (see Appendix G) that the gain of a tuned circuit can be expressed as

$$A_v(j\omega) = -g_m R_P \frac{1}{1 + jQ\left(\dfrac{\omega}{\omega_0} - \dfrac{\omega_0}{\omega}\right)} \tag{11-6}$$

where ω_0 is the resonant frequency.

It can also be shown that the bandwidth of the tuned circuit is

$$BW = f_1 - f_2 = \frac{f_0}{Q} \tag{11-7}$$

These equations are valid only for high Q circuits ($Q > 5$), but they apply to most practical circuits.

The frequency response of a tuned-circuit amplifier is shown in Figure 11-4. As the Q of a circuit increases, its gain becomes larger and its bandwidth narrows.

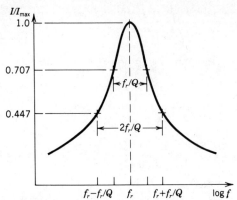

FIGURE 11-4
Response of a series resonant circuit as a function of frequency.

EXAMPLE 11-3

Find the center frequency, the gain, and the bandwidth for the circuit of Figure 11-5.

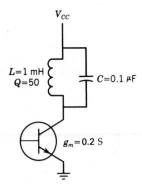

FIGURE 11-5
Circuit for Example 11-3.

a. If $1/h_{oe} = \infty$.
b. If $1/h_{oe} = 20{,}000 \; \Omega$.

Solution

a. The center frequency is

$$f = \frac{1}{2\pi\sqrt{LC}} = \frac{1}{2\pi\sqrt{10^{-10}}} = \frac{10^5}{2\pi} = 15{,}924 \text{ Hz}$$

b. To find the gain R_P must be found.

$$R_P = Q\omega L = 50 \times 2\pi \times 15{,}924 \text{ Hz} \times 10^{-3}\text{H} = 5900 \; \Omega$$

The gain of the circuit is

$$A_v = -g_m R_L = -0.2 \text{ S} \times 5{,}000 \text{ }\Omega = -1000$$

at the resonant frequency. The bandwidth is $f_0/Q = 15{,}924 \text{ Hz}/50 = 315 \text{ Hz}$. Therefore the gain would be 3 dB down, or 707 at $f = 15{,}924 \pm 157.5 \text{ Hz}$.

c. If $1/h_{oe} = 20{,}000 \text{ }\Omega$, the effective R_P is $5{,}000 \text{ }\Omega \parallel 20{,}000 \text{ }\Omega = 4{,}000 \text{ }\Omega$.

$$Q' = \frac{R_P}{\omega L} = \frac{4000}{2\pi \times 15{,}924 \text{ Hz} \times 10^{-3}\text{H}} = 40$$

The mid-frequency gain of the circuit is

$$A_v = -g_m R_L = -800$$

The bandwidth is

$$BW = \frac{15{,}924 \text{ Hz}}{40} = 398 \text{ Hz}$$

11-3.3 The Variation of Q with Frequency

Equation 11-3 indicates that Q increases linearly with frequency. For real coils, this is not necessarily true. High-frequency effects tend to increase the impedance of the coil. One of these is called the *skin effect*. At high frequencies, ac current tends to migrate to the outside of a wire. This reduces the effective cross-sectional area of the wire and increases its resistance.

The Q of a coil can increase, decrease, or peak as a function of frequency. If a coil is being used in wide-band applications, a Q versus frequency plot might help. In narrow-band applications, as discussed in this section, the Q is generally taken as constant, and coils are often specified by their Q and inductance at a particular frequency.

11-4 MORE SELECTIVE AMPLIFIERS

An amplifier is often required to amplify a *band* of frequencies and reject any frequencies beyond the band. An AM radio signal is an example. It operates at an assigned frequency and is limited by FCC (Federal Communications Commission) regulations to a bandwidth of approximately 15 kHz. The information in an AM radio signal is therefore contained in the frequencies $f_0 \pm 15$ kHz, where f_0 is the transmission frequency of the AM station. Thus a radio station that transmits at 1 MHz (about the middle of the AM band), actually transmits between 985 kHz and 1.015 MHz. This also means that any sounds or music that have frequencies higher than 15 kHz cannot be transmitted on AM radio.

The ideal amplifier for this signal would have a flat top and steep sides, as shown in Figure 11-6. In actual AM radio receivers this signal is heterodyned down to the intermediate frequency of 455 kHz, but the wave shape remains essentially the same.

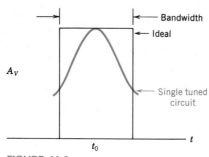

FIGURE 11-6
The actual and ideal bandpass characteristics of a tuned circuit.

The frequency response of a single-tuned circuit is also shown in Figure 11-6. Circuits that are more *selective*, that have flatter tops and steeper sides, are an improvement over the single-tuned circuit. Selectivity is often measured by the *shape factor* or *skirt ratio*. A typical skirt ratio is the 60 dB/6 dB value, or the ratio of the bandwidth at 60 dB down from midfrequency to the bandwidth at 6 dB down. A small ratio indicates that the amplifier has sharp sides and is highly selective. A more detailed discussion is given in the References.

11-4.1 The Double-tuned Transformer Amplifier

Some amplifiers, especially those in the intermediate frequency (IF) stages of a radio or TV, use a *double-tuned transformer amplifier*. The circuit of a double-tuned transformer is shown in Figure 11-7. The capacitors, C_p and C_s, tune the primary and secondary sides of the transformers to the same frequency, f_0. The primary and secondary sides of the transformer are coupled by their *mutual inductance, M*.

$$M = k\sqrt{L_p L_s}$$

where *k* is the *coefficient of coupling*. If the transformer windings are far apart,

A double-tuned transformer circuit.

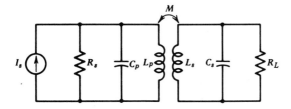

An approximate equivalent circuit for the double-tuned transformer.

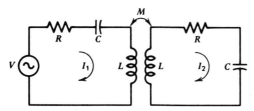

FIGURE 11-7
A double-tuned transformer circuit. (From *Solid State Radio Engineering* by Herbert L. Krauss, Charles W. Bostian, and Federik H. Raab. Copyright © 1980 by John Wiley & Sons, Inc. Reprinted by John Wiley & Sons, Inc.)

there is no coupling between them. As they are moved closer together, the coupling increases. This affects the waveshape of the output as shown in Figure 11-8. There is a *critical value* for the coefficient coupling, k_c, given by

Frequency response of the double-tuned transformer for three values of k.

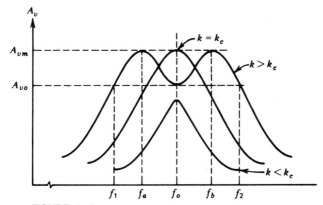

FIGURE 11-8
Frequency response of the double-tuned transformer for three values of k. (From *Solid State Radio Engineering* by Herbert L. Krauss. Charles W. Bostian, and Federik H. Raab. Copyright © 1980 by John Wiley & Sons, Inc.)

$$k_c = \frac{1}{\sqrt{Q_p Q_s}}$$

Where Q_p is the Q of the primary side and Q_s is the Q of the secondary side. If the actual value of k is less than k_c, the output is low. It increases until $k = k_c$. If k is adjusted so that it is greater than k_c, the output no longer increases, but a *double-peak* and *valley* occur, as shown in the figure.

Most double-tuned transformers are adjusted for critical coupling. They have a better waveshape (flatter top and steeper sides) than the single-tuned amplifier. If an IF amplifier, the double-tuned transformers are usually contained within an *IF can*. IF cans are critically coupled, and their inductance is adjustable over a small frequency range, so they can be properly aligned.

Figure 11-9 shows a radio receiver with its covers off. The IF cans are the square metal cans. They contain a screw driver adjustment to allow the user to set the inductance.

FIGURE 11-9
A radio receiver showing the IF cans. (Courtesy of Scientific Radio Systems, Inc., Rochester, N.Y.)

11-4.2 Stagger-tuned Amplifiers

The gain of a tuned amplifier can be increased by adding stages, as shown in Figure 11-10. JFETs (see Chapter 5) have been used in this amplifier, because their high input impedance is needed to isolate the tank circuits from each other and from the transistor impedances. The gain of the amplifier is then the product of the gains of each individual stage. If each stage is tuned to the same frequency, however, the bandwidth decreases as each stage is added, so the improvement due to the additional stages is limited.

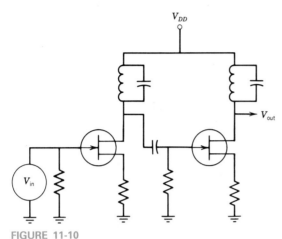

FIGURE 11-10
A two-stage tuned amplifier.

In a *stagger-tuned amplifier*, each stage is tuned to a slightly different frequency, which improves its gain–bandwidth curve. The most common stagger-tuned amplifier is the *Butterworth* or *maximally flat* amplifier. Figure 11-11 is a gain versus frequency plot for this type of amplifier, where n is the number of stages. The curve for $n = 1$ is the simple single-tuned circuit previously discussed. It can be seen that as the number of stages increases, the top gets flatter and the sides get steeper.

In Figure 11-11, x is the ratio of the frequency deviation from resonance to *half* the bandwidth. Thus, if an amplifier has a center frequency of 10 MHz and a bandwidth of 1 MHz, x will be 1 at 950 KHz and 1050 KHz. Figure 11-11 shows that the gain of the amplifier is down 3 dB when $x = 1$, as we would expect.

EXAMPLE 11-4

An amplifier has a center frequency of 15 MHz and a bandwidth of 2 MHz. By how many dB is its gain reduced at 13 MHz?
a. For a single-stage amplifier
b. For a three-stage amplifier

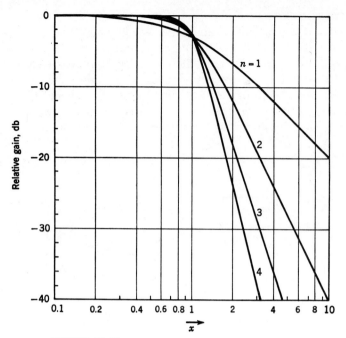

FIGURE 11-11
Gain vs. frequency for maximally flat functions and various n's. (From *Electronic Amplifier Circuits* by Pettit and McWhorter. Published by McGraw-Hill Inc., 1961.)

Solution
The required frequency deviates from the center frequency of the amplifier by 2 MHz. Half the bandwidth is 1 MHz. Therefore:

$$x = \frac{\text{frequency deviation}}{\text{bandwidth}/2} = \frac{2 \text{ MHz}}{1 \text{ MHz}} = 2$$

Figure 11-11 shows that when $x = 2$ the gain for a single-stage amplifier is down by 6 dB and for the three-stage amplifier the gain is down by 18 db.

In order to design a maximally flat amplifier, the user must first determine the number of stages of the amplifier (n), its mid-frequency (f_0), and its bandwidth (BW). The simplest and most common case is where the bandwidth is small compared to f_0. This is the *narrow-band approximation*. The user must then decide where to place the poles (see section 7-3.3) of the tank circuits. As a first approximation there is one pole for each stage. The imaginary part of each pole is the resonant frequency of the tank circuit (which is not f_0 for most stages) and the real part depends on the Q of the tank circuit.

The pole placement for the narrow-band approximation is shown in Figure 11-12. The poles are placed as follows:

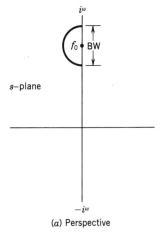

s-plane

(a) Perspective

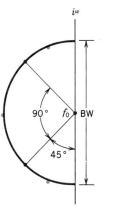

(b) Magnified view of the poles

FIGURE 11-12
Pole placement in the s-plane.

1. A circle is drawn on the *s-plane*. The center of the circle is f_0 and the diameter is the required bandwidth.
2. The first pole is placed on this circle at an angle of $180°/2n$ degrees with the vertical, where n is the number of poles in the amplifier.
3. The remaining poles are placed on the circle, separated by an angle of $180°/n$ from the first pole.

Figure 11-12a shows the circle on the *s-plane* is perspective. Figure 11-12b is a magnified view of the circle. It shows the pole placement for the two-pole amplifier, in black, and for the three-pole amplifier in blue.

EXAMPLE 11-5

Determine the placement of the higher pole in a two-pole Butterworth amplifier.

Table 11-1
Narrow-band Stagger Tuning (Maximally Flat)

1. Staggered pair ($n = 2$)
   Two stages tuned to $f_0 \pm 0.35BW$, each having a bandwidth $0.707BW$
2. Staggered triple ($n = 3$)
   One stage tuned to f_0, with bandwidth BW
   Two stages tuned to $f_0 \pm 0.43BW$ with bandwidth $0.50BW$
3. Staggered quadruple ($n = 4$)
   Two stages tuned to $f_0 \pm 0.46BW$, with bandwidth $0.38BW$
   Two stages tuned to $f_0 \pm 0.10BW$, with bandwidth $0.92BW$

Note: f_0 is the center frequency of the overall amplifier, and BW is the overall 3-dB bandwidth.

Solution
Because $n = 2$, the first pole is at an angle of 45° with the vertical. Its radius is $BW/2$. The imaginary part of the pole is therefore at $f_0 + BW/2 \sin 45°$ or at $f_0 + 0.35\ BW$. The real part of the pole is at $-0.35\ BW$.

Table 11-1 shows the pole placement for stagger-tuned amplifiers with 2, 3, and 4 poles. It helps the engineer design stagger-tuned amplifiers as Example 11-6 shows.

EXAMPLE 11-6

Design the tuned circuits for a three-stage Butterworth amplifier with a center frequency of 10 MHz and a bandwidth of 1 MHz. Use an inductor of 10 µH, with a Q of 50.

Solution
The design for the highest frequency pole will be presented. The design of the other two poles is similar. Table 11-1 shows that the highest frequency pole is tuned to $f_0 + 0.43B$. For this amplifier, the high-frequency pole is therefore at 10.43 MHz. From the table we can also determine that the bandwidth of the pole is 500 kHz.

$$f = \frac{1}{2\pi\sqrt{LC}}$$

$$C = \frac{1}{4\pi^2 f^2 L}$$

$$= \frac{1}{4\pi^2 \times (10.43 \times 10^6)^2 \times 10^{-5}}$$

$$C = \frac{1}{4290 \times 10^7} = 23.3 \text{ pF}$$

The Q of the circuit is

$$Q = \frac{f_0}{BW} = \frac{10.43 \text{ MHz}}{500 \text{ kHz}} = 20.86$$

The required parallel resistance is, from equation 11-4

$$R_P = Q\omega L = 20.86 \times 2\pi \times 10.43 \times 10^{+6} \times 10^{-5} \text{ H} = 13,663 \text{ }\Omega$$

The R_P due to the coil is

$$R_P = Q\omega L = 50 \times 2\pi \times 10.43 \times 10^6 \times 10^{-5} \text{ H} = 32,750$$

A resistor must be placed across the coil to reduce the parallel resistance of the circuit from 32,750 Ω to 13,663 Ω. The value of this resistor should also take other impedances across the load into account, such as the input impedance of the next stage.

11-4.3 Ceramic Filters

Many modern amplifiers use a wide-band amplifier and follow it with a *ceramic filter* for selectivity. Ceramic filters are rugged, simplify circuit construction, and have excellent selectivity characteristics. Figure 11-13a shows a circuit using a ceramic filter, and Figure 11-13b shows the attenuation characteristic for a ceramic filter that is designed for an IF amplifier in an AM radio. The curve shows that it has a resonant frequency of 455 kHz, and a bandwidth of about 10 kHz.

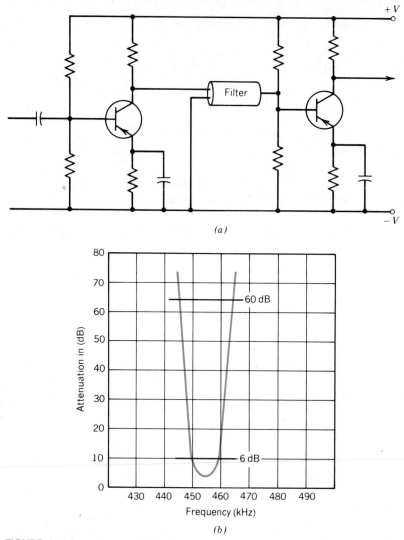

FIGURE 11-13
Ceramic filter application. (a) Typical circuit. (b) Typical response. (From *Essentials of Solid State Electronics* by Rodney B. Faber. Copyright © 1985 by John Wiley & Sons, Inc. Reprinted by permission of John Wiley & Sons, Inc.)

11-5 OSCILLATOR THEORY

Oscillations are repetitive waveforms that cycle at a particular frequency. Three types of oscillations are shown in Figure 11-14. An *oscillator* is a circuit that is designed to produce these oscillations. Many types of oscillators exist, but we will divide them into three categories:

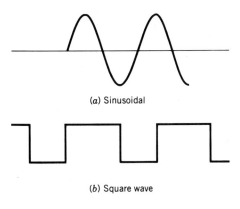

(a) Sinusoidal

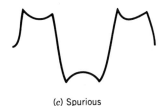

(b) Square wave

(c) Spurious

FIGURE 11-14
Types of oscillations.

1. High-frequency sine wave oscillators
2. Low-frequency sine wave oscillators
3. Square-wave or digital oscillators

The first two types produce sinusoidal waveforms, as shown in Figure 11-14a, and digital oscillators generate square waves, as shown in Figure 11-14b.

An oscillator is a circuit that seems to provide something for nothing—an output when no input, other than dc power, is applied. A *useful* oscillator provides a specific waveform at a particular frequency, and it is capable of *controlling the frequency* of the output. Sine wave oscillators are used to generate test waveforms, to provide the frequencies for radio transmitters, and in radio receivers. The local oscillator in a radio is an example. Square-wave oscillators often provide the *clock* signals used to synchronize and control the operation of digital circuits and computers.

Spurious oscillations, shown in Figure 11-14c, are unwanted, uncontrolled oscillations that occur in circuits that are *not* designed to be oscillators. Their waveforms are not necessarily sinusoidal, as Figure 11-14c shows. Spurious oscillations most commonly occur in high-gain amplifiers, and the designers of these amplifiers take great pains to eliminate them. The two-stage BJT amplifier of Figure 6-3 has a gain of 25,000 if the reduction due to the input is omitted. If this circuit is built in the laboratory, it will almost surely oscillate, and therefore be useless as an amplifier. Engineers have often been heard to lament "If I build an oscillator, it amplifies; if I build an amplifier, it oscillates."

11-5.1 The Barkhausen Criterion

An oscillator depends upon feedback to produce its oscillations. The *Barkhausen criterion* can be used to determine whether, and under what conditions, a circuit will oscillate. The equation for a feedback circuit has been developed in Chapter 8:

$$A(f) = \frac{A}{1 + A\beta} \qquad (8\text{-}1)$$

If, at any frequency or under any conditions, $A\beta = -1$, the gain of the circuit is infinite. This is the Barkhausen criterion for oscillations. In most circuits both A and β are functions of frequency, but if $A\beta = -1$ at any frequency, it means that any miniscule voltage at that frequency will produce an output, or that a circuit will break into oscillations.

An alternate way to look at the Barkhausen criterion is to reason that it is satisfied if $A\beta = 1$ and the *phase shift* is 180°.

Most of the circuits studied in Chapter 8 contained amplifiers that had an inherent 180° phase shift, but the feedback voltage was introduced in such a way as to *oppose* the applied voltage. This negative feedback reduced the gain. If the feedback circuit produces an *additional* 180° of phase shift, the total phase shift will be 360°; the voltage fed back will *aid* the applied voltage. This is *positive feedback*. As stated in Chapter 8, it leads to instability and oscillations.

A more complex criterion for oscillation is the *Nyquist criterion*, but it can be simplified as follows:

1. The Nyquist criterion states that the phase shift of the feedback voltage, $A\beta$, varies as a function of frequency.
2. At some frequency, f_0, the phase shift of $A\beta$ will be 180°.
3. If $|A\beta|$ is less than 1 at f_0, the circuit is stable; if $|A\beta|$ is 1, the circuit will oscillate.

The fact that the phase shift must be 180° is very useful in determining the frequency of oscillation of a circuit, and the required relationship between A and β.

11-5.2 Conditions for Spurious Oscillations

The Barkhausen and Nyquist criteria indicate why spurious oscillations tend to occur in high-gain amplifiers. If the gain of the amplifier, A, is high, there is a greater probability that $A\beta$ will be ≥ 1. In addition, the feedback path for spurious oscillations is generally through stray capacitance between the wires and other circuit components. These small capacitances are very difficult to control, but they are more conductive at high frequencies. This is why spurious oscillations are usually at high frequencies.

11-6 HIGH-FREQUENCY SINUSOIDAL OSCILLATORS

Most high-frequency oscillators are built around a tank circuit. The simplest of these is the inductively coupled LC oscillator. A typical LC oscillator is shown in Figure 11-15. FETs (see chapter 5) are often used in oscillator circuits because of their high input impedance. Under dc conditions the gate of the JFET is at ground and the bias is developed across the source resistor, R_S.

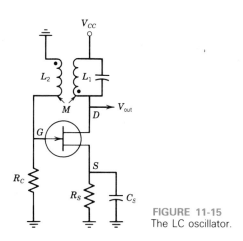

FIGURE 11-15
The LC oscillator.

As explained in the previous section, oscillators require positive feedback. In the LC oscillator, when the drain current is rising, the drain voltage is falling [$V_D = V_{CC} - L(di/dt)$]. The changing voltage across the inductor is coupled to the gate by transformer action in such a way as to raise the gate voltage and increase the current in the JFET. This provides the positive feedback and the 180° of phase shift necessary for oscillation. When the JFET saturates, the drain current stops rising. At this time the gate is well below the source voltage due to the large drop across R_S, and the current starts to decrease. The falling current is coupled to the gate so as to make it more negative and accelerate the decrease. This continues until the drain current is zero. Then the coupled voltage again goes to zero, the current starts to rise, and the cycle repeats itself. The frequency of oscillation is controlled by the resonant frequency of the tank circuit, which also filters harmonics and other nonlinearities, and provides a sinusoidal output.

For oscillations to occur, $A\beta$ must equal 1. In the LC oscillator $A = g_m R$ where $R = R_P \parallel r_d$ and β, the portion of the output voltage fed back to the gate, is equal to the turns ratio between L_1 and L_2.

EXAMPLE 11-7

The circuit of Figure 11-15 uses an inductor of 1 mH with a Q of 50. If $C = 1000$ pF and $L_1:L_2 = 10:1$, find the frequency of oscillation, and the minimum g_m of the JFET for oscillations. Assume the JFET has a drain resistance, r_d, of 50 kΩ.

Solution
The frequency of oscillation is determined by the LC tank circuit.

$$f = \frac{1}{2\pi\sqrt{LC}} = \frac{1}{2\pi\sqrt{10^{-12}}} = \frac{10^6}{2\pi} = 159{,}000 \text{ Hz}$$

$$R_P = Q\omega L = 50 \times 2\pi \times 159{,}000 \text{ Hz} \times 10^{-3} \text{H} = 50{,}000 \text{ }\Omega$$

$$R = r_d \parallel R_P = 25{,}000 \text{ }\Omega$$

$$\beta = \frac{L_2}{L_1} = 0.1$$

Because $A\beta = 1$, we have

$$1 = g_m \times 25{,}000 \times 0.1$$

or

$$g_m = \frac{1}{2500 \text{ }\Omega} = 400 \text{ }\mu S$$

This is the minimum value of g_m for oscillation. Most JFETs have larger values of g_m, so the circuit will almost surely oscillate.

LC oscillators can also be built using BJT transistors. Some of the References discuss them.

11-6.1 The Theoretical Basis for Reactive Oscillators

Figure 11-16 shows a theoretical oscillator. Practical applications of this circuit are given in the next two sections. In Figure 11-16, X_1, X_2, and X_3 are all reactances. X_1 and X_2 are the same type of reactances (inductive or capacitive) and X_3 is the opposite type. The blue dashed line in the figure indicates that X_1 and X_2 can be connected together, and they usually are. They are in parallel with X_3 and form a tank circuit. The circuit assumes a JFET amplifier, although BJT transistors could also be used. Again, R is the parallel impedance of the transistor output impedance, r_d, and the parallel resistance of the inductor, R_P.

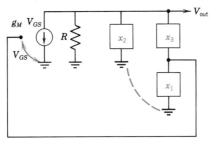

FIGURE 11-16
The basic circuit for a reactive oscillator.

The gain of this circuit is

$$A_v = -g_m Z$$

where

$$Z = R \parallel X_2 \parallel (X_1 + X_3)$$

After some algebra, this becomes

$$A_v = \frac{-g_m R\, X_2(X_1 + X_3)}{R(X_1 + X_2 + X_3) + X_2(X_1 + X_3)} \qquad (11\text{-}8)$$

For this circuit, β is the voltage at the gate divided by V_{out}.

$$\beta = \frac{X_1}{X_1 + X_3} = \frac{-X_1}{X_2} \qquad (11\text{-}9)$$

Assuming, as we have, that X_1, X_2, and X_3 are purely imaginary, equation 11-9 shows that β is a real number. For oscillation, $A\beta = 1 < 180°$, so that A must also be a real number. The term in the numerator and the second term in the denominator of A are real, but the first term in the denominator is imaginary. For A to be real, this term must be zero. Therefore

$$X_1 + X_2 + X_3 = 0$$

This allows us to determine the frequency of oscillation.

11-6.2 The Colpitts Oscillator
The Colpitts oscillator is the most common practical application of the reactive oscillator of Figure 11-16. Its basic circuit is shown in Figure 11-17, with the reactive components shown in blue. In the Colpitts oscillator, X_1 and X_2 are capacitors and X_3 is an inductor. The RFC is a radio frequency choke coil used

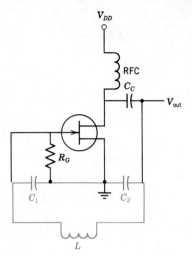

FIGURE 11-17
The basic Colpitts oscillator.

to isolate the power supply from the tank circuit, C_C is a coupling capacitor used to block the dc drain voltage, and R_G is the gate resistance.

Equation 11-8 requires that $X_1 + X_2 + X_3 = 0$, or that

$$X_1 + X_2 = -X_3 \qquad (11\text{-}10)$$

Equation 11-10 shows that the resonant frequency of the Colpitts oscillator is

$$f = \frac{1}{2\pi L C_{eq}} \qquad (11\text{-}11)$$

where C_{eq} is the equivalent capacity of the two capacitors in series.

$$C_{eq} = \frac{C_1 C_2}{C_1 + C_2} \qquad (11\text{-}12)$$

EXAMPLE 11-8

Given a coil, $L = 10\ \mu H$, design a Colpitts oscillator to oscillate at 5 MHz, with $\beta = 0.1$.

Solution
From equation 11-11 we have

$$C_{eq} = \frac{1}{4\pi^2 L f^2} = \frac{1}{4\pi^2 \times 10^{-5} \times 25 \times 10^{12}} = \frac{10^9}{\pi^2} = 101\ \text{pf}$$

From equation 11-9 we have

$$\beta = 0.1 = \frac{X_1}{X_2} = \frac{C_2}{C_1}$$

Combining these in equation 11-12 gives

$$C_{eq} = 101 \text{ pf} = \frac{C_1 C_2}{C_1 + C_2} = \frac{0.1\, C_1^2}{1.1\, C_1}$$

The results are $C_1 = 1111$ pf and $C_2 = 111$ pf.
This circuit will oscillate as long as its gain is greater than 10.

Figure 11-18 shows a more practical form of the Colpitts oscillator. Here, the FET is connected to a source follower. This circuit has a low gain, but it also has a low output impedance, so it can drive a heavy load.

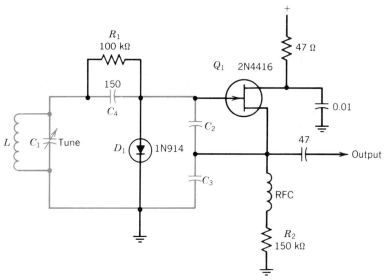

FIGURE 11-18
A practical Colpitts oscillator circuit. (Adapted from *Radio Handbook*. Reproduced with permission of the publisher Howard W. Sams & Co., Indianapolis, Indiana, *Radio Handbook* by Wm. Orr, Copyright © 1986.)

Because the gain of the FET is approximately 1, β must be greater than 1. Equation 11-9 shows that this will occur if $C_2 > C_3$. C_1 is used to fine-tune the circuit and set the frequency of oscillation. Placing C_4 in the loop allows the designer to increase the values of C_2 and C_3, which makes them less sensitive to stray capacitance or changes in capacity due to temperature or other factors. The equivalent capacity across the inductor is

$$C_{eq} = C_1 + \frac{1}{1/C_2 + 1/C_3 + 1/C_4}$$

11-6.3 The Hartley Oscillator

In the Colpitts oscillator, X_1 and X_2 were capacitors and X_3 was an inductor (see Figure 11-16). The Hartley oscillator is the complement of the Colpitts; X_1 and X_2 are inductors and X_3 is a capacitor. In most Hartley oscillators X_1 and X_2 are part of a single tapped inductor, as shown in Figure 11-19. The frequency of oscillation is

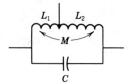

FIGURE 11-19
The Hartley oscillator tank circuit.

$$f = \frac{1}{2\pi \sqrt{L_{eq} C}}$$

where

$$L_{eq} = L_1 + L_2 + 2M$$

For a tightly coupled inductor, such as a tapped coil, L_{eq} is simply the inductance of the entire coil.

Figure 11-20 shows a typical Hartley oscillator, using a BJT transistor.

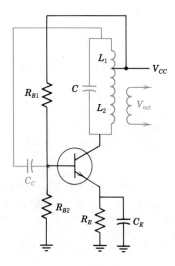

FIGURE 11-20
A Hartley oscillator circuit.

For this circuit, $\beta = L_1/L_2$ and C_c is a capacitor that couples the feedback signal to the base. The output is coupled to the circuit by transformer action.

EXAMPLE 11-9

The circuit of Figure 11-20 is to oscillate at 1 MHz. For the transistor, assume $h_{fe} = 100$ and $h_{ie} = 1000\ \Omega$. If the coil is 100 μH, tapped at 90 μH and 10 μH, with a Q of 20, find the value of the capacitor and the minimum value of the reflected impedance of the load.

Solution
To find the resonant frequency

$$f = \frac{1}{2\pi \sqrt{LC}}$$

$$C = \frac{1}{4\pi^2 f^2 L} = \frac{1}{4\pi^2 \times 10^{12} \times 10^{-4}} = \frac{10^8}{4\pi^2} = 254\ \text{pF}$$

For oscillation

$$\beta = \frac{L_1}{L_2} = \frac{1}{9}$$

The gain of the circuit must therefore be greater than 9.

$$g_m = \frac{h_{fe}}{h_{ie}} = 0.1\ \text{S}$$

This means that the total equivalent load impedance must be greater than 90 Ω.

$$R_P = Q\omega L = 20 \times 2\pi \times 10^6 \times 10^{-4} = 12{,}560$$

The reflected impedance in parallel with R_P must be at least 90 Ω.

$$R_F = \frac{R_P R_T}{R_P - R_T} = \frac{12{,}560\ \Omega \times 90\ \Omega}{12{,}470\ \Omega} = 90.65\ \Omega$$

11-6.4 Crystal Oscillators

Colpitts and Hartley oscillators are used when variable-frequency oscillators are required, such as in the local oscillators in radio receivers. When a highly stable, fixed-frequency oscillator is required, a *crystal oscillator* is usually used. Crystal oscillators have a wide range of uses, from controlling the frequency of a radio transmitter to generating the "clock" signals for a microprocessor. One commonly available crystal oscillates at 3.59 MHz and is used in color television.

A crystal is a small piece of quartz that exhibits *piezoelectric effects*; it produces a voltage when it is mechanically deformed, and it deforms in response to an applied voltage. When an alternating voltage is applied to a crystal, it oscillates at a resonant frequency that depends on its physical dimensions. Thus crystals can be cut to resonate at any required frequency. Practical crystals oscillate at high frequencies (1 MHz or more). At lower frequencies crystals become too large, but lower frequencies can often be obtained by dividing down a high frequency.

Figure 11-21a shows a crystal packaged for use in an oscillator. Figure 11-21b shows its symbol, and Figure 11-21c is the electrical equivalent circuit. It is a tank circuit with a very high Q; 2000 is typical.

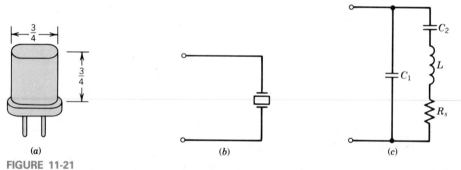

FIGURE 11-21
The quartz crystal. (From *Essentials of Solid State Electronics* by Rodney B. Faber. Copyright © 1985 by John Wiley & Sons, Inc. Reprinted by permission of John Wiley & Sons, Inc.)

The impedance of a crystal as the frequency varies is shown in Figure 11-22. The crystal exhibits a series resonant frequency, at which its impedance is very low, and a parallel resonance at which its impedance is very high. The two frequencies are generally within 1% of each other. Oscillators can be built with the crystal resonating in either mode.

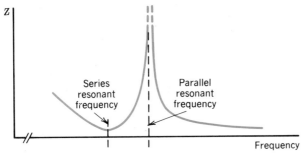

FIGURE 11-22
Crystal impedance versus frequency. (From *Essentials of Solid State Electronics* by Rodney B. Faber. Copyright © 1985 by John Wiley & Sons, Inc. Reprinted by permission of John Wiley & Sons, Inc.)

The *Pierce* oscillator shown in Figure 11-23 is one of the most popular crystal-controlled oscillators. The crystal operates in the series resonant mode and provides maximum feedback when its impedance is low. The circuit can be tuned over a small range of frequencies by varying C_G or C_D or by placing a small capacitor across the crystal.

The *tuned-drain crystal-gate oscillator*, also called the Miller oscillator, is shown in Figure 11-24. The Miller capacitance between the gate and the drain

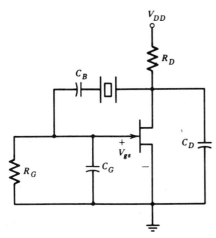

FIGURE 11-23
Pierce oscillator circuit. (From *Solid State Radio Engineering* by Herbert L. Krauss, Charles W. Bostian, and Frederik H. Raab. Copyright © 1980 by John Wiley & Sons, Inc. Reprinted by John Wiley & Sons, Inc.)

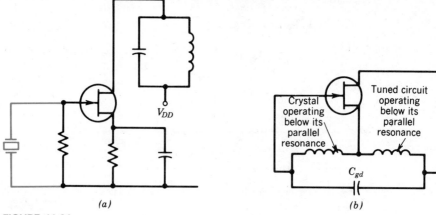

FIGURE 11-24
A tuned-drain crystal-gate oscillator. (From *Essentials of Solid State Electronics* by Rodney B. Faber. Copyright © 1985 by John Wiley & Sons, Inc. Reprinted by permission of John Wiley & Sons, Inc.)

provides coupling between the crystal and the tank circuit. If the tank circuit is tuned anywhere near the resonant frequency of the crystal, the crystal will determine the frequency of oscillation.

There are many variations on the Colpitts, Hartley, and crystal-controlled oscillators. Obviously we cannot describe all of them, but more information is available in the References.

11-7 PHASE-SHIFT OSCILLATORS

Oscillators that depend on tank circuits or crystals are generally used for frequencies above 100 kHz. Crystals, in particular, become large and unwieldy when designed to oscillate at low frequencies.

For low-frequency oscillations, *phase-shift oscillators* are often used. As with any oscillator, a portion of the output must be fed back *in phase* with the input for oscillations to occur. This also means that a 360° phase shift will allow oscillations. The basic amplifier provides 180° of phase shift, and an additional 180° must be introduced by other means. Phase-shift oscillators are controlled by RC networks that provide the proper phase shift.

11-7.1 The Wien Bridge Oscillator

The Wien bridge, shown in Figure 11-25, has been in use for many years. It can be used to measure the value of an unknown capacitor. If C_1 is a known, adjustable capacitor and C_2 is an unknown capacitor, and an ac input is applied, the circuit can be made to *null* (no current flowing in the center arm) when C_1 is adjusted to equal C_2.

The Wien bridge can also be used as the basis of an oscillator. It requires two stages of amplification to provide the necessary 360° phase shift. The

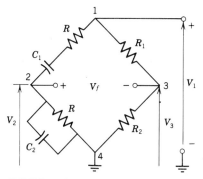

FIGURE 11-25
A Wien bridge frequency-selective network.

output is then fed back to the input through the reactive arm of the bridge, as shown in Figure 11-26.

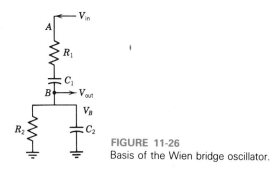

FIGURE 11-26
Basis of the Wien bridge oscillator.

If point A is connected to the output of an amplifier with a 360° phase shift and point B is connected to its input, then

$$\beta = \frac{V_B}{V_A} = \frac{\dfrac{R_2/j\omega C_2}{R_2 + \dfrac{1}{j\omega C_2}}}{\dfrac{R_2/j\omega C_2}{R_2 + \dfrac{1}{j\omega C_2}} + R_1 + \dfrac{1}{j\omega C_1}} \qquad (11\text{-}13)$$

In many circuits, $R_1 = R_2 = R$ and $C_1 = C_2 = C$. Then equation 11-13 reduces to

$$\beta = \frac{R}{3R + j\left(\omega C R^2 - \dfrac{1}{\omega C}\right)}$$

11-7 PHASE-SHIFT OSCILLATORS   553

For oscillations, β must be a real number. This occurs when $\omega = 1/RC$. Then $\beta = 1/3$, so the gain of the two-stage amplifier must be at least 3. This requirement is easily met.

When $R_1 \neq R_2$ and $C_1 \neq C_2$, the algebra is a bit more complex, but the results are

$$f = \frac{1}{2\pi\sqrt{R_1 R_2 C_1 C_2}}$$

$$\beta = \frac{1}{1 + (C_2/C_1) + (R_1/R_2)}$$

A basic Wien bridge oscillator circuit is shown in Figure 11-27.

EXAMPLE 11-10

Using the circuit of Figure 11-27, design an oscillator for 1 kHz.

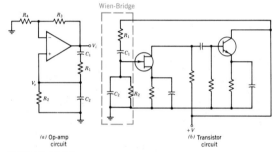

FIGURE 11-27
Basic Wien bridge oscillator circuits. (From *Essentials of Solid State Electronics* by Rodney B. Faber. Copyright © 1985 by John Wiley & Sons, Inc. Reprinted by permission of John Wiley & Sons, Inc.)

Solution

The designer is at liberty to select the values of R and then find the corresponding values of C. The simplest case is to let $R_1 = R_2$ and $C_1 = C_2$. If $R = 10\ \text{k}\Omega$ is selected

$$C = \frac{1}{2\pi f R} = \frac{1}{2\pi \times 10^3 \times 10^4} = \frac{10^7}{2\pi} = 1.59\ \mu\text{F}$$

11-7.2 The RC Phase-Shift Oscillator

An oscillator can be built using a single transistor and an RC network to develop the additional 180° phase shift required. The RC network is shown in

Figure 11-28. The resistors and capacitors in the network are equal. The equations for it are

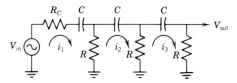

FIGURE 11-28
The RC phase-shift network.

$$V_{in} = (R_C + R + X_C)I_1 - RI_2$$
$$0 = RI_1 + (2R + X_C)I_2 - RI_3$$
$$0 = -RI_2 + (2R + X_C)I_3$$

Solving for I_3 we have

$$I_3 = \frac{V_{in} R^2}{\begin{vmatrix} R_C + R + X_C & -R & 0 \\ -R & 2R + X_C & -R \\ 0 & -R & 2R + X_C \end{vmatrix}}$$

$$V_{out} = I_3 R = \frac{V_{in} R^3}{R^3 + 6R^2 X_C + 5RX_C^2 + X_C^3 + {}^3eR_C R^2 + 4R_C R X_C + R_C X_C^2}$$

(11-14)

Oscillations will occur when the denominator is a real number. Equating the imaginary terms in the denominator gives

$$6 R^2 X_C + 4 R_C R X_C + X_C^3 = 0$$

or

$$X_C = R \sqrt{6 + \frac{4R_C}{R}} \qquad (11\text{-}15)$$

This leads to an oscillation frequency of

$$f = \frac{1}{2\pi RC \sqrt{6 + \dfrac{4R_C}{R}}} \qquad (11\text{-}16)$$

Replacing X_C^2 with $-(6R^2 + 4R_C R)$ in equation 11-14 and eliminating the imaginary terms gives

$$\frac{V_{out}}{V_{in}} = -\left(\frac{1}{29 + \dfrac{23R_C}{R} + \dfrac{4R_C^2}{R^2}}\right) \qquad (11\text{-}17)$$

where the minus sign indicates the 180° phase shift.

Figure 11-29 shows a JFET circuit using the RC network. With JFETs it is

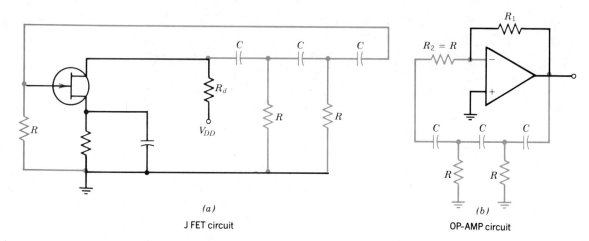

(a) J FET circuit

(b) OP-AMP circuit

FIGURE 11-29
Phase-shift oscillator. (From *Essentials of Solid State Electronics* by Rodney B. Faber. Copyright © 1985 by John Wiley & Sons, Inc. Reprinted by permission of John Wiley & Sons, Inc.)

possible to make $R \gg R_C$, and feed the entire voltage back. If $R \gg R_C$, equations 11-16 and 11-17 reduce to

$$f = \frac{1}{2\pi RC \sqrt{6}}$$

$$\beta = \frac{V_{out}}{V_{in}} = \frac{1}{29}$$

Thus the JFET circuit must have a gain of at least 29 for oscillation.

An oscillator circuit using a BJT is shown in Figure 11-30. The resistor, R_F, has been added to make R_F plus the input impedance of the transistor equal to R. In this circuit it is necessary to keep R small, just a bit larger than h_{ie}. Example 11-11 shows why.

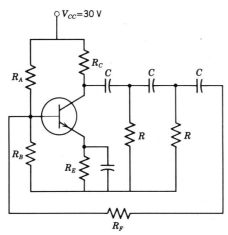

FIGURE 11-30
The BJT phase-shift oscillator.

EXAMPLE 11-11

For the BJT oscillator circuit of Figure 11-30, assume the transistor has an $h_{ie} = 400\,\Omega$, $R_C = 1500\,\Omega$, and the parallel resistance of the two biasing resistors is $1600\,\Omega$.

a. Choose R as $500\,\Omega$ and find C if the circuit is to oscillate at 8 kHz.
b. Find R_F.
c. Find the minimum value of h_{fe} for oscillations.

Solution
a. From Equation 11-16 we have

$$C = \frac{1}{2\pi f R \sqrt{6 + \dfrac{4R_C}{R}}}$$

Here

$$\frac{R_C}{R} = \frac{1500\,\Omega}{500\,\Omega} = 3$$

$$C = \frac{1}{2\pi \times (8 \times 10^3)\, 500 \times \sqrt{18}} = \frac{10^6}{106.5} = 0.00938\ \mu F$$

b. The input impedance of the transistor is $1600\,\Omega \parallel 400\,\Omega = 320\,\Omega$. R_F must be $180\,\Omega$ to bring R up to $500\,\Omega$.

c. From equation 11-17 we have

$$\frac{V_{out}}{V_{in}} = \frac{-1}{29 + 23 \times 3 + 4 \times 9} = \frac{-1}{134}$$

The portion of V_{out} that appears across the base will be $V_{out} \times 320/500$. Thus the gain of the transistor must be at least

$$134 \times \frac{500}{320} = 209.375$$

$$A_{v(tr)} = h_{fe} \frac{R_C}{h_{ie}}$$

$$h_{fe(min)} = \frac{209.375 \times 400}{1500} = 55.833$$

Making R larger will reduce the portion of V_{out} that appears at the base.

11-8 DIGITAL OSCILLATORS

Digital oscillators are used for many applications because they are small, inexpensive, and readily available. They produce a square wave output instead of a sine wave output, and often the square wave does not have a 50% duty cycle.

The output of a digital oscillator is often called a *clock* and is used in many circuits to synchronize operations. Both variable- and fixed-frequency digital oscillators exist.

The frequency of a variable-frequency oscillator is usually determined by varying the R or C of an RC network that controls the output. Highly stable digital oscillators generally use crystals to establish their frequency. Most microprocessor systems, for example, use crystal-controlled oscillators to set their basic timing. The crystal-controlled oscillator in an Apple computer oscillates at 14.31818 MHz.

11-8.1 The 555 Timer

The **555** is a very popular and commonly used digital oscillator. As shown in Figure 11-31, the **555** comes in an 8-pin dual-in-line (DIP) package and is

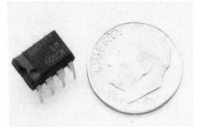

FIGURE 11-31
The 555 IC Timer. (Photograph courtesy of Howard LeVant.)

about half the size of an ordinary digital integrated circuit. The frequency of the **555** is controlled by external resistors and capacitors that must be tied to the pins on the **555**. The **556** is another version of the **555**. It comes in a 14-pin DIP, but contains two timers within it.

Figure 11-32 shows a simplified equivalent circuit for the **555**. Internally it has five basic parts:

1. The lower comparator.
2. The upper comparator.
3. The internal FF.
4. The discharge transistor.
5. The output driver.

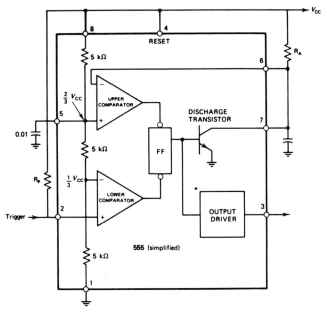

FIGURE 11-32
A simplified equivalent circuit of the **555** timer. (Taken from Motorola **MC 1455/1555** data sheets and Motorola Semiconductor Data Library, Vol. 7, pp. 2-25. Courtesy of Motorola Integrated Circuits Division.)

The lower comparator compares the voltage on its input (pin 2) with $\frac{1}{3}V_{CC}$. V_{CC} can be any voltage from 4.5 to 18 V. If the voltage on pin 2 is less than $\frac{1}{3}V_{CC}$, the lower comparator produces a *low* output voltage, which *clears* the FF.

The upper comparator compares the voltage on its input (pin 6) with $\frac{2}{3}V_{CC}$. If its input is greater than $\frac{2}{3}V_{CC}$, it *sets* the FF. Pin 5 of the **555** can be used to control or change the threshhold voltage of the upper comparator. Normally this feature is not used and pin 5 is connected to ground through a 0.01 μF capacitor.

If the internal FF is *set*, it saturates the discharge transistor and also causes the output driver to produce a 1 output. Conversely, if the FF is *clear*, the discharge transistor is cut off and the output driver produces a 0 output.

The **555** can be used as an astable circuit or clock generator, by connecting it as shown in Figure 11-33a. Here, both comparator inputs (pins 2 and 6) are connected to the capacitor. The circuit operates as follows:

1. Assume the discharge transistor is off. Then the capacitor charges toward V_{CC} through R_A and R_B.
2. When the capacitor voltage reaches $\frac{2}{3}V_{CC}$ the upper comparator *sets* the FF. This turns on the discharge transistors and the capacitor discharges toward ground through R_B.
3. When the capacitor voltage becomes as low as $\frac{1}{3}V_{CC}$, the lower comparator *clears* the FF. The discharge transistor turns *off* and the capacitor starts to charge again. This process continues indefinitely, resulting in an oscillator or clock generator.

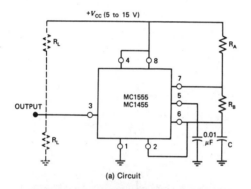

(a) Circuit

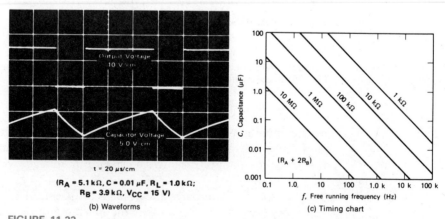

(b) Waveforms

(c) Timing chart

FIGURE 11-33
The **555** connected as an oscillator. (Taken from Motorola **MC 1455/1555** data sheets and Motorola Semiconductor Data Library, Vol. 7, pp. 2-25. Courtesy of Motorola Integrated Circuits Division.)

4. The output frequency can be obtained from the timing chart (Figure 11-33c) or

$$f = \frac{1.44}{(R_A + 2R_B)C} \qquad (11\text{-}18)$$

5. The output and capacitor voltage waveforms for an 11 kHz oscillator are shown in Figure 11-33b. Note that $R_A + 2R_B \approx 13\ k\Omega$, and a frequency of 11 kHz results using either equation 11.8 or the chart.

EXAMPLE 11-12

Design a 100 kHz oscillator using a **555**.

Solution
Using equation 11-18 we find

$$f(R_A + 2R_B)C = 1.44$$

If we choose $C = 10^{-9}$ (0.001 µF), then $R_A + 2R_B = 14{,}400\ \Omega$. This is a good choice as it keeps R_A and R_B in the kΩ range. There are many ways to select R_A and R_B. One choice, which makes them approximately equal, is to set $R_A = 5\ k\Omega$ and $R_B = 4.7\ k\Omega$.

To check, we use the timing chart with $R_A + 2R_B = 14{,}400\ \Omega$ and $C = .001$ µF. This combination does yield a frequency of approximately 100 kHz.

Other digital oscillators can be made using one-shots, Schmitt triggers, or astable multivibrators. A more thorough discussion of digital one-shots and oscillators is given in Chapter 7 of *Practical Digital Design Using ICs* (see References).

11-8.2 Digital Crystal Oscillators

Digital oscillators that require a constant, highly stable output use crystals to control their frequency. Figure 11-34 shows a typical digital crystal oscillator.

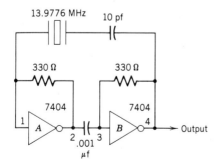

A crystal-controlled digital oscillator.

It consists of two digital inverters, connected in a loop, with the crystal between them. The resistors and capacitors stabilize the circuit and the crystal controls the frequency of oscillation. This circuit produces a square wave or clock output of 13.9776 MHz.

11-9 SUMMARY

This chapter started with a discussion of tuned-circuit amplifiers, which amplify only at the resonant frequency of the tank circuit in their output. Then amplifiers that have a better, more selective waveshape, such as double-tuned and stagger-tuned amplifiers, were considered. The chapter then discussed the design and construction of oscillators. Tank-circuit oscillators, such as the Colpitts, Hartley, and crystal-controlled oscillator, were described. The operation of phase-shift and Wien bridge oscillators was also explained. The chapter concluded with a brief introduction to digital oscillators.

11-10 GLOSSARY

Butterworth amplifier. An amplifier whose poles are set at frequencies to provide a flat output over a particular frequency range.

Clock. The fixed frequency output of a digital oscillator.

Coefficient of coupling. A measure of the coupling between the windings of a transformer.

Colpitts oscillator. An oscillator characterized by a tuned circuit consisting of an inductor and two capacitors connected in series.

Crystal. A pierce of quartz or other material that resonates at a specific frequency.

Digital oscillator. An oscillator that produces a square wave output. It is usually constructed from digital circuits.

Double-tuned transformer. A transformer whose primary and secondary are both tuned to a specific frequency by capacitors.

Hartley oscillator. An oscillator characterized by a tank circuit consisting of a capacitor and a tapped inductor. Part of the inductor voltage is fed back to produce the oscillations.

Maximally flat amplifier. See Butterworth.

Oscillator. A circuit that generates a sine wave or square wave output of fixed frequency with no input other than V_{CC}.

Piezoelectric effect. The effect in a crystal whereby a mechanical deformation produces a voltage, and vice versa.

Phase-shift oscillator. A circuit that oscillates by introducing an additional 180° of phase shift in the output and feeding it back to the input.

Q. The quality factor of a coil. It is the ratio of the inductive reactance to the coil's resistance.

Selectivity. The ability of an amplifier to be selective; to amplify a certain range of frequencies and reject all other ranges.

Stagger-tuned amplifier. An amplifier consisting of several stages, with each stage tuned to a different frequency.

Tank circuit. An LC circuit set to resonate at a specific frequency.

Tuned-circuit amplifier. An amplifier that uses a tuned circuit or tank circuit as the load. Amplification takes place over a small frequency range.

Wien bridge. A bridge circuit sometimes used to determine the value of an unknown capacitor. It is also used as a Wien bridge oscillator, which is a phase-shift type.

11-11 REFERENCES

ROBERT BOYLESTAD and LOUIS NASHELSKY, *Electronic Devices and Circuit Theory*, 4th Edition, Prentice-Hall, Englewood Cliffs, N.J., 1987.

RODNEY B. FABER, *Essentials of Solid State Electronics*, John Wiley, New York, 1985.

JOSEPH D. GREENFIELD, *Practical Digital Design Using ICs*, 2nd Edition, John Wiley, New York, 1983.

HERBERT L. KRAUSS, CHARLES W. BOSTIAN, FREDERICK H. RAAB, *Solid State Radio Engineering*, John Wiley, New York, 1980.

JACOB MILLMAN, *Microelectronics*, 2nd Edition, McGraw-Hill, New York, 1987.

JOSEPH MAYO PETTIT and MALCOLM MYERS MCWHORTER, *Electronic Amplifier Circuits*, McGraw-Hill, New York, 1961.

DONALD L. SCHILLING and CHARLES BELOVE, *Electronic Circuits, Discrete and Integrated*, 2nd Edition, McGraw-Hill, New York, 1979.

11-12 PROBLEMS

11-1 The Q of a coil used in a 1.0 MHz tuned amplifier is 50 and its inductive reactance is 100 Ω. What is the resonant impedance of the circuit and the bandwidth?

11-2 For the circuit of Figure P11-2, assume the coil has a Q of 30. Find the resonant frequency and bandwidth of the output.

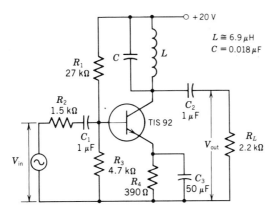

FIGURE P11-2

11-3 For the circuit of Figure P11-3 find the gain and bandwidth of the circuit
   a. If $1/h_{oe} = \infty$.
   b. If $1/h_{oe} = 18{,}000 \, \Omega$.
   c. For part a, find the gain of the circuit at 18 kHz. (*Hint:* Figure 11-11 might help.)

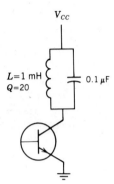

FIGURE P11-3

11-4 Given a coil $L = 10^{-4}$ H, $Q = 50$, design a circuit to resonate at 50,000 Hz with a bandwidth of 2 kHz. Put the circuit in the collector of a transistor with $g_m = 0.1$ S, and find the mid-frequency gain.

11-5 A tuned transistor amplifier operates with a center frequency of 40 MHz and a bandwidth of 8 MHz, a coil with a Q of 100 and an inductance of 0.2 µH. What resistance is required across the coil to produce the desired bandwidth and what value of capacitance is required for resonance?

11-6 Find the capacitor and resistor required for the other two poles in Example 11-5.

11-7 A two-pole, stagger-tuned amplifier is to have a resonant frequency of 31.8 MHz and a bandwidth of 2 MHz. If the available coils are 1 µH and have a Q of 100, find the resistor and capacitor needed for each circuit.

11-8 An amplifier is to amplify a 15 MHz signal with a bandwidth of 1 MHz. Find the pole locations for
   a. A three-stage maximally flat amplifier.
   b. A four-stage maximally flat amplifier.

11-9 An amplifier must amplify the band of frequencies between 9.5 and 10.5 MHz using a three-stage Butterworth amplifier. Find the frequency and bandwidth of each pole. If each stage individually gains 16 dB, what is the gain of the overall amplifier at
   a. 10 MHz
   b. 9.5 MHz
   c. 9.0 MHz

11-10 Design a Colpitts oscillator to oscillate at 1 MHz with $\beta = 0.15$ and $C_T = 1000$ pF. If a JFET is used with $g_m = 0.0015$, find the inductance and the minimum Q of the coil.

11-12 For the circuit of Figure 11-18, $C_1 = 50\text{–}500$ pF, $C_2 = 0.04$ µF, $C_3 = 0.01$ µF, and $C_4 = 200$ pF. If $L = 100$ µH, find the range of oscillation frequencies.

11-13 For a Hartley oscillator, $L_1 = 5$ µH, $L_2 = 50$ µH and $C = 100$ pF. Find the frequency of oscillation and β.

11-14 For the Wien bridge oscillator of Figure 11-27, find the frequency of oscillation if $R_1 = R_2 = 15$ kΩ and $C_1 = C_2 = 0.1$ µF.

11-15 Repeat problem 11-14 if R_1 only is changed to 10 kΩ. What is the new value of β?

11-16 Design a Wien bridge oscillator for 5 kHz.

11-17 The phase-shift oscillator of Figure 11-30 has $C = 0.01$ µF, $R_C = 2000$ Ω, $R = 1000$ Ω, and the input impedance of the transistor circuit is 600 Ω. Find the frequency of oscillation, the value of R_F and the minimum h_{fe} of the transistor for oscillation.

11-18 A 555 timer has $R_A = R_B = 100$ kΩ, and $C = 1$ µF. Find its frequency of oscillation from the curves (Figure 11-33) and verify it by the equation.

CHAPTER TWELVE

Operational Amplifiers

12-1 Instructional Objectives

This chapter completes this book by introducing the operational amplifier (op-amp), which is a very popular and commonly used circuit element. After reading it, the student should be able to:

1. Find the gains of inverting and noninverting op-amps.
2. Design inverting and noninverting op-amps for specific gains.
3. Design summing amplifiers using op-amps.
4. Compensate an op-amp.
5. Design integrator and differentiator circuits.
6. Set up op-amp circuits to solve differential equations.
7. Build log and antilog amplifiers.
8. Build filters with a specific frequency response.

12-2 Self-Evaluation Questions

Watch for the answers to the following questions as you read the chapter. They should help you to understand the material presented.

1. What is a virtual ground? What are its characteristics?
2. Why is an op-amp needed in a summing network?
3. Why are integrators preferred to differentiators?

4. What is the difference between a difference amplifier and an instrumentation amplifier?
5. Why can't the output of a log amplifier be more than 0.7 V?
6. What is the difference between an active and a passive filter? What are the advantages of active filters?
7. What is the function of a weighted resistor network in a D/A converter?

12-3 INTRODUCTION

The purpose of this section is to introduce the student to the operational amplifier, to show how it operates and what it does. A more detailed discussion of the characteristics of an operational amplifier, such as input impedance, offset, and slew rate, is deferred until section 12-4.

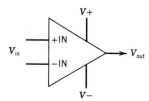

FIGURE 12-1
The basic op-amp.

Figure 12-1 shows the basic symbol for an operational amplifier (op-amp). It shows that an op-amp has two inputs: a *noninverting* input (+IN), and an *inverting input* (−IN). A voltage applied to the noninverting input drives the output positive, while a voltage applied to the inverting input drives the output negative. The power supply for an op-amp is *double-ended;* it supplies identical positive and negative voltages, V+ and V−. The output voltage, V_{out} can swing either positive or negative, but its excursion is limited by the power supply voltages. A typical op-amp uses a power supply of ± 15 V, and the output swing might be limited to ± 10 V.

The voltage gain of an op-amp is very high, typically 200,000, or higher than 100 dB. This intrinsic gain of the op-amp is called the *open-loop gain*, A_{ol}. In actual operation, negative feedback is introduced, closing the loop and greatly reducing the gain.

The voltage gain of an op-amp, A_{ol}, is given by equation 12-1:

$$V_{out} = A_{ol}(V_{+IN} - V_{-IN}) \tag{12-1}$$

where V_{+IN} is the voltage at the +IN terminal and V_{-IN} is the voltage at the −IN terminal.

Because V_{out} is limited by the supply voltage and A_{ol} is high, the voltage difference, $V_{+IN} - V_{-IN}$, must be very small. If it is large, it will *saturate* the op-amp by driving it to the limits of its positive or negative excursion.

EXAMPLE 12-1

The output voltage of an op-amp is limited to $+12$ V. If its open loop gain is 150,000, find the maximum voltage difference between the positive and negative input voltages.

Solution
From equation 12-1 we have

$$V_{+IN} - V_{-IN} = \frac{V_{out}}{A_{ol}} = \frac{12 \text{ V}}{150,000} = 80 \text{ }\mu\text{V}$$

This voltage is so small that digital voltmeters reading to three decimal places cannot detect it.

12-3.1 Virtual Ground

The *virtual ground* concept is a great help in understanding op-amps. Example 12-1 showed that the voltage difference at the inputs must be very small, or that $V_{+IN} \approx V_{-IN}$. This means that if one of the inputs is tied to ground, the other input is *virtually* at ground. The virtual ground concept means that the two inputs are approximately a short circuit in the sense that their voltages must be the same. The inputs are *not* a short circuit in the sense that there is any current flow between them. The impedance between V_{+IN} and V_{-IN} is kept very high so that there is almost no current flow between them.

12-3.2 The Inverting Amplifier

Figure 12-2 shows the basic inverting amplifier using an op-amp. The input voltage is applied to the negative terminal through resistor R_1, and the feedback resistor, R_f, is connected between the output and the input. The

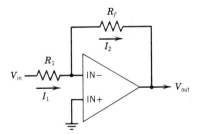

FIGURE 12-2
The inverting amplifier.

reader should not confuse the input voltage to the circuit, V_{in}, which is connected to R_1, with the voltage at the op-amp terminals, V_{+IN} or V_{-IN}. In most circuits, including this one, they are different.

Using the virtual ground concept, it is very easy to calculate the gain of this circuit. Because the voltage at the positive terminal is grounded, the voltage at the negative terminal must also be at ground. Furthermore, the current flowing in R_1 must also flow in R_f, if we assume the input impedance of the op-amp is very high. This leads to the following equations:

$$V_{in} = i_1 R_1$$
$$V_{out} = i_2 R_f$$
$$i_1 = i_2$$

and it follows that

$$A_v = \frac{V_{out}}{V_{in}} = \frac{R_f}{R_1} \quad (12\text{-}2)$$

Equation 12-2 shows that the gain of this circuit depends only on the resistors and not on the characteristics of the op-amp. This is true unless the op-amp fails completely. The negative feedback in this circuit is very high, so the output is highly stable.

EXAMPLE 12-2

The resistors in Figure 12-2 are $R_1 = 5$ kΩ and $R_2 = 50$ kΩ. The open-loop gain of the op-amp is 200,000. If V_{in} is 0.5 V, find
a. The gain.
b. The currents i_1 and i_2.
c. The exact voltage at the input.
d. The feedback factor.

Solution
a. The gain is given by equation 12-2.

$$A_v = \frac{R_f}{R_1} = \frac{50 \text{ k}\Omega}{5 \text{ k}\Omega} = 10$$

The output voltage must therefore be 5 V.
b. With the voltage at the input terminal at ground, the currents are

$$i_1 = \frac{0.5 \text{ V}}{5 \text{ k}\Omega} = 0.1 \text{ mA}$$

$$i_2 = \frac{5 \text{ V}}{50 \text{ k}\Omega} = 0.1 \text{ mA}$$

or

$$i_2 = i_1 = 0.1 \text{ mA}$$

c. The exact voltage at the input is

$$\frac{V_{out}}{A_{ol}} = \frac{5 \text{ V}}{200{,}000} = 25 \text{ }\mu\text{V}$$

If this voltage is considered, it will alter the values of i_1 and i_2 as calculated in part b, but only in the ninth decimal place.

d. Here

$$\beta = \frac{R_1}{R_1 + R_f} = \frac{1}{11}$$

$$A\beta = \frac{200{,}000}{11} = 18{,}182$$

The feedback factor is very high, as predicted.

12-3.3 A Precise Analysis of the Inverting Amplifier[1]

A more precise analysis of the inverting amplifier can be made by considering the voltage V_{-IN} instead of neglecting it. As before, we will assume no current flows between the $+IN$ and $-IN$ terminals.

Because V_{+IN} is grounded, V_{-IN} must be slightly above ground to produce the negative V_{out} (refer to Figure 12-2). It then follows that

$$\frac{V_{in} - V_{-IN}}{R_1} = \frac{V_{out} + V_{-IN}}{R_f}$$

$$\frac{V_{in}}{R_1} = \frac{V_{out}}{R_f} + V_{-IN}\left(\frac{1}{R_1} + \frac{1}{R_f}\right)$$

$$A_v = \frac{V_{out}}{V_{in}} = \frac{R_f}{R_1} - V_{-IN}\left(\frac{1}{R_1} + \frac{1}{R_f}\right)$$

---

[1]This section may be omitted at first reading.

But

$$V_{-IN} = \frac{V_{out}}{A_{ol}}$$

$$A_v = \frac{R_f}{R_1} - \frac{V_{out}}{A_{ol}}\left(\frac{1}{R_1} + \frac{1}{R_f}\right) \qquad (12\text{-}3)$$

Equation 12-3 is a more precise gain formula. It shows how the gain depends on A_{ol}. As A_{ol} becomes large, however, the second term drops out and equation 12-3 reduces to equation 12-2.

12-3.4 The Noninverting Amplifier

A noninverting amplifier cannot be built by simply grounding the minus input and connecting the input signal to the plus terminal because if both V_{in} and V_{out} are positive, there is no way to get the voltage at the +IN terminal down to ground. A noninverting amplifier can be built, however, by connecting resistors between the input and output, as shown in Figure 12-3. To

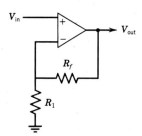

FIGURE 12-3
The noninverting amplifier.

analyze this circuit, assume the output voltage is V_{out}. Then

$$V_{-IN} = V_{out}\left(\frac{R_1}{R_1 + R_f}\right)$$

But $V_{-IN} \approx V_{+IN}$ because of the virtual ground so that

$$A_v = \frac{V_{out}}{V_{in}} = \frac{R_1 + R_f}{R_1} \qquad (12\text{-}4)$$

EXAMPLE 12-3

A voltage V_{in} of 0.5 V is applied to the input of the noninverting op-amp of Figure 12-3. If $R_1 = 1$ kΩ and $R_f = 4$ kΩ, find the output voltage and the voltage at the (−IN) terminal.

Solution
From equation 12-4 we have

$$A_v = \frac{R_1 + R_f}{R_1} = \frac{1\ \text{k}\Omega + 4\ \text{k}\Omega}{1\ \text{k}\Omega} = 5$$

Therefore the output voltage is 2.5 V.

$$V_{\text{out}} = V_{+\text{IN}} A_v = 0.5\ \text{V} \times 5 = 2.5\ \text{V}$$

The voltage at the ($-$IN) terminal is

$$V_{\text{out}} \left(\frac{R_1}{R_1 + R_f}\right) = 2.5\ \text{V} \left(\frac{1\ \text{k}\Omega}{5\ \text{k}\Omega}\right) = 0.5\ \text{V}$$

Check: Due to the virtual ground, $V_{+\text{IN}} \approx V_{-\text{IN}} = 0.5$ V.

12-3.5 The Summing Amplifier
The circuit of Figure 12-4 is a resistive *summing* network. The voltage V_A is

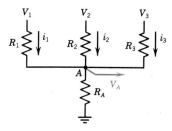

FIGURE 12-4
A resistive summing network.

$$V_A = (i_1 + i_2 + i_3) R_A = R_A \left(\frac{V_1 - V_A}{R_1} + \frac{V_2 - V_A}{R_2} + \frac{V_3 - V_A}{R_3}\right) \quad (12\text{-}5)$$

If $R_1 = R_2 = R_3 = R$ and V_A is kept small by selecting $R_A \ll R$, then equation 12-5 reduces to

$$V_A = \frac{R_A}{R}(V_1 + V_2 + V_3) \quad (12\text{-}6)$$

Thus V_A is proportional to the sum of the applied voltages. Equation 12-6 becomes less accurate as higher voltages or more inputs increase V_A. The circuit also suffers the drawback of having V_A small. The addition of an op-amp improves the accuracy of this circuit tremendously.

The summing op-amp circuit is shown in Figure 12-5. The op-amp holds the input voltage, which corresponds to V_A in Figure 12-4, at ground. The

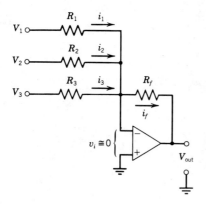

FIGURE 12-5
The summing amplifier. (From *An Introduction to Operational Amplifiers with Linear IC Applications* Second Edition by Luces M. Faulkenberry. Copyright © 1982 by John Wiley & Sons, Inc. Reprinted by permission of John Wiley & Sons, Inc.)

output current is the sum of the input currents:

$$i_f = i_1 + i_2 + i_3$$

If the resistors are equal and equal R_f, the output voltage is simply $V_1 + V_2 + V_3$.

EXAMPLE 12-4

In the circuit of Figure 12-5, $R_1 = R_2 = R_3 = R_f = 1 \text{ k}\Omega$, $V_1 = 1 \text{ V}$, $V_2 = 2 \text{ V}$ and $V_3 = 4 \text{ V}$. Find V_{out}.

Solution
The output voltage is simply the inverse of the sum of the input voltages.

$$V_{out} = -(1 \text{ V} + 2 \text{ V} + 4 \text{ V}) = -7 \text{ V}$$

Check: Since $V_{-IN} \approx 0$, the currents i_1, i_2, and i_3 are 1, 2, and 4 mA respectively. Thus 7 mA flows in R_f and causes V_{out} to be -7 V. Note that the sum of the inputs cannot exceed the limit on the range of output voltages for the op-amp.

If the resistors in Figure 12-5 are not all equal, the circuit is called a *scaling* or *weighted adder*. The outputs are scaled by the weight of each resistor. The equations for the scaling adder are

$$V_{out} = i_f R_f$$
$$= (i_1 + i_2 + i_3) R_f$$
$$= R_f \left(\frac{V_1}{R_1} + \frac{V_2}{R_2} + \frac{V_3}{R_3} \right) \qquad (12\text{-}7)$$

The scaling adder is often used in a Digital-to-Analog (D/A) converter that converts a digital input to an analog output by adding the bits scaled by their proper weight. Perhaps the simplest way to do this is to have each resistor twice as large as the previous resistor.

EXAMPLE 12-5

For the circuit of Figure 12-5, $R_1 = 1$ kΩ, $R_2 = 2$ kΩ, $R_3 = 4$ kΩ, and $R_f = 1$ kΩ. If $V_1 = V_3 = 1$ V and $V_2 = 0$ V, find V_{out}.

Solution
Equation 12-7 gives

$$V_{out} = R_f \left(\frac{V_1}{R_1} + \frac{V_2}{R_2} + \frac{V_3}{R_3} \right) = 1 \text{ k}\Omega \,(1 \text{ mA} + 0 + 0.25 \text{ mA})$$

$$V_{out} = 1.25 \text{ V}$$

From the digital standpoint, the numbers given ($V_1 = 1$, $V_2 = 0$, $V_3 = 1$) correspond to the number 5 $(101)_2$. The unit of current is the smallest value or 0.25 mA, which corresponds to 0.25 V. The output of this circuit is 5 times 0.25 V or 1.25 V. Changing the digital number of the input will change the analog output accordingly.

12-3.6 The Voltage Follower

The voltage follower is an op-amp circuit that is used to isolate an input from an output or load. It serves as a super emitter-follower. The characteristics of the voltage follower are

1. The gain is very close to 1.
2. The input impedance is very high.
3. The output impedance is very low.

The voltage-follower circuit is shown in Figure 12-6. If a voltage is

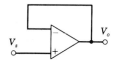

FIGURE 12-6
The voltage follower. (From *Essentials of Solid State Electronics* by Rodney B. Faber. Copyright © 1985 by John Wiley & Sons, Inc. Reprinted by permission of John Wiley & Sons, Inc.)

applied to the +IN terminal, the same voltage must appear at the −IN terminal, because of the virtual ground. But −IN is connected to V_{out}. Therefore

$$V_{out} \approx V_{in}$$

We should also observe that the input impedance of this circuit is the input impedance of the op-amp and is therefore very high.

It is possible to buy op-amps connected internally as voltage-followers. The **LM310** is a voltage-follower op-amp manufactured by both Texas Instruments and National Semiconductor. The typical characteristics of an **LM310** are

A_v voltage gain   0.9999
r_i input resistance 10^{12}
r_o output resistance   0.75

Note how closely the voltage-follower approximates the ideal emitter-follower.

This section introduced the basic op-amp circuits. More sophisticated circuits, such as differentiators and integrators will be discussed in section 12-5.

12-4 CHARACTERISTICS OF THE OP-AMP

In order to introduce the op-amp in the preceding section, and to show some of its more commonly used circuits, we assumed that the op-amp had the ideal characteristics of high open-loop gain and high input impedance. In this section the characteristics of existing op-amps are examined in greater detail.

Figure 12-7 is a simplified drawing of a basic op-amp circuit. Transistors Q_1 and Q_2 form a difference amplifier, and Q_3 is a constant-current source. Difference amplifiers and constant-current sources have beeen discussed in sections 6-7 and 6-9. The presence of the emitter resistors and the constant-current source causes the op-amp to have a high input impedance. Transistors Q_4 and Q_5 form an intermediate stage of amplification to increase the gain of the op-amp. Transistors Q_7 and Q_8 provide a complementary output that can be either positive or negative. They are driven by Q_6. The circuit is similar to the output stage of a power amplifier (see Figure 9-26), but its function is to provide a complementary output instead of a power output.

The **uA741** (commonly called the **741**) is one of the least expensive and most readily available op-amps. Its characteristics are given in the specifications of Figure 12-8. Although it is not the best or most modern op-amp, it is sufficient for many applications and is widely used.

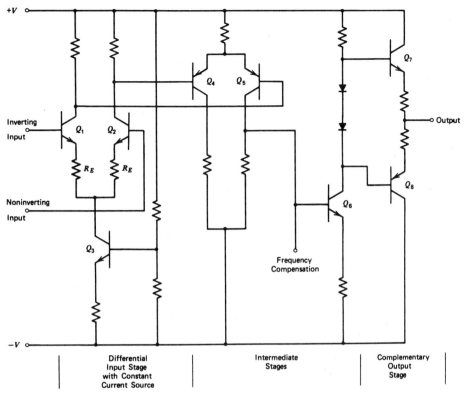

FIGURE 12-7
A simple operational amplifier circuit. (From *An Introduction to Operational Amplifiers with Linear IC Applications* Second Edition by Luces M. Faulkenberry. Copyright © 1982 by John Wiley & Sons, Inc. Reprinted by permission of John Wiley & Sons, Inc.)

Figure 12-8a shows the packaging of the **741**. The **741C**, which comes in an 8-pin dual-in-line (DIP), package is the most commonly used. It looks like the **555** timer (see Figure 11-31). Figure 12-8b, the second page of the specifications, gives the circuit diagram (too complex to be covered here), and the maximum ratings of the op-amp. The third page (Figure 12-8c) gives the electrical characteristics that will be discussed in the following paragraphs. Figure 12-8d gives the frequency characteristics and shows some test circuits, and Figure 12-8e shows some of the most important characteristics graphically.

12-4.1 The Open-Loop Voltage Gain

As previously stated, the open-loop voltage gain of an op-amp, A_{ol}, must be high for proper operation. Some manufacturers specify A_{ol} directly, whereas others specify it as volts-per-millivolt (V/mV). When specified in this way the actual gain is 10^3 times the gain in V/mV. Figure 12-8c shows that the

LINEAR INTEGRATED CIRCUITS

TYPES uA741M, uA741C
GENERAL-PURPOSE OPERATIONAL AMPLIFIER

D920, NOVEMBER 1970 – REVISED AUGUST 1983

- Short-Circuit Protection
- Offset-Voltage Null Capability
- Large Common-Mode and Differential Voltage Ranges
- No Frequency Compensation Required
- Low Power Consumption
- No Latch-up
- Designed to be Interchangeable with Fairchild μA741M, μA741C

description

The uA741 is a general-purpose operational amplifier featuring offset-voltage null capability.

The high common-mode input voltage range and the absence of latch-up make the amplifier ideal for voltage-follower applications. The device is short-circuit protected and the internal frequency compensation ensures stability without external components. A low potentiometer may be connected between the offset null inputs to null out the offset voltage as shown in Figure 2.

The uA741M is characterized for operation over the full military temperature range of −55°C to 125°C; the uA741C is characterized for operation from 0°C to 70°C.

symbol

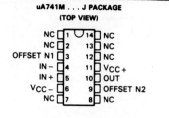

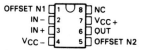

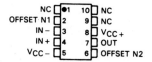

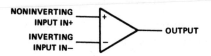

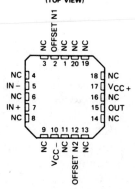

NC – No internal connection

FIGURE 12-8
The **uA741** specifications. (Courtesy of Texas Instruments, Inc.)

TYPES uA741M, uA741C
GENERAL-PURPOSE OPERATIONAL AMPLIFIERS

schematic

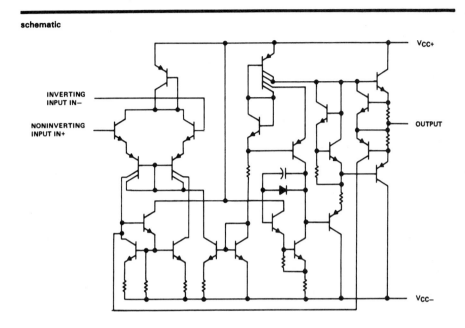

absolute maximum ratings over operating free-air temperature range (unless otherwise noted)

	uA741M	uA741C	UNIT
Supply voltage V_{CC+} (see Note 1)	22	18	V
Supply voltage V_{CC-} (see Note 1)	−22	−18	V
Differential input voltage (see Note 2)	±30	±30	V
Input voltage any input (see Notes 1 and 3)	±15	±15	V
Voltage between either offset null terminal (N1/N2) and V_{CC-}	±0.5	±0.5	V
Duration of output short-circuit (see Note 4)	unlimited	unlimited	
Continuous total power dissipation at (or below) 25°C free-air temperature (see Note 5)	500	500	mW
Operating free-air temperature range	−55 to 125	0 to 70	°C
Storage temperature range	−65 to 150	−65 to 150	°C
Lead temperature 1,6 mm (1/16 inch) from case for 60 seconds FH, FK, J, JG, or U package	300	300	°C
Lead temperature 1,6 mm (1/16 inch) from case for 10 seconds D, N or P package		260	°C

NOTES: 1. All voltage values, unless otherwise noted, are with respect to the midpoint between V_{CC+} and V_{CC-}.
2. Differential voltages are at the noninverting input terminal with respect to the inverting input terminal.
3. The magnitude of the input voltage must never exceed the magnitude of the supply voltage or 15 volts, whichever is less.
4. The output may be shorted to ground or either power supply. For the uA741M only, the unlimited duration of the short-circuit applies at (or below) 125°C case temperature or 75°C free-air temperature.
5. For operation above 25°C free-air temperature, refer to Dissipation Derating Curves, Section 2. In the J and JG packages, uA741M chips on alloy mounted; uA741C chips are glass mounted.

FIGURE 12-8b

**TYPES uA741M, uA741C
GENERAL-PURPOSE OPERATIONAL AMPLIFIERS**

electrical characteristics at specified free-air temperature, $V_{CC+} = 15$ V, $V_{CC-} = -15$ V

PARAMETER		TEST CONDITIONS†		uA741M			uA741C			UNIT
				MIN	TYP	MAX	MIN	TYP	MAX	
V_{IO}	Input offset voltage	$V_O = 0$	25°C		1	5		1	6	mV
			Full range			6			7.5	
$\Delta V_{IO(adj)}$	Offset voltage adjust range	$V_O = 0$	25°C		±15			±15		mV
I_{IO}	Input offset current	$V_O = 0$	25°C		20	200		20	200	nA
			Full range			500			300	
I_{IB}	Input bias current	$V_O = 0$	25°C		80	500		80	500	nA
			Full range			1500			800	
V_{ICR}	Common-mode input voltage range		25°C	±12	±13		±12	±13		V
			Full range	±12			±12			
V_{OM}	Maximum peak output voltage swing	$R_L = 10$ kΩ	25°C	±12	±14		±12	±14		V
		$R_L \geq 10$ kΩ	Full range	±12			±12			
		$R_L = 2$ kΩ	25°C	±10	±13		±10	±13		
		$R_L \geq 2$ kΩ	Full range	±10			±10			
A_{VD}	Large-signal differential voltage amplification	$R_L \geq 2$ kΩ, $V_O = \pm10$ V	25°C	50	200		20	200		V/mV
			Full range	25			15			
r_i	Input resistance		25°C	0.3	2		0.3	2		MΩ
r_o	Output resistance	$V_O = 0$, See Note 6	25°C		75			75		Ω
C_i	Input capacitance		25°C		1.4			1.4		pF
CMRR	Common-mode rejection ratio	$V_{IC} = V_{ICR}$ min	25°C	70	90		70	90		dB
			Full range	70			70			
k_{SVS}	Supply voltage sensitivity ($\Delta V_{IO}/\Delta V_{CC}$)	$V_{CC} = \pm9$ V to ±15 V	25°C		30	150		30	150	μV/V
			Full range			150			150	
I_{OS}	Short-circuit output current		25°C		±25	±40		±25	±40	mA
I_{CC}	Supply current	No load, $V_O = 0$	25°C		1.7	2.8		1.7	2.8	mA
			Full range			3.3			3.3	
P_D	Total power dissipation	No load, $V_O = 0$	25°C		50	85		50	85	mW
			Full range			100			100	

†All characteristics are measured under open-loop conditions with zero common-mode input voltage unless otherwise specified. Full range for uA741M is −55°C to 125°C and for uA741C is 0°C to 70°C.
NOTE 6: This typical value applies only at frequencies above a few hundred hertz because of the effects of drift and thermal feedback.

FIGURE 12-8c

minimum gain for a **741C** is 20,000 and for a **741M** it is 50,000. It is rare to find op-amps with gains this low. A typical **741** has a gain of 200,000, and this is what the user would expect.

12-4.2 The Input Impedance

The input impedance of an op-amp is generally the impedance between the +IN and −IN terminals, although it is sometimes considered as the impedance between one of these terminals and ground. The **741** specifications give a minimum value of 300,000 Ω and a typical value of 2 MΩ. The input impedance is high due to the difference amplifier input and the constant-current

TYPES uA741M, uA741C
GENERAL-PURPOSE OPERATIONAL AMPLIFIERS

operating characteristics, V_{CC+} = 15 V, V_{CC-} = −15 V, T_A = 25°C

PARAMETER		TEST CONDITIONS	uA741M			uA741C			UNIT
			MIN	TYP	MAX	MIN	TYP	MAX	
t_r	Rise time	V_I = 20 mV, R_L = 2 kΩ, C_L = 100 pF, See Figure 1		0.3			0.3		µs
	Overshoot factor			5%			5%		
SR	Slew rate at unity gain	V_I = 10 V, R_L = 2 kΩ, C_L = 100 pF, See Figure 1		0.5			0.5		V/µs

PARAMETER MEASUREMENT INFORMATION

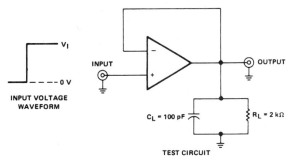

FIGURE 1 – RISE TIME, OVERSHOOT, AND SLEW RATE

TYPICAL APPLICATION DATA

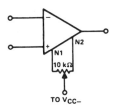

FIGURE 2 – INPUT OFFSET VOLTAGE NULL CIRCUIT

FIGURE 12-8d

source. These values are typical for a *bi-polar* op-amp, constructed using BJT transistors. Other op-amps, using *BIFET* technology (see section 12-4.10), have higher input impedances. BIFET op-amps use JFET transistors on their inputs and take advantage of their intrinsically higher input impedance. Typical BIFET op-amps have input impedances of 10^{12} Ω.

EXAMPLE 12-6

For the inverting amplifier of Example 12-2 assume the op-amp is a typical **741**. How much current flows between the −IN and +IN terminals?

12-4 CHARACTERISTICS OF THE OP-AMP **581**

TYPES uA741M, uA741C
GENERAL-PURPOSE OPERATIONAL AMPLIFIERS

TYPICAL CHARACTERISTICS

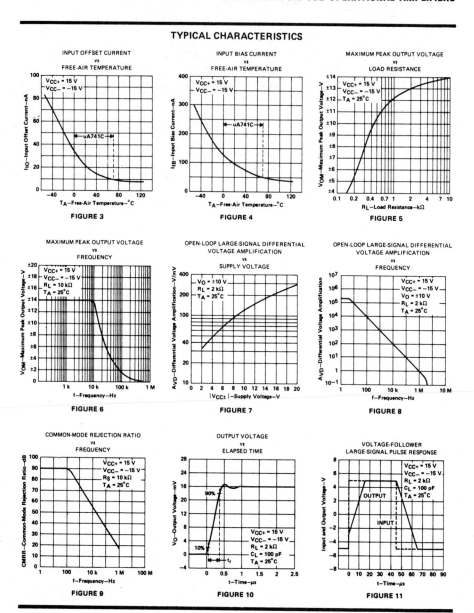

FIGURE 12-8e

Solution
In Example 12-2 we found the voltage at the $-\text{IN}$ terminal was 25 µV. The current is therefore

$$i = \frac{V_{-\text{IN}}}{R_{\text{typical}}} = \frac{25\ \mu V}{2 \times 10^6} = 12.5\ \text{pA}$$

Of course, the input impedance of the circuit may be different from the input impedance of the op-amp. For the noninverting circuit of Figure 12-3, the input is applied directly to the $+\text{IN}$ terminal, and the input impedance of the circuit and the op-amp are the same and very high. For the inverting op-amp of Figure 12-2, however, the input impedance of the circuit is basically R_1 because $V_{-\text{IN}}$ is at ground. Thus the inverting amplifier has a much lower input impedance.

12-4.3 The Output Impedance

The output impedance of an op-amp should be very low so that the load does not affect its output. The output resistance of a typical **741** is given in Figure 12-8c as 75 Ω. This is the intrinsic output impedance of the op-amp, however, and in most circuits the feedback greatly reduces this impedance.

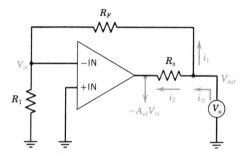

FIGURE 12-9
Determining the output impedance of an inverting op-amp.

The output impedance of an op-amp *circuit* can be found by applying a voltage to the output and shorting the input to ground (see section 4-7.1) For the inverting op-amp this is shown in Figure 12-9, where R_o is the intrinsic output impedance of the op-amp. The input impedance is assumed to be so large that no current enters the $-\text{IN}$ terminal. The current due to V_o is

$$i_o = i_1 + i_2$$
$$= \frac{V_o}{R_f + R_1} + \frac{V_o + A_{ol} V_{in}}{R_o}$$

But R_1 and R_f form a voltage divider and

$$V_{in} = V_o \frac{R_1}{R_1 + R_f}$$

Therefore

$$i_o = \frac{V_o}{R_1 + R_f} + \frac{V_o\left(1 + \dfrac{A_{ol} R_1}{R_1 + R_f}\right)}{R_o} \quad (12\text{-}8)$$

Because A_{ol} is large, the first term in equation 12-8 is much smaller than the second term and will be disregarded. If r_o is the output impedance of the circuit, then

$$r_o = \frac{V_o}{i_o} = \frac{R_o}{1 + \dfrac{A_{ol} R_1}{R_1 + R_f}} \quad (12\text{-}9)$$

EXAMPLE 12-7

Find the output impedance of the circuit of Example 12-2.

Solution

Example 12-2 is an inverting amplifier with $R_1 = 5\ \text{k}\Omega$ and $R_f = 50\ \text{k}\Omega$. If a typical **741** is assumed ($A_{ol} = 200{,}000$, $R_o = 75\ \Omega$), then, by equation 12-9

$$r_o = \frac{75\ \Omega}{1 + 200{,}000 \times \dfrac{5\ \text{k}\Omega}{55\ \text{k}\Omega}} = \frac{75\ \Omega}{18{,}183} = 0.004\ \Omega$$

The output impedance of the circuit is very small.

12-4.4 Input Offset Voltage

If both the +IN and −IN terminals of an op-amp are grounded, the ideal output voltage would be 0 V. Real op-amps generally produce an output voltage that only is a few mV, but in some applications this cannot be tolerated. The output voltage can be adjusted to be 0 V by applying a small voltage, called the *input offset voltage* (V_{IO}), to one of the terminals and adjusting this voltage until the output is 0 V. For a **741**, Figure 12-8c shows that V_{IO} is typically 1 mV.

If the no-signal output of an op-amp must be exactly 0 V, a *compensating potentiometer* is usually used instead of an input voltage adjustment. The **741** has two terminals, N1 and N2, to accept this potentiometer, as shown in the null circuit of Figure 12-8d. The potentiometer causes a small redistribution of the current in the emitter legs of the difference amplifier on the input, which allows the user to null the output.

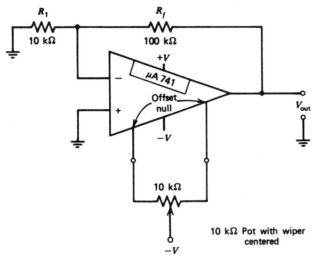

FIGURE 12-10
Connection for offset compensation. (From *An Introduction to Operational Amplifiers with Linear IC Applications* Second Edition by Luces M. Faulkenberry. Copyright © 1982 by John Wiley & Sons, Inc. Reprinted by permission of John Wiley & Sons, Inc.)

Faulkenberry (see References) recommends the circuit of Figure 12-10, an inverting amplifier with a gain of 10, for setting the compensation. When set up in the laboratory, we found the output voltage without the potentiometer was 2 to 3 mV, and it was easily nulled when the potentiometer was connected.

12-4.5 Input Bias Current and Input Offset Current

For an op-amp to function properly, a small amount of *bias current* must flow in the base of the transistors connected to the +IN and −IN terminals. In laboratory experiments, where both terminals were left open so that no bias current flowed, the output voltage went to either the positive or negative saturation voltage. The input bias current can cause a voltage of a few millivolts to appear at the output. The input bias current, as given in the specifications, is the average of the two currents flowing into the +IN and −IN terminals.

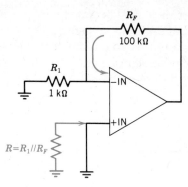

FIGURE 12-11
The circuit for Example 12-8, and a resistor for balancing the input bias currents.

EXAMPLE 12-8

The op-amp in the circuit of Figure 12-11 is a **741**. What is the worst-case output voltage caused by the bias current?

Solution

The specifications (Figure 12-8c) list the maximum input bias current, I_{IB}, as 500 nA. The voltage at the $-$IN terminal can be assumed to be 0 V, because of both the virtual ground and the fact that the output voltage will be very small, causing the input voltage to be V_{out}/A_{ol}, which is of the order of 10^{-8} V. Because the voltage at the $-$IN terminal is 0 V, we assume no current is flowing in R_1, and the entire bias current flows through R_f. The worst-case output voltage is therefore

$$V_{out} = I_{IB} \times R_f = 500 \times 10^{-9} \times 10^5 = 0.5 \text{ mV}$$

The offset voltage due to the input bias current is quite small, as Example 12-8 indicates, but it can be troublesome in some circuits. Some authors recommend keeping R_f as small as possible, consistent with gain and power requirements, to reduce this effect. Another way to minimize the effect of the bias current is to place a resistor in the +IN leg of the circuit, as shown in blue in Figure 12-11. If the value of this resistor is chosen as the parallel combination of R_1 and R_f, approximately equal bias currents will flow in each leg of the difference amplifier and minimize the offset voltage (see problem 12-8).

The input bias currents entering the two terminals cannot be precisely controlled, and in most op-amps there is a difference between them called the *input offset current*, I_{IO}. This current is not predictable and cannot be compensated by a resistor. Fortunately, $I_{IO} = I_{B(+IN)} - I_{B(-IN)}$, so that I_{IO} is less than I_{IB}. If the op-amp can accommodate a compensating potentiometer (see section 12-4.4), it compensates for all the offset effects when it nulls the output.

12-4.6 The Common-Mode Rejection Ratio

The common-mode rejection ratio (CMRR) was introduced in section 6-9.3 as the ratio between the *difference-mode* gain, where a signal is applied to only one terminal, and the *common-mode* gain, which is the change in V_{out} when the same signal is applied to both terminals. Because a good op-amp should reject output changes due to identical signals applied to both the +IN and −IN terminals, a high CMRR is desirable. Most op-amps have difference amplifiers on their inputs and achieve a high CMRR.

In most specifications the CMRR is given in decibels (dBs).

$$\text{CMRR}_{(dB)} = 20 \log_{10} \frac{A_d}{A_c} \qquad (12\text{-}10)$$

where A_c is the common-mode gain and A_d is the difference mode gain.

EXAMPLE 12-9

What is the CMRR of a typical **741**?

Solution

The specifications of Figure 12-8 give the typical CMRR as 90 dB. This translates to a CMRR of 31,600.

The CMRR is a function of frequency. The curve of CMRR versus frequency for a **741** is given as one of the curves in Figure 12-8e. It shows that the CMRR remains flat up to 100 Hz, and then decreases rapidly.

12-4.7 The Frequency Response of an Op-amp

The frequency response of an op-amp is flat only over a small range of frequencies. Figure 12-12, taken from part 8 of Figure 12-8c, is the frequency

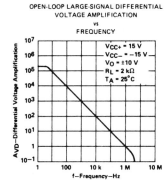

FIGURE 12-12
Open-loop large-signal differential voltage amplification versus frequency. (Courtesy of Texas Instruments, Inc.)

response of the **741**. It shows that the response of the **741** is flat only to about 6 Hz; beyond 6 Hz it falls off at 20 dB per decade.

The most commonly used indicator of an op-amp's frequency response is its *gain–bandwidth product*, sometimes called the *unity-gain bandwidth*, B_1. The gain–bandwidth product was introduced in section 7-6.3; it is the frequency at which the gain is unity. Although B_1 is not specified for the **741**, Figure 12-12 shows that this frequency is 1 MHz.

EXAMPLE 12-10

If the gain of a **741** is flat to 6 Hz, what is its low-frequency open-loop gain?

Solution
Since B_1 is MHz for a **741**

$$A_{ol} = \frac{B_1}{f} = \frac{1 \text{ MHz}}{6 \text{ Hz}} = 167{,}000$$

This is close to the manufacturer's specifications for A_{ol}.

Although a frequency response of 6 Hz may seem to be a severe limitation, the reader must remember that most op-amps operate with feedback that increases the frequency response. If a **741** is used as an amplifier with a gain of 10, for example, its frequency response will be 100 kHz.

Because an op-amp has a high gain, it may oscillate in some circuits. *Compensating capacitors* are used in most op-amps. They limit the frequency response, but reduce the possibilities of oscillation. The **741** is an *internally compensated* op-amp, because its capacitor is within the IC. Other op-amps, like the **709**, are *externally compensated;* they have terminals where the user can attach a compensating capacitor. In this way the user can control the frequency response of the op-amp and tailor it for the particular application.

12-4.8 Slew Rate
The input to an op-amp can change instantly; the output cannot. When an input pulse is applied that changes by several volts almost instantly, the output will *ramp* from its former voltage to the new voltage it must assume in response to the change in the input. The rate of change (dV/dt) of this ramp is called the *slew rate*. It is related to the frequency response of the amplifier. The higher the frequency response, the more rapidly the output can change. This results in a higher slew rate.

The slew rate for a **741** is given in Figure 12-8d as 0.5 V/μs. Figure 12-13a shows the response of the **741** to a 10 V change, and Figure 12-13b shows the circuit for obtaining the slew rate. It is a voltage-follower. Figure 12-13a shows

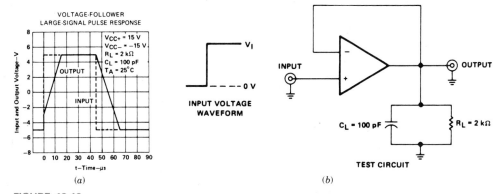

FIGURE 12-13
Slew rate response and test circuit. (Courtesy of Texas Instruments, Inc.)

that, in response to the positive transition, the output jumps instantly by 2 V and then slews, but it slews continuously in response to the negative transition.

EXAMPLE 12-11

Using the negative transistion of Figure 12-13a, find the slew rate of a **741**.

Solution
Figure 12-13a shows that the output voltage goes from $+5$ V to -5 V in 20 µs (from 45 to 65 µs). Therefore the slew rate is

$$SR = \frac{dV}{dt} = \frac{10 \text{ V}}{20 \text{ µs}} = 0.5 \text{ V/µs}$$

This agrees with the specifications for the **741**.

12-4.9 Drift

Drift is the name given to the change of op-amp parameters as the ambient temperature changes. The change of offset current and offset voltage are of particular interest. Part 3 of Figure 12-8e shows the change of offset current as a function of temperature. It shows that as the temperature changes from 20° C to 80° C, the offset current will change by 10 nA. If the parallel resistance connected to the op-amp is 10 kΩ, this will produce an offset voltage of 0.1 mV.

The change of offset voltage with temperature is not given, but a reasonable guess would be 15 µV/°C. Thus the temperature would have to change by 66° C to produce a 1 mV change in offset voltage.

Table 12-1
Typical Parameters for the 741 and TL080 Op-amps

	Parameter	Op-amp 741	Op-amp TL080	
A_{ol}	Open-loop voltage gain	200,000	200,000	
r_i	Input resistance	2×10^6	10^{12}	Ω
V_{IO}	Input offset voltage	1	3	mV
I_{IO}	Input offset current	20,000	5	pA
I_{IB}	Input bias current	80,000	30	pA
CMRR	Common-mode rejection ratio	90	86	dB
B_1	Unity-gain bandwidth	1	3	MHz
SR	Slew rate	0.5	13	V/μ

12-4.10 Other Op-amps

The **741** is a BJT (bipolar junction transistor) op-amp. It has been used throughout this section as an example because it is readily available and very widely used as a general-purpose op-amp. Of course, many other op-amps are available, and the reader should consult the manufacturers' catalogs to find the most appropriate op-amp for the application.

The **TL080** is a newer and faster op-amp. It is a BIFET op-amp, constructed primarily of BJT transistors, but it has JFET transistors on its input. Table 12-1 compares the parameters of the **741** and the **TL080**. Because of its JFET inputs, the **TL080** has a much higher input impedance and much lower bias and offset currents. The **TL080** also has a higher gain–bandwidth product and a faster slew rate. The gains and CMRR of the ICs are comparable. The **TL080** is the better op-amp, but it may be more expensive and not as available. These advantages are unimportant in many applications, and a **741** may be sufficient.

12-5 INTEGRATORS AND DIFFERENTIATORS

The circuits considered in the previous sections had only resistors in their input and feedback paths. By placing capacitors in these paths as well as resistors, the designer can construct a variety of other useful circuits such as integrators, differentiators, and filters. The integrator is a particularly interesting circuit because it is the basis of the *analog computer*.

12-5.1 The Op-amp Integrator

To review, the basic equation for a capacitor is

$$i = C\frac{dV}{dt}$$

where V is the voltage across the capacitor.

This translates into

$$V = \frac{1}{C}\int i\, dt + V_c$$

where V_c is a constant of integration.

An electronic integrating circuit can be constructed using an op-amp as shown in Figure 12-14.

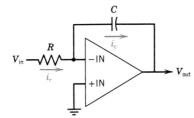

FIGURE 12-14
The basic op-amp integrator.

Because the virtual ground holds the voltage at $-$IN to ground, we have

$$i_R = \frac{V_{in}}{R} = i_c = C\frac{dv}{dt}$$

and

$$V = -\frac{1}{RC}\int V_{in}\, dt + V_c \tag{12-11}$$

where V_c is the initial voltage across the capacitor. The minus sign is caused by the inversion of the op-amp.

In equation 12-11, the voltage across the capacitor, V, is the same as V_{out} because one side of the capacitor is connected to the virtual ground at the $-$IN terminal.

One way to establish an initial voltage across a capacitor is to charge the capacitor to the proper voltage and then open a mechanical or electrical switch when integration starts. This is illustrated in Example 12-12.

EXAMPLE 12-12

For the circuit of Figure 12-15a, if switch S_1 is opened and S_2 is closed at the same time, find V_{out} as a function of time.

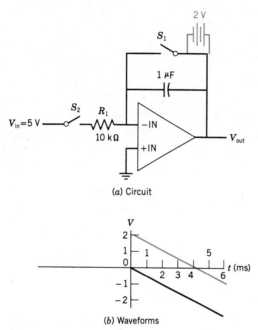

FIGURE 12-15
The circuit for Example 12-12.

a. Assume S_1 shorts C as shown in black.
b. Assume there are 2 V across C when S_2 is closed, as shown in blue.

Solution
a. The integral of a constant is a ramp. When S_2 closes and S_1 opens, $V_c = 0$ V (the capacitor was short-circuited), and $V_{in} = 5$ V. From 12-11, we have

$$V_{out} = \frac{1}{RC}\int V_{in}\, dt = -\frac{1}{RC} \times 5t = -5 \times 10^2\, t$$

This shows that V_{out} ramps down at a rate of 0.5 V/ms as shown in black in Figure 12-15b.

b. If the capacitor is initially charged to 2 V, it will ramp down from 2 V, as shown in blue in Figure 12-15b. Note that the initial voltage offsets the waveform of the output but does not change it.

12-5.2 Practical Integrator Considerations

An *ideal* capacitor will not allow any leakage current to flow through it. If an ideal capacitor is used in the integrating circuit of Figure 12-14, there will be no dc path from input to output. Unfortunately, this means that the circuit will integrate both the offset voltage and the offset current and will produce a ramp output even without an input. Eventually this ramp will drive the op-amp into saturation.

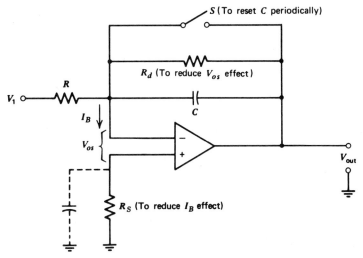

FIGURE 12-16
An integrator with dc error correction. (From *An Introduction to Operational Amplifiers with Linear IC Applications* Second Edition by Luces M. Faulkenberry. Copyright © 1982 by John Wiley & Sons, Inc. Reprinted by permission of John Wiley & Sons, Inc.)

Figure 12-16 is an integrating circuit that shows three ways of minimizing the effects of the offset voltage and currents. The switch, which can be either mechanical or electrical, can be used to periodically discharge the capacitor, or to set up initial conditions. In many circuits the switch is closed to discharge the capacitor and set the initial output voltage to 0 V. Integration begins when the switch is opened.

A resistor, R_d, is sometimes used to reduce the effect of the offset voltage. Without the resistor, the dc output voltage is $A_{ol}V_{os}$; with R_d in the circuit it becomes $V_{os}R_d/R_1$.

EXAMPLE 12-13

For the circuit of Figure 12-16, assume $V_{os} = 3$ mV, and $A_{ol} = 200{,}000$. If $R_1 = 10$ kΩ and V_{in} is 0 V, find the output if
a. R_d is not present.
b. $R_d = 1$ MΩ.

Solution

a. Without R_d

$$V_{out} = A_{ol}V_{os} = 200{,}000 \times 3 \times 10^{-3} \text{ V} = 600 \text{ V}$$

Obviously the output will never reach 600 V, but the op-amp will saturate somewhere between 10 and 12 V.

b. With R_d

$$V_{out} = \frac{R_d}{R_1} \times V_{os} = \frac{1 \text{ M}\Omega}{10 \text{ k}\Omega} \times 3 \times 10^{-3} \text{ V} = 300 \text{ mV}$$

The use of R_d places a pole in the output at $f = 1/2\pi R_d C$. This reduces the accuracy of the integrator at low frequencies. Consequently R_d is generally a high resistance (1 MΩ is typical) to reduce this frequency and improve the accuracy of the integrator. A general rule of thumb used by many engineers is to select R_d so that f is one-tenth of the lowest frequency to be integrated. If the capacitor is not of high-quality, it may have enough leakage resistance to make R_d unnecessary.

An integrator circuit using $R = 10$ kΩ and $C = 0.22$ μF was tested in the laboratory. A square was applied and the output was a *triangular* wave as shown in Figure 12-17. If R_d was not in the circuit, the output waveform drifted until either its positive or negative corner reached saturation. When an R_d of 1 MΩ was placed across the capacitor, it was possible to keep the output out of saturation.

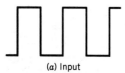

(a) Input

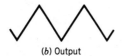

(b) Output

FIGURE 12-17
Response of an integrator to a square wave input.

EXAMPLE 12-14

What is the lowest frequency of integration for the circuit just described?

Solution
Since $R_d = 1\ M\Omega$ and $C = 0.22\ \mu F$, then

$$f = \frac{1}{2\pi R_d C} = \frac{1}{2\pi \times 10^6 \times 0.22 \times 10^{-6}} = 0.72\ Hz$$

If accurate integration is required, the lowest frequency of the wave to be integrated should be 10 times this or 7.2 Hz. This circuit is accurate down to very low frequencies.

The resistor R_S can be used to reduce the effects of the input bias and leakage currents. As before, R_S should be equal to the parallel combination of R_1 and R_d. Practically, however, R_d is high, so R_S can be chosen as equal to R_1. This selection is also valid if R_d is not in the circuit.

12-5.3 The Differentiator

Op-amps can also be used to construct *differentiator* circuits, where the output is proportional to the derivative of the input. The circuit of Figure 12-18 is an ideal differentiator circuit. Its equations are

$$i = C \frac{dV_{in}}{dt}$$

$$V_{out} = iR = RC \frac{dV_{in}}{dt}$$

Thus the output voltage is proportional to the derivative of the input.

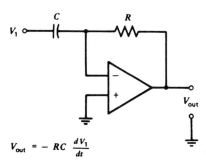

FIGURE 12-18
The op-amp differentiator. (From *An Introduction to Operational Amplifiers with Linear IC Applications* Second Edition by Luces M. Faulkenberry. Copyright © 1982 by John Wiley & Sons, Inc. Reprinted by permission of John Wiley & Sons, Inc.)

Unfortunately, differentiating circuits provide high gain at high frequencies. If a sine wave, $A \sin \omega t$, is applied to an integrator, for example, the output will be $(-A/\omega) \cos \omega t$. This output decreases as the frequency increases. If the same sine wave is applied to a differentiator, its output will be $A\omega \cos \omega t$, which increases as the frequency increases. Because a differentiator amplifies high-frequency signals, it is susceptible to noise and oscillations. As a result integrators are often used in preference to differentiators.

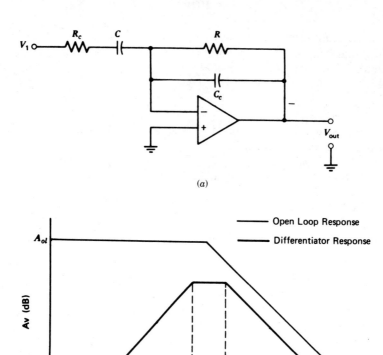

FIGURE 12-19
The compensated (a) differentiator and (b) differentiator response. (From *An Introduction to Operational Amplifiers with Linear IC Applications* Second Edition by Luces M. Faulkenberry. Copyright © 1982 by John Wiley & Sons, Inc. Reprinted by permission of John Wiley & Sons, Inc.)

Faulkenberry (see References) shows that a differentiator can be compensated so that it attenuates high-frequency signals. The circuit of Figure 12-19 is a *compensated differentiator*. It differentiates up to the frequency f_1, where

596 OPERATIONAL AMPLIFIERS

$$f_1 = \frac{1}{2\pi R_c C}$$

but it integrates at frequencies above f_2, where

$$f_2 = \frac{1}{2\pi RC_c}$$

The values of f_1 and f_2 can be controlled by the resistors and capacitors in the circuit. Generally f_2 is set equal to or slightly higher than f_1. This circuit is a stable differentiator for frequencies up to approximately f_1.

12-5.4 Solving Differential Equations

Differential equations can be solved using op-amp integrators and differentiators. Because of the noise problems associated with differentiators, integrators are generally used. They also allow the user to set in any initial conditions that may be required.

A typical second-order differential equation is

$$K_1 f_1(t) = K_2 x + K_3 \frac{dx}{dt} + \frac{d^2 x}{dt^2} \tag{12-12}$$

where $f_1(t)$ is the *forcing function* or driving function applied to the system, and K_1, K_2, and K_3 are constants. The problem is to find $x(t)$, the response of the system to the forcing function.

The RLC circuit of Figure 12-20 is a familiar electrical circuit that can be described by a second-order differential equation. Its equation is

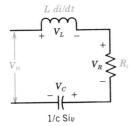

FIGURE 12-20
An RLC circuit.

$$V_{in}(t) = \frac{1}{C}\int i\, dt + Ri + L\frac{di}{dt} \tag{12-13}$$

where $V_{in}(t)$ is the forcing voltage.

If the charge q is used in equation 12-13, ($i = dq/dt$) it becomes

$$V_{in}(t) = \frac{q}{C} + R\frac{dq}{dt} + L\frac{d^2 q}{dt^2} \tag{12-14}$$

Equation 12-14 is of the same form as equation 12-12. To solve Equation 12-12, it is first transposed so that d^2x/dt^2 is alone on the left side.

$$\frac{d^2x}{dt^2} = K_1 f_1(x) - K_2 x - K_3 \frac{dx}{dt}$$

This shows that the second derivative is the sum of the forcing function plus terms that depend on x and its first derivative.

The op-amp circuit for solving this equation is shown in Figure 12-21. The summing and scaling amplifier has inputs of $f_1(t)$, $x(t)$, and dx/dt. It scales

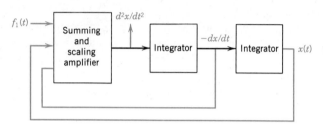

FIGURE 12-21
Using op-amps to solve a differential equation.

and adds them (see section 12-3.5) to produce d^2x/dt^2. This is then fed into op-amp integrators to get dx/dt and $x(t)$. Sometimes op-amp inverters must be added to the circuit of Figure 12-21 to correct the sign of a function.

EXAMPLE 12-15

Set up a circuit to solve the differential equation

$$\frac{d^2v}{dt^2} + 5\frac{dv}{dt} + \frac{v}{2} = 2 V_{in} \sin \omega t$$

Solution
The solution to this problem requires several steps.
a. Rewrite the given equation so that the d^2v/dt^2 is by itself:

$$\frac{d^2v}{dt^2} = 2 V_{in} \sin \omega t - 5\frac{dv}{dt} - \frac{v}{2}$$

b. Now concentrate on the dv/dt term.

$$\frac{dv}{dt} = \int \frac{(d^2v)}{dt^2} dt + V_C$$

Assuming the constant of integration is 0, this becomes

$$\frac{dv}{dt} = \int \left(2 V_{in} \sin \omega t - 5 \frac{dv}{dt} - \frac{v}{2}\right) dt \qquad (12\text{-}15)$$

Because an op-amp integrator provides an output of $(-1/RC) \int V_{in}\, dt$, the circuit for $-dv/dt$ can be set up as shown in Figure 12-22.

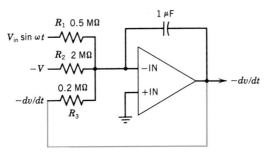

FIGURE 12-22
The circuit for finding dv/dt for Example 12-15.

c. In Figure 12-22, the capacitor was chosen to be 1 µF. The resistors R_1, R_2, and R_3 were then selected to give the required constants in equation 12-15. For example, the term $2 V_{in} \sin \omega t$ requires that the circuit be driven by an input of $V_{in} \sin \omega t$. The 2 can be obtained by setting $1/RC = 2$. With $C = 10^{-6}$, R must be 0.5 MΩ. The other resistors are determined similarly.

d. Because the circuit of Figure 12-22 provides $-dv/dt$, this signal can be fed back to the input via resistor R_3, as shown by the blue line in the figure.

e. To complete the solution, the voltage V must be found. This can be done by integrating $-dv/dt$. Another integrator with $1/RC = 1$ will produce v. Unfortunately, a voltage $-v$ is required as the input in Figure 12-22, so an inverter is needed. The entire circuit is shown in Figure 12-23. It requires two integrators and an inverter. The feedback lines are shown in blue.

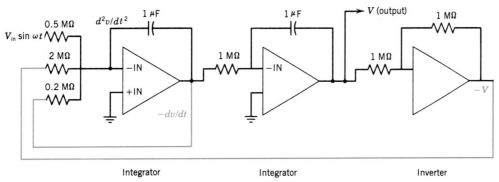

FIGURE 12-23
Solution to the differential equation of Example 12-15.

f. Once the circuit has been set up and the capacitors discharged, if the voltage $V_{in} \sin \omega t$ is applied to the input, the waveform of the response $v(t)$ can be observed. This waveform is the solution of the differential equation.

Circuits to solve the RLC circuit equation 12-13 can also be set up, but for practical values of R, L, and C, serious scaling problems arise. The reader should consult more advanced books for further information.

12-6 APPLICATIONS OF OP-AMPS

Op-amps are used in many circuits and have a large variety of applications. Some of these, such as the use of the summing amplifier in A/D converters and of integrators to solve differential equations, have already been discusssed briefly. There are many other uses for op-amps; the most commonly used will be introduced in this section. Unfortunately space limitations preclude a thorough discussion of any of them, but several other authors have written entire books on op-amps (see References) and the reader should refer to them for more details on any specific application.

12-6.1 The Difference Amplifier

The difference amplifier is used to amplify the *difference* between two input voltages. In many cases this difference is small and may exist in the presence of a large common-mode signal, so a high common-mode rejection ratio is essential.

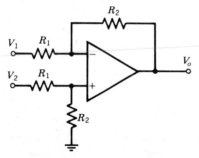

FIGURE 12-24
Differential-input amplifier. (From *Essentials of Solid State Electronics* by Rodney B. Faber. Copyright © 1985 by John Wiley & Sons, Inc. Reprinted by permission of John Wiley & Sons, Inc.)

The op-amp version of the difference amplifier is shown in Figure 12-24. The voltage at the +IN terminal is (by voltage division)

$$V_{+IN} = V_2 \frac{R_2}{R_1 + R_2}$$

Because of the virtual ground this is also the voltage at the −IN terminal. The current in the resistors that are connected to the −IN terminal is therefore

$$i = \frac{V_1 - V_{-IN}}{R_1} = \frac{V_1 - V_{+IN}}{R_1}$$

$$i = \frac{1}{R_1}\left(V_1 - V_2\frac{R_2}{R_1 + R_2}\right)$$

and

$$V_{out} = V_{-IN} - iR_2$$

$$V_{out} = \frac{V_2 R_2}{R_1 + R_2} - \frac{R_2}{R_1}\left(V_1 - V_2\frac{R_2}{R_1 + R_2}\right)$$

$$V_{out} = \frac{V_2 R_2}{R_1 + R_2} + \frac{V_2 R_2^2}{R_1(R_1 + R_2)} - \frac{R_2}{R_1}V_1$$

$$V_{out} = \frac{V_2 R_1 R_2}{R_1(R_1 + R_2)} + \frac{V_2 R_2^2}{R_1(R_1 + R_2)} - \frac{R_2}{R_1}V_1$$

$$V_{out} = \frac{V_2 R_2 \,\cancel{(R_1 + R_2)}}{R_1 \,\cancel{(R_1 + R_2)}} - \frac{R_2}{R_1}V_1 = \frac{R_2}{R_1}(V_2 - V_1) \qquad (12\text{-}16)$$

Equation 12-16 shows that the output of the difference amplifier is the multiplication factor (R_2/R_1) times the difference of the input voltage.

12-6.2 Instrumentation Amplifiers

An instrumentation amplifier is usually required to amplify small difference-mode signals in the presence of a large common-mode signal. It must also have a very high input impedance so it does not affect the source of the signals.

Instrumentation amplifiers are often used to amplify the output of transducers. In many cases transducers are connected as a leg of a *bridge* circuit, as shown in Figure 12-25. The resistance of the transducer varies as its temperature or pressure changes. This change in resistance causes a change in the voltage between points A and B. This voltage change could be amplified by the difference amplifier of Figure 12-24, but its input impedance is limited to R_1. If a higher input impedance is required, the instrumentation amplifier shown in Figure 12-26 should be used. This is a difference amplifier preceded by a pair of voltage followers. The input impedance is now the impedance of the voltage followers and can be very high, especially if FET input op-amps are used for the voltage followers.

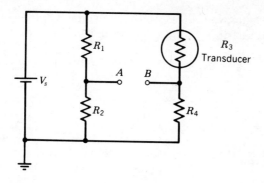

FIGURE 12-25
Bridge circuit—example of an underground signal source. (From *Essentials of Solid State Electronics* by Rodney B. Faber. Copyright © 1985 by John Wiley & Sons, Inc. Reprinted by permission of John Wiley & Sons, Inc.)

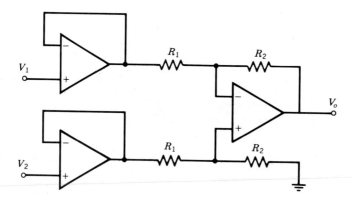

FIGURE 12-26
High-input-impedance instrumentation amplifier. (From *Essentials of Solid State Electronics* by Rodney B. Faber. Copyright © 1985 by John Wiley & Sons, Inc. Reprinted by permission of John Wiley & Sons, Inc.)

A practical version of the instrumentation amplifier is shown in Figure 12-27. Input amplifiers A1 and A2 will have some differential gain but the common-mode input voltages will experience only unity gain. These voltages will not appear as differential signals at the input of amplifier A3 because, when they appear at equal levels on both ends of resistor R2, they are effectively canceled. The potentiometer R7 is adjusted for the best common-mode rejection and the potentiometer at A3 is used to null the output.

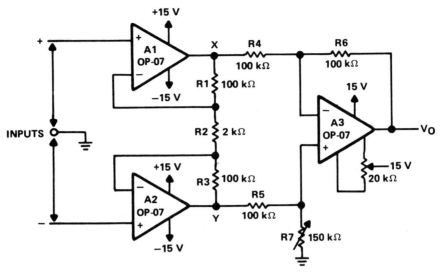

FIGURE 12-27
A practical instrumentation amplifier. (Courtesy of Texas Instruments, Inc.)

EXAMPLE 12-16

Assume that all the resistors in Figure 12-25, including the resistance of the transducer, are 500 Ω and that V_s is 10 V. The voltage V_{AB} is then 0 V, and the output is to be 0 V. Now assume that the resistance of the transducer changes to 525 Ω due to a change in the temperature or pressure it is measuring. Also assume that R_2 is limited to 1 MΩ. If the output voltage of the amplifier is to be 5 V under these conditions, and a difference amplifier is to be used, find its gain and input impedance.

Solution
Under the given conditions voltage division shows that $V_A = 5$ V and $V_B = 4.878$ V. Therefore $V_{AB} = 0.122$ V, and the required gain of the difference amplifier is

$$A_v = \frac{5 \text{ V}}{0.122 \text{ V}} = 41$$

The resistance, R_1 can then be found:

$$R_1 = \frac{R_2}{A_v} = \frac{1 \text{ M}\Omega}{41} = 24{,}390 \text{ }\Omega$$

If this input impedance is too low, the instrumentation amplifier of Figure 12-26 should be used.

12-6 APPLICATIONS OF OP-AMPS

12-6.3 Logarithmic Amplifiers

A logarithmic amplifier produces an output that is proportional to the natural logarithm of the input. To do this, a logarithmic element must be included in the circuit. Figure 12-28 shows two logarithmic amplifiers. Figure 12-28a uses a simple diode as the logarithmic element. An ordinary **1N914** diode will operate logarithmically over several orders of magnitude. Figure 12-28b

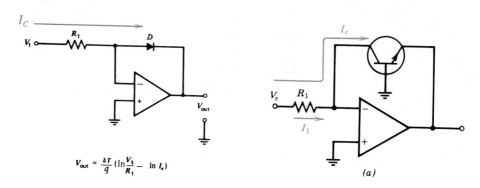

FIGURE 12-28
The logarithmic amplifier. (a) (From *An Introduction to Operational Amplifiers with Linear IC Applications* Second Edition by Luces M. Faulkenberry. Copyright © 1982 by John Wiley & Sons, Inc. Reprinted by permission of John Wiley & Sons, Inc.) (b) (From *Essentials of Solid State Electronics* by Rodney B. Faber. Copyright © 1985 by John Wiley & Sons, Inc. Reprinted by permission of John Wiley & Sons, Inc.)

shows a common-base transistor being used instead of the diode. The base-to-emitter voltage is the logarithmic element in this circuit. The transistor is often used because it is easier to provide temperature compensation circuits. In either circuit the output diode will saturate at about 0.6 V, so the output must be less than this for proper operation. If higher output voltages are required, an amplifier may have to follow the logarithmic circuit.

For either the diode or the transistor the basic equation is

$$I_C = I_S(e^{qV_E/\eta kT} - 1) \qquad (12\text{-}17)$$

in which

I_S = emitter saturation current (emitter-to-base leakage current)
V_E = emitter-base voltage
k = Boltzman's constant, = 1.38062×10^{-23}
T = temperature, in kelvins (K = °C + 273.15)
q = charge of electron, = 1.60219×10^{-19} coulomb (C)
η = the carrier recombination constant

η is 1 for germanium and slightly more than 1 for silicon. The value $\eta kT/Q$ is 26 mV for germanium and approximately 30 mV (to be consistent with equation 4-3) for silicon. The value of I_S is on the order of 10 nA or less for most diodes or transistors.

For silicon, equation 12-17 simplifies to

$$I_C = I_S (e^{V_E/30\,\text{mV}} - 1) \tag{12-18}$$

In most cases V_E is larger than 30 mV, and 12-18 becomes

$$I_C = I_S\, e^{V_E/30\,\text{mV}} \tag{12-19}$$

The logarithmic form of equation 12-19 is

$$V_E = \frac{\eta kT}{q}\ln\left(\frac{I_C}{I_S}\right) = \frac{\eta kT}{q}(\ln I_C - \ln I_S) \tag{12-20}$$

but, due to the virtual ground

$$I_C = \frac{V_S}{R_1}$$

and equation 12-20 becomes:

$$V_E = \frac{\eta kT}{q}\left(\ln \frac{V_S}{R_1} - \ln I_S\right) \tag{12-21}$$

EXAMPLE 12-17

If I_C is to be 1 mA and the leakage current of the silicon transistor is 10 nA, find V_E.

Solution

For these values, equation 12-21 becomes

$$V_E = 30\text{ mV }(\ln 10^{-3} - \ln 10^{-8})$$
$$= 30\text{ mV }\left(\ln \frac{10^{-3}}{10^{-8}}\right)$$
$$= 30\text{ mV }(\ln 10^5)$$
$$= 30\text{ mV} \times 11.51$$
$$V_E = 345\text{ mV}$$

Note that at an I_C of 1 mA, it starts to approach the saturation voltage of 700 mV. Generally I_C should not be larger than this in logarithmic amplifier circuits. Fortunately, I_C can be limited by the proper choice of R_1.

There are two problems with equation 12-21. The equation is directly dependent on the temperature, T, and while I_S is often very small, its logarithm is not; it is a large negative number. The circuit of Figure 12-29 eliminates both problems. The voltage V_{ref} is selected to eliminate the ln (I_S) term and the output of amplifier A_2 also changes with temperature. Because A_3 subtracts the outputs of amplifiers A_1 and A_2, it compensates for any temperature change.

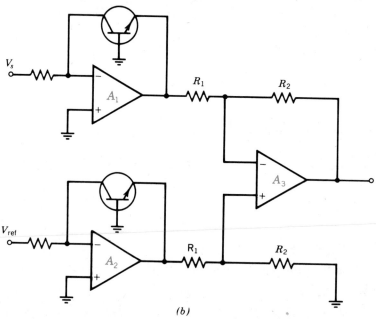

FIGURE 12-29
A compensated logarithmic amplifier. (From *Essentials of Solid State Electronics* by Rodney B. Faber. Copyright © 1985 by John Wiley & Sons, Inc. Reprinted by permission of John Wiley & Sons, Inc.)

12-6.4 The Antilog Amplifier

The antilog amplifier can be used to convert logarithmic values into natural numbers, or to take the antilog of a number. Antilog amplifiers are constructed by using a diode or transistor as the input impedance to an op-amp, as shown in Figure 12-30. The equations for this amplifier are

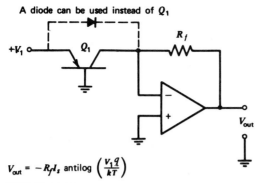

$V_{out} = -R_f I_s \text{ antilog}\left(\frac{V_1 q}{kT}\right)$

FIGURE 12-30
Antilog amplifier. (From *An Introduction to Operational Amplifiers with Linear IC Applications* Second Edition by Luces M. Faulkenberry. Copyright © 1982 by John Wiley & Sons, Inc. Reprinted by permission of John Wiley & Sons, Inc.)

$$V_{out} = R_f I_C$$
$$I_C = I_S \, e^{qV_1/\eta kT}$$

or

$$V_{out} = -R_f I_S \, e^{qV_1/\eta kT}$$

Therefore if V_1 is a logarithm, the antilog amplifier transforms it to a natural number. Op-amp multipliers can be constructed by taking the logarithms of the numbers to be multiplied using logarithmic amplifiers, adding them in a summing op-amp (see section 12-3.5), and then using an antilog amplifier.[2] Divider circuits can be similarly constructed by subtracting the logarithms.

12-6.5 Filters

Filters are circuits that pass some frequencies and reject others. Four types of filters and their frequency responses are shown in Figure 12-31. The high-pass filter rejects low frequencies. It was discussed in Chapter 7 in conjunction with the low-frequency response of amplifiers. The low-pass filter was also discussed there in conjunction with the high-frequency response of amplifiers. The band-pass filter can be made up of a combination of a low-pass and a high-pass circuit.

Filters can be either *passive* or *active*. Passive filters are built of passive components: resistors, capacitors, and inductors. Active filters use amplifiers,

---

[2]These circuits are shown in Faulkenberry (see References).

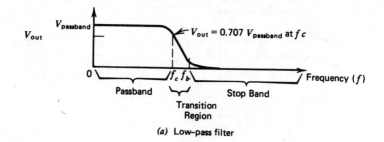

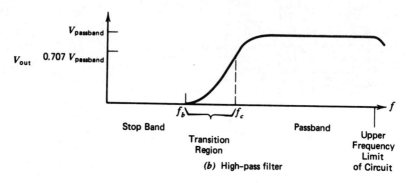

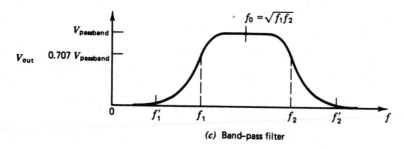

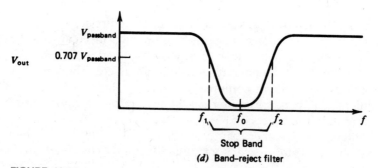

FIGURE 12-31
Filter characteristics (From *An Introduction to Operational Amplifiers with Linear IC Applications* Second Edition by Luces M. Faulkenberry. Copyright © 1982 by John Wiley & Sons, Inc. Reprinted by permission of John Wiley & Sons, Inc.)

often op-amps, to enhance their performance. Active filters have the following advantages:

1. They provide a high input impedance, so they do not degrade the performance of the filter.
2. They provide *isolation* so that the various sections of the filter do not interact with each other.
3. They can provide gain.
4. Inductors can be replaced by capacitors, especially in op-amp circuits. Capacitors are usually less expensive and easier to obtain.
5. Low-frequency filters can be constructed using modest component values.

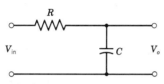

(a) Passive one pole low pass filter section

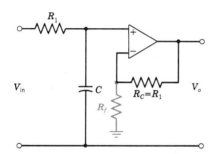

(b) Active one pole low pass filter section

FIGURE 12-32
Low-pass filters.

Figure 12-32 shows both a passive and an active low-pass filter. The active filter uses a noninverting op-amp with a low-pass section on its input. The upper-half-power frequency of this circuit is given by

$$f_h = \frac{1}{2\pi RC}$$

If R_f, shown in blue, is not used, the gain of the filter is 1. If R_f is included, the filter has the noninverting op-amp gain of

$$A_v = 1 + \frac{R_f}{R_1}$$

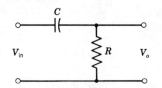

(a) One-pole passive high-pass filter section

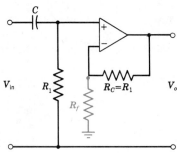

(b) One-pole active high-pass filter section

FIGURE 12-33
One-pole high-pass filters.

If the position of the resistor and capacitor at the input are interchanged, a high-pass circuit results, as shown in Figure 12-33. These are simple one-pole filters and beyond the pass-band they roll off at 20 dB/decade. By adding capacitors and a more complicated input circuit, two-pole filters can be constructed. The *voltage-controlled voltage source* (VCVS), or Sallen and Key filters, shown in Figure 12-34, are very popular two-pole or second-order filters. Beyond the pass-band they roll off at 40 dB/decade.

Butterworth (see section 11-4.2) and other more sophisticated types of filters, such as Bessel and Chebychev, can also be built using op-amps. Entire books have been written on active filters, and the reader who wants to pursue the matter further must consult more advanced material.

12-6.6 Digital-to-Analog Converters

A digital-to-analog (D/A)[3] converter is used to produce an analog output that is equivalent to the value of the digital input. Because computers express numbers digitally, a D/A converter must be used whenever a computer is controlling a circuit to convert its output into the proper voltage. D/A converters are also often used in the construction of *analog-to-digital* (A/D) converters. In an A/D converter, the input analog voltage, which is to be converted to digital form before being sent to the computer, is often compared to the output of a D/A converter. When the two are most nearly equal, the input to the D/A converter is the proper digital value.

---

[3]A more thorough discussion of A/D and D/A converters is presented in *Practical Digital Design Using ICs* (see References).

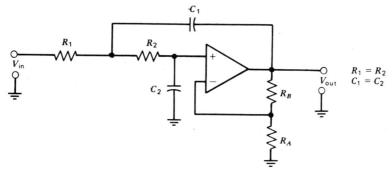

(a) Second-order low-pass filter

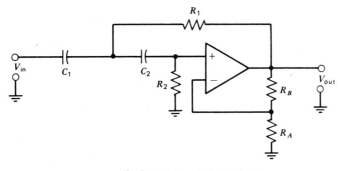

(b) Second-order high-pass filter

FIGURE 12-34
Sallen and Key active filters (From *An Introduction to Operational Amplifiers with Linear IC Applications* Second Edition by Luces M. Faulkenberry. Copyright © 1982 by John Wiley & Sons, Inc. Reprinted by permission of John Wiley & Sons, Inc.)

The simplest D/A converter is shown in Figure 12-35 and uses a *weighted resistor* network. This is a summing network in which each resistor is twice as large as the preceding resistor. Consequently, the current that flows in R_1 is twice as large as the current that flows in R_2, and so on. Because each binary bit is half the value of the preceding bit, the currents are in proportion to the value of the bits. The op-amp sums these currents and develops the correct analog output. It also keeps the voltage at point *a* at 0 V, due to the virtual ground. This prevents the currents in the various resistors from interacting with each other.

The R-2R network, shown in Figure 12-36, accomplishes the same result as the weighted-resistor network. Because it requires only two values of resistors, it is more commonly used.

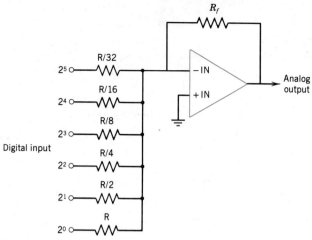

FIGURE 12-35
A weighted resistor D/A converter. (Adapted from *Digital Engineering* by George K. Kostopoulos. Copyright © 1975 by John Wiley & Sons, Inc. Reprinted by permission of John Wiley & Sons, Inc.)

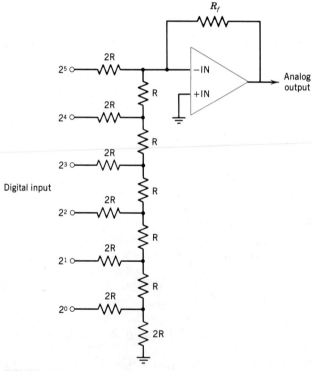

FIGURE 12-36
The R-2R network D/A converter. (From *Digital Engineering* by George K. Kostopoulos. Copyright © 1975 by John Wiley & Sons, Inc. Reprinted by permission of John Wiley & Sons, Inc.)

12-7 SUMMARY

Except for the transistor itself, the op-amp is probably the most commonly used analog circuit component. The basic operation of the op-amp and the most commonly used op-amp circuits were introduced first. Then the characteristics of op-amps were discussed so the reader could make an intelligent selection of an op-amp for his particular use. Finally several common uses of op-amps, such as integrators, differentiators, log amplifiers, and others, were presented.

12-8 GLOSSARY

Active filter. A filter that uses amplifiers.

Antilog amplifier. An amplifier whose output is proportional to the exponential of its input.

Compensating capacitor. Capacitors added to op-amps to prevent oscillation.

Compensating potentiometer. A potentiometer that is used to null the output voltage.

Difference amplifier. An op-amp circuit whose output is proportional to the difference between its inputs.

Differentiator. An op-amp circuit whose output voltage is the derivative of its input voltage.

Drift. The change in op-amp parameters as ambient conditions change.

Filter. A circuit that passes some frequencies and rejects or attenuates other frequencies.

Input offset voltage. A small voltage that must be applied to the input to null the output voltage.

Instrumentation amplifier. A difference amplifier with a high input impedance.

Integrator. An op-amp circuit whose output voltage is the integral of its input voltage.

Logarithmic amplifier. An amplifier whose output voltage is proportional to the natural log (base e) of the input.

Slew rate. The rate of change of the output of an op-amp in response to a step change in its input.

Summing circuit. A circuit that produces an output proportional to the sum of its inputs.

Virtual ground. The fact that the $V_{+\text{IN}}$ and $V_{-\text{IN}}$ terminals of an op-amp are usually at the same voltage, although there is a high impedance between them.

12-9 REFERENCES

Robert Boylestad and Louis Nashelsky, *Electronic Devices and Circuit Theory*, 4th Edition, Prentice-Hall, Englewood Cliffs, N.J., 1987.

Rodney B. Faber, *Essentials of Solid State Electronics*, John Wiley, New York, 1985.

Luces M. Faulkenberry, *An Introduction to Operational Amplifiers With Linear IC Applications*, 2nd Edition, John Wiley, New York, 1982.

Joseph D. Greenfield, *Practical Digital Design Using ICs*, 2nd Edition, John Wiley, New York, 1983.

Robert G. Irvine, *Operational Amplifier Characteristics and Applications*, Prentice-Hall, Englewood Cliffs, N.J., 1981.

JACOB MILLMAN and HERBERT TAUB, *Pulse, Digital, and Switching Waveforms*, McGraw-Hill, New York, 1965.

D. E. PIPPENGER and E. J. TOBABEN, *Linear and Interface Circuits Applications, Volume 1, Amplifiers, Comparators, Timers, and Voltage Regulators*, Texas Instruments, Inc., 1985.

DONALD L. SCHILLING and CHARLES BELOVE, *Electronic Circuits, Discrete and Integrated*, 2nd Edition, McGraw-Hill, New York, 1979.

WILLIAM STANLEY, *Operational Amplifiers With Linear Integrated Circuits*, Charles E. Merrill Co., Columbus, Ohio, 1984.

12-10 PROBLEMS

12-1 The swing of an op-amp is limited to $+12$ V. If its open-loop gain is 200,000 and V_{+IN} is connected to ground, what is the possible swing on V_{-IN} for linear operation?

12-2 An inverting op-amp has $R_1 = 5$ kΩ and $R_f = 25$ kΩ. What is its gain? Repeat this problem for a noninverting op-amp.

12-3 For the inverting amplifier of problem 12-2, assume the input is 2 V. If $A_{ol} = 200{,}000$, find the voltage at the $-IN$ terminal. Repeat if the input is -2 V.

12-4 Assume the output voltages of the op-amps of Figure 12-2 are limited to $+12$ V. What are the limitations on their input voltages?

12-5 For the circuit of Figure P12-5

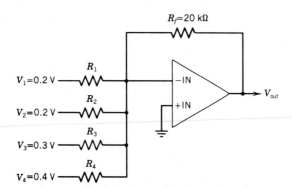

FIGURE P12-5

a. Find the output if $R_1 = R_2 = R_3 = R_4 = 2$ kΩ.
b. Find the output if V_3 is changed from 0.3 V to -0.3 V.

12-6 Repeat problem 12-5 if $R_1 = 1$ kΩ, $R_2 = 2$ kΩ, $R_3 = 3$ kΩ, and $R_4 = 4$ kΩ.

12-7 For a **741**, what are the typical and worst-case values of I_{IO}. What percent of I_{IB} are they?

12-8 In the circuit of Figure 12-10, R_f is changed to 50 kΩ. If the bias current flowing into both terminals is 400 nA
  a. What is the output voltage if the $+IN$ terminal is grounded?
  b. What is the proper value of the resistor that should be placed in the $+IN$ leg?

c. If this resistor is put into the +IN leg, find
   (1) The voltage at the +IN terminal.
   (2) The voltage at the −IN terminal.
   (3) The current in R_1.
   (4) The current in R_f.
   (5) The output voltage.

12-9 Repeat problem 12-8 if the bias current into the −IN terminal is 500 nA and the bias current into the +IN terminal is 300 nA.

12-10 An inverting op-amp has R_1 = 15 kΩ and is set up for a gain of 5. If its output resistance is 100 Ω and A_{ol} = 100,000, find the output impedance of the circuit with feedback.

12-11 Figure P12-11 shows the frequency response of a **TL080** op-amp. Find its gain–bandwidth product

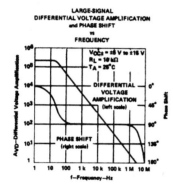

FIGURE P12-11
(Courtesy of Texas Instruments, Inc.)

a. At the corner frequency.
b. At 1 kHz.
c. When the gain is 1.

12-12 Figure P12-12 shows the large signal pulse response of a **TL080**. Find its slew rate.

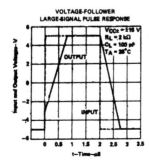

FIGURE P12-12
(Courtesy of Texas Instruments, Inc.)

12-13 An op-amp has a slew rate of 5 V/μs. How long will the output take to go from +3 V to −3 V?

12-14 Design an integrator similar to Figure 12-15 that will ramp from +5 V to −10 V in 10 ms.

12-15 For the circuit of Figure P12-15a, sketch the output voltage if the input voltage is given by Figure P12-15b. Assume the capacitor is initially uncharged and has no leakage. Be sure to place the critical voltage values on your sketch.

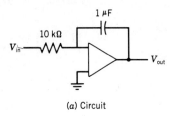

(a) Circuit

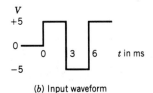

(b) Input waveform

FIGURE P12-15

12-16 For the circuit of Figure P12-16, V_{os} = 4 mV. Find

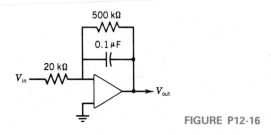

FIGURE P12-16

a. The output voltage due to V_{os}.
b. The general expression of the output voltage as a function of the input.
c. The lowest frequency of integration.

12-17 Design an op-amp integrator so that $V_{out} = 200 \int V_{in} \, dt$. What is the value of R_d if the minimum frequency of integration is to be 20 Hz?

12-18 For the circuit of Figure 12-19 assume $R_c = R = 100\ \text{k}\Omega$. If the circuit is to differentiate signals up to 100 Hz and integrate signals above 300 Hz, find the values of C and C_c.

12-19 For the circuit of Figure P12-19 $R_1 = 100\ \text{k}\Omega$, $R_2 = 250\ \text{k}\Omega$ and $R_3 = 50\ \text{k}\Omega$. If $C = 1\ \mu\text{f}$, express the output as a function of the inputs.

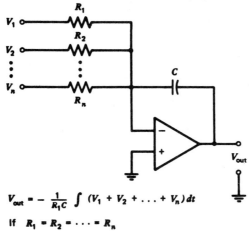

$$V_{out} = -\frac{1}{R_1 C} \int (V_1 + V_2 + \ldots + V_n)\, dt$$

If $R_1 = R_2 = \ldots = R_n$

FIGURE P12-19
Summary integrator. (From *An Introduction to Operational Amplifiers with Linear IC Applications* Second Edition by Lucas M. Faulkenberry. Copyright © 1982 by John Wiley & Sons, Inc. Reprinted by permission of John Wiley & Sons, Inc.)

12-20 Set up a circuit to solve the differential equation

$$6\frac{dv}{dt} + \frac{v}{3} = 4V_{in} \sin \omega t$$

Use 1 μf capacitors.

12-21 Set up a circuit to solve the differential equation

$$\frac{d^2v}{dt^2} - \frac{3dv}{dt} + \frac{v}{4} = 3V_{in} \sin \omega t$$

12-22 For the circuit of Figure 12-24 assume $R_1 = 10\ \text{k}\Omega$ and $R_2 = 50\ \text{k}\Omega$.
  a. If V_2 is held constant at 4 V, sketch V_{out} if V_1 ramps from 3 to 5 V.
  b. What is the input impedance of this amplifier?

12-23 Design a difference amplifier whose output is

$$V_{out} = 20(V_1 - V_2)$$

12-24 For the circuit of Figure P12-24 assume all the resistors, including the resistance of the transducer, are 1000 Ω. Find the output if the resistance of the transducer becomes

12-10 PROBLEMS   **617**

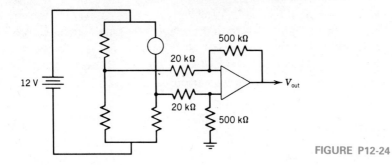

FIGURE P12-24

    a. 1050 Ω
    b. 950 Ω

12-25 A temperature transducer has a resistance of 2000 Ω at 25°C and a resistance of 2200 Ω at 85°C. Design a circuit that will have an output voltage of 0 V at 25°C, and 10 V at 85°C.

12-26 Repeat Example 12-17 if the leakage current of the transistor is 5 nA.

12-27 A transistor has a leakage current of 20 nA. Using this transistor, design a log amplifier so that its output voltage is 0.4 V if its input voltage is 10 V.

APPENDIX A

Characteristics of the 2N3903 and 2N3904 Transistors

Appendix A tables are courtesy of Motorola. Copyright by Motorola, Inc. Used by permission.

2N3903
2N3904

NPN SILICON ANNULAR♦ TRANSISTORS

... designed for general purpose switching and amplifier applications and for complementary circuitry with types 2N3905 and 2N3906.

- High Voltage Ratings — BV_{CEO} = 40 Volts (Min)
- Current Gain Specified from 100 μA to 100 mA
- Complete Switching and Amplifier Specifications
- Low Capacitance — C_{ob} = 4.0 pF (Max)

NPN SILICON SWITCHING & AMPLIFIER TRANSISTORS

MAXIMUM RATINGS

Rating	Symbol	Value	Unit
*Collector-Base Voltage	V_{CB}	60	Vdc
*Collector-Emitter Voltage	V_{CEO}	40	Vdc
*Emitter-Base Voltage	V_{EB}	6.0	Vdc
*Collector Current	I_C	200	mAdc
Total Power Dissipation @ T_A = 60°C	P_D	250	mW
**Total Power Dissipation @ T_A = 25°C Derate above 25°C	P_D	350 2.8	mW mW/°C
**Total Power Dissipation @ T_C = 25°C Derate above 25°C	P_D	1.0 8.0	Watts mW/°C
**Junction Operating Temperature	T_J	150	°C
**Storage Temperature Range	T_{stg}	−55 to +150	°C

THERMAL CHARACTERISTICS

Characteristic	Symbol	Max	Unit
Thermal Resistance, Junction to Ambient	$R_{\theta JA}$	357	°C/W
Thermal Resistance, Junction to Case	$R_{\theta JC}$	125	°C/W

*Indicates JEDEC Registered Data
**Motorola guarantees this data in addition to the JEDEC Registered Data.
♦Annular Semiconductors Patented by Motorola Inc.

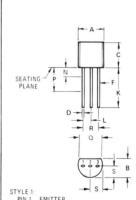

STYLE 1:
PIN 1. EMITTER
    2. BASE
    3. COLLECTOR

DIM	MILLIMETERS		INCHES	
	MIN	MAX	MIN	MAX
A	4.450	5.200	0.175	0.205
B	3.180	4.190	0.125	0.165
C	4.320	5.330	0.170	0.210
D	0.407	0.533	0.016	0.021
F	0.407	0.482	0.016	0.019
K	12.700	−	0.500	−
L	1.150	1.390	0.045	0.055
N	−	1.270	−	0.050
P	6.350	−	0.250	−
Q	3.430	−	0.135	−
R	2.410	2.670	0.095	0.105
S	2.030	2.670	0.080	0.105

CASE 29-02
TO-92

© MOTOROLA INC. 1973   DS 5127 R2

APPENDIX A **621**

*ELECTRICAL CHARACTERISTICS ($T_A = 25°C$ unless otherwise noted)

Characteristic		Fig. No.	Symbol	Min	Max	Unit
OFF CHARACTERISTICS						
Collector-Base Breakdown Voltage ($I_C = 10$ μAdc, $I_E = 0$)			BV_{CBO}	60	–	Vdc
Collector-Emitter Breakdown Voltage (1) ($I_C = 1.0$ mAdc, $I_B = 0$)			BV_{CEO}	40	–	Vdc
Emitter-Base Breakdown Voltage ($I_E = 10$ μAdc, $I_C = 0$)			BV_{EBO}	6.0	–	Vdc
Collector Cutoff Current ($V_{CE} = 30$ Vdc, $V_{EB(off)} = 3.0$ Vdc)			I_{CEX}	–	50	nAdc
Base Cutoff Current ($V_{CE} = 30$ Vdc, $V_{EB(off)} = 3.0$ Vdc)			I_{BL}	–	50	nAdc
ON CHARACTERISTICS						
DC Current Gain (1) ($I_C = 0.1$ mAdc, $V_{CE} = 1.0$ Vdc)	2N3903 2N3904	15	h_{FE}	20 40	– –	–
($I_C = 1.0$ mAdc, $V_{CE} = 1.0$ Vdc)	2N3903 2N3904			35 70	– –	
($I_C = 10$ mAdc, $V_{CE} = 1.0$ Vdc)	2N3903 2N3904			50 100	150 300	
($I_C = 50$ mAdc, $V_{CE} = 1.0$ Vdc)	2N3903 2N3904			30 60	– –	
($I_C = 100$ mAdc, $V_{CE} = 1.0$ Vdc)	2N3903 2N3904			15 30	– –	
Collector-Emitter Saturation Voltage (1) ($I_C = 10$ mAdc, $I_B = 1.0$ mAdc) ($I_C = 50$ mAdc, $I_B = 5.0$ mAdc)		16, 17	$V_{CE(sat)}$	– –	0.2 0.3	Vdc
Base-Emitter Saturation Voltage (1) ($I_C = 10$ mAdc, $I_B = 1.0$ mAdc) ($I_C = 50$ mAdc, $I_B = 5.0$ mAdc)		17	$V_{BE(sat)}$	0.65 –	0.85 0.95	Vdc
SMALL-SIGNAL CHARACTERISTICS						
Current-Gain–Bandwidth Product ($I_C = 10$ mAdc, $V_{CE} = 20$ Vdc, $f = 100$ MHz)	2N3903 2N3904		f_T	250 300	– –	MHz
Output Capacitance ($V_{CB} = 5.0$ Vdc, $I_E = 0$, $f = 100$ kHz)		3	C_{ob}	–	4.0	pF
Input Capacitance ($V_{BE} = 0.5$ Vdc, $I_C = 0$, $f = 100$ kHz)		3	C_{ib}	–	8.0	pF
Input Impedance ($I_C = 1.0$ mAdc, $V_{CE} = 10$ Vdc, $f = 1.0$ kHz)	2N3903 2N3904	13	h_{ie}	0.5 1.0	8.0 10	k ohms
Voltage Feedback Ratio ($I_C = 1.0$ mAdc, $V_{CE} = 10$ Vdc, $f = 1.0$ kHz)	2N3903 2N3904	14	h_{re}	0.1 0.5	5.0 8.0	X 10^{-4}
Small-Signal Current Gain ($I_C = 1.0$ mAdc, $V_{CE} = 10$ Vdc, $f = 1.0$ kHz)	2N3903 2N3904	11	h_{fe}	50 100	200 400	–
Output Admittance ($I_C = 1.0$ mAdc, $V_{CE} = 10$ Vdc, $f = 1.0$ kHz)		12	h_{oe}	1.0	40	μmhos
Noise Figure ($I_C = 100$ μAdc, $V_{CE} = 5.0$ Vdc, $R_S = 1.0$ k ohms, $f = 10$ Hz to 15.7 kHz)	2N3903 2N3904	9, 10	NF	– –	6.0 5.0	dB
SWITCHING CHARACTERISTICS						
Delay Time	($V_{CC} = 3.0$ Vdc, $V_{BE(off)} = 0.5$ Vdc, $I_C = 10$ mAdc, $I_{B1} = 1.0$ mAdc)	1, 5	t_d	–	35	ns
Rise Time		1, 5, 6	t_r	–	35	ns
Storage Time	($V_{CC} = 3.0$ Vdc, $I_C = 10$ mAdc, $I_{B1} = I_{B2} = 1.0$ mAdc)	2, 7	t_s	– –	175 200	ns
Fall Time		2, 8	t_f	–	50	ns

(1) Pulse Test: Pulse Width = 300 μs, Duty Cycle = 2.0%.
*Indicates JEDEC Registered Data

FIGURE 1 – DELAY AND RISE TIME EQUIVALENT TEST CIRCUIT

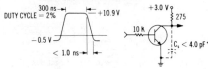

FIGURE 2 – STORAGE AND FALL TIME EQUIVALENT TEST CIRCUIT

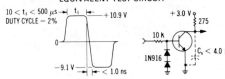

*Total shunt capacitance of test jig and connectors

MOTOROLA Semiconductor Products Inc.

2N3903 • 2N3904

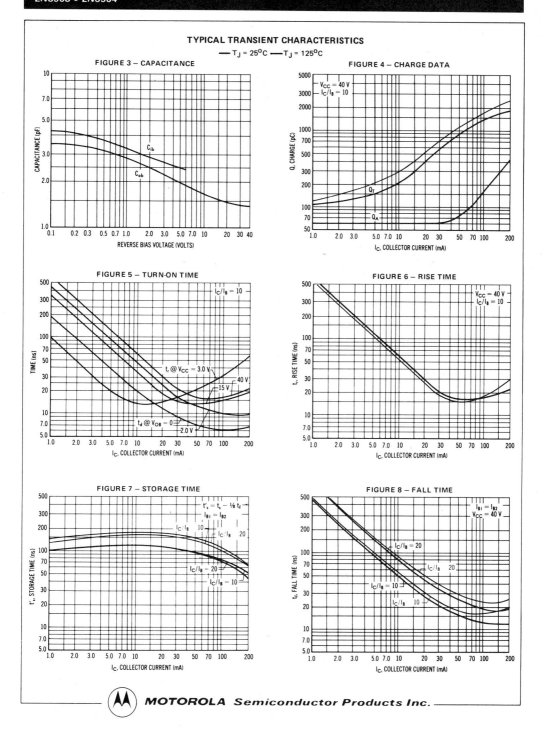

TYPICAL TRANSIENT CHARACTERISTICS
— $T_J = 25°C$ — $T_J = 125°C$

FIGURE 3 — CAPACITANCE

FIGURE 4 — CHARGE DATA

FIGURE 5 — TURN-ON TIME

FIGURE 6 — RISE TIME

FIGURE 7 — STORAGE TIME

FIGURE 8 — FALL TIME

MOTOROLA *Semiconductor Products Inc.*

APPENDIX A **623**

2N3903 • 2N3904

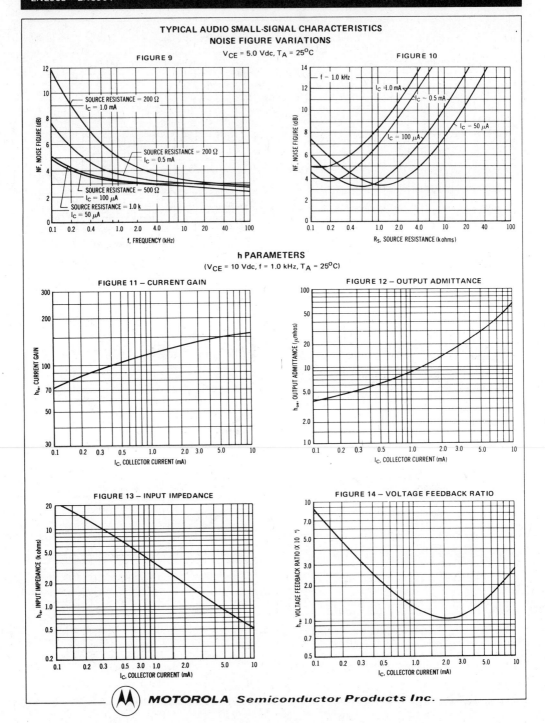

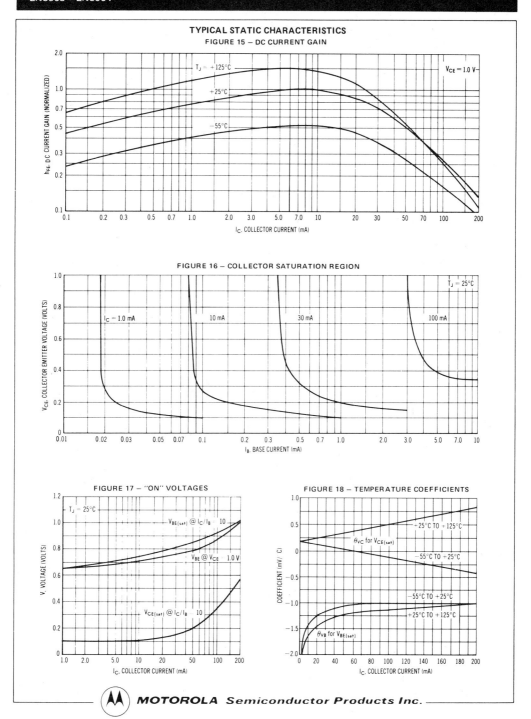

TYPICAL STATIC CHARACTERISTICS
FIGURE 15 – DC CURRENT GAIN

FIGURE 16 – COLLECTOR SATURATION REGION

FIGURE 17 – "ON" VOLTAGES

FIGURE 18 – TEMPERATURE COEFFICIENTS

APPENDIX B

Derivations of the Exact Hybrid Parameter Equations for a Transistor

To find the gain we can start with the basic hybrid parameter transistor equations.

$$v_1 = h_{11} i_1 + h_{12} v_2$$
$$i_2 = h_{21} i_1 + h_{22} v_2 \tag{B-1}$$

For the typical common emitter circuit $h_{11} = h_{ie}$, $h_{12} = h_{re}$, $h_{21} = h_{fe}$ and $h_{22} = h_{oe}$.

Assume the transistor is connected to a collector resistor, R_C, as shown in Figure B-1. Then we have

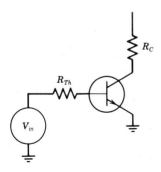

FIGURE B-1

$$i_2 = \frac{-v_2}{R_C}$$

and equations B-1 can be expressed as

$$v_1 = h_{11} i_1 + h_{12} v_2 \tag{B-2}$$

$$0 = h_{21} i_1 + \left(h_{22} + \frac{1}{R_C}\right) v_2 \tag{B-3}$$

To find the voltage gain, the simultaneous equations B-2 and B-3 can be solved for v_2. Using determinants we obtain

$$v_2 = \frac{\begin{vmatrix} h_{11} & v_1 \\ h_{21} & 0 \end{vmatrix}}{\begin{vmatrix} h_{11} & h_{12} \\ h_{21} & h_{22} + \frac{1}{R_C} \end{vmatrix}}$$

$$v_2 = \frac{-h_{21} v_1}{\Delta^h + h_{11/R_C}}$$

where

$$\Delta^h = h_{11} h_{22} - h_{21} h_{12}$$

$$A_v = \frac{v_2}{v_1} = \frac{-R_C h_{21}}{R_C \Delta^h + h_{11}} \tag{B-4}$$

This is the same as equation 4-4 in the text. Note that the approximate equations in the text are derived from the exact equation by assuming

$$h_{11} \gg R_C \Delta^h$$

The current gain can be found by using equation B-3 and substituting $-i_2 R_C$ for v_2. This gives

$$0 = h_{21} i_1 - \left(h_{22} + \frac{1}{R_C}\right) i_2 R_C$$

$$A_i = \frac{i_2}{i_1} = \frac{h_{21}}{R_C h_{22} + 1}$$

This is equation 4-5 in the text.
If $h_{22} R_C \ll 1$ as is usually the case, $A_i \approx h_{21}$.

The input impedance is found by solving equations B-2 and B-3 for i_1.

$$i_1 = \frac{\begin{vmatrix} v_1 & h_{12} \\ 0 & h_{22} + \dfrac{1}{R_C} \end{vmatrix}}{\begin{vmatrix} h_{11} & h_{12} \\ h_{21} & h_{22} + \dfrac{1}{R_C} \end{vmatrix}}$$

$$i_1 = \frac{v_1\left(h_{22} + \dfrac{1}{R_C}\right)}{\Delta^h + h_{11/R_C}}$$

$$R_{in} = \frac{v_1}{i_1} = \frac{\Delta^h + h_{11/R_C}}{h_{22} + \dfrac{1}{R_C}} = \frac{h_{11} + R_C \Delta^h}{1 + h_{22} R_C}$$

This is equation 4-6.

If $h_{11} \gg R_C \Delta^h$ and $1 \gg h_{22} R_C$, as is usually the case, then $R_{in} \approx h_{11}$.

The output impedance can be found by shorting the input voltage generator, but not its input impedance R_{Th}, and applying a voltage at v_2. Under these conditions $v_1 = -i_1 R_{Th}$ and the equations become

$$0 = (h_{11} + R_{Th})i_1 + h_{12} v_2$$
$$i_2 = h_{21} i_1 + h_{22} v_2$$

Solving for v_2 by determinants we have

$$v_2 = \frac{\begin{vmatrix} h_{11} + R_{Th} & 0 \\ h_{21} & i_2 \end{vmatrix}}{\begin{vmatrix} h_{11} + R_{Th} & h_{12} \\ h_{21} & h_{22} \end{vmatrix}} = \frac{i_2(h_{11} + R_{Th})}{\Delta^h + R_{Th} h_{22}}$$

$$R_o = \frac{v_2}{i_2} = \frac{h_{11} + R_{Th}}{\Delta^h + R_{Th} h_{22}}$$

This is the same as equation 4-7.

APPENDIX C

A Program for Calculating the Gain of a JFET with a Capacitively Coupled Load

The program for calculating the gain of a JFET with a capacitively coupled, resistive load is given in this appendix. It is assumed that the JFET and its pertinent characteristics (I_{DSS}, V_P) and the ac coupled load resistor, R_L is also specified. The program then steps through all the possible values of V_{GS}. For each value of V_{GS} it calculates the corresponding drain resistor, R_D, the gain, and other parameters of the circuit. The program can be used for either fixed- or self-biased circuits.

To obtain the program, the JFET curves were triangularized and idealized as shown in Figure C-1. The $V_{GS} = 0$ curve was assumed to rise linearly

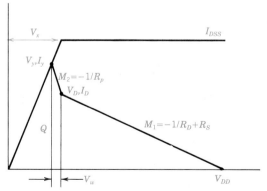

FIGURE C-1
Load lines on an idealized JFET characteristic.

631

until $I_D = I_{DSS}$ and then flatten out. V_X is the voltage at the knee point where it flattens out; for a **2N5459** we estimate V_X to be 8 V.

V_W is the required output voltage swing. Once the output voltage swing is specified, I_D can no longer equal I_{DSS}, nor $V_{GS} = 0$. Instead, the situation is as shown in Figure C-1. The circuit operates at some quiescent point that determines I_D and V_{DS}. The quiescent point is at the intersection of the ac and dc load lines. From the quiescent point, the operating point can swing until it reaches the I_{DSS} curve at V_Y, I_Y.

To run the program the user must specify the required voltage swing and the load resistor R_L. Then V_{DD} and the JFET parameters must be entered. Finally the variable Y must be entered. For a self-biased circuit $Y = 1$, and for a fixed-bias circuit $Y = 0$. The program steps through the allowable values of V_{GS} (in tenth volt steps) and finds the gain and the resistors at each point. After running it the user can determine the best operating point.

Mathematical Derivation

The mathematics behind the program are given below. From Figure C-1 we have

$$V_Y = V_X \frac{I_Y}{I_{DSS}}$$

For a self-biased circuit

$$I_D = \frac{V_{DD} - V_W - V_Y}{R_D + R_S} = \frac{V_{DD} - V_W - \frac{V_X I_Y}{I_{DSS}}}{R_D + R_S} \quad \text{(C-1)}$$

But $I_D R_S = V_{GS}$ for a self-biased JFET, so

$$I_D R_D = V_{DD} - V_W - V_{GS} - V_X \frac{I_Y}{I_{DSS}}$$

We will let $K_1 = V_{DD} - V_W - V_{GS}$.

For a fixed-biased circuit equation C-1 can be used by setting $R_S = 0$. Then we have

$$I_D R_D = V_{DD} - V_W - V_X \frac{I_Y}{I_{DSS}}$$

Here we let $K_2 = V_{DD} - V_W$.

Both values of K can be combined if we define K as

$$K = V_{DD} - V_W - Y V_{GS}$$

where $Y = 1$ for a self-biased circuit and $Y = 0$ for a fixed biased circuit. It then follows that

$$I_D R_D = K - \frac{V_X I_Y}{I_{DSS}}$$

$$R'_L = \frac{R_D R_L}{R_D + R_L} = \frac{V_W}{I_Y - I_D}$$

$$I_Y = \frac{V_W + R'_L I_D}{R'_L} = \frac{V_W}{R'_L} + I_D$$

$$I_D R_D = K - \frac{V_X V_W}{I_{DSS} R'_L} - \frac{I_D V_X}{I_{DSS}}$$

$$I_D R_D R'_L = K R'_L - \frac{V_X V_W}{I_{DSS}} - \frac{I_D R'_L V_X}{I_{DSS}}$$

Substituting $R_D R_L/(R_D + R_L)$ for R'_L and multiplying by $R_D + R_L$ we get

$$I_D R_D^2 R_L = K R_D R_L - \frac{I_D V_X R_D R_L}{I_{DSS}} - (R_D + R_L)\frac{V_X V_W}{I_{DSS}}$$

Finally we have

$$I_D R_L R_D^2 - R_D\left(KR_L - \frac{I_D R_L V_X}{I_{DSS}} - \frac{V_X V_W}{I_{DSS}}\right) + \frac{R_L V_X V_W}{I_{DSS}} = 0$$

This is a quadratic in R_D. The program selects a value of V_{GS}. All other values are specified. If the value of V_{GS} is a possible value, it solves the quadratic for R_D. Once R_D is known the gain and operating points are easily found.

The program is given as Program C-1. This program can also be used to calculate the gain of a JFET with no capacitively coupled load by making R_L very large (perhaps 1 MΩ).

```
10 REM  This is a program to find the best gain
20 REM  for a self biased JFET
30 INPUT "TYPE THE SWING VOLTAGE AND RL";VW,RL
40 INPUT "TYPE VDD,VP,IDSS,AND VX";VC,VP,IS,VX
41 INPUT "TYPE 1 IF THIS IS A SELF BIASED CIRCUIT OR 0 FOR FIXED BIAS";Y
42 IP = IS/1000
50 REM Step VGS
55 PRINT
56 PRINT "  GAIN      ID(MA)      VGS       RS        RD       VDS         IY "
57 PRINT
60 FOR M = 1 TO 10*VP-1
70 VG = M/10
80 ID= IP*(1-(VG/VP))^2
90 IF VW/(IP-ID)>RL GO TO 210
100 REM Solve the quadratic for RD
110 A = RL*ID
120 K = VC-VW-VG*Y
130 B = K*RL - ID*RL*VX/IP - VX*VW/IP
140 C = RL*VW*VX/IP
145 IF B^2-4*A*C < 0 GO TO 210
150 D = SQR(B^2-4*A*C)
160 RD =( B + D)/(2*A)
165 RP = RL*RD/(RL + RD)
170 RS = VG*Y/ID
175   REM  Find the gain.
180 G = 2*IP/VP
190 AV = G*(1-VG/VP)*RP
191 REM Find IY
192 IY =   (VW+(RP*ID))/(RP)
193 IF IY>IP GO TO 210
194 IF IY - 2*(IY-ID)<0 GO TO 210
196 VD = VX*IY/IP + VW
200 PRINT AV;TAB(10);1000*ID;TAB(20);VG;TAB(30);INT(RS);TAB(40);INT(RD);TAB(50);VD;TAB(60);IY
210 NEXT M
220 END
```

PROGRAM C-1
A program to find the gain of a JFET with a capacitively coupled load.

PART C-1: PROBLEMS USING PROGRAM C-1

This Appendix contains three problems that were run using the Program C-1.

Problem 1

Find the gain of a **2N5459** JFET using fixed-bias with $V_{DD} = 20$ V, a required swing of 2.5 V, and $R_L = 1000\Omega$. This is the same as Example 5-15.

The input and output files for the problem are shown in Figure C-2.

```
2.5,1000
20,6,9,8
0
```

GAIN	ID(MA)	VGS	RS	RD	VDS	IY
1.4735	5.0625	1.5	0	1897	10.3933	.887994E-02
1.47701	4.84	1.6	0	2042	10.1122	.856373E-02
1.47758	4.6225	1.7	0	2197	9.8424	.826019E-02
1.4754	4.41	1.8	0	2362	9.58299	.796837E-02
1.47061	4.2025	1.9	0	2538	9.3333	.768746E-02
1.46335	4	2	0	2726	9.09273	.741682E-02
1.45375	3.8025	2.1	0	2929	8.86081	.715591E-02
1.44191	3.61	2.2	0	3147	8.63711	.690425E-02
1.42793	3.4225	2.3	0	3383	8.4213	.666146E-02
1.4119	3.24	2.4	0	3637	8.21307	.64272E-02

FIGURE C-2
Results of running Problem 1.

Problem 2

Find the gain of a **2N5459** using self-bias with $V_{DD} = 20$ V, a required swing of 2.5 V, and $R_L = 1000\Omega$.

The input and output files for the problems are shown in Figure C-3.

```
2.5,1000
20,6,9,8
1
```

GAIN	ID(MA)	VGS	RS	RD	VDS	IY
1.37312	4.84	1.6	330	1660	10.3626	.884547E-02
1.37608	4.6225	1.7	367	1778	10.0809	.852853E-02
1.37647	4.41	1.8	408	1902	9.81033	.822412E-02
1.37443	4.2025	1.9	452	2034	9.55005	.793131E-02
1.37011	4	2	499	2175	9.29942	.764935E-02
1.36359	3.8025	2.1	552	2325	9.05788	.737761E-02
1.35499	3.61	2.2	609	2486	8.82495	.711557E-02

FIGURE C-3
Results of running Problem 2.

Problem 3

Find the gain of a **2N5459** using self-bias with $V_{DD} = 25$ V, a swing of 2.5 V, and $R_L = 1000\Omega$.

The input and output files for the problems are shown in Figure C-4.

```
2.5,1000
25,6,9,8
1
```

GAIN	ID(MA)	VGS	RS	RD	VDS	IY
1.64519	5.29	1.4	264	2512	10.3089	.878503E-02
1.6344	5.0625	1.5	296	2654	10.0592	.850413E-02
1.62205	4.84	1.6	330	2806	9.81624	.823077E-02
1.60818	4.6225	1.7	367	2968	9.57981	.796478E-02
1.59283	4.41	1.8	408	3140	9.34979	.770602E-02
1.57603	4.2025	1.9	452	3325	9.12609	.745435E-02
1.5578	4	2	499	3522	8.90858	.720965E-02
1.53818	3.8025	2.1	552	3735	8.69718	.697182E-02
1.5172	3.61	2.2	609	3963	8.49178	.674076E-02
1.49489	3.4225	2.3	672	4209	8.29233	.651637E-02
1.47127	3.24	2.4	740	4475	8.09873	.629857E-02
1.44638	3.0625	2.5	816	4763	7.91093	.608729E-02

FIGURE C-4
Results of running Problem 3.

PART C-2: PROGRAM CHECKS

This section checks the results of the previous program at one point. Other points can be checked similarly.

As a check, we will use line 5 of the output file for Problem 2. It gives

$$I_C = 4 \text{ mA} \quad V_{GS} = -2 \text{ V} \quad R_S = 499\Omega \quad R_D = 2175\Omega$$
$$V_{DS} = 9.3 \text{ V} \quad I_Y = 7.649 \text{ mA}$$

The dc slope of the load line must be the same as the sum of R_D and R_S. The dc load line goes from $V_{DS} = 20$ V, $I_D = 0$ to the quiescent point, $V_{DS} = 9.3$ V, $I_D = 4$ mA

$$R = \frac{V_{DS}}{I_D} = \frac{20 \text{ V} - 9.3 \text{ V}}{4 \text{ mA}} = \frac{10.7 \text{ V}}{4 \text{ mA}} = 2675\Omega$$
$$R_D + R_S = 2175\Omega + 499\Omega = 2674\Omega$$

This shows that the quiescent point, as given by the computer and using the computer-generated values of R_D and R_S, is indeed on the dc load line.

The ac swing is 2.5 V along the ac load line. The slope of the ac load line is

$$R_P = R_L \| R_D = \frac{1000\Omega \times 2175\Omega}{3175\Omega} = 685\Omega$$

Because the voltage swings by 2.5 V

$$\Delta I_D = \frac{\Delta V}{R_P} = \frac{2.5 \text{ V}}{685} = 3.649 \text{ mA}$$

Adding this to the quiescent value of I_D gives I_D at the point where it intersects the I_{DSS} curve.

$$I_Y = \Delta I_D + I_{DQ} = 3.649 \text{ mA} + 4 \text{ mA} = 7.649 \text{ mA}$$

This checks with the computer result.

Having found I_Y, we can now find V_Y.

$$V_Y = \frac{I_Y}{I_{DSS}} V_X = \frac{7.649}{9} \times 8 = 6.8 \text{ V}$$

This shows that the ac voltage swing between the quiescent point and the I_{DSS} curve is 9.3 V $-$ 6.8 V $= 2.5$ V, as specified.

APPENDIX D

Derivation of Equations for the Difference Amplifier

To obtain the equations for a difference amplifier, refer to Figure D-1. Assume a Thevenized voltage source, ΔV, is driving the difference amplifier through the Thevenin's resistance, R_{Th}.

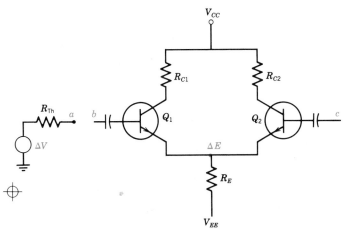

FIGURE D-1

To find the common mode gain, assume the source is driving both bases simultaneously, by connecting point a to both points b and c. Any increase of base voltage due to ΔV will cause the emitter voltage, ΔE to increase. Then it follows that

$$\Delta I_{B1} = \Delta I_{B2} = \frac{\Delta V - \Delta E}{h_{ie} + R_{Th}}$$

$$\Delta I_{C1} = \Delta I_{C2} = h_{fe} \Delta I_B = \frac{h_{fe}(\Delta V - \Delta E)}{h_{ie} + R_{Th}}$$

Assuming the base current is small compared to the collector current, as it is for reasonable values of h_{fe}

$$\Delta E = (\Delta I_{C1} + \Delta I_{C2})R_E = \frac{2h_{fe}R_E(\Delta V - \Delta E)}{h_{ie} + R_{Th}}$$

$$\Delta E \left[1 + \frac{2 h_{fe}R_E}{h_{ie} + R_{Th}} \right] = \frac{2h_{fe}R_E}{h_{ie} + R_{Th}} \Delta V$$

Multiplying by $h_{ie} + R_{Th}$, we have

$$\Delta E(h_{ie} + R_{Th} + 2 h_{fe}R_E) = 2 h_{fe}R_E \Delta V$$

$$\Delta E = \frac{2 h_{fe}R_E}{h_{ie} + R_{Th} + 2h_{fe}R_E} \Delta V$$

$$\Delta I_{C2} = \frac{h_{fe}(\Delta V - \Delta E)}{h_{ie} + R_{Th}} = \frac{h_{fe} \Delta V \left[1 - \dfrac{2 R_E h_{fe}}{h_{ie} + R_{Th} + 2h_{fe}R_E} \right]}{h_{ie} + R_{Th}}$$

$$= \frac{h_{fe} \Delta V \left[\dfrac{h_{ie} + R_{Th} + 2h_{fe}R_E - 2R_E h_{fe}}{h_{ie} + R_{Th} + 2h_{fe}R_E} \right]}{h_{ie} + R_{Th}}$$

$$= \frac{h_{fe} \Delta V}{h_{ie} + R_{Th} + 2h_{fe}R_E}$$

$$A_C = \frac{\Delta I_{C2} R_{C2}}{\Delta V} \left(\text{or } \frac{\Delta I_{C1} R_{C1}}{\Delta V} \right) = \frac{h_{fe} R_C}{h_{ie} + R_{Th} + 2h_{fe}R_E}$$

This is equation 6-6.

To find the difference mode gain, as in the laboratory, assume ΔV is applied to point b in Figure D-1, while point c is grounded. A voltage ΔV will again produce a change at the emitter of ΔE. Then

$$\Delta I_{B1} = \frac{\Delta V - \Delta E}{R_{Th} + h_{ie}}$$

638 APPENDIX D

$$\Delta I_{B2} = -\frac{\Delta E}{R_{Th} + h_{ie}}$$

$$\Delta E = h_{fe}(\Delta I_{B1} + \Delta I_{B2})R_E$$

But

$$\Delta I_{B1} + \Delta I_{B2} = \frac{\Delta V - 2\Delta E}{R_{Th} + h_{ie}}$$

$$\Delta E = \frac{h_{fe}R_E}{R_{Th} + h_{ie}}(\Delta V - 2\Delta E)$$

$$\Delta E \left[1 + \frac{2 h_{fe}R_E}{R_{Th} + h_{ie}} \right] = \frac{h_{fe}R_E \, \Delta V}{R_{Th} + h_{ie}}$$

In most cases, however

$$\frac{2 h_{fe}R_E}{R_{Th} + h_{ie}} \gg 1$$

and

$$1 + \frac{2 h_{fe}R_E}{R_{Th} + h_{ie}} \approx \frac{2 h_{fe}R_E}{R_{Th} + h_{ie}}$$

Therefore

$$\Delta E \approx \frac{\Delta V}{2}$$

$$I_{B1} = \frac{\Delta V}{2(R_{Th} + h_{ie})}$$

$$\Delta V_{out} = h_{fe}R_{C1} \Delta I_{B1} = \frac{-h_{fe}R_{C1} \Delta V}{2(R_{Th} + h_{ie})}$$

$$A_d = \frac{\Delta V_{out}}{\Delta V} = \frac{-h_{fe}R_C}{2(R_{Th} + h_{ie})}$$

This is equation 6-7.

  This equation can also be derived, even more simply, if it is assumed that point *a* increases by $\Delta V/2$, point *b* decreases by $\Delta V/2$, and that the resulting current changes compensate for each other so that $\Delta E = 0$.

BALANCING RESISTORS

With balancing resistors added, as in Figure 6-21, the common mode currents become

$$\Delta I_{B1} - \Delta I_{B2} = \frac{\Delta V - \Delta E}{h_{ie} + R_{Th} + h_{fe}R_B}$$

where R_B is the value of the balancing resistor placed between the base and the emitter-resistor, R_E.

The steps of the derivation follow the common-mode derivation exactly if $h_{ie} + R_{Th}$ is replaced by $h_{ie} + R_{Th} + h_{fe}R_B$. The results are

$$A_C = \frac{h_{fe}R_C}{h_{ie} + R_{Th} + h_{fe}(R_B + 2R_E)}$$

Generally the balancing resistor is small so that $R_B \ll 2R_E$ and the equation reduces to the standard equation for A_C.

For the difference mode gains

$$I_{B1} = \frac{\Delta V - \Delta E}{R_S + h_{ie} + h_{fe}R_B}$$

$$I_{B2} = \frac{\Delta E}{R_{Th} + h_{ie} + h_{fe}R_B}$$

The steps in the derivation follow with $R_{Th} + h_{ie}$ replaced by $R_{Th} + h_{ie} + h_{fe}R_B$. With the assumption that

$$\frac{2h_{fe}R_E}{R_{Th} + h_{ie} + h_{fe}R_B} \gg 1$$

the final equation becomes

$$A_d = \frac{h_{fe}R_C}{2(R_{Th} + h_{ie} + h_{fe}R_B)}$$

This is equation 6-8, and it is significantly different from equation 6-7, the equation for A_d without balancing resistors.

APPENDIX E

The Frequency Response Due to the Emitter Bypass Capacitor

The circuit of Figure E-1 can be used to find the frequency response due to the emitter bypass capacitor. From it we have:

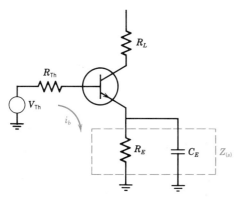

FIGURE E-1
The low-frequency-dependent circuit.

$$i_b = \frac{V_{Th}}{R_{Th} + h_{ie} + (1 + h_{fe})Z(s)} \qquad \text{(E-1)}$$

where

$$Z(s) = R_E \| X_C = \frac{R_E X_C}{R_E + X_C} = \frac{R_E \cdot \frac{1}{sC}}{R_E + \frac{1}{sC}}$$

$$Z(s) = \frac{R_E}{1 + sR_E C_E}$$

Multiplying equation E-1 by $(1 + sR_E C_E)$

$$i_b = \frac{V_{Th}(1 + sR_E C_E)}{(1 + sR_E C_E)[R_{Th} + h_{ie}] + (1 + h_{fe})R_E}$$

$$v_{out} = h_{fe}R_L i_b$$

so that

$$A_{v(ck)} = \frac{v_{out}}{v_{Th}} = \frac{h_{fe}R_L(1 + sR_E C_E)}{[R_{Th} + h_{ie}][1 + sR_E C_E] + (1 + h_{fe})R_E}$$

$$= \frac{h_{fe}R_L(1 + sR_E C_E)}{(R_{Th} + h_{ie})(sR_E C_E) + R_{Th} + h_{ie} + (1 + h_{fe})R_E}$$

$$= \frac{h_{fe}R_L(1 + sR_E C_E)}{(R_{Th} + h_{ie})\left[sR_E C_E + \dfrac{R_{Th} + h_{ie} + (1 + h_{fe})R_E}{R_{Th} + h_{ie}}\right]}$$

Dividing by $R_E C_E$

$$A_{v(ck)} = \frac{h_{fe}R_L\left(s + \dfrac{1}{R_E C_E}\right)}{(R_{Th} + h_{ie})\left[s + \dfrac{R_{Th} + h_{ie} + (1 + h_{fe})R_E}{R_E C_E(R_{Th} + h_{ie})}\right]}$$

This is equation 7-7.

APPENDIX F

Derivation of Feedback Circuit Equations

PART F-1

To show that the output impedance of a series voltage amplifier is

$$R_{of} = \frac{R'_o}{1 + \beta A_2}$$

where R'_o is the parallel combination of the output impedance without feedback and R_β, the impedance of the β network.

The circuit is shown in Figure F-1 with a voltage generator v_o applied.

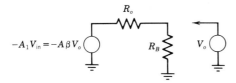

FIGURE F-1

The resulting current is

$$i_o = \frac{v_o(1 + A_1\beta)}{R_o} + \frac{V_o}{R_\beta}$$

$$R_{of} = \frac{v_o}{i_o} = \frac{1}{\frac{1+A_1\beta}{R_o} + \frac{1}{R_\beta}}$$

$$R_{of} = \frac{R_o R_\beta}{R_\beta + R_\beta \beta A_1 + R_o}$$

But A_2, the gain with the β network in place, is

$$A_2 = A_1 \frac{R_\beta}{R_o + R_\beta}$$

$$A_1 = \frac{A_2(R_o + R_\beta)}{R_\beta}$$

$$R_{of} = \frac{R_o R_\beta}{R_o + R_\beta + A_2\beta(R_o + R_\beta)}$$

Dividing top and bottom by $(R_o + R_\beta)$ we have

$$R_{of} = \frac{\frac{R_o R_\beta}{R_o + R_\beta}}{1 + A_2\beta}$$

Of course

$$\frac{R_o R_\beta}{R_o + R_\beta} = R_o \parallel R_\beta$$

PART F-2

This part of the appendix shows that the current and voltage gains of a shunt feedback circuit are both reduced by the factor $(1 + A_i)$.

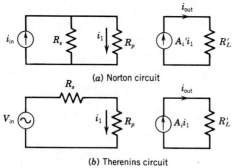

(a) Norton circuit

(b) Therenins circuit

FIGURE F-2
Circuit for Appendix F-2.

Consider the Norton's circuit of Figure F-2a where:

$$A_i = \frac{i_{out}}{i_{in}} \quad A_i' = \frac{i_{out}}{i_1}$$

R_L' is the output load resistance (including the feedback network, but with no feedback).

$$i_1 = i_{in} \times \frac{R_S}{R_S + R_P}$$

$$A_i' = \frac{i_{out}}{i_1} = \frac{i_{out}}{i_{in}} \frac{R_S + R_P}{R_S} = A_i \frac{(R_S + R_P)}{R_S}$$

$$v_{out} = A_i' i_1 R_L' = A_i \frac{(R_S + R_P)}{R_S} \times \frac{i_{in} R_S}{R_S + R_P} \times R_L' = A_i i_{in} R_L'$$

If the Thevenin's equivalent circuit of Figure F-2b is used

$$v_{in} = i_1 (R_S + R_P) = i_{in} R_S$$

$$v_{out} = A_i i_{in} R_L' = \frac{A_i v_{in} R_L'}{R_S}$$

$$A_v = \frac{v_{out}}{v_{in}} = \frac{A_i R_L'}{R_S}$$

If feedback is added, R_S and R_L' do not change, but

$$A_{if} = \frac{A_i}{1 + \beta A_i}$$

Therefore

$$A_{vf} = \frac{A_{if} R_L'}{R_S} = \frac{A_i}{1 + \beta A_i} \times \frac{R_L'}{R_S} = \frac{A_v}{1 + \beta A_i}$$

PART F-3

This part of the appendix shows that the current flowing in the feedback resistor of Example 8-11 due to the emitter-resistor is negligible, compared to the current flowing due to the feedback.

In Figure F-3, the emitter current is 101 i_{b1} and the collector current is 5000 i_{b1}. The current flowing in the feedback resistor, R_f, can be found by superposition.

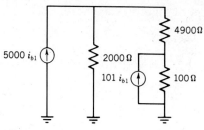

FIGURE F-3
Circuit for Appendix F-3.

If the emitter current generator, $101\ i_{b1}$, is open-circuited

$$i_{R_f} = 5000\ i_{b1} \times \frac{2000\ \Omega}{7000\ \Omega} = 1428\ i_{b1}$$

If the collector current generator, $5000\ i_{b1}$, is open circuited

$$i_{R_f} = 101\ i_{b1} \times \frac{100\ \Omega}{7000\ \Omega} = 1.4\ i_{b1}$$

These equations show that the current in R_{f1} due to the emitter current is 0.1% of the current in R_{f1} due to the collector current.

PART F-4

This part shows that the voltage gain of the circuit of Figure 9-26 is

$$A_{vf} = \frac{h_{fe}\ R_C}{h_{ie} + R_S\left(1 + \dfrac{h_{fe}\ R_C}{R_F}\right)}$$

Method 1

Assume that ac current in the base of Q_3 is I_b. It then follows that

$$i_c = h_{fe}\ i_b \qquad v_c = h_{fe}\ R_C\ i_b$$

Assuming the voltage gain of an emitter-follower is 1 (this is approximately true whether a single transistor or a Darlington is used)

$$v_{out} \approx v_c = h_{fe}\ R_C\ i_b$$

If it is also assumed that the ac voltage at the base of transistor Q_3 is close to 0, then

$$i_{R_F} = \frac{v_{out}}{R_F} = \frac{h_{fe} R_C i_b}{R_F}$$

The source voltage, V_S, is the base voltage at Q_3, plus the drop across R_S.

$$v_S = h_{ie} i_b + i_b R_S \left(1 + \frac{h_{fe} R_C}{R_F}\right)$$

$$A_{vf} = \frac{v_{out}}{v_S} = \frac{h_{fe} R_C}{h_{ie} + R_S\left(1 + \frac{h_{fe} R_C}{R_F}\right)}$$

Method 2
The circuit can be considered as a current feedback circuit. If an input current, I_b, is applied to the base of Q_3, with no feedback

$$v_{out} = h_{fe} R_C i_b$$

$$i_{out} = \frac{h_{fe} R_C i_b}{R_L}$$

$$A_i \text{ (without feedback)} = \frac{h_{fe} R_C}{R_L}$$

But

$$\beta = \frac{R_L}{R_F}$$

$$A\beta = \frac{h_{fe} R_C}{R_F}$$

From equation 8-12, we have

$$\frac{A_{vf}}{A_v} = \frac{1}{1 + \beta A_i} \frac{R_S + R_i}{R_S + R_{if}} \tag{8-12}$$

A_v, the voltage gain without feedback is

$$\frac{h_{fe} R_C}{h_{ie} + R_S}$$

$$R_i = h_{ie}$$

$$R_{if} = \frac{h_{ie}}{1 + \beta A_i}$$

$$A_{vf} = \frac{A_v(R_S + R_i)}{(1 + \beta A_i)R_S + h_{ie}} = \frac{\dfrac{h_{fe} R_C}{h_{ie} + R_S}(R_S + h_{ie})}{\left(1 + \dfrac{h_{fe} R_C}{R_F}\right) R_S + h_{ie}}$$

$$A_{vf} = \frac{h_{fe} R_C}{h_{ie} + \left(1 + \dfrac{h_{fe} R_C}{R_F}\right) R_S}$$

Both methods give the same results.

APPENDIX G

The Equivalence of a Series and a Parallel Coil

A coil can be represented as an inductance in series with a resistance, as shown in Figure 11-2a. Its input impedance is then

$$Z_{in} = R_S + j\omega L_S$$

where R_S is the series resistance of the coil and L_S is its series inductance.

The same coil can also be represented as a resistor in parallel with an inductor, L_P, as shown in Figure 11-2b. The input impedance of this circuit is

$$Z'_{in} = \frac{j\omega L_P R_P^2 + \omega^2 L_P^2 R_P}{R_P^2 + \omega^2 L_P^2}$$

as shown in section 11-3.1

For the circuits to be equivalent

$$Z_{in} = Z'_{in}$$

Equating the resistors (the real parts of Z_{in} and Z'_{in}) we have

$$R_S = \frac{\omega^2 L_P^2 R_P}{R_P^2 + \omega^2 L_P^2}$$

Equating the imaginary parts we find

$$L_S = \frac{L_P R_P^2}{\omega^2 L^2 + R_P^2}$$

If the Q's of the circuit are to be equal

$$Q = \frac{\omega L_S}{R_S} = \frac{\dfrac{\omega L_P R_P^2}{\omega^2 L^2 + R_P^2}}{\dfrac{\omega^2 L_P^2 R_P}{\omega^2 L^2 + R_P^2}} = \frac{R_P}{\omega L_P}$$

Now we have

$$L_S = \frac{L_P R_P^2}{R_P^2 + \omega^2 L^2} = \frac{L_P}{1 + \dfrac{\omega^2 L^2}{R_P^2}} = \frac{L_P}{1 + \dfrac{1}{Q^2}}$$

If $Q \gg 1$ this reduces to $L_S \approx L_P$.
If the Q of the circuit is 5, for example, the approximation is only off by 4% and becomes better as the Q increases.
Similarly

$$R_S = \frac{\omega^2 L^2 R_P}{\omega^2 L^2 + R_P^2} = \frac{\dfrac{\omega^2 L^2}{R_P^2} R_P}{1 + \dfrac{\omega^2 L^2}{R_P^2}} = \frac{\dfrac{R_P}{Q^2}}{1 + \dfrac{1}{Q^2}}$$

Again, if Q is high the denominator becomes 1 and we have

$$R_P = Q^2 R_S$$

APPENDIX H

The Derivation of Tank Circuit Equations

A tank circuit is the parallel combination of a coil, a capacitor, and a resistor. Its admittance is

$$Y = \frac{1}{R} + \frac{1}{j\omega L} - j\omega C \tag{H-1}$$

$$Y = \frac{1}{R}\left[1 - j\left(\omega CR - \frac{R}{\omega L}\right)\right] \tag{H-2}$$

At resonance $\omega_0^2 = LC$.
The Q of the circuit is $Q = R/\omega_0 L = R\omega_0 C$
At any frequency, ω, equation H-2 becomes

$$Y = \frac{1}{R}\left[1 - j\left(\frac{Q\omega}{\omega_0} - \frac{Q\omega_0}{\omega}\right)\right]$$

$$Y = \frac{1}{R}\left[1 - jQ\left(\frac{\omega}{\omega_0} - \frac{\omega_0}{\omega}\right)\right]$$

The 3 dB points are at ω_1 and ω_2 where

$$Q\left(\frac{\omega}{\omega_0} - \frac{\omega_0}{\omega}\right) = +1 \text{ or } -1$$

For ω_1 we have

$$Q\left(\frac{\omega_1}{\omega_0} - \frac{\omega_0}{\omega_1}\right) = 1$$

$$\frac{\omega_1}{\omega_0} - \frac{\omega_0}{\omega_1} = \frac{1}{Q}$$

$$\omega_1^2 - \omega_0^2 = \frac{\omega_1 \omega_0}{Q}$$

or

$$f_1^2 - f_0^2 = \frac{f_1 f_0}{Q} \tag{H-3}$$

For ω_2 we have

$$Q\left(\frac{\omega_2}{\omega_0} - \frac{\omega_0}{\omega_2}\right) = -1$$

$$\frac{\omega_0}{\omega_2} - \frac{\omega_2}{\omega_0} = \frac{1}{Q}$$

$$\omega_0^2 - \omega_2^2 = \frac{\omega_2 \omega_0}{Q}$$

$$f_0^2 - f_2^2 = \frac{f_2 f_0}{Q} \tag{H-4}$$

Combining equations H-3 and H-4 gives

$$f_1^2 - f_2^2 = \frac{f_0 (f_1 + f_2)}{Q}$$

$$(f_1 - f_2)(f_1 + f_2) = \frac{f_0 (f_1 + f_2)}{Q}$$

$$f_1 - f_2 = \frac{f_0}{Q}$$

But $f_1 - f_2$ is the frequency between the two 3 dB points or the bandwidth of the circuit.

Answers to Selected Problems

CHAPTER 1

1-1 4.36 μohms
1-2b 1.03 ohms (30 gauge wire)
1-4 45.4 ohm-cm
1-6 a. 1.7×10^{10}
     b. 6.8×10^{10}
1-7 The forward resistance is 20 ohms.
1-9b 3.33×10^9 ohms
1-11 At 45°C, $I = 4$

1-12 a, b See Figure A1-12.

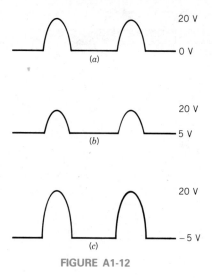

FIGURE A1-12

1-14

$-50\ V \leq V_{in} \leq 10\ V$	D_1 ON D_2 OFF	$V_{out} = \dfrac{V_{in} + 5}{3}$
$10\ V \leq V_{in} \leq 40\ V$	D_1, D_2 OFF	$V_{out} = \dfrac{V_{in}}{2}$
$40\ V \leq V_{in} \leq 50\ V$	D_2 ON	$V_{out} = 20\ V$

1-16 See Figure A1-16.

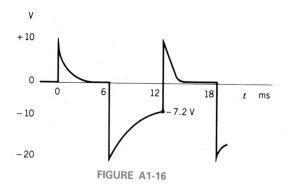

FIGURE A1-16

1-17 See Figure A1-17.

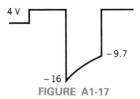

FIGURE A1-17

1-19 See Figure A1-19.

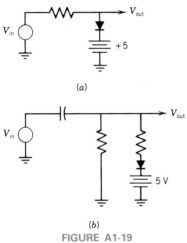

FIGURE A1-19

1-20 a. 60 mA
b. 57.7 mA

1-22 See Figure A1-22.

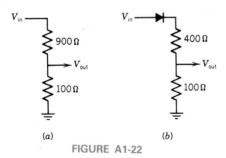

FIGURE A1-22

1-23 a. $I = 0$
b. $I = 100$ mA

1-24 420 ohms

1-25 a. $R = 4$ kohms
  b. The range of R is 1 kohm and higher.
1-26 0.5 V

CHAPTER 2

2-1 $I_E = 2.985$ mA
  $I_B = 15$ μA
  $V_C = 5.075$ V

2-3 $A_{i(ck)} = A_{i(tr)} = 0.995$
  $A_{v(tr)} = 39.8$
  $A_{v(ck)} = 7.96$

2-4 $\alpha = 0.985$

2-5b $h_{FE} = 199$

2-7b $\alpha = 0.993$

2-8 $I_B = 9.3$ uA
  $I_C = 1.4$ mA
  $V_C = 23.31$ V

2-9a $h_{FE} = 118$
  $h_{fe} = 140$

2-11 $V_C = 9$ V
  $I_C = 4.5$ mA

2-14 At 30 mA, $h_{fe} = 85.7$ and V_{CE} at $I = 1.1$ mA is 0.19 V.

2-15b 1.4 mW

2-16 $V_{CE} = 15$ V

2-18 $V_{CC} = 30$ V, $R_C = 6$ kΩ

2-20 $h_{fe(forced)} = 22.6$

2-22b $I_B = 243$ μA, $I_C = 12.5$ mA
  $h_{FE(forced)} = 51.4$

CHAPTER 3

3-1 $R_B = 494$ kΩ
  $R_C = 333$ kΩ

3-2a 5 V

3-3b 8.5 V

3-7 a. 12.7 V
  b. 0.1 V (saturation)

3-9 $I_B = 65$ μA
  $I_C = 6.44$ mA
  $V_C = 12.3$ V

$I_2 = 1.94$ mA
$I_1 = 2.10$ mA

3-11 $R_{B1} = 26{,}575$
$R_{B2} = 3382$

3-14 $V_{CC} = -30$ V
$R_E = 1$ kΩ
$R_C = 4$ kΩ
$R_{B1} = 14{,}062$
$R_{B2} = 2165$

3-18 $I_C = 3.27$ mA
$V_C = 9.81$ V

3-19 $I_C \cong I_E = 4.3$ mA
$V_C = 15.7$ V

3-21 $V_B = 9.92$ V

3-24 $I_E = 8.58$ mA
$V_E = 8.58$ V

3-26 $S_I = 5.712$
$S_V = 0.0052$

CHAPTER 4

4-1a 5.25 V

4-2 $A_{V(\text{tr})} = 75$
$A_{V(\text{ck})} = 7.5$

4-3 $h_f = 0.6$
$h_i = 60$
$h_r = 0.6$
$h_o = 0.1$

4-4 $\Delta^h = 0.96$
$V_2 = 7.01$ V

4-5 At 2 mA
$h_{fe} = 135$
$h_{oe} = 15 \times 10^{-6}$ S
$h_{ie} = 2000$ ohms
$h_{re} = 10^{-4}$

4-8 $\Delta^h = 15 \times 10^{-3}$
$R_i = 971.7$ ohms
$R_o = 40$ kΩ
$A_i = 94.3$
$A_v = -194.2$

4-10 Using the approximation that the base voltage is 4 V
$h_{ie} = 581$ ohms
$A_{i(\text{tr})} = 100$

$A_{V(tr)} = -206.5$
$A_{v(ck)} = -63.9$
$A_{i(ck)} = 77.5$

4-13 $h_{ie} = 2500 \, \Omega$
$A_{v(tr)} = 177.7$

4-14 $V_{out} = 2.83 \, V$

4-15 $A_{v(tr)} = 200$
$A_{v(ck)} = 9.52$

4-18 $h_{ib} = 6.85$ ohms
$h_{fb} = -0.980$

4-20 $\Delta^h = 3.065 \times 10^{-4}$
$A_v = 29.28$
$A_i = -0.985$

4-21 $A_{v(tr)} = 1000$
$A_{v(ck)} = 5$

4-23 $A_{v(ck)} = 0.4868$
$R_o = 490$ ohms

4-24 $A_{v(ck)} = 0.346$

4-26 $A_{v(tr)} = 1.84$
$A_{v(ck)} = 0.906$

4-28b (Using approximate equations)
$V_{out1} = 4 \, V$
$V_{out2} = 0.5 \, V$

4-29 a. 250 mV
b. 2.5 mV
c. 500 mV

CHAPTER 5

5-1b $-3.5 \, V$

5-3 At $V_{GS} = -3 \, V$, for example, $I_D = 2.2 \, mA$.

5-5 At $V_{GS} = -2 \, V$, $I_D = 0.92 \, mA$

5-7 $I_D = 2.35 \, mA$. This is slightly high.

5-9 For the self-biased circuit, $R_S = 1.68 \, k\Omega$.

5-10 $I_D = 4 \, mA$, $V_D = 17 \, V$

5-11 $V_{GS} = -1.6$, $I_D = 1.6 \, mA$

5-12 $A_V = 7.5$

5-13 $R_S = 275 \, \Omega$, $R_D = 4132 \, \Omega$, $g_m = 2074 \times 10^{-6} \, S$, $A_V = 8.57$

5-14b $g_m = 2041 \times 10^{-6}$
5-15 $A_V = 10.77$
5-16 From the computer the gains are:
fixed bias: 15.5925
self bias: 16.8561
5-17 For both cases the gain is about 1.6.
5-19b The best gain is 20.7686.
5-20b The best gain is 2.06.
5-21 $A_V = 2.67$
5-22 $A_V = 1.01$
5-26 $A_V = 0.614$
5-27 $R_D = 3$ kohms.
5-29 The gains are 8 and 4.85.
5-30c $g_m = 2520 \times 10^{-6}$
5-31c $A = 8.32$
5-32 $g_m = 0.3$ S
$A_V = 7.5$

CHAPTER 6

6-1 $A_{V(total)} = 125{,}000$
6-2 $A_{V(total)} = 3440$
6-3 $A_{V(total)} = 107.52$
6-4 $A_{V2} = 38.37$
$A_{V1} = 33.5$
$A_{Vin} = 0.387$
$A_{V(total)} = 498.1$
6-6 $R_{in} = 1087.5$
$A_{V(total)} = 7.89$
$R_O = 24.56\ \Omega$
6-7 $V_{C2} = 6$ V
6-9 $A_{V(total)} = 15.36$
6-10 $A_{V(total)} = 66.6$
6-13 $R_E = 4.3$ k
$A_{V(total)} = 6.62$
6-15 $A_i = 4800 = 4800$
$R_i = 240{,}000\ \Omega$
6-18 The constant current is 64 mA.
The maximum resistance is 461 ohms.

6-19

V_{ia}	I_{C1}	V_{C1}	I_{C2}	V_{C2}
.7 V	0	25 V	6 mA	13 V
4.7 V	3 mA	19 V	3 mA	19 V
6.7 V	6.5 mA	12 V	0	25 V

6-23 $V_{ca1} = 8$ V

6-25 $A_c = .49875$
$A_d = 100$

6-26 $A_c = 0.0747$
$A_d = 10$

6-27 $A_d = 3.75$

6-28 87.95 dB

6-29 CMRR = 5623

6-32 $V_{out(dc)} = -3.3$ V

6-34 $I_{RB1} = 1003$ μA
$I_{RB2} = 993$ μA

CHAPTER 7

7-2b As f increases by a factor of 10
dB = 20 log 10, = 20 dB

7-4 $f_L = 26.53$ Hz

7-6 $C = 82$ μF
The zero is at 8 Hz.

7-7 The pole is at 297 Hz. The zero is at 13.26 Hz.

7-8b 5 dB

7-9 $f_H = 3.18$ MHz
At 5 MHz the gain is 0.536.

7-11 $f_\beta = 1.66$ MHz
$C_{b'e} = 95$ pF
$C_M = 543$ pF
$f_H = 525$ kHz

7-12 $A_{V(ck)} = 26$
At 100 kHz, $A = 8.06$

7-14 $f_H = 7.96$ MHz

7-19 $f_H = 945$ kHz

7-20 The pole is at 199 Hz. The zero is at 39.8 Hz.

7-22 $C = 7.96$ μF.

7-23b The input pole is at 53 kHz. The output pole is at 3.18 MHz.

7-24 For $R_i = 100\ \Omega$, the input pole is at 149 MHz. The output pole is at 19.08 MHz.

7-25 $f = 2.04$ MHz

7-26 At 100 Hz, the attenuation factor is 0.316.

7-27 $A_V(200\text{ Hz}) = 1247$

7-28 The poles are at 578 kHz and at 247 kHz.

7-29 $h_{ie} = 500\ \Omega$
$A_{V(tr)} = 220$
$A_{V(ck)} = 42.3$
The low-frequency pole is at 1886.7 Hz. The zero is at 36.7 Hz. The high-frequency pole is at 603 kHz.

7-32 The high-frequency poles are at 3.18 MHz and 31.8 MHz.

7-33 $P = .133 = 13.3\%$
$C = 0.3\ \mu\text{F}$

7-34 $P = 0.314$

7-36 $t = 0.7\ \mu\text{s}$

7-38 $t = 0.32\ \mu\text{s}$

7-40 The output rises to 9.57 V.

7-41 For 10:1 attenuation
$R = 4.5\ \text{M}\ \Omega$
$C = 6.66\ \text{pF}$

CHAPTER 8

8-1 $V_S = 4.2$ V

8-3 $A_{VF} = 14.28$
$R_{iF} = 3500$

8-4 $\beta = 0.3$

8-7 $A1 = 300$
$A2 = 250$
$A_{VF} = 16.80$
$R_{it} = 13{,}400$
$R_{ot} = 100.75$

8-9 $A = \dfrac{1}{\beta} = 10$

8-10 $A_V = 5$
$A_{VF} = 1.667$

8-14 $A_{Q1} = 4.13$
$A_{Q2} = 214.2$
$A_{Vf} = 47.3$

8-15 $V_{in} = 0.0526$ V
$V_r = 0.0476$ V

8-16 $R_F = 24750 \, \Omega$
$A_{VF} = 78.2$

8-17 $A_{VF} = 11.94$
$A_{if} = 9.55$

8-19 $A_{iF} = 27.7$
$A_{V(ck)} = 74.33$
$\beta = 0.122$

CHAPTER 9

9-1 $P_D = 14.04$ W
$P_{RE} = 2.36$ W

9-3 $P_{ac}(10a) = 1.152$ W
$P_{ac}(15a) = 0.768$ W
$P_{tr} = 9.08$ W

9-6 $R_{dc} = 10 \, \Omega$
$R_{ac} = 6 \, \Omega$
$\eta = 5\%$

9-7 $V_i = 2.7$ V

9-8 $T_J = 60.2°C$

9-10 $P_D = 61.5$ W

9-13 $T = 150°C$
$Q_{ic} = 3.125$

9-14b $P_D = 1.85$ W

9-15b $P_D = 27.4$ W

9-16b $T_{sink} = 90°C$

9-17b $T_A = 100°C$

9-19 $\eta = 38.46\%$

9-20 $V_{CC} = 19.375$ V

9-22 $P_{CC} = 20$ W

9-24 $N = 4:1$

9-26 $V_{CC} = 31.5$ V
$\eta = 34.2\%$

9-27 $P_{RE} = 2.25$ W
$P_{(transformer\ ac)} = 9.6$ W

$P_{(transistor)} = 10.41$ W
$P_{(transformer\ dc)} = 1.12$ W

9-28 $I_{ca} = 1$ A
$V_{CEQ} = 18$ V
$N = 3:1$
$V_{cm} = 18$ V
$I_{cm} = 1$ A

9-32 $R = 64\ \Omega$
$N = 4:1$
$I_{lm} = 1$ A
$P_{cc} = 3.18$ W
$\eta = 62.9\%$

9-33 $P_{load} = 30$ W
$N = 1.548$

9-34 $R = 227\ \Omega$

9-35 $V_{cm} = 9$ V
$I_{cm} = 1$ A
$P_L = 6.38$ W

9-37 $A_{VF} = 104$

CHAPTER 10

10-3 $V_{rms} = 4.062$ V
$RF = .17675$

10-5 $V_{rms} = 0.289$ V
$RF = .0118$

10-7 a. $N = 165:128$
b. $N = 165:64$
c. $N = 165:65$

10-8a $N = 165:40.7$

10-9 $V_{dc} = 26.2$ V
$RF = 0.3$

10-10b $N = 165:13.2:13.2$
$C = 66.3\ \mu F$

10-12 $C = 83\ \mu F$

10-13 a. $C = .241$ Farads
b. $L = 10$ Hy
$C_1 = C_2 = 510\ \mu F$

10-14 $R_1 = 1\ \Omega$
The resistor must dissipate 400 W.
The Zener diode must dissipate 300 W.

10-17 $2998.6 = 3331V_L + 3.05I_L$

When $I_L = 0$ A, $V_L = 9.06$ V.
When $I_L = 5$ A, $V_L = 9.013$ V.

10-18 $2278.6 = 81V_L + 6I_L$

10-19 $28.75 = V_L + .05I_L$

10-22 30 W

10-26 $L = 6.4$ mHy
$C = 50$ μF

CHAPTER 11

11-1 $R_p = 5000$ Ω
$BW = 20000$ Hz

11-2 $f_r = 452$ kHz
$BW = 15.08$ kHz

11-4 $R_p = 1570$
$C = .101$ μF

11-5 $R_p = 264.17$
$C = 79.2$ pF

11-7 For the pole at 32.5 MHz
$R_p = 3960$
$C = 24$ pF

11-8 For the four-stage amplifier, one stage is at 15.46 MHz with a BW of 380 kHz.

11-10 $L = 1.94 \times 10^{-4}$ Hy
The minimum Q is 5.47.

11-12 The range is 98.6 kHz to 174 kHz.

11-13 $f = 2.145$ MHz

11-16 If $R_1 = R_2 = 10$ kΩ, then $C_1 = C_2 = .00318$ μF.

11-17 $R_F = 400$ Ω
$f = 15.9$ kHz
h must be at least 76.

CHAPTER 12

12-1 $V_{in} = 60$ μV

12-2 a. 15 (inverting)
b. +6 (noninverting)

12-5 a. −11 V
b. −5 V

12-7 I_{IO} typical = 20 nA, worst case = 200 nA.

12-8 a. 20 mV
b. 8.3 kΩ
12-10 6×10^{-3} Ω
12-11 b. 3 MHz
c. 3 MHz
12-13 $T = 1.17$ μs
12-16 a. $V_{out} = -0.1$ V
c. $f = 3.18$ Hz
12-18 $C_C = .005$ μF
$C = .016$ μF
12-19 $V_{out} = -10 \int V_1\, dt - 4 \int V_2\, dt - 20 \int V_3\, dt$
12-22 $R = 60$ kΩ

Index

Index

A

Acceptor impurities, 11–13
ac gain:
  Butterworth amplifier, 535–539
  cascaded BJT amplifier, 239–247, 322–327
  cascaded common-emitter, 365–367, 375–381
  cascaded JFET amplifier, 247–248, 322–323, 327
  CE-CC amplifier, 245–247
  common-base, 74–76, 166–169
  common-collector, 127–128, 169–178
  common-emitter, 77–79, 151–160, 363–365, 370–375, 381–382, 386–387
  common-mode, 587
  common-source JFET amplifier, 209–218, 362–363
  complementary symmetry amplifier, 444, 449
  Darlington, 251, 255–257
  defined, 74, 142
  difference amplifier, 266–267
  difference-mode, 587
  direct-coupled amplifier, 250–251
  and emitter bypass resistors, 299–304
  emitter-follower, 127–128, 169–178
  and feedback amplifier stability, 360–361
  frequency dependence of, *see* Frequency response of amplifiers
  and H-bias circuits, 160–166
  maximally flat amplifier, 535–539
  mid-frequency, 142, 322–327
  MOSFET, 229–230
  nomenclature for, 74
  op-amp, 250–251, 568–580
  phase splitter, 177–178
  series feedback amplifier, 356–360
  shunt feedback amplifier, 367–369
  single-tuned amplifier, 526
  source-follower JFET, 218–222
  stagger-tuned amplifier, 535–539
  VMOS-FET, 232
ac nomenclature for transistors, 73–74
Active filters, 607–611
Active probes for oscilloscopes, 341
Active region of BJTs, 87–88, 93–94, 105, 505
A/D (analog-to-digital) converters, 610
Adjustable-voltage regulators, 502–503
Admittance, 148, 524–526
Admittance (y) parameters, 145
alpha (α) for BJTs, 70–71, 77, 78
AM (amplitude modulation) demodulators, 25–26

AM (amplitude modulation) receivers, 524, 531–532, 539–540
Amplifiers:
  antilog, 606–607
  bandwidth of, 290–291, 316, 382–384, 390
  biasing of, see Biasing entries
  bootstrap circuits in, 447–449
  buffer, 127–128, 169–177
  Butterworth, 535–539
  cascaded BJT, 239–247, 248–251
  cascaded common-emitter, 240–245, 365–367, 375–381
  cascaded JFET, 221–222, 247–248
  cascode, 277–279
  CE-CC, 245–247
  choke-coupled, 420–426
  class A, see Class A amplifiers
  class AB, 438
  class B, see Class B amplifiers
  clipping in, 105–106, 330, 419–420
  common-base, 71–77, 124–127, 166–169
  common-collector, 127–128, 169–178
  common-drain JFET, 218–222
  common-emitter, 77–82, 151–160, 240–245, 363–365, 370–375, 381–382, 386–387
  common-emitter common-collector, 245–247
  common-source JFET, 209–218, 362–363
  compensated logarithmic, 604–606
  complementary symmetry, see Complementary symmetry amplifiers
  current, 388–390
  Darlington, 251–257, 275–276, 445–449
  defined, 1
  difference, see Difference amplifiers
  direct-coupled, 248–251
  distortion in, 105–108, 327–333, 390, 419–420, 437–438, 453
  double-tuned, 532–534
  efficiency of, see Efficiency
  emitter-follower, 127–128, 169–178
  feedback, see Feedback amplifiers
  frequency response of, see Frequency response of amplifiers
  gain-bandwidth product of, 309–311, 316, 384
  half-power frequencies of, 290–291
  IC, 449–454
  IF, see Tuned-circuit amplifiers
  instrumentation, 601–603
  "integrated" output of, unintentional, 332–333
  isolating, 127–128
  JFET, common-drain, 218–222
  JFET, common-source, 209–218, 362–363
  JFET, multistage, 221–222, 247–248
  linear, 224
  load lines for, see Load lines
  logarithmic, 604–606
  maximally flat, 535–539
  MOSFET, 224, 230–231
  "motor-boating" in, 454
  operating (active) region of, 87–88, 93–94
  operating point of, see Quiescent (Q) point
  operational, see Operational amplifiers (op-amps)
  oscillations in, 242, 243, 360, 454, 541–543, 588
  oscillators as, 541–543
  oscilloscope, 339–343
  power, see Power amplifiers
  preamplifiers, 398
  push-pull, 432–437
  Q point of, see Quiescent (Q) point
  quasi-complementary, 445–446
  RC coupled, 240–248
  resistively-coupled, 404–407
  RF, 169
  single-source common-base, 126–127
  single-tuned, 524–526, 529–532
  single-tuned, ideal, 532
  source-follower JFET, 218–222
  spurious oscillations in, 242, 243, 360, 541–543, 588
  square-wave response of, 327–333
  stability of, 360–361
  stages of, 238
  stagger-tuned, 535–539
  summing and scaling, 598
  swing in, see Swing
  3-dB points of, 290–291
  transconductance, 388–390
  transformer, see Transformer amplifiers
  transformer-coupled, 426–431
  transformerless, see Complementary symmetry amplifiers
  transient response of, 327–333
  transresistance, 388–390
  tuned-circuit, see Tuned-circuit amplifiers
  video, 524
  VMOS-FET, 231–232
  voltage, 388–390
  wide-band, 524
Amplitude modulation (AM) demodulators, 25–26
Amplitude modulation (AM) receivers, 524, 531–532, 539–540
Analog computers, 590–595, 597–600
Analog-to-digital (A/D) converters, 610
Analog voltage comparators, 261–263

Angles, conduction, 476–477
Antilog amplifiers, 606–607
Apple computer, 2, 117, 213, 558
Application notes for ICs, 453
Astable multivibrators, 560, 561
Atoms, 6
Attenuated oscilloscope probes, 339–341
Attenuation, 292. *See also* Gain
Attenuation factor, 322
Attenuators:
   capacitively loaded, 334–335
   compensated, 333–334, 336–340
Audio systems, 398–399, 420, 440, 448–453
Avalanche breakdown:
   diode, 16, 39–40
   JFET, 191
Average values of waveforms, 466–469

B

Balancing resistors, 271–273
Band-pass filters, 607, 608. *See also* Tuned-circuit amplifiers
Band-reject filters, 608
Bands of frequencies, 531
Bandwidth:
   amplifier, 290–291, 316, 382–384, 390
   oscilloscope, 341–343
   tuned-circuit amplifier, 529–533, 535–540
Bar code readers, 50–51
Barkhausen criterion for oscillation, 542–543
Barrels of oscilloscope probes, 339
Base region, 66
Base-spreading resistance, 309
Base-to-emitter junction, 66–67
BASIC programming language, 117
BASIC programs, *see* Computer programs
BB139 varactor diode, 56
Bell Telephone Laboratories, 2, 195
Bessel filters, 610
beta (β) for BJTs, 78–79, 91–92, 108, 129–133, 310–311
beta (β) feedback factor, 388
beta (β) feedback networks, 354–356, 359, 360–361, 388
Bias current of op-amps, 585–586, 590
Biased clamping circuits, 37–38
Biased clippers, 26–31
Biasing bipolar junction transistors (BJTs):
   common-base amplifiers, 124–127
   common-collector circuits, 127–128
   emitter-follower circuits, 127–128
   emitter-resistor bias circuit, 110–120
   fixed-bias circuit, 105–110
   introduction to, 102–104
   self-bias circuit, 120–123
   stability issues, 128–133
Biasing class B amplifiers, 438–440
Biasing complementary symmetry amplifiers, 440–444
Biasing difference amplifiers, 265–266
Biasing junction field effect transistors (JFETs), 197–206
Biasing metal oxide semiconductor field effect transistors (MOSFETs), 227–228
Biasing multistage amplifiers, 239
BIFET operational amplifiers, 581, 590
Bipolar junction transistors (BJTs):
   active region of, 87–88, 93–94
   alpha (α) for, 70–71, 77, 78
   base region in, 66
   base-to-emitter junction in, 66–67
   beta (β) for, 78–79, 91–92, 108, 129–133, 310–311
   biasing of, *see* Biasing bipolar junction transistors (BJTs)
   breakdown voltages in, 88–89
   collector-base junction in, 69–70
   collector region in, 66
   common-base configuration for, 71–77
   common-emitter configuration for, 77–82
   construction of, 67–69
   curve tracers for measuring, 83–85, 89–90
   cutoff region of, 87–90, 93–94
   emitter-base junction in, 70–71
   emitter region in, 66
   equivalent circuits for, *see* Equivalent circuits
   forced beta (β) for, 91–92
   at high frequencies, 304–316
   h parameters for, *see* Hybrid (h) parameters
   load lines for, 84–87
   at low frequencies, 293–304
   manufacturers' specifications for, 94–95
   nomenclature for, 73–74
   in op-amps, 581, 590
   power dissipation in, 94
   saturation region of, 91–94
Bi-polar operational amplifiers, 581, 590
BJTs, *see* Bipolar junction transistors (BJTs)
Bleeder resistors, 481
Blocking capacitors, 240–247, 293–300, 307–308, 317–318, 365–367
Bode plots, 291–293, 299–300
Boltzman's constant, 604
Boost regulators, 503–504
Boost switching regulators, 513
Bootstrap circuits, 447–449
Breakdown voltages:
   BJT, 88–90

diode, 20, 39–40, 46, 465–466, 469
  maximum, for **2N3055** and **2N3904** BJTs, 409
Bridge circuits, 601, 602
Bridge rectifiers, 469–472
Buffer amplifiers, 127–128, 169–177
Buses in computers, 504
Butterworth filters, 535–539, 610
Bypass capacitors:
  emitter, 110–111, 176–177, 243–245, 291, 299–304, 307–308, 363–367
  source, 200, 216–218, 317–318

C

Calibrated output of oscilloscopes, 340
Capacitance, *see also* Capacitors
  drain-to-gate, 320
  gate-to-drain, 320
  gate-to-ground, 320
  interelectrode, 308
  Miller, 311–316, 320
  probe, 333–340
  stray, 291, 305–308, 320, 542–543, 547–548
  thermal, 419
Capacitive filters, 472–478, 482
Capacitively coupled loads, 163–166, 203, 211–215, 230, 291
Capacitively loaded attenuators, 334–335
Capacitors, *see also* Capacitance
  blocking, 240–247, 293–300, 307–308, 317–318, 365–367
  compensating, 588
  coupling, 291
  in differentiators, 595–597
  diffusion, 309
  emitter bypass, 110–111, 176–177, 243–245, 291, 299–304, 307–308, 363–367
  ideal, 111, 290, 593
  in integrators, 590–595, 597–600
  Miller, 311–312
  source bypass, 200, 216–218, 317–318
Carbon, 6–7
Carrier density, intrinsic, 9
Carrier recombination coefficient, 604–605
Cascaded amplifiers:
  BJT, 239–247, 248–251, 322–327, 365–367, 375–381
  common-emitter, 240–245
  JFET, 221–222, 247–248, 322–323, 327
  rise times in, 341–343
Cascode amplifiers, 277–279
Case-to-air thermal resistance, 412–414
Cathode-ray oscilloscopes (CROs):
  amplifiers in, 339–343
  bandwidth of, 341–343
  capacitively loaded attenuators in, 334–335
  compensated attenuators in, 333–334, 336–340
  curve tracers, 83–85, 89–90
  frequency response limitations of, 341–343
  probes for, 333–334, 339–341
CA3000 difference amplifier, 275–276
CE-CC (common-emitter common-collector) amplifiers, 245–247
Celsius temperature scale, 604
Ceramic filters, 539–540
Channels in FETs, 188–191
Characteristic curves:
  active region of, 87–88
  for BJTs in the switching mode, 93–94
  for breakdown voltages, 89–90
  common-base, 76–77
  common-emitter, 80–81
  cutoff region of, 87–90, 93–94
  for JFETs, 191, 192, 199, 201–205, 209–212, 220
  load lines and, *see* Load lines
  for MOSFETs, 226, 228, 230
  for op-amps, 582
  saturation region of, 87–88, 92–93
  for VMOS-FETs, 231–232
Charge of an electron, 604
Chebychev filters, 610
Choke-coils, 420, 545–546
Choke-coupled amplifiers, 420–426
Clamping circuits, *see also* Clipping circuits
  biased, 37–38
  defined, 31
  simple, 18
  sine wave, 35–38
  unbiased, 31–37
Class A amplifiers:
  choke-coupled, 420–426
  compared with Class B amplifiers, 431–432
  defined, 419–420
  transformer-coupled, 426–431
Class AB amplifiers, 438
Class AB distortion, 438
Class B amplifiers:
  biasing of, 438–440
  collector characteristics of, 433–434
  compared with class A amplifiers, 431–432
  distortion in, 437–438
  efficiency of, 434–437
  power dissipation in, 434–437
  push-pull, 432–437
Clipping in amplifiers, 105–106, 330
Clipping circuits, *see also* Clamping circuits
  biased, 26–31

corner points of, 30–31
defined, 24
half-wave rectifiers as, 24–26
Clock signals, 550, 558
CMOS (complementary metal oxide semiconductor FETs), 188, 223, 224
CMRR (common-mode rejection ratio), 266–267, 587, 590, 600
Coaxial cables, 169, 339–340
Coefficient of coupling, 532–534
Coils, see also Transformers
analysis of, 527–529
choke, 420, 545–546
Q factor of, 527–529
slugs in, 527
Collector-base junction, 69–70
Collector pole, 314–315
Collector region, 66
Color television, crystal in, 550
Colpitts oscillators, 545–548
Common-anode diode circuits, 53, 56–57
Common-base amplifiers, 71–77, 124–127, 166–169
Common-cathode diode circuits, 53–54, 56–57
Common-collector amplifiers, 127–128, 169–178
Common-drain JFET circuits, 218–222
Common-emitter amplifiers, 77–82, 151–160, 240–245. See also Capacitively coupled loads; H-bias circuits
Common-emitter common-collector (CE-CC) amplifiers, 245–247
Common mode, 264–267, 587
Common-mode rejection ratio (CMRR), 266–267, 587, 590, 600
Common-mode voltage, 263
Common-source JFET amplifiers, 209–218
Comparators, 259–263, 487–489, 559
Compensated attenuators, 333–334, 336–340
Compensated differentiators, 596–597
Compensated logarithmic amplifiers, 604–606
Compensated operational amplifiers, 588
Compensating capacitors, 588
Compensating potentiometers, 585
Complementary metal oxide semiconductor FETs (CMOS), 188, 223, 224
Complementary symmetry amplifiers:
bootstrap circuits in, 447–448
Darlingtons in, 445–449
diodes and double-ended power supplies in, 442–443, 446–449
input transistor in, 443–445
introduced, 440–442
Complex frequencies, 292–293
Complex numbers, 292–293

Complex plane, 292–293, 298, 536–538
Computer programs:
amplifier frequency response, 295–296, 303–304, 319
gain of JFET amplifiers, 213–216
Q point of BJT H-bias circuits, 117
Q point of JFET self-bias circuits, 200–202
Computers:
analog, 590–595, 597–600
Apple, 2, 117, 213, 588
buses in, 504
historical background of, 2
programs for, see Computer programs
TRS-80, 2, 117
UNIVAC I, 2
VAX, 117, 213
Conduction angle, 476–477
Conduction region, see Active region of BJTs; Saturation region
Conduction in semiconductors, 8–9
Conductivity, 9. See also Resistivity
Conductors, 4–6
Constant-current sources, 257–259
Convection, 417–419
Conventional current flow, 66
Converters:
A/D, 610
D/A, 610–612
Copper, 6
Corner points of clipping circuits, 30–31
Coupling:
coefficient of, 532–534
critical, 533–534
Coupling capacitors, 291
Covalent bonds, 7–8, 11–12
Critical coupling, 533–534
CROs, see Cathode-ray oscilloscopes (CROs)
Crossover distortion, 437–438
Crowbar circuits, 495–497
Crystal lattice, 7–8, 11–12
Crystal oscillators, 550–552, 561–562
Current amplifiers, 388–390
Current feedback amplifiers, see Shunt feedback amplifiers
Current feedback equations, 369
Current flow:
conventional, 66
of majority carriers, 188
Current-limiting circuits, 497–498
Current loop converters, 44
Current probes for oscilloscopes, 341
Current-series feedback, 388–391
Current-shunt feedback, 388–391
Curve tracers, 83–85, 89–90
Cut-in point, 476–478

Cutoff region:
    amplifier clipping in, 105, 419
    of BJTs, 87–90, 93–94
    in class B amplifiers, 433–434
Cut-out point, 476–477

D

D/A (digital-to-analog) converters, 610–612
Dark current in photodiodes, 47
Darlington pairs, 251–257, 275–276, 445–449, 489–498
dB (decibel) units, 273–274, 291–293
dc gain:
    cascaded BJT amplifier, 250
    op-amp, 250–251, 568–580
dc nomenclature for transistors, 73–74
Dead time in waveforms, 515
Decibel (dB) units, 273–274, 291–293
Demodulators, 25–26
Depletion-mode MOSFETs, 222–226
Depletion region, 14, 22, 190–191, 308
Derating curves, 407, 408, 457
Detectors:
    AM, 25–26
    intrusion, 49–50
Determinants, 149, 153, 555
Difference-amplifier regulators, 488–489
Difference amplifiers:
    analysis of, 266–271
    balancing resistors in, 271–273
    biasing of, 265–266
    commercial, 274–276
    common mode of, 264–267, 587
    difference mode of, 264–267, 587
    introduced, 263–265
    op-amps as, 600–601
Difference mode, 264–267, 587
Differential equations, 597–600
Differential probes for oscilloscopes, 341
Differentiators, 595–597
Diffusion capacitor, 309
Digital crystal oscillators, 561–562
Digital Equipment Corporation, 117
Digital inverters, 561–562
Digital operation of BJTs, 93–94
Digital oscillators, 558–562
Digital-to-analog (D/A) converters, 610–612
Diodes:
    avalanche breakdown in, 16
    in bar code readers, 50–51
    breakdown voltages in, 20, 39–40, 46
    in capacitive filters, 476–478
    characteristics of, 16–24
    in class B amplifiers, 439
    in complementary symmetry amplifiers, 442–443, 446–449
    current equation for, 604–605
    in detector circuits, 25–26
    dynamic resistance of, 17, 39–43, 486
    forward characteristics of, 17, 20, 39–43, 46
    GaAs, 49
    GaAsP, 49–56
    germanium, 23–24
    hot carrier, 45–46
    ideal, 16
    JEDEC types, 41–43
    leakage current in, 16, 20–24, 39–43, 46
    LEDs, 49–56, 58
    in optical couplers, 50
    photodetectors, 47–51
    photodiodes, 46–56
    photoemitters, 49–56
    PIV for, 20, 46, 465–466, 469
    p-n junctions as, 13–16
    power dissipation in, 24, 41, 58
    rectifier, see Rectifiers
    reverse current in, 16, 20–24, 39–43, 46
    reverse recovery time of, 20–23, 46
    reverse resistance of, 17, 20
    Schottky, 45–46
    seven-segment displays, 54–56, 58
    silicon, 23–24
    state change of, 29
    storage time of, 20–23
    surface barrier, 45–46
    surge currents in, 479
    switching, 17–24
    transition time of, 22–23
    tunnel, 57
    turn-on voltage for, 16–17
    varactor (varicap), 56–57
    Zener, 39–44, 259, 483–487
DIPs (dual-in-line packages), 68–69, 558–559, 577, 578
Direct-coupled amplifiers, 248–251
Displays:
    dot matrix, 55
    seven-segment, 54–56, 58
Distortion in amplifiers, 105–108, 327–333, 390, 419–420, 437–438, 453
Distortion analyzers, 437
Dominant poles, 320–323
Donor impurities, 11–13
Dopant, 11
Doped semiconductors, 11–13, 189, 222–223, 231
Dot matrix displays, 55
Double-ended power supplies, 568
Double-peaked response, 533–534

Double-tuned transformer amplifiers, 532–534
Drain:
   FET, 188–191
   MOSFET, 222–224
   VMOS-FET, 231
Drain-to-gate capacitance, 320
Drift in op-amps, 589
Dual-in-line packages (DIPs), 68–69, 558–559, 577, 578
Duty cycle, 506–507
Dynamic resistance of diodes, 17, 39–43, 486

E

Efficiency:
   choke-coupled amplifier, 420–426
   class B amplifier, 434–437
   complementary symmetry amplifier, 442
   regulator, 483
   resistively-coupled amplifier, 402–407
   transformer-coupled amplifier, 430
   Zener regulator, 485, 487
Electrical energy, 398
Electric eyes, 49–50
Electron charge, 604
Electron-hole pairs, 10
Electronics, 1
Electronic voltage regulators, see Transistor voltage regulators
Electrons, 6, 189
Emitter-base junction, 70–71
Emitter bypass capacitors, 110–111, 176–177, 243–245, 291, 299–304, 307–308, 363–367
Emitter carriers, 67
Emitter-follower amplifiers, 127–128, 169–178
Emitter region, 66
Emitter-resistor bias circuits, 110–120
Emitter saturation current, 604–605
Emitter-to-base leakage current, 604–605
Energy, electrical, 398
Enhancement-mode MOSFETs, 222–224, 230–231
Equivalent circuits, see also Hybrid parameters
   ac, 143
   BJT amplifiers, at high frequencies, 306–309, 311–314
   BJT amplifiers, at low frequencies, 293–297
   blocking capacitors at low frequencies, 293–297, 317–318
   cascaded amplifiers, 238–239
   coils, 527–529
   crystals, 550
   Darlingtons, 251–254
   difference amplifier bias, 265
   555 timer, 559
   hybrid-pi, 309–311
   JFET amplifiers, at high frequencies, 320–322
   JFET amplifiers, at low frequencies, 317–319
   JFET amplifiers, at mid frequencies, 206–207, 218–219
   Miller capacitance, 311–312
   Norton, 146, 306, 367, 372, 388, 389
   series feedback amplifiers, 354–355
   shunt feedback amplifiers, 367–368, 370–372
   Thevenin, 145–149, 174, 267, 306, 325, 334, 389
   Zener diodes, 486
Externally compensated operational amplifiers, 588

F

Fairchild, Inc., 56, 509
Fall time, 330
Faraday's law, 426–427
Federal Communications Commission (FCC), 531
Feedback:
   in integrators, 599
   negative, 360, 390
   positive, 360, 542
Feedback amplifiers:
   analysis of, summarized, 388–389
   beta (β) networks in, 354–356, 359, 360–361, 388
   complementary symmetry, see Complementary symmetry amplifiers
   current, see Shunt feedback amplifiers
   current-series, 388–391
   current-shunt, 388–391
   feedback circuits in, 354–356, 359, 360–361, 388
   frequency response of, 382–387, 390
   IC, 453–454
   input impedance of, 389–390
   multistage, 365–367, 375–379
   output impedance of, 389–390
   series, see Series feedback amplifiers
   shunt, see Shunt feedback amplifiers
   voltage, see Series feedback amplifiers
   voltage-series, 388–391
   voltage-shunt, 388–391
Feedback equation, 542–543
Field effect transistors (FETs):
   JFETs, see Junction field effect transistors (JFETs)

MOSFETs, see Metal oxide semiconductor FETs (MOFETs)
VMOS-FETs, 231–232
Filters:
   active, 607–611
   band-pass, 607, 608. See also Tuned-circuit amplifiers
   band-reject, 608
   Bessel, 610
   Butterworth, 535–539, 610
   capacitive, 472–478, 482
   ceramic, 539–540
   Chebychev, 610
   functions of, 464
   high-pass, 293–297, 328–329, 607–611
   isolation of sections in, 609
   low-pass, 306–307, 329–331, 334–335, 472–482, 607–611
   L-section, 480–481, 482
   maximally flat, 535–539, 610
   passband of, 608
   passive, 607–610
   pi, 481, 482
   RC, 481, 482. See also High-pass filters; Low-pass filters
   Sallen and Key, 610, 611
   shunt-capacitance, 472–478, 482
   stop band of, 608
   transition region of, 608
   VCVS, 610, 611
Firing of SCRs, 497
555 and **556** timers, 558–561
Fixed-bias circuits:
   BJT, 105–110
   JFET, 197–206
Fixed-voltage power supplies, 464
Fixed-voltage regulators, 499–501
FM (frequency modulation) modulators, 56–57
FM (frequency modulation) receivers, 188
Focusing of photodiodes, 50
Forced beta (β) for BJTs, 91–92
Forced convection, 417–419
Forcing functions, 597–600
Forward-biased p-n junction, 14–15
Forward characteristics:
   BJTs in the switching mode, 93–94
   diodes, 17, 20, 39–43, 46
Four-terminal networks, 145–146
Free electrons, 6–12
Frequencies:
   complex, 292–293
   half-power, 290–291, 382–387
   heterodyned, 524
   IF, 524, 532
   pulse, 330
   radian, 298
   resonant, 524–526, 529–533, 550–551
   of switching regulator oscillators, 509
   3-dB, 290–291, 382–387
Frequency modulation (FM) modulators, 56–57
Frequency modulation (FM) receivers, 188
Frequency response of amplifiers:
   and bandwidth, 290–291
   bandwidth increased by feedback, 382–384, 390
   BJT, at high frequencies, 304–316
   BJT, at low frequencies, 293–304, 300
   BJT, at mid frequencies, 142, 300
   Bode plots of, 291–293, 299–300
   cascaded amplifiers, 322–327
   frequency-response curves, 290–293
   general equation for, 292–293
   JFET, at high frequencies, 320–322
   JFET, at low frequencies, 317–320
   JFET, at mid frequencies, 206–207, 218–219
   op-amp, 587–588, 590
Frequency-response curves, 290–293
Frequency-response limitations of oscilloscopes, 341–343
Full-load current, 478
Full-load voltage, 478–479
Full-wave rectifiers, 468–469
Fuses, 495–497

G

GaAs (gallium arsenide) diodes, 49
GaAsP (gallium arsenide phosphorus) diodes, 49–56
Gain:
   ac, see ac gain
   dc, see dc gain
   defined, 74, 142
   general equation for, 292–293
   and H-bias circuits, 160–166
   high-frequency, see High-frequency response
   low-frequency, see Low-frequency response
   mid-frequency, see Mid-frequency response
Gain-bandwidth product, 309–311, 316, 384, 587, 588
Gallium arsenide (GaAs) diodes, 49
Gallium arsenide phosphorus (GaAsP) diodes, 49–56
Gate:
   FET, 188–191
   MOSFET, 222–224
   VMOS-FET, 231

Gate-to-drain capacitance, 320
Gate-to-ground capacitance, 320
Germanium:
    carrier recombination coefficient for, 604–605
    n-type, 11–16
    properties of, 9
    p-type, 11–16
    as a semiconductor, 6–7
Germanium diodes, 23–24
Glitches, 497
Gold, 6
Ground, virtual, 569

H

Half-power frequencies, 290–291, 382–387
Half-wave rectifiers, 24–26, 465–466
Harmonic distortion, 437–438, 453
Harmonics, 437
Hartley oscillators, 548–549
H-bias circuits, 112–120, 142–145, 160–166
Heat sinks, 414–419, 449, 450
HEDS-1000 reflective sensor, 51
Heterodyned frequencies, 425
Hewlett-Packard Corporation, 51, 52
High-frequency region, 290–293
High-frequency response:
    BJT amplifier, 300
    cascaded amplifers, 322–323
    coil, 531
    JFET amplifier, 320–323, 327
    square-wave amplifier, 329–331
High-frequency sinusoidal oscillators, 543–552
High-pass filters, 293–297, 328–329, 607–611
High-voltage probes for oscilloscopes, 341
Holes in semiconductors, 7–12, 189
Hot carrier diodes, 45–46
HP 415A curve tracer, 76, 80, 83–85
Hybrid (h) parameters:
    and beta (β), 78–79, 91–92
    common-base amplifier, 166–169
    common-collector amplifier, 169–178
    common-emitter amplifier, 151–160
    defined, 145–146
    emitter-follower amplifier, 169–178
    resistive circuit, 147–149
Hybrid-pi equivalent circuits, 309–311

I

IC (integrated-circuit) power amplifiers, 449–454
IC (integrated circuit) regulators, see Integrated-circuit (IC) voltage regulators
ICs (integrated circuits), 2, 224, 231, 453, 495, 504
Ideal capacitors, 111, 290, 593
Ideal diodes, 16
Ideal single-tuned amplifiers, 532
Ideal transformers, 426–427
IEEE-696 bus, 504
IF (intermediate frequency), 524, 532
IF (intermediate-frequency) amplifiers, see Tuned-circuit amplifiers
IF (intermediate-frequency) cans, 534
Imaginary axis, 293
Impedance:
    input, see Input impedance
    output, see Output impedance
    reflected, 427
Impurities in semiconductors, 11–13
Inductance, mutual, 532–534
Inductive reactance, 527–529
Inductors, see Coils; Transformers
Infrared photodiodes, 49
Input bias current of op-amps, 585–586, 590
Input impedance:
    cascaded BJT amplifier, 239–247
    common-emitter circuit, 151–157
    Darlington, 251, 255
    emitter-follower circuit, 171
    feedback amplifier, 389–390
    FET, 188, 190
    JFET, 190, 206, 218–219
    MOSFET, 227
    op-amp, 569, 580–581, 590
    series feedback amplifier, 356–360, 389–390
    shunt feedback amplifier, 369–372, 389–390
Input offset current of op-amps, 585–586, 590
Input offset voltage of op-amps, 584–585, 590
Input voltage of IC voltage regulators, 499
Instrumentation amplifiers, 601–603
Insulators, 4–5
Integrated-circuit (IC) power amplifiers, 449–454
Integrated-circuit (IC) voltage regulators:
    adjustable-voltage, 502–503
    boost, 503–504
    characteristics of, 498–499
    fixed-voltage, 499–501
    practical, 504–505
Integrated circuits (ICs), 2, 224, 231, 453, 495, 504
Integrators, 590–595, 597–600
Interelectrode capacitance, 308
Intermediate frequency (IF), 524, 532
Intermediate-frequency (IF) amplifiers, see Tuned-circuit amplifiers
Intermediate-frequency (IF) cans, 534

Intermodulation distortion, 438
Internally compensated operational amplifiers, 588
Intrinsic carrier density, 9
Intrinsic resistivity, 4–5, 9–11
Intrusion detectors, 49–50
Inverters, 561–562, 599
Inverting inputs in op-amps, 568–569
Inverting operational amplifiers, 569–572
Inverting switching regulators, 513–514
Isolating amplifiers, 127–128
Isolation of filter sections, 609
Isosceles right triangles, 298, 299

J

JEDEC cases, 412–416, 449–450
JEDEC diodes, 41–43
Junction field effect transistors (JFETs):
  ac gain of, 209–210, 218–222
  biasing of, 197–206
  cascaded amplifiers using, 221–222, 247–248
  characteristics of, 196–197
  common-drain amplifiers using, 218–222
  common-drain characteristics of, 192–193
  common-source amplifiers using, 209–218, 362–363
  common-source characteristics of, 192–193
  construction of, 189–190
  feedback amplifiers using, 362–363
  at high frequencies, 320–322
  introduced, 188–189
  at low frequencies, 317–320
  in op-amps, 581, 590
  operation of, 190–191
  small signal analysis of, 206–209
  source-follower circuits, 218–222
  specifications for, 193–195
Junctions:
  base-to-emitter, 66–67
  collector-base, 69–70
  emitter-base, 70–71

K

Kelvin temperature scale, 604
Kirchhoff's current law, 70
Kirchhoff's voltage law, 111, 120
Knee:
  JFET drain curve, 193
  Zener curve, 39–40

L

Large-scale integrated (LSI) circuits, 188
LCDs (liquid-crystal displays), 58
Leakage current:
  in diodes, 16, 20–24, 39–43, 46
  emitter-to-base, 604–605
LEDs (light-emitting diodes), 49–56, 58
Level shifters, 277–279
Light-emitting diodes (LEDs), 49–56, 58
Line, point-slope equation for, 123
Linear amplifiers, 224
Line regulation, 493, 499
Liquid-crystal displays (LCDs), 58
Litronix, 49
LM106 comparator, 262–263
LM123/LM223/LM323 voltage regulators, 499–501
LM310 voltage follower op-amp, 576
Load lines:
  ac, 120–123
  BJT, 84–87, 102–108, 120–123, 404–405, 420–421
  dc, 120–123
  JFET, 198–199, 204–206, 211–212, 220
  MOSFET, 227–228
Load on power supplies, 464
Load regulation, 499
Logarithmic amplifiers, 604–606
Loudspeakers, 398–399, 420, 440, 448–453
Lower half-power frequency, 290–291, 382–387
Low-frequency region, 290–293
Low-frequency response:
  BJT amplifier, 293–304
  bypass capacitor, 110–111
  cascaded amplifiers, 322–323
  JFET amplifier, 317–320
  square-wave amplifier, 328–329
Low-pass filters, 306–307, 329–331, 334–335, 472–482, 607–611
L-section filters, 480–481, 482
LSI (large-scale integrated) circuits, 188

M

Magnitude of a complex number, 292
Majority carriers, 188
Maximally flat filters, 535–539, 610
Maximum efficiency of resistively-coupled amplifiers, 404–407
MC1455/1555 timers, 558–561
MC1723C *npn* boost regulator, 502–504
MC3240 pulse-width modulator, 515

MC3423 crowbar IC, 497
MDA970A1 through MDA970A6 bridge rectifiers, 471–472
Measurement errors for rise times, 341–343
Mechanical motion, 398
Medical electronics, 263–264
Metal oxide semiconductor FETs (MOSFETs):
   amplifiers using, 224, 230–231
   biasing of, 227–228
   depletion-mode, 222–226
   enhancement-mode, 222–224, 230–231
   gain of, 229–230
   introduced, 188, 222–223
   operation of, 224
   specifications for, 224–226
Metals, 6
Mho (unit of admittance), 148
Microstrip line, 169
Mid-frequency region, 290–293
Mid-frequency response:
   BJT amplifier, 142, 300
   cascaded amplifiers, 322–327
   JFET amplifier, 206–207, 218–219
Miller capacitance, 311–316, 320, 551–552
Miller capacitor, 311–312
Miller oscillators, 551–552
Minority carriers, 16
MJ2955 power BJT, 407–408
Mobility in semiconductors, 9, 189
Modulators:
   FM, 56–57
   pulse-width, 514, 515
MOSFETs, *see* Metal oxide semiconductor FETs (MOSFETs)
Motion, mechanical, 398
"Motor-boating" oscillations, 454
Motorola, Inc., 41, 42, 256–257, 471, 486, 497
Multistage amplifiers:
   Butterworth, 535–539
   cascaded, *see* Cascaded amplifiers
   cascode, 277–279
   maximally flat, 535–539
   narrow-band stagger-tuned, 536–539
Multivibrators, astable, 560, 561
Mutual inductance, 532–534

N

Narrow-band approximation for maximally flat amplifiers, 536–538
Narrow-band stagger-tuned amplifiers, 536–539
National Semiconductor, Inc., 499

Natural convection, 417–419
n-channel FETs, 188–191
n-channel MOSFETs, 222–226
n-channel VMOS-FETs, 231–232
Negative feedback, 360, 390
Negative peak of a square wave, 330
Negative resistance region, 57
Networks:
   beta (β) feedback, 354–356, 359, 360–361, 388
   four-terminal, 145–146
   RC, in phase-shift oscillators, 554–558
   R-2R, 611–612
   summing, 573–575, 611–612
   three-terminal, 145–146
   two-port, 145–146
   unilateral, 354
   weighted resistor, 611, 612
Neutrons, 6
Noise figure of FETs, 188
Noise susceptibility of differentiators, 596
No-load voltage, 478–479
Nomenclature for transistors, 73–74
Noninverting inputs in op-amps, 568–569
Noninverting operational amplifiers, 572–573
Nonmetals, 6
Nonrepetitive peak surge current, 479
Norton equivalent circuits, 146, 306, 367, 372, 388, 389
npn BJTs, *see* Bipolar junction transistors (BJTs)
npn boost regulators, 502–504
npn Darlington pairs, 256–257, 445–449
npn power transistors, 407–411, 440–445
n-type semiconductors, 11–16, 188–191, 222–223, 231–232
Nucleus of an atom, 6
Null of a bridge circuit, 552
Nyquist criterion for oscillation, 542–543

O

Off-line switching regulators, 514–515
Offset current of op-amps, 585–586, 590
Offset voltage of op-amps, 584–585, 590
1M3.3ZS10 through 1M200ZS10 Zener diodes, 41–43
1N914 diode, 547, 604
1N2808 Zener diode, 486
1N4004 rectifier diode, 19–20, 23
1N4148 switching diode, 17–24
1N4728 through 1N4764 Zener diodes, 41–43
One-shot circuits, 561

Op-amps, *see* Operational amplifiers (op-amps)
Open-loop voltage gain of op-amps, 568–569, 577, 580, 590
Operating point, *see* Quiescent (Q) point
Operating region for power transistors, 409–410
Operational amplifiers (op-amps):
  as antilog amplifiers, 606–607
  BIFET, 581, 590
  bi-polar, 581, 590
  characteristics of, 576–582, 590
  CMRR of, 587, 590
  compared with difference amplifiers, 275
  compensated, 588
  as D/A converters, 610–612
  as difference amplifiers, 600–601
  as differentiators, 595–597
  drift in, 589
  externally compensated, 588
  filters in, 607–611
  frequency response of, 587–588, 590
  gain of, 250–251, 568–580
  input bias current of, 585–586, 590
  input impedance of, 580–581, 590
  input offset current of, 585–586, 590
  input offset voltage of, 584–585, 590
  as instrumentation amplifiers, 601–603
  as integrators, 590–595, 597–600
  internally compensated, 588
  introduced, 568–569
  inverting, 569–572
  as logarithmic amplifiers, 604–606
  noninverting, 572–573
  open-loop voltage gain of, 577, 580, 590
  output impedance of, 583–584
  as RC phase-shift oscillators, 556
  slew rate of, 588–589, 590
  summing, 573–575, 611–612
  virtual ground in, 569
  voltage follower, 575–576
Optical couplers, 50
Opto-isolators, 50, 514
OP-07 op-amp, 603
Oscillations:
  in amplifiers, 242, 243, 360, 454, 541–543, 588
  defined, 540
  in differentiators, 596
Oscillators:
  amplifiers as, 541–543
  Barkhausen criterion for, 542–543
  Colpitts, 545–548
  crystal, 550–552, 561–562
  digital, 558–562
  Hartley, 548–549
  high-frequency sinusoidal, 543–552
  Miller, 551–552
  Nyquist criterion for, 542–543
  phase-shift, 552–558
  Pierce, 551
  positive feedback in, 360, 542
  pulse-width modulated square-wave, 506–507
  RC phase-shift, 554–558
  reactive, 544–545
  spurious oscillations in, 541–543
  in switching regulators, 509
  theoretical basis for, 544–545
  tuned-drain crystal-gate, 551–552
  types of, 540–541
  variable-frequency, 545–550, 558
  Wien bridge, 552–554
Oscilloscope amplifiers, 339–343
Oscilloscopes, *see* Cathode-ray oscilloscopes (CROs)
Output impedance:
  cascaded BJT amplifier, 239–247
  common-collector circuit, 171, 173–174
  common-emitter circuits, 151–157
  emitter-follower circuit, 171, 173–174
  feedback amplifier, 389–390
  IC voltage regulator, 499
  JFET, 218–219
  op-amp, 583–584
  series feedback amplifier, 356–360, 389–390
  shunt feedback amplifiers, 389–390
Output swing:
  BJT amplifier, 102–108, 249, 404–405, 420–421
  class B power amplifier, 433–434
  JFET amplifier, 201–206
Overcompensation, 338
Overvoltage protection, 495–497

P

Parallel equivalent circuits for coils, 527–529
Parallel resonant frequency of crystals, 550–551
Passband of filters, 608
Passive filters, 607–610
Pass transistors, 503–504, 505–506
p-channel FETs, 188–191
Peak inverse voltage (PIV) for diodes, 20, 46, 465–466, 469
Peaks of amplifier square-wave response, 330
Pentode vacuum tubes, 207
Percentage tilt, 328–329

Percent regulation, 478–479, 492–493, 495, 501
Perfect compensation, 338–340
Phase angle of a complex number, 292
Phase shift, 292, 526, 542
Phase-shift oscillators, 552–558
Phase splitters, 177–178
Photodetectors, 47–51
Photodiodes, 46–56
Phototransistors, 49
Pierce oscillators, 551
Piezoelectric effects, 550
Pi filters, 481, 482
Pinch-off region, 191
PIV (peak inverse voltage) for diodes, 20, 46, 465–466, 469
p-n junctions, 13–16, 188–191, 222, 224. *See also* Diodes
pnp BJTs, *see* Bipolar junction transistors (BJTs)
pnp Darlington pairs, 256–257, 445–449
pnp power transistors, 407–408, 440–445
Point-slope equation for a line, 123
Poles:
    cascaded amplifier, 322–323
    collector, 314–315
    dominant, 320–323
    high-frequency effects of, 307, 320–323
    introduced, 292–293
    low-frequency effects of, 297–299, 322–323
    in maximally flat amplifiers, 536–538
Positive feedback, 360, 542
Positive peak of a square wave, 330
Potentiometer, compensating, 585
Power amplifiers, *see also* Power transistors
    choke-coupled, 420–426
    class A, *see* Class A amplifiers
    class AB, 438
    class B, *see* Class B amplifiers
    complementary symmetry, *see* Complementary symmetry amplifiers
    Darlington, 445–449
    distortion in, 437–438, 453
    IC, 449–454
    power dissipation in, 399–404, 407–411, 437–438, 442, 457
    quasi-complementary, 445–446
    transformer-coupled, 426–431
    transformerless, *see* Complementary symmetry amplifiers
Power change (dB units), 273–274, 291–293
Power Darlingtons, 254–257
Power dissipation:
    BJT, 94
    class B amplifier, 434–437
    complementary symmetry amplifier, 442
    Darlington, 254–257
    diode, 24, 41, 58
    IC voltage regulator, 499, 500, 505
    maximum, for **2N3055** and **2N3904** BJTs, 409
    MOSFET, 231
    power transistor, 399–404, 407–411, 457
    regulator, 483
    transistor voltage regulator, 493
    VMOS-FET, 231, 232
    worst case, 404
    Zener regulator, 485, 487
Power supplies:
    average values of waveforms in, 466–469
    block diagram of, 464
    bridge rectifiers in, 469–472
    double-ended, 568
    filters in, 472–482
    full-wave rectifiers in, 468–469
    half-wave rectifiers in, 24–26, 465–466
    introduced, 464–465
    protection circuits in, 495–498, 499
    regulation and, 478–479
    regulators for, *see* Regulators
    ripple factor in, 467–469, 482
    rms values of waveforms in, 466–469
    surge currents in, 479
    voltage doublers in, 482–483
Power transistors, *see also* Power amplifiers
    cases for, 412–416
    defined, 407
    derating curves for, 407, 408, 457
    efficiency of, 402–407
    heat sinks for, 414–419, 449, 450
    power dissipation in, 399–404, 407–411, 457
    pulsed operation of, 410–411
    safe operating region of, 409–410, 457
    thermal resistance of, 408, 409, 411–419
Preregulators, 494–495
Primary windings, 426–427
Probe capacitance, 333–340
Probes for oscilloscopes, 333–334, 339–341
Programs, *see* Computer programs
Protection circuits, 495–498, 499
Protons, 6
p-type semiconductors, 11–16, 188–191, 222–223, 231
Pulsed operation of power transistors, 410–411
Pulse frequency, 330
Pulse period, 330
Pulse width, 330
Pulse-width modulated square-wave oscillators, 506–507

Pulse-width modulators, 514, 515
Push-pull amplifiers, 432–437

Q

Q factor, *see* Quality (Q) factor
Q point, *see* Quiescent (Q) point
Q2T2905 quad *pnp* BJT package, 68–69
Quality (Q) factor:
   of coils, 527–529
   effect on bandwidth, 529–532
Quasi-complementary amplifiers, 445–446
Quiescent (Q) point:
   BJT amplifier, 102–108, 120–123, 128–133, 249, 420–421
   class B power amplifier, 433–434
   complementary symmetry amplifier, 440–444
   JFET amplifier, 197–206, 248
   power transistor, 399–404

R

Radian frequencies, 298
Radio Corporation of America (RCA), 255, 257, 275
Radio frequency choke (RFC) coils, 420, 545–546
Radio frequency interference (RFI), 509
Radio frequency (RF) amplifiers, 169
Ramp waveforms, 509–511, 588–589, 593
RCA (Radio Corporation of America), 255, 257, 275
RC coupled amplifiers, 240–248
RC filters, 481, 482. *See also* High-pass filters; Low-pass filters
RC phase-shift oscillators, 554–558
RC time constants, 33–35, 334–335, 337–339
Reactive oscillators, 544–545
Recovery time of diodes, 20–23, 46
Rectifiers, *see also* Power supplies
   bridge, 469–472
   defined, 24
   diodes as, 16, 19–20, 23
   full-wave, 468–469
   functions of, 464
   half-wave, 24–26, 465–466
   SCRs, 495–497
   surge currents in, 479
Reflected impedance, 427
Regulation:
   line, 493, 499
   of power supplies, 478–479
Regulators:
   boost, 503–504
   difference-amplifier, 488–489
   functions of, 464
   IC, *see* Integrated-circuit (IC) voltage regulators
   percent regulation for, 492–493, 495, 501
   preregulators, 494–495
   protection circuits in, 495–498, 499
   series, 488–493
   switching, *see* Switching regulators
   tracking, 503
   transistor, *see* Transistor voltage regulators
   Zener, 483–487
Resistance, *see also* Resistors
   base-spreading, 309
   calculated from resistivity, 4–5
   reverse, 17, 20
   thermal, 408, 409, 411–419
   thermal, case-to-air, 412–414
Resistively-coupled amplifiers, 404–407
Resistive (r) parameters, 145
Resistivity, 4–5, 9–11
Resistors, *see also* Resistance
   balancing, 271–273
   bleeder, 481
   voltage-controlled, 191
Resonant frequencies, 524–526, 529–533, 550–551
Reverse-biased p-n junctions, 14–16, 190, 222, 224
Reverse current in diodes, 16, 20–24, 39–43
Reverse recovery time of diodes, 20–23, 46
Reverse resistance of diodes, 17, 20
RFC (radio frequency choke) coils, 420, 545–546
RFI (Radio Frequency Interference), 509
RF (radio frequency) amplifiers, 169
Rings (electron), 6
Ripple factor, 467–469, 482
Rise time, 329–331, 334–335, 341–343
RIT (Rochester Institute of Technology), 161
RLC circuits, 597–600. *See also* Tuned-circuit amplifiers
rms (root-mean-square) values of waveforms, 466–469
Rochester Institute of Technology (RIT), 161
Root-mean-square (rms) values of waveforms, 466–469
r (resistive) parameters, 145
R-2R network D/A converters, 611–612

S

Safe operating region for power transistors, 409–410
Sallen and Key filters, 610, 611
Saturation current, emitter, 604–605

Saturation region:
   amplifier clipping in, 105–106, 419–420
   BJT, 91–94
   constant-current source, 257–259
   logarithmic amplifier, 604, 606
   op-amp, 569, 593, 594
Schmitt triggers, 561
Schottky diodes, 45–46
Scientific Radio Systems Corporation, 453
Scopes, *see* Cathode-ray oscilloscopes (CROs)
SCRs (silicon controlled rectifiers), 495–497
Secondary windings, 426–427
Second-order differential equations, 597–600
Selectivity of tuned circuits, 532
Self-bias circuits:
   BJT, 120–123
   JFET, 197–206
Semiconductors, *see also* Diodes; Transistors
   conduction in, 8–9
   crystal lattice of, 7–8, 11–12
   defined, 6–7
   doped, 11–13, 189, 222–223, 231
   n-type, 11–16, 188–191, 222–223, 231–232
   p-type, 11–16, 188–191, 222–223, 231
Series equivalent circuits for coils, 527–529
Series feedback amplifiers:
   BJT, 363–365
   BJT, multistage, 365–367
   feedback equations for, 356–360
   JFET, 362–363
   negative feedback in, 360, 390
   positive feedback in, 360
   series feedback circuits in, 354–356
   stability of, 360–361
Series feedback equations, 356–357
Series regulators, 488–493
Series resonant frequency of crystals, 550–551
709 BJT op-amp, 588
741C and **741M** BJT op-amps, 576–582, 584–590
Seven-segment displays, 54–56, 58
7404 digital inverters, 561–562
78S40 switching regulator, 508, 509
SG1524 pulse-width modulator, 515
Shape factors for tuned circuits, 532, 540
Shockley, W., 195
Shockley's equation, 195–196
Short-circuit current gain, 310
Shunt-capacitance filters, 472–478, 482
Shunt feedback amplifiers:
   BJT, 370–375, 381–382, 386–387
   BJT, multistage, 375–381
   equivalent circuits for, 367–368, 370–372
   feedback equations for, 369
   negative feedback in, 390

Siemen (unit of admittance), 148
Silicon:
   carrier recombination coefficient for, 604–605
   crystal lattice of, 7–8, 11–12
   n-type, 11–16
   properties of, 9
   p-type, 11–16
   as a semiconductor, 6–7
   turn-on voltage for, 16–17
Silicon controlled rectifiers (SCRs), 495–497
Silicon diodes, 23–24
Silicon dioxide (SiO_2), 222, 223
Silicon General, Inc., 515
Single-phase full-wave bridge rectifiers, 471–472
Single-source common-base amplifiers, 126–127
Single-tuned transformer amplifiers, 524–526, 529–532
SiO_2 (silicon dioxide), 222, 223
Skin effect, 531
Skirt ratios for tuned circuits, 532, 540
Slew rate of op-amps, 588–589, 590
Slugs in coils, 527
"Solid-state electronics", 2
S-100 bus, 504
Source:
   FET, 188–191
   MOSFET, 222–224
   VMOS-FET, 231
Source bypass capacitors, 200, 216–218, 317–318
Source-follower JFET circuits, 218–222
Southwest Technical Products Corporation, 448
Speakers in audio systems, 398–399, 420, 440, 448–453
Spectral response of photodiodes, 49
Spectrum analyzers, 437
Spikes, 497
Spurious oscillations in amplifiers, 242, 243, 360, 541–543, 588
Square-wave oscillators, pulse-width modulated, 506–507
Square-wave response of amplifiers, 327–333
Stability:
   BJT bias, 128–133
   feedback amplifier, 360–361, 542
   JFET bias, 199
   oscillator, 542–543
Stability criteria, 542–543
Stability factor, 129–133
Stages of amplifiers, 238, 536–539
Stagger-tuned amplifiers, 535–539

State change of diodes, 29
Step-down switching regulators, 509–513
Step-up switching regulators, 513
Stop band of filters, 608
Storage time of diodes, 20–23
Stray capacitance, 291, 305–308, 320, 542–543, 547–548
Summing networks, 573–575, 611–612
Summing operational amplifiers, 573–575, 611–612
Summing and scaling amplifiers, 598
Superposition theorem, 74
Surface barrier diodes, 45–46
Surge currents, 479
Swing:
   BJT amplifier, 102–108, 249, 404–405, 420–421
   class B power amplifier, 433–434
   JFET amplifier, 201–206
Switching diodes, 17–24
Switching mode operation of BJTs, 93–94
Switching regulators:
   analysis of, 509–513
   boost, 513
   control of, 506–509
   introduced, 505–506
   inverting, 513–514
   off-line, 513–514
   oscillator frequencies in, 509
   step-down, 509–513
   step-up, 513

T

Tank circuits, 524–525, 529
TDA2002 power amplifier, 449–454
Tektronix Corporation, 83, 90, 340
Tektronix **576** and **577** curve tracers, 83–84
Television (color), crystal in, 550
Temperature-compensated Zener diodes, 487
Temperature scales, 604
Temperature variations, *see also* Heat sinks; Thermal resistance
   BJT bias stability and, 130–131
   Colpitts oscillators and, 547–548
   Darlington characteristics and, 256–257
   diode forward characteristics and, 17
   immunity of compensated logarithmic amplifiers to, 604–606
   immunity of feedback amplifiers to, 361
   op-amps and, 589
   power transistors and, 407–411
   resistivity and, 10–11
   Zener diodes and, 486–487

Texas Instruments, Inc., 68, 262
THD (total harmonic distortion), 453
Thermal capacitance, 419
Thermally generated electron-hole pairs, 10
Thermally generated minority carriers, 16
Thermal resistance, 408, 409, 411–419
Thevenin equivalent circuits, 145–159, 174, 267, 306, 325, 334, 389
Thevenin's theorem, 113, 115
3-dB points of frequency response, 290–291, 382–387
Three-terminal devices:
   comparators, 259–263, 487–489, 559
   IC voltage regulators, *see* Integrated-circuit voltage (IC) regulators
Three-terminal networks, 145–146
Three-terminal regulators, *see* Integrated-circuit voltage (IC) regulators
Tilted square waves, 328–329
Time constants (RC), 33–35, 334–335, 337–339
Tips of oscilloscope probes, 339
TIP31 power transistor, 457
TL080 BIFET op-amp, 590
Total harmonic distortion (THD), 453
TO-type cases, 412–416, 449–450
Tracking regulators, 503
Transconductance, 207, 312
Transconductance amplifiers, 388–390
Transducers, 601–603
Transformer amplifiers:
   double-tuned, 532–534
   single-tuned, 524–526, 529–532
   stagger-tuned, 535–539
Transformer-coupled amplifiers, 426–431
Transformerless amplifiers, *see* Complementary symmetry amplifiers
Transformers, *see also* Coils
   coefficient of coupling in, 532–534
   critical coupling in, 533–534
   functions of, 464
   ideal, 426–427
   mutual inductance of, 532–534
Transient response of amplifiers, 327–333
Transistors:
   BJTs, *see* Bipolar junction transistors (BJTs)
   cases for, 412–416, 449–450
   CMOS, 188, 223, 224
   current equation for junctions in, 604–605
   Darlington pairs, 251–257, 275–276, 445–449, 489–498
   heat sinks for, 414–419, 449, 450
   introduced, 2
   JFETs, *see* Junction field effect transistors (JFETs)

MOSFETs, *see* Metal oxide semiconductor FETs (MOSFETs)
  nomenclature for, 73–74
  pass, 503–504, 505–506
  phototransistors, 49
  power, *see* Power transistors
  VMOS-FETs, 231–232
Transistor-Transistor Logic (TTL) circuits, 18, 44, 53–54
Transistor voltage regulators:
  analysis of series regulators in, 489–494
  current limiting in, 497–498
  introduced, 487–489
  overvoltage protection in, 495–497
  preregulators in, 494–495
Transition region of filters, 608
Transition time of diodes, 22–23
Transmission lines, 264
Transresistance amplifiers, 388–390
Triangular waveforms, 594
Triggers, Schmitt, 561
TRS-80 computer, 2, 117
TTL (Transistor-Transistor Logic) circuits, 18, 44, 53–54
Tubes (vacuum), 1–2, 207, 440
Tuned-circuit amplifiers:
  bandwidth of, 529–533, 535–540
  ceramic filters in, 539–540
  coils in, 527–529
  double-tuned, 532–534
  introduced, 524–526
  Q factors in, 527–532
  selectivity of, 531–532
  single-tuned, 524–526, 529–532
  stagger-tuned, 535–539
Tuned-drain crystal-gate oscillators, 551–552
Tunnel diodes, 57
Turn-on voltage for diodes, 16–17
Turns ratio of transformers, 427
2N3055 power BJT, 83, 407–413, 417–419, 423–424, 430–431, 503–504
2N3796 and **2N3797** MOSFETs, 224–231
2N3903 BJT, 94–95
2N3904 amplification and switching BJT:
  biasing of, 108
  characteristics of, 76–86, 90–95
  construction of, 68–69
  in difference amplifiers, 268–271
  at high frequencies, 308–311
  h parameters of, 150
  maximum ratings of, 409
  not suitable as second Darlington, 250
  temperature stability of, 130
2N3905 and **2N3906** BJTs, 94–95

2N4351 MOSFET, 230–231
2N5457 and **2N5458** JFETs, 193–195
2N5459 JFET, 192–197, 200–206, 211–216, 220–221, 248
2N6034 through **2N6039** Darlingtons, 256–257
2N6383 Darlington, 255
2N6660 VMOS-FET, 231–232
Two-port networks, 145–146
Two-stage amplifiers:
  BJT, 241–247, 248–251, 323–327, 375–381
  Butterworth, 535–539
  cascode, 277–279
  common-emitter, 240–245
  JFET, 221–222, 247–248, 327, 535–539
  maximally flat, 535–539
  narrow-band stagger-tuned, 536–539
  rise times in, 341–343

U

uA709 BJT op-amp, 588
uA741C and **uA741M** BJT op-amps, 576–582, 584–590
ULN-3701Z/TDA2002 power amplifier, 449–454
Unbiased clamping circuits, 31–37
Unbiased *p-n* junctions, 13–14
Undercompensation, 338
Unilateral networks, 354
Unity-gain bandwidth, *see* Gain-bandwidth product
UNIVAC I computer, 2
Upper half-power frequency, 290–291, 382–387

V

Vacuum tubes, 1–2, 207, 440
Valence bands, 6–9, 11–12
Valence electrons, 7–9, 11–12
Valleys in response curves, 533–534
Varactor diodes, 56–57
Variable-frequency oscillators (VFOs), 545–550, 558
Variable-voltage power supplies, 464
Varicap diodes, 56–57
VAX computer, 117, 213
VCVS (voltage-controlled voltage source) filters, 610, 611
Vertical metal oxide semiconductor FETs (VMOS-FETs), 231–232
VFOs (variable-frequency oscillators), 545–550, 558
Video amplifiers, 524

Virtual ground, 569
VMOS-FETs (vertical metal oxide semiconductor FETs), 231–232
Voltage amplifiers, 388–390
Voltage-controlled resistors, 191
Voltage-controlled voltage source (VCVS) filters, 610, 611
Voltage doublers, 482–483
Voltage feedback amplifiers, *see* Series feedback amplifiers
Voltage follower operational amplifiers, 575–576
Voltage regulators, *see* Regulators
Voltage-series feedback, 388–391
Voltage-shunt feedback, 388–391

W

Waveforms:
  average values of, 466–469
  dead time in, 515
  ramp, 509–511, 588–589, 593
  rms values of, 466–469
  in switching regulators, 509–511
  triangular, 594
Weighted resistor D/A converters, 611–612
Wide-band amplifiers, 524
Wien bridge oscillators, 552–554
Windings of transformers, 426–427
Worst case power dissipation, 404

Y

y (admittance) parameters, 145

Z

Zener breakdown, 39–40
Zener diodes, 39–44, 259, 483–487
Zener regulators, 483–487
Zeros:
  of cascaded amplifiers, 322–323
  high-frequency effects of, 307, 322–323
  introduced, 292–293
  low-frequency effects of, 297–299, 322–323